Atmospheric Science: Principles and Applications

Atmospheric Science: Principles and Applications

Edited by Evie Hughes

MURPHY & MOORE
www.murphy-moorepublishing.com

Published by Murphy & Moore Publishing,
1 Rockefeller Plaza,
New York City, NY 10020, USA
www.murphy-moorepublishing.com

Atmospheric Science: Principles and Applications
Edited by Evie Hughes

International Standard Book Number: 978-1-63987-064-6 (Hardback)

Cataloging-in-Publication Data

Atmospheric science : principles and applications / edited by Evie Hughes.
p. cm.
Includes bibliographical references and index.
ISBN 978-1-63987-064-6
1. Atmospheric science. 2. Atmosphere. 3. Earth sciences. I. Hughes, Evie.
QC861.3 .A86 2022
551.5--dc23

TABLE OF CONTENTS

PREFACE

This book aims to highlight the current researches and provides a platform to further the scope of innovations in this area. This book is a product of the combined efforts of many researchers and scientists, after going through thorough studies and analysis from different parts of the world. The objective of this book is to provide the readers with the latest information of the field.

The study of the physics and chemistry of clouds, gases, and aerosols surrounding the planetary bodies of the solar system is known as atmospheric science. Research in atmospheric science deals with a wide range of areas such as climatology, dynamic meteorology, cloud physics, atmospheric chemistry, atmospheric physics, aeronomy and oceanography. It studies the atmosphere of the Earth as well as the atmospheres of the planets and moons in our solar system. Atmospheric science is applied in instrumentation and data acquisition, data analysis and modeling, and laboratory studies of the chemical and physical processes that occur in the atmosphere. It also finds application in the study of photochemical reactions, cloud microphysics, and absorption and emission of radiation by atmospheric gases and particles. This book traces the progress of this field and highlights some of its key concepts and applications. It presents researches and studies performed by experts across the globe. Scientists and students actively engaged in this field will find this book full of crucial and unexplored concepts.

I would like to express my sincere thanks to the authors for their dedicated efforts in the completion of this book. I acknowledge the efforts of the publisher for providing constant support. Lastly, I would like to thank my family for their support in all academic endeavors.

Editor

1

Typhoon-induced sea surface cooling during the 2011 and 2012 typhoon seasons: observational evidence and numerical investigations of the sea surface cooling effect using typhoon simulations

Akiyoshi Wada[1*], Tomohiro Uehara[2] and Shiro Ishizaki[2]

Abstract

Understanding oceanic responses to typhoons and the impacts those responses have on the typhoons themselves is important so that typhoon predictions performed using numerical models and typhoon forecasts can be improved. However, *in situ* oceanic observations underneath typhoons are still limited. To gain a deep understanding of the oceanic response and estimate the magnitude of its impact, three profiling floats were deployed in the western North Pacific during the 2011 and 2012 typhoon seasons. The daily observations showed that the sea surface cooled by more than 2°C in typhoons Ma-on and Muifa in 2011, and typhoons Bolaven and Parapiroon in 2012. The response was different at different float locations relative to the typhoon center, that is, the response within 100 km of the typhoon center was different to the response more than 100 km from the center on the right- or left-hand sides of the typhoon track, even though the response was affected by pre-existing oceanic conditions, precipitation, and the typhoon intensity. The salinity and temperature profiles were also considerably different before, during, and after the passage of a typhoon. To determine the impacts of typhoon-induced sea surface cooling on typhoon predictions, the impacts of the four typhoons were numerically evaluated using an atmosphere-wave-ocean coupled model. The coupled model simulated sea surface cooling and the resultant increases in the central pressures caused by the passages of the typhoons reasonably well. When axisymmetrically simulated, the mean sea surface cooling beneath a typhoon decreased the latent heat fluxes by 24% to 47%. A larger cooling effect gave a larger decrease in the latent heat flux only during the intensification phase. The decrease in the latent heat flux affected the inner core structure, particularly in the inflow boundary layer and around the eyewall. The cooling effect significantly affected the track simulation only for Typhoon Muifa, which had the weakest zonal steering flow of the four typhoons. These results suggest that making more frequent typhoon observations using profiling floats, further developing oceanic analysis techniques, and improving our understanding of typhoon-ocean interactions are required to produce more accurate typhoon predictions.

Keywords: Typhoon; Profiling float; Sea surface cooling; Latent heat flux; Atmosphere-wave-ocean coupled model

Background

Advances in observational technologies, data assimilation systems, and numerical modeling have made it possible to predict tropical cyclones (TCs) more precisely than was previously the case. Little progress has been made, however, in predicting TC intensities, despite notable improvements in TC track predictions in recent decades. In particular, the dynamics and thermodynamics associated with rapid intensification are not well understood, and simulations using numerical prediction systems sometimes overestimate or underestimate TC intensities and intensification rates. It has previously been stated that understanding interactions between TCs and the ocean is crucial to decreasing the extent of the overestimations that are currently made of TC intensities (Bender et al. 1993; Shade and Emanuel 1999) and intensification rates (Wada 2009).

* Correspondence: awada@mri-jma.go.jp
[1]Meteorological Research Institute, Japan Meteorological Agency, 1-1 Nagamine, Tsukuba, Ibaraki 305-0052, Japan
Full list of author information is available at the end of the article

A TC will induce sea surface cooling in its wake (Price 1981; Wada 2002; Walker et al. 2005; D'Asaro et al. 2011; Pun et al. 2011; Jeong et al. 2013). This and other oceanic responses to a TC have been the subjects of numerous studies (e.g., Shay 2010), and highly complex physical processes associated with typhoon-ocean interactions have recently been investigated in a large typhoon-ocean field experiment ('Impact of Typhoons on the Ocean in the Pacific') (D'Asaro et al. 2011; Pun et al. 2011). In the western North Pacific, TCs with maximum sustained wind speeds exceeding approximately 33 m s^{-1} are called 'typhoons,' and the Japanese term *taifu* means a TC with a maximum sustained wind speed exceeding approximately 17 m s^{-1}. It has previously been reported that TC-induced sea surface cooling is mainly caused by Ekman upwelling and vertical turbulent mixing (Price 1981; Wada et al. 2009a; Sanford et al. 2011). The rate of entrainment at the base of the mixed layer caused by the passage of a TC is primarily determined by different horizontal current speeds at different depths across the base (Price 1981; Sanford et al. 2007). Wada et al. (2010) suggested that vertical turbulent mixing caused by breaking surface waves caused by a TC also plays an important role in sea surface cooling. The effect of vertical turbulent mixing on TC-induced sea surface cooling depends very much on the speed at which the TC moves (Wada et al. 2009a; Lin 2012).

Sea surface cooling induced by a TC is also affected by pre-existing oceanic conditions such as the depth of the mixed layer and the temperature gradient within the underlying thermocline (Price 1981; Wada 2002; Moon and Kwon 2012). In particular, the passage of a TC over a warm eddy will decrease the magnitude of the sea surface cooling that occurs (Wu et al. 2007; Wada et al. 2010). The presence of a warm eddy beneath a TC will, therefore, enable the TC to become intensified more rapidly than would be the case otherwise, because the air-sea sensible and latent heat fluxes increase rapidly (Lin et al. 2005, 2008, 2009; Wada and Usui 2007; Shay 2010). It can be seen that the mechanisms involved in TC-induced sea surface cooling are complex and depend on various factors, such as the TC intensity, translation speed, and size, and on pre-existing oceanic conditions such as the depth of the mixed layer and the temperature gradient within the underlying seasonal thermocline. *In situ* observations underneath a TC are, therefore, needed to allow the mechanisms involved in TC-induced sea surface cooling to be determined.

Decreased enthalpy (sensible and latent heat) fluxes underneath a TC are caused by TC-induced sea surface cooling. Cione and Uhlhorn (2003) found that sea surface cooling of approximately 1°C within the inner core of a TC can effectively alter the maximum total enthalpy flux by 40% or more. A lower translation speed will cause a much higher enthalpy flux to be required to maintain the intensity of an intense TC (Lin et al. 2009). However, the dependence of the decrease in enthalpy (particularly latent heat) flux within the inner core on TC-induced sea surface cooling has not yet been effectively determined.

The atmospheric response to TC-induced sea surface cooling and the impact such a response has on an idealized TC-like vortex has been investigated using atmosphere-ocean coupled models (Chang and Anthes 1979; Zhu et al. 2004; Wu et al. 2005; Bender et al. 2007; Wada 2009), and an atmosphere-wave-ocean coupled model has recently been developed to numerically simulate TCs (Chen et al. 2007, 2013; Wada et al. 2010, 2013; Warner et al. 2010). It has been shown in numerical experiments that TC-induced sea surface cooling leads to the suppression of the intensity of an idealized TC-like vortex, resulting in decreased latent heat fluxes within the inner core of the vortex and increased inner-core asymmetries (Zhu et al. 2004; Wu et al. 2005). However, the effects of interactions between the atmospheric and oceanic environments on TC evolution were not examined using these idealized numerical experiments. Numerical simulations of TCs using atmosphere-wave-ocean models have, so far, provided only preliminary results. The degree to which TC-induced sea surface cooling and the associated decreased latent heat flux affects the intensity of a TC in realistic atmospheric and oceanic environments is still uncertain, as is how the sea surface cooling and decreased latent heat flux affect the inner core structure of a TC.

This study aims to present the following: first, to determine the oceanic response to the passage of a typhoon using daily *in situ* observations made by profiling floats deployed in the western North Pacific Ocean and second, to quantitatively evaluate the effect of the oceanic response on the typhoon predictions produced using an atmosphere-wave-ocean coupled model, especially when the observed extent of sea surface cooling exceeded 2°C. The observations made by the profiling floats will be outlined in the next subsection.

Observations made by profiling floats during the 2011 and 2012 typhoon seasons

Profiling floats are autonomous instruments that drift with ocean currents while measuring the water temperature, salinity (as conductivity), and pressure (i.e., depth) as they ascend through the water column. Three profiling floats, each with an Iridium communication system, were deployed around 20°N (World Meteorological Organization (WMO) number 5901989, identification number (ID) 6399), 22°N (WMO number 5901988, ID 7560), and 24°N (WMO number 2901670, ID 6429) along the 137°E meridian by the R/V Keifu Maru, which is operated by the Japan Meteorological Agency (JMA), on 22 and 23 June 2011 (Figure 1). The profiling floats were used to monitor rapid variations in the water temperature and salinity in the upper ocean caused by the passage of typhoons during the 2011 and 2012 seasons. A bidirectional telecommunication system was used, and the three profiling floats recorded vertical profiles of the water temperature and salinity between approximately 5- and 500-m deep every day during the typhoon season (June to October) and between approximately 5- and 2,000-m deep every 10 days outside the typhoon season.

In this study, TCs with maximum wind speeds exceeding approximately 17 m s^{-1} (called *taifu* in Japanese, as mentioned above) are categorized as typhoons. Each typhoon is referred to by its number, Tyynn, where yy are the last two digits of the year and nn is the sequential number for the typhoon in that typhoon season. Five typhoons, Ma-on (T1106), Muifa (T1109), Talas (T1112), Kulap (T1114), and Roke (T1115), passed over the area monitored by the floats (around 17°N to 27°N and 130°E to 140°E) during the 2011 typhoon season, and three typhoons, Bolaven (T1215), Sanba (T1216), and Prapiroon (T1221), passed over the area monitored by the floats (around 20°N to 27°N and 125°E to 135°E) during the 2012 typhoon season (Figure 1).

The names and numbers of these eight typhoons, the analysis period for each typhoon, and the maximum amount of sea surface cooling that was observed by the three profiling floats during each analysis period are shown in Table 1. The analysis period is the period during which the typhoon passed over the area monitored by the profiling floats (Figure 1), and the maximum amount of sea surface cooling, which ranged from 0.8°C to 4.3°C (Table 1), was defined as the difference between the maximum and minimum water temperature at a depth of 5 m during each analysis period.

Regarding the relationship between the maximum amount of sea surface cooling and the TC, it has been suggested that TC-induced sea surface cooling of more than 2.5°C will not cause the intensification of a TC because this degree of cooling should be sufficient to shut down the entire energy production process of the TC (Emanuel 1999; Emanuel et al. 2004; Lin et al. 2008). The amount of cooling that occurred during typhoons T1106, T1109, and T1221 exceeded this threshold (Table 1), and the sea surface was cooled by 2°C, which is close to the threshold, during typhoon T1215. We will later assess the changes in water temperature and salinity profiles during the passage of four typhoons (T1106, T1109, T1215, and T1221) and perform numerical simulations of the typhoons using an atmosphere-wave-ocean coupled model.

The three profiling floats drifted westward in the 2 years wherein the observations were made (Figure 1), mainly following the wind-driven North Equatorial Current (Qiu and Chen 2010), and they sometimes became trapped in mesoscale eddies. The areas in which the observations were made in the 2011 and 2012 seasons were, therefore, different (Figure 1). However, the

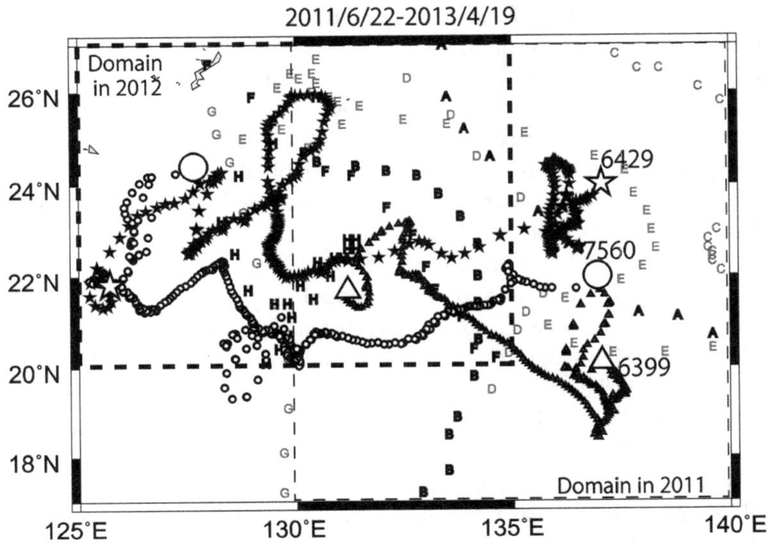

Figure 1 Trajectories of the three profiling floats from 22 June 2011 to 19 April 2013. All of the profiling floats were deployed along the 137°E meridian on 22 and 23 June 2011. The first symbols accompanying the float IDs (stars for 6429, circles for 7560, and triangles for 6399) indicate the locations where the floats were initially deployed, and the second symbols indicate the locations of the floats on 19 April 2013. The letters in bold indicate the central positions of the typhoons listed in Table 1 (A, T1106; B, T1109; C, T1112, D, T1114; E, T1115; F, T1215; G, T1216; and H, T1221) every 6 h. The bold characters show when a profiling float was within 300 km of a typhoon track. The thin dashed box indicates the domain wherein the three profiling floats were monitored in 2011, and the thick dashed box indicates the domain wherein the three profiling floats were monitored in 2012.

Table 1 Typhoons and the maximum sea surface cooling observed during the typhoons by the profiling floats

Name	Number	Analysis period	Maximum sea surface cooling (°C) (float ID)
Ma-on	T1106	15 to 18 July 2011	4.3 (6429)
Muifa	T1109	31 July to 3 August 2011	3.5 (7560)
Talas	T1112	24 to 30 August 2011	1.1 (6429)
Kulap	T1114	6 to 8 September 2011	1.0 (6399)
Roke	T1115	10 to 19 September 2011	0.8 (7560)
Bolaven	T1215	23 to 26 August 2012	2.0 (6399)
Sanba	T1216	14 to 16 September 2012	0.8 (6429)
Prapiroon	T1221	12 to 17 October 2012	3.3 (6399)

horizontal distributions of weekly sea level height anomalies indicated that the locations of warm eddies and cold wakes changed little over the time taken for a typhoon to pass (a week), although the amplitudes of the cold wakes became stronger after a typhoon had passed over than before it had arrived (Figure 2). Weekly sea level height anomaly observations made by a satellite-borne altimeter were obtained from the Archiving, Validation and Interpretation of Satellite Oceanographic data (AVISO) website (http://www.aviso.oceanobs.com/). On the basis of the results shown in Figure 2, we assumed that weekly mesoscale oceanic variations caused by planetary Rossby waves were negligible compared with variations caused by the response of the ocean to a typhoon. The cold wake area, with a negative sea surface height anomaly, corresponded with the area of relatively low sea surface temperatures (sea surface cooling) that was formed after the passage of each typhoon (Figure 3). The horizontal sea surface temperature distributions were obtained from the daily dataset produced by the Tropical Rainfall Measuring Mission (TRMM) microwave imager (TMI), and the Aqua/Advanced Microwave Scanning Radiometer for Earth observing system (AMSRE) satellite radiometers, which have a horizontal resolution of 0.25° (http://www.ssmi.com/). A remarkable amount of sea surface cooling occurred along the path of each typhoon after it had passed, except for T1221 (Figure 3i,j,k,l).

Figure 4 shows the trajectories of the three profiling floats and the best tracks (from the Regional Specialized Meteorological Center-Tokyo (RSMC-Tokyo) best-track data) for T1106, T1109, T1215, and T1221, along with three-hourly accumulated precipitation measurements collected during the analysis period (Table 1) for each typhoon (when the typhoon entered the domain shown in Figure 4). The three-hourly global precipitation estimates at a resolution of 0.25° latitude and 0.25° longitude produced by the US National Oceanic and Atmospheric Administration (NOAA) Climate Prediction Center morphing method (Joyce et al. 2004) were used to determine the total precipitation distribution during each analysis period. The horizontal total precipitation distribution indicated that the total precipitation was higher in T1221 than in T1106, T1109, and T1215, and this was caused by the relatively slow and irregular track followed by T1221.

Methods

Profiling floats

Vertical profiles of the water temperature and salinity were obtained from three profiling floats during their ascents. Real-time quality control was performed before the data were made available on the Japan Argo Real Time Data Base (http://argo.kishou.go.jp/index.html), which is operated by the Japan Meteorological Agency, from which the data used in this study were taken. Daily water temperature and salinity observations made by the three profiling floats were interpolated from the subsurface (at a depth of nearly 5 m) to a depth of 200 m using the method developed by Akima (1970). The temperature and salinity between the surface and a depth of 5 m were assumed to be constant, because no surface observations were made by the profiling floats. The maximum sea surface cooling values (Table 1) were calculated from the water temperatures observed at a depth of 5 m during each analysis period. This method is valid when surface wind speeds are higher than approximately 7 m s^{-1} (Soloviev and Lukas 2006; Kawai and Wada 2007). Soloviev and Lukas (2006) showed that the water was well mixed (in terms of temperature) between the subsurface and a depth of 5 m under relatively strong winds (approximately 7 m s^{-1}) in the Pacific warm pool during the Coupled Ocean-Atmosphere Response Experiment.

The mixed layer was defined as the water down to a depth at which the density was 0.25 g m^{-3} higher than the density at the surface, whereas the isothermal layer was defined as the water down to a depth at which the temperature was 0.5°C different from the temperature at the surface.

Model

The nonhydrostatic model (NHM) that was used in this study was developed jointly by the JMA Numerical Prediction Division and the Meteorological Research Institute (MRI). The model is a regional mesoscale atmospheric model that includes physical processes (see Table 2) (such as clouds, which are described using an explicit three-ice bulk microphysics scheme that is based on the work published by Lin et al. (1983)), an assumed resistance law for the air-sea momentum and for enthalpy fluxes in the atmospheric surface boundary layer, exchange coefficients for the air-sea momentum and for enthalpy transfers over the sea (based on bulk formulae (Kondo 1975) or when the ocean-wave model is coupled (Wada et al. 2010), based on the roughness lengths proposed by Taylor and

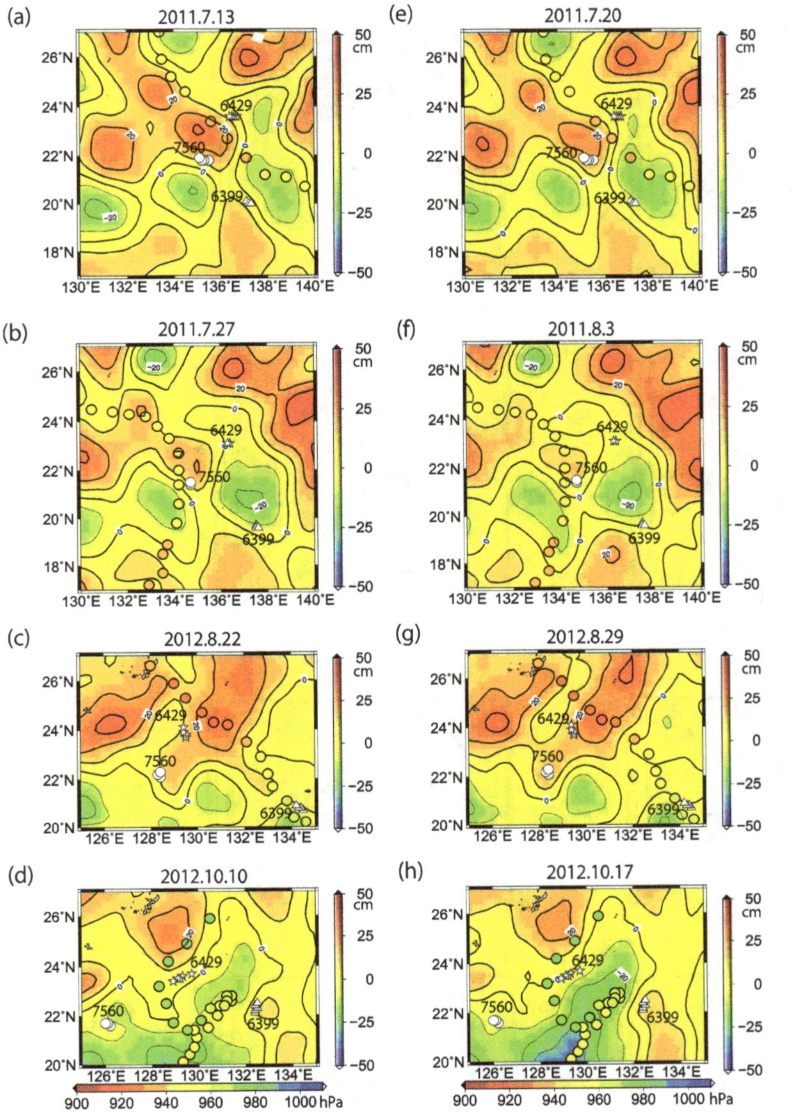

Figure 2 Horizontal distributions of the AVISO weekly sea surface height anomaly. Horizontal distributions of the AVISO weekly sea surface height anomaly (cm) (http://www.aviso.oceanobs.com/) on **(a)** 13 July 2011, **(b)** 27 July 2011, **(c)** 22 August 2012, and **(d)** 10 October 2012 before the passage of a typhoon and on **(e)** 20 July 2011, **(f)** 3 August 2011, **(g)** 29 August 2012, and **(h)** 17 October 2012 after the passage of the typhoons. The colors within the circles indicate the central pressure of the typhoon at each position along the RSMC-Tokyo best track. The open stars, circles, and triangles show the positions of the profiling floats, as in Figure 1, except during each analysis period.

Yelland (2001)), a turbulent closure model in the atmospheric boundary layer (based on the work of Klemp and Wilhelmson (1978) and Deardorff (1980)), and an atmospheric radiation scheme (based on the work of Sugi et al. (1990)). The NHM used was an older version of the current nonhydrostatic mesoscale model (Saito 2012), and it was coupled with both a multilayer ocean model and a third-generation ocean-wave model (Wada et al. 2010). No cumulus parameterization was used.

The JMA third-generation ocean-wave model was coupled with the NHM to estimate changes in the surface roughness lengths, drag coefficients, and enthalpy coefficients. The wave spectrum in the third-generation ocean-wave model had 900 components, each associated with one of 25 frequencies and one of 36 directions. The frequency of the wave spectrum was divided logarithmically from 0.0375 to 0.3000 Hz. The ocean was assumed to be motionless at the initial time. Wada et al. (2010) described the wave-ocean coupling procedure in detail.

The multilayer ocean model used a decreased gravity approximation and a hydrostatic approximation, and it was assumed that the water was a Boussinesq fluid. The

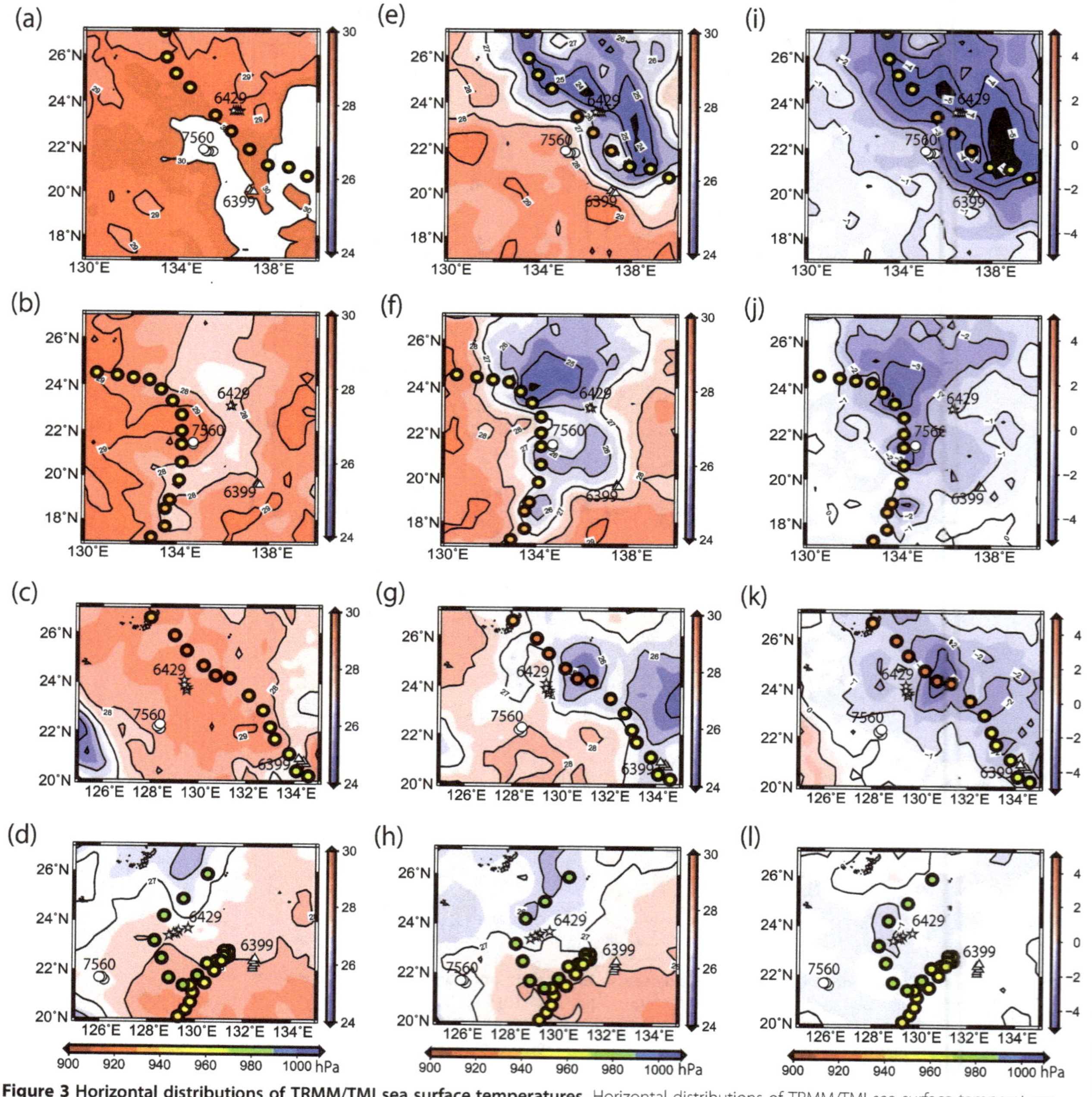

Figure 3 Horizontal distributions of TRMM/TMI sea surface temperatures. Horizontal distributions of TRMM/TMI sea surface temperatures (°C) on **(a)** 15 July 2011, **(b)** 31 July 2011, **(c)** 23 August 2012, and **(d)** 12 October 2012 before the passage of a typhoon and on **(e)** 18 July 2011, **(f)** 3 August 2011, **(g)** 26 August 2012, and **(h)** 17 October 2012 after the passage of the typhoons. **(i to l)** Differences in the TRMM/TMI sea surface temperature between before and after the passage of the typhoon (before subtracted from after). The colors within the circles indicate the central pressure of the typhoon at each position along the RSMC-Tokyo best track. The open stars, circles, and triangles show the positions of the profiling floats, as in Figure 1, except during each analysis period.

model had three layers and four levels. The uppermost layer was the mixed layer, where the density was vertically uniform. The middle layer was the seasonal thermocline, where the vertical temperature gradient was greatest. The bottom layer was assumed to be undisturbed by entrainment. The four levels were the sea surface, the mixed layer base, the thermocline base, and the sea bottom. The degree of entrainment was calculated using the multi-limit entrainment formula proposed by Deardorff (1983) and modified by Wada et al. (2009a). The model calculated the water temperature and salinity at the surface and at the base of the mixed

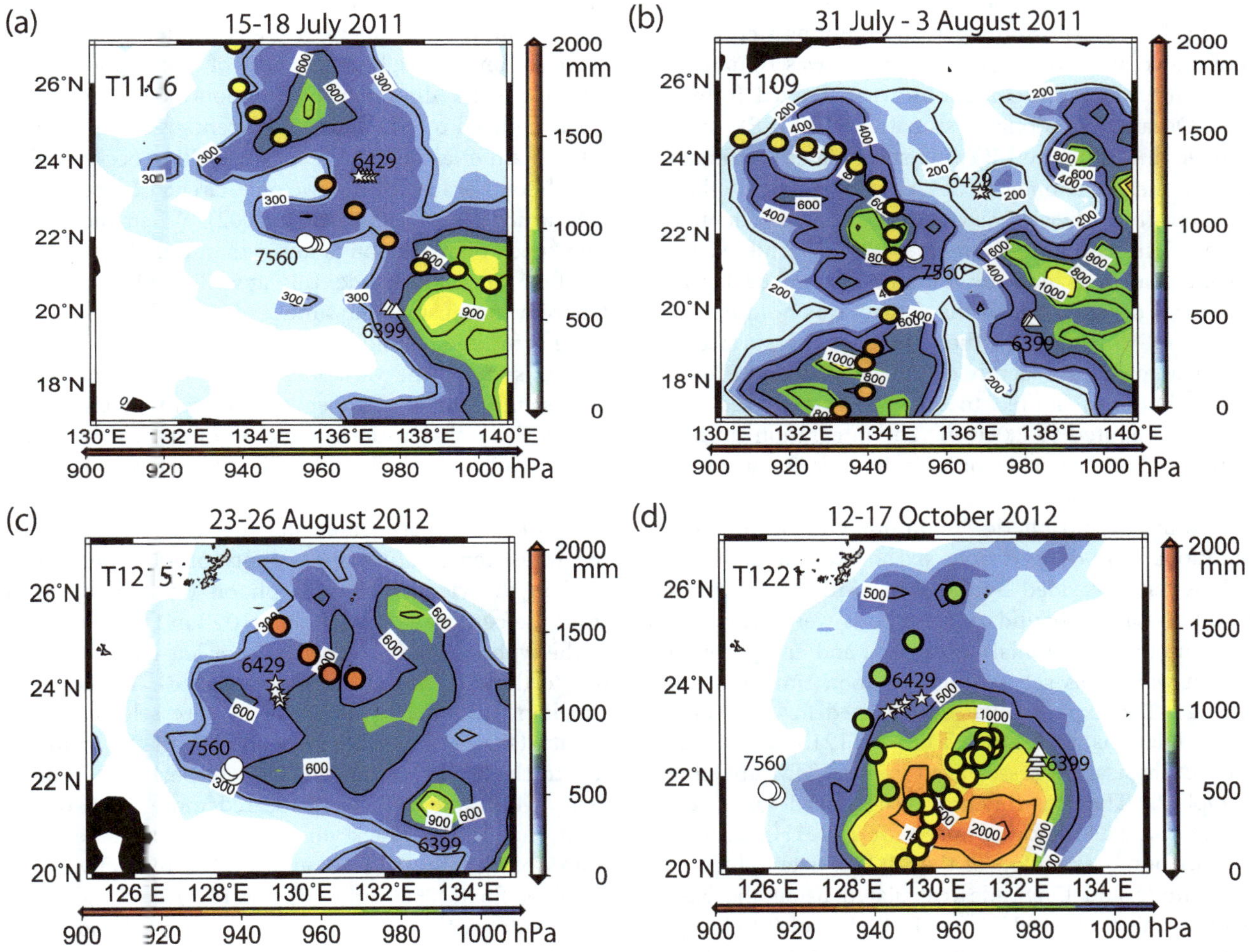

Figure 4 Horizontal distributions of three-hourly accumulated precipitation and the RSMC-Tokyo best track. Horizontal distributions of three-hourly accumulated precipitation (color shading) and the RSMC-Tokyo best track (filled circles) during the periods **(a)** 15 to 18 July 2011 for T1106, **(b)** 31 July to 3 August 2011 for T1109, **(c)** 23 to 26 August 2012 for T1215, and **(d)** 12 to 17 October 2012 for T1221. The colors within the circles indicate the central pressure of the typhoon at each position along the RSMC-Tokyo best track. The open stars, circles, and triangles show the positions of the profiling floats, as in Figure 1, except during each analysis period.

layer, and the thickness of each layer and the two-dimensional flows in the layers. The model could simulate near-inertial currents behind a TC reasonably well (Wada 2002).

The atmosphere-ocean coupling procedure is described next. Short- and long-wavelength radiation, sensible and latent heat fluxes, wind stresses, and precipitation were calculated by the atmospheric model and supplied to the ocean model at every time step in the ocean model. Land and sea distributions, extracted from the Global TOPOgraphic 30 digital elevation data produced by the US Geological Survey, were provided by the atmosphere model to the ocean model only at the initial time, to ensure that the land and sea distributions in the atmosphere and ocean models matched. The topography of the ocean bottom at 5′ latitude and longitude intervals was used in the ocean model, and these data were taken from the Earth Topography Digital Dataset 5 provided by the NOAA National Geophysical Data Center. The sea surface temperature calculated by the ocean model was provided to the atmospheric model at every time step in the ocean model.

Experimental design

The experimental design used in the numerical simulations of T1106, T1109, T1215, and T1221 are summarized in Table 2. The horizontal resolution of each typhoon simulation performed by the NHM was applied to the ocean-wave model and the multilayer ocean model. The vertical coordinates used in all of the NHM and coupled model experiments had 40 vertical levels, with different intervals ranging from 40 m for the lowermost (near-surface) layer to 1,180 m for the uppermost layer. The time step used in the ocean model was six times the time step used in the atmosphere model, and the time step used in all of the ocean-wave model experiments was 10 min.

The initial oceanic conditions were derived from the MRI Ocean Variational Estimation (MOVE) system (Usui et al. 2006). The MOVE system directly assimilates sea level height anomalies observed by a satellite-borne altimeter (http://www.aviso.oceanobs.com/). The initial depth of the oceanic mixed layer was determined from oceanic reanalysis data by defining the mixed layer as reaching a depth at which the density was no more than 0.25 kg m^{-3} higher than the density at the surface and having a maximum depth of 200 m. The base of the thermocline was limited to 600 m, whereas the water depth was limited to 2,000 m. The integration time was relatively short, so open oceanic boundaries were used in a series of numerical simulations.

The initial atmospheric conditions and the boundary conditions were derived from six-hourly global objective analysis data provided by the JMA, with a grid spacing of 20 km. Lateral boundary conditions were provided to the computational domain every 6 h, and an appropriate domain width was set for each typhoon simulation (see Table 2). The experimental design used for the numerical simulations of T1106, T1109, T1215, and T1221, and the physical processes used in the NHM are shown in Table 2. The experimental design used for T1221L was the same as the design used for T1221H, except for the horizontal resolution of the MOVE data, which was 0.5° latitude and longitude (calculated using the North Pacific version of MOVE) for T1221L and 0.1° latitude and longitude (calculated using the western North Pacific version of MOVE) for T1221H (Usui et al. 2006).

Results and discussion

Observational evidence for the oceanic response to typhoons

In situ observations made by autonomous oceanic profiling floats have been used to investigate the oceanic response to a typhoon in previous studies. Wada et al. (2009b) found that vertical water temperature profiles obtained at 10-day intervals were clearly affected by the passage of a typhoon and that the change was dependent on the position of the profiling float relative to the typhoon track. Baranowski et al. (2011) found that the passage of two consecutive typhoons, Hagupit and Jangmi, in 2008, caused changes in the oceanic mixed layer. The *in situ* observations that showed these changes were made by an autonomous oceanic profiling float operating on a cycle with a period of approximately 1 day, and the authors showed that more frequent profiling (at least once a day) is required to allow variations in the water temperature and salinity in the upper ocean caused by the passage of a TC to be studied.

Two essential processes must be taken into account when assessing the oceanic response to a typhoon. First, vertical turbulent mixing occurs beneath a typhoon. In the Northern Hemisphere, this entrainment occurs more strongly on the right-hand side of the typhoon track (relative to the direction the typhoon moves), where the shear induced by near-inertial currents is stronger (Price 1981; Nam et al. 2012). Second, upwelling occurs behind a typhoon because of the cyclonic circulation of the surface winds (Price 1981; Wada 2002). When the translation speed of a typhoon is lower than the phase speed of the first baroclinic mode, this upwelling plays an essential role in causing sea surface cooling because a cold wake is formed behind the typhoon (Shay et al. 1998).

In this study, the variations in the water temperature and salinity in the upper ocean were studied using *in situ* daily observations made by three profiling floats. In particular, we will assess how the upper ocean responses vary according to the position of the float relative to the typhoon center and track. The evolutions of the water temperature and salinity profiles were examined in the vicinity of the typhoon (less than 100 km from the center) and more than 100 km from the center of the typhoon track on the right- or left-hand side (relative to the direction the typhoon moved). The following profiling floats and typhoon cases were selected: profiling floats 7560 and 6399 were located in the vicinity of the tracks of T1109 and T1215, respectively; profiling floats 6429 and 6399 were located more than 100 km from the nearest track on the right-hand side of the tracks of typhoons T1106 and T1221, respectively; and profiling floats 7560 and 6429 were located more than 100 km from the nearest track on the left-hand side of the tracks of typhoons T1106 and T1215, respectively (Figures 2, 3, and 4).

Figures 5, 6 and 7 show the vertical water temperature and salinity profiles that were found by the profiling floats within 100 km of the typhoon centers (Figure 5) and more than 100 km from the typhoon centers on the right-hand (Figure 6) and left-hand (Figure 7) sides of the tracks. The vertical water temperature and salinity profiles for T1109 (float 7560) and T1215 (float 6399) (Figure 5) showed that the water in the vicinity of each typhoon center was warm and that the salinity in the mixed layer was low and uniform before the passage of T1109 on 29 July 2011 and the passage of T1215 on 21 August 2012. Both the mixed and isothermal layers became deeper during the passage of both typhoons, and the water temperature abruptly decreased and the salinity abruptly increased in the mixed layer on 2 August 2011 for T1109 and 23 August 2012 for T1215. The isothermal layer was deeper than the mixed layer in both typhoons. The difference between the isothermal and mixed layer depths was caused by the upwelling of saline water in the seasonal thermocline. The seasonal thermocline became steeper during the passage of each typhoon

Table 2 Specifications of the coupled model and the experimental design for the simulations

	T1106	T1109	T1215	T1221L(H)
Horizontal resolution	2 km	2 km	2 km	3 km
Horizontal grid number	1,391 × 1,201	1,101 × 1,101	1,291 × 1,021	701 × 701
Vertical grid number	40			
Top height	22,801 m			
Time step (atmosphere model)	6 s	6 s	6 s	8 s
Initial time	2011/7/13/0000UTC	2011/7/30/0000UTC	2012/8/21/0000UTC	2012/10/8/0000UTC
Integration time	144 h	84 h	96 h	168 h
Lateral boundary relaxation sponge layers	70	100	60	15
Oceanic initial condition	MOVE 0.1°	MOVE 0.5°	MOVE 0.1°	MOVE 0.5° (0.1°)
Surface roughness length	Kondo (1975) [uncoupled model]/Taylor and Yelland (2001) [coupled model]			
Surface boundary layer	Louis et al. (1982)			
Atmospheric boundary process	Klemp and Wilhelmson (1978) and Deardorff (1980)			
Cloud physics	Explicit three-ice bulk microphysics (Lin et al. 1983)			
Radiation	Sugi et al. (1990)			

MOVE, MRI Ocean Variational Estimation.

than it had been beforehand. After the typhoons had passed, the water temperature had decreased from the surface to a depth of 200 m, and this decrease was attributed to upwelling occurring behind the typhoon. However, the amount by which the temperature decreased was clearly different for each of the two typhoons. The salinity within the mixed layer increased more after T1215 had passed than after T1109 had passed, even though the amount of precipitation was similar for both typhoons. The magnitude of the salinity increase was found to be related to the pre-existing salinity profile, i.e., the salinity was higher below the seasonal thermocline than near the surface before the passage of the typhoon.

The vertical water temperature profiles for T1106 (float 6429) and T1221 (6399) (Figure 6) showed that both the mixed and isothermal layers became deeper more than 100 km from the typhoon center on the right-hand side of the typhoon track when T1106 passed on 16 and 17 July 2011 and when T1221 passed on 13 and 14 October 2012. The deepening of the mixed layer led to an increase in the water temperature (compared with the water temperature before the passage) below the base of the mixed layer during and after the passage of the typhoon (after 18 July 2011 for T1106 and after 15 October 2012 for T1221). After T1221 had passed, in particular, the mixed layer became deeper than 100 m.

The vertical salinity profiles (Figure 6) showed that salinity in the mixed layer decreased after the passage of T1106. The salinity in the mixed layer was relatively high on 13 July 2011, before the typhoon passed, but decreased during and after the passage of the typhoon (after July 17), when heavy precipitation occurred. The mixed layer depth was almost the same as the isothermal depth, indicating that the effect of vertical turbulent mixing overwhelmed the effect of the precipitation. It should be noted that float 6429 was located at the eastern edge of the warm eddy on July 13 (Figure 2a), whereas it was located within the cold eddy on July 20 (Figure 2e). The westward propagation of mesoscale oceanic circulation and the resultant changes in environmental oceanic conditions possibly caused changes to occur in the temperature and salinity profiles at depth greater than 80 m before the passage of T1106.

The salinity profile of the upper ocean was quite different for T1221 than for T1106. The salinity was high in the seasonal thermocline on 10 October 2012 before T1221 passed. The salinity became low near the surface when T1106 passed over, whereas the salinity became low around the seasonal thermocline when for T1221 passed over. Figure 4 shows that the precipitation accompanying T1221 was much heavier (the three-hourly accumulated precipitation reached 1,000 mm around the location of float 6399) than the precipitation accompanying T1106, and this was caused by the relatively slow and irregular track of T1221. Water with a low salinity near the surface was transported downward by strong vertical turbulent mixing around the location of float 6399. Strong vertical turbulent mixing associated with T1221 not only deepened the mixed layer but also homogenized the water by decreasing the salinity in the lower part of the mixed layer and increasing the salinity in the upper part of the mixed layer, resulting in the water at depths of 60 to 100 m becoming fresher from October 15 to 20 (Figure 6b). A similar decrease in the salinity after the typhoon had passed was not observed

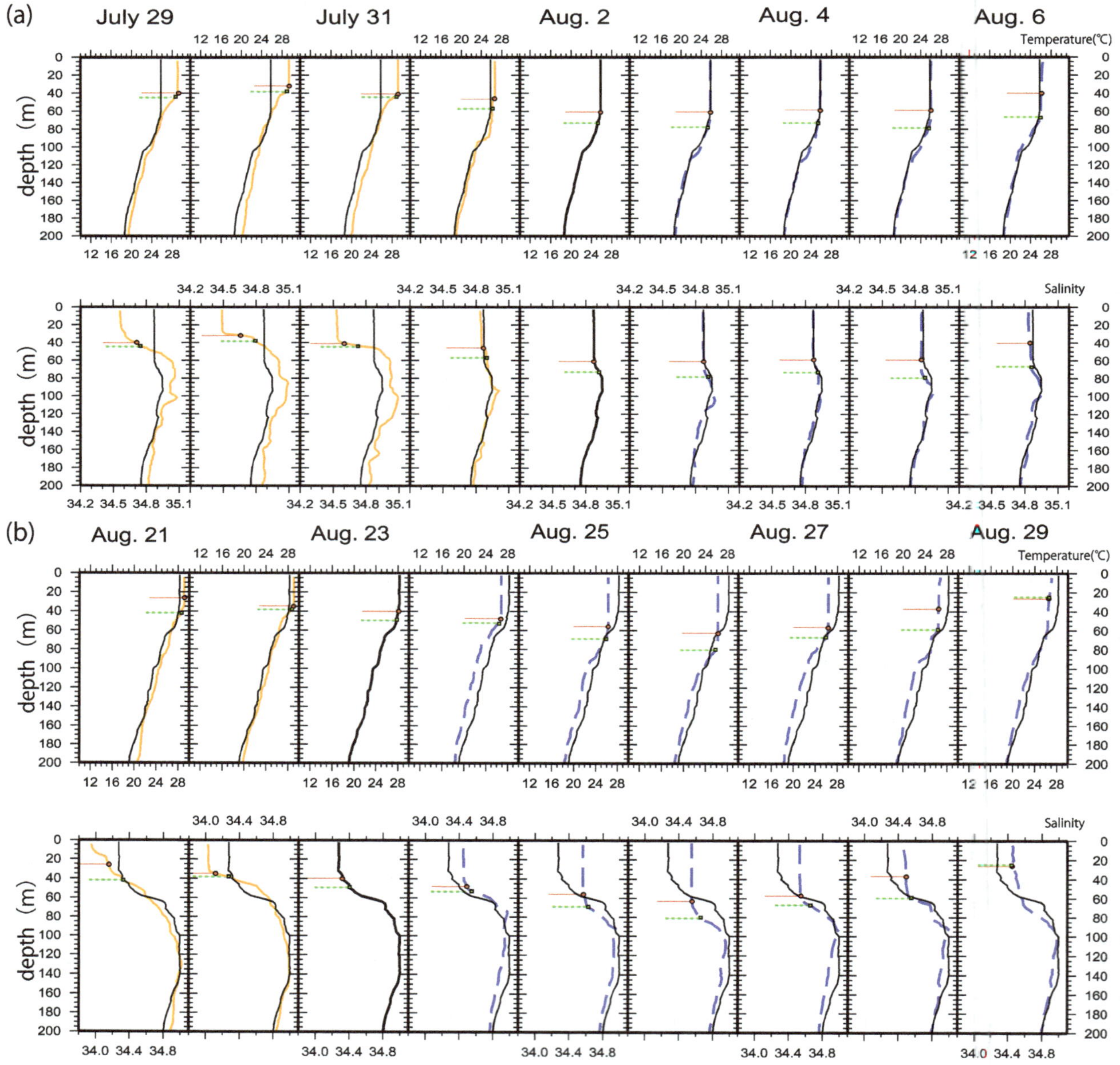

Figure 5 Vertical profiles of the water temperature and salinity in the vicinity of the typhoon track. Vertical profiles of the water temperature (upper panel) and salinity (lower panel) in the vicinity of the typhoon track during the periods **(a)** 31 July to 3 August 2011 for T1109, observed by profiling float 7560, and **(b)** 23 to 26 August 2012 for T1215, observed by profiling float 6399. Thin solid orange lines indicate the profiles before the passage of a typhoon. The thick line indicates the profile during the passage of the typhoon. Dashed blue lines indicate the profiles after the passage of a typhoon. Red circles with a solid line indicate the depth of the mixed layer. Green squares with a dashed line indicate the depth of the isothermal layer.

in the vicinity of a typhoon, less than 100 km from the center (Figure 5). The slow and irregular track of T1221 and the accompanying heavy precipitation played crucial roles in changing the vertical water temperature and salinity profiles.

The vertical water temperature profiles for T1106 (float 7560) and T1215 (float 6429) showed that both the mixed layer and isothermal layer depths more than 100 km from the typhoon center on the left-hand side of the typhoon track were deepest during the passage of T1106 on 16 July 2011 and T1215 on 24 and 25 August 2012 (Figure 7). The differences between the mixed and isothermal layer depths in T1106 and T1215 were relatively small compared with the differences between the mixed and isothermal layer depths found in the vicinity of a typhoon (Figure 5) and on the right-hand side of a

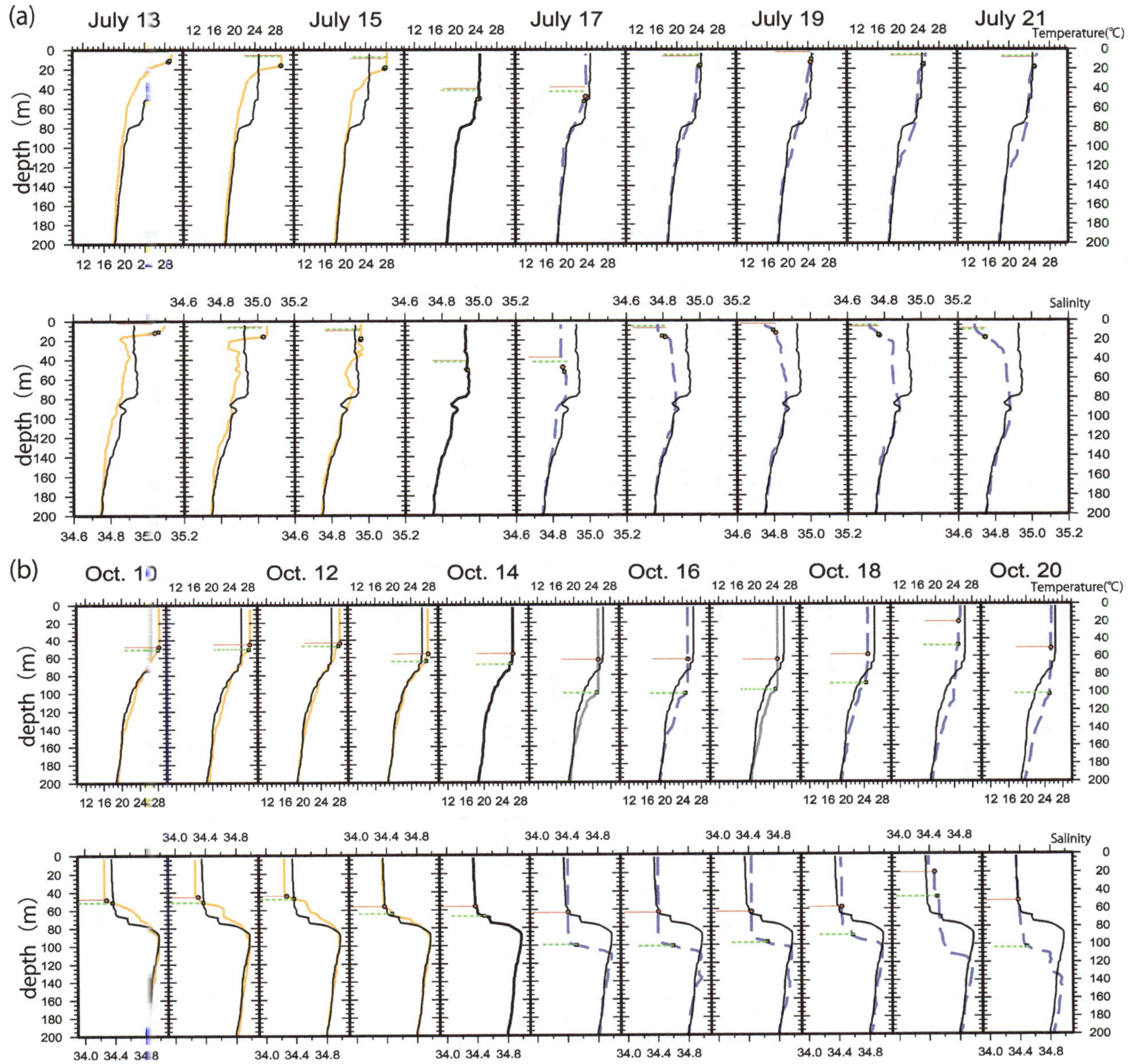

Figure 6 Vertical profiles of the water temperature and salinity on the right-hand side of the typhoon. Vertical profiles of the water temperature (upper panel) and salinity (lower panel) on the right-hand side of the typhoon, relative to its direction of motion, during the periods **(a)** 15 to 18 July 2011 for T1106, observed by profiling float 6429, and **(b)** 12 to 17 October 2012 for T1221, observed by profiling float 6399. Thin solid orange lines indicate the profiles before the passage of a typhoon. The thick line indicates the profile during the passage of the typhoon. Dashed blue lines indicate the profiles after the passage of a typhoon. Red circles with a solid line indicate the depth of the mixed layer. Green squares with a dashed line indicate the depth of the isothermal layer.

typhoon (Figure 6). Similar to the salinity profiles found more than 100 km from the center of a typhoon on the right-hand side the vertical salinity profiles on the left-hand side showed that both the mixed layer and the isothermal layer became deeper during the passage of each typhoon. However, the salinity was lower both before and after the passage of the typhoon than during its passage, even though the total amount of precipitation was not remarkably high. Moreover, the decreases in the water temperatures caused by the passage of the typhoons were relatively small compared with the decreases found in the vicinity of a typhoon and more than 100 km from the center on the right-hand side. Figure 2 shows that vertical temperature and salinity profiles were both located over the warm core eddies. Both the mixed and isothermal layer depths were approximately 30 to

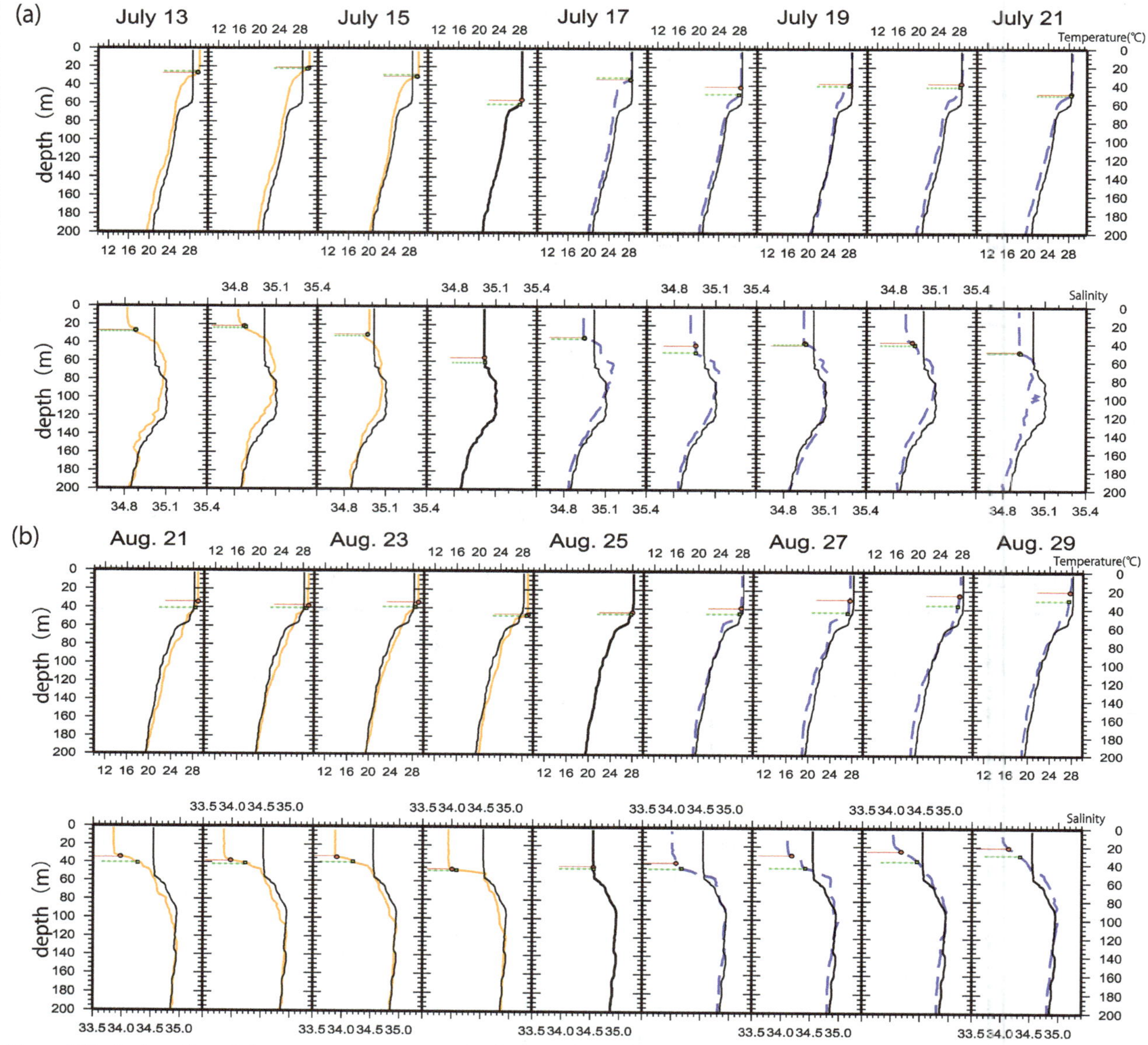

Figure 7 Vertical profiles of the water temperature and salinity on the left-hand side of the typhoon. Vertical profiles of the water temperature (upper panel) and salinity (lower panel) on the left-hand side of the typhoon, relative to its direction of motion, during the periods **(a)** 15 to 18 July 2011 for T1106, observed by profiling float 7560 and **(b)** 23 to 26 August 2012 for T1215, observed by profiling float 6429. Thin solid orange lines indicate the profiles before the passage of a typhoon. The thick line indicates the profile during the passage of the typhoon. Dashed blue lines indicate the profiles after the passage of a typhoon. Red circles with a solid line indicate the depth of the mixed layer. Green squares with a dashed line indicate the depth of the isothermal layer.

40 m, which was relatively deep compared with the depths before the passage of T1106 on 13 July 2011 and of T1215 on 21 August 2012 (Figure 7). These conditions are not favorable for the passage of a typhoon causing sea surface cooling.

Cione and Uhlhorn (2003) found differences between inner-core and ambient sea surface temperatures of approximately 0°C to 2°C. These differences are much smaller than the cold-wake-like decrease in the sea surface temperature of 4°C to 5°C, which is typically observed. Our findings were consistent with those of Cione and Uhlhorn (2003). During the passage of a typhoon, the inner-core sea surface temperature is mainly determined by vertical turbulent mixing, irrespective of the position relative to the typhoon center. In fact, the vertical water temperature and salinity profiles we found demonstrate that the mixed layer, determined from both the water temperature and the salinity, becomes deeper almost the entire time that a typhoon passes over. However, in our analysis, we did not take into account the effect of the translation speed of the typhoon on vertical turbulent mixing, even though the translation speed is known to be important to the oceanic response to a typhoon (Lin et al. 2009) and certainly affected changes in the vertical water temperature and salinity profiles for T1221 (Figures 5b and 6b). It will be necessary to further explore the relationship between the translation speed of a typhoon and the different oceanic responses to the typhoon at different positions relative to the center.

The results of this study imply that the oceanic response to a typhoon is influenced by pre-existing oceanic conditions. In fact, Figure 2 shows that the water on the right-hand side of the tracks of some of the typhoons was much colder than the water on the left-hand side before the typhoons passed, which is confirmed by Figures 6 and 7. In that sense, the contribution of cool conditions to strong sea surface cooling is not negligible. A strong cooling effect depends, to some extent, on the location relative to the typhoon, especially when warming or cooling occurs in the seasonal thermocline; however, the cooling effect is also partly caused by differences in pre-existing oceanic vertical profiles. To explore this subject more thoroughly, many more *in situ* observations underneath typhoons, with different pre-existing oceanic conditions, will be required. That is beyond the scope of this study, but such a study should be performed in the future.

Typhoon track, central pressure, and sea surface temperature

In previous studies, sea surface cooling has been found to have led to the weakening of the typhoons themselves (Chang and Anthes 1979; Zhu et al. 2004; Wu et al. 2005; Bender et al. 2007; Wada 2009). The weakening of a typhoon means that surface wind weakened, leading to the vertical turbulent mixing in the oceanic mixed layer being modified, and this modification affecting the cooling of the sea surface. Therefore, in this subsection, we will investigate the impact of typhoon-induced sea surface cooling on the simulations of the tracks and intensities of T1106, T1109, T1215, and T1221 using the atmosphere-wave-ocean coupled model. The maximum sea surface cooling determined from the *in situ* observations exceeded 2°C for these four typhoons (Table 1). It should be noted that the maximum sea surface cooling that occurred during the other four typhoons (T1112, T1114, T1115, and T1116) could have exceeded 2°C; however, we did not assess these four typhoons because no *in situ* observational evidence that the maximum sea surface cooling exceeded 2°C was available (Table 1). The position of each simulated typhoon during the integration time was defined as the location where the sea-level pressure reached a minimum value at a certain time. The intensity of a typhoon was defined as its central pressure. The TRMM/TMI maximum sea surface cooling effects shown in Table 3 were calculated as the maximum differences between the sea surface temperature on a certain day within each analysis period (Table 1) and the sea surface temperature on the initial analysis day of that analysis period (Figure 1). The maximum sea surface cooling effect simulated using the coupled model was calculated in a similar way; however, it was the maximum difference between the sea surface temperature at a certain integration time within the entire integration period (Table 1) and the sea surface temperature at the initial time.

The RSMC-Tokyo best tracks for T1106 (Figure 8a), T1109 (Figure 8b), T1215 (Figure 8c), and T1221 (T1221L and T1221H; Figure 8d) were compared with their simulated tracks, and calculated using the NHM (uncoupled) and the full coupled atmosphere-wave-ocean models. Both of the simulated T1106 tracks showed excessive westward translation only at around 20°N and 136°E (Figure 8a). The impact of simulated sea surface cooling on the track simulation was small for T1106, and ocean coupling did not decrease the westward bias of the simulation. In contrast, the T1109 track simulated using the coupled model was closer to the best track than was the track simulated using the uncoupled model (Figure 8b). Like the T1106 track simulations, both of the T1215 and T1221 track simulations showed that simulated sea surface cooling did not significantly affect the simulated tracks of T1215 (Figure 8c), T1221L, or T1221H (Figure 8d). The T1215 tracks simulated using both the uncoupled and coupled models showed a northward bias at around 21°N and 134°E and a northwestward bias east of 132°E without changing the direction of movement to the north-northwest. The T1221 tracks simulated using both the uncoupled and coupled models showed a southward bias when the typhoon was moving westward and a northward bias around the recurvature point (20°N and 128°E) when the typhoon was moving northeastward. The translation of the simulated typhoon was relatively slow compared with the translation determined using the RSMC-Tokyo best-track data.

The differences between the tracks simulated using the uncoupled and coupled models reflected the configuration of the lateral sponge layer in the regional atmospheric model (Wada 2012). It is possible that a small number of lateral sponge layers caused the track simulation of the regional 'child' model to differ from that of the 'parent' analysis (e.g., by global objective analysis) because the effect of the parent analysis on the track simulations using the regional uncoupled and coupled models became small. In these numerical simulations, we set the sponge layer number to 70 for T1106 with 1,391 × 1,201 grids, 100 for T1109 with 1,101 × 1,101 grids, 60 for T1215 with 1,291 × 1,021 grids, and 15 for

Table 3 Maximum sea surface cooling effects

Location of float ID	T1106 (°C)	T1109 (°C)	T1215 (°C)	T1221L (°C)	T1221H (°C)
6429	4.0	2.8	3.2	3.3	3.2
7560	3.1	2.8	3.2	2.5	2.6
6399	3.0	2.6	2.7	3.9	3.9
Max. cooling (model)	5.4	3.2	3.2	2.9	3.3
Max. cooling (TMI/AMSRE)	6.3	3.9	4.8	5.7	

T1221 with 701 × 701 grids (Table 2). The sponge layer number of 70 for T1106 was comparable to the sponge layer number of 100 for T1109; thus, the differences between the tracks simulated using the coupled and uncoupled models could not be accounted for by the sponge layer number for T1109. Therefore, these results suggest that the differences between the impacts of the simulated sea surface cooling on the track simulation depended on the typhoon and the atmospheric environments. Later in this subsection, we will investigate the impact of ocean coupling on the track simulation of T1109 further.

The central pressure time series for T1106 (Figure 9a), T1109 (Figure 9b), T1215 (Figure 9c), and T1221 (T1221L and T1221H; Figure 9d) simulated using the uncoupled and coupled models were compared with the time series for the RSMC-Tokyo best-track central pressures for each typhoon. The intensification, mature, and decaying phases for all of the typhoons except for T1215, which includes only the intensification phase according to the best-track data, are shown in Figure 9.

The simulated sea surface cooling induced by the typhoon began to have an impact on the simulated central pressure during the intensification phase for all four typhoons. The central pressures for T1106 (Figure 9a) and T1221 (Figure 9d) simulated using the uncoupled model were lower than the RSMC-Tokyo best-track central pressures, particularly during the mature phase. The

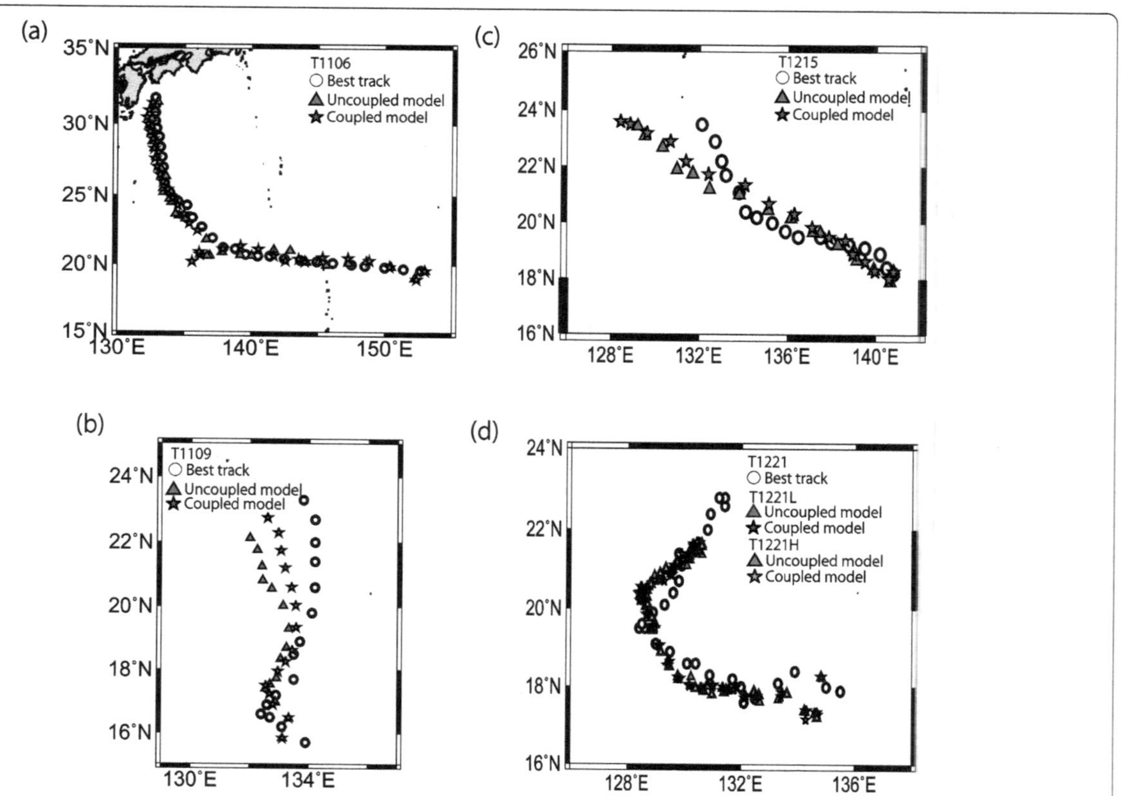

Figure 8 RSMC-Tokyo best tracks and simulated best tracks. (a) T1106, **(b)** T1109, **(c)** T1215, and **(d)** T1221L and T1221H.

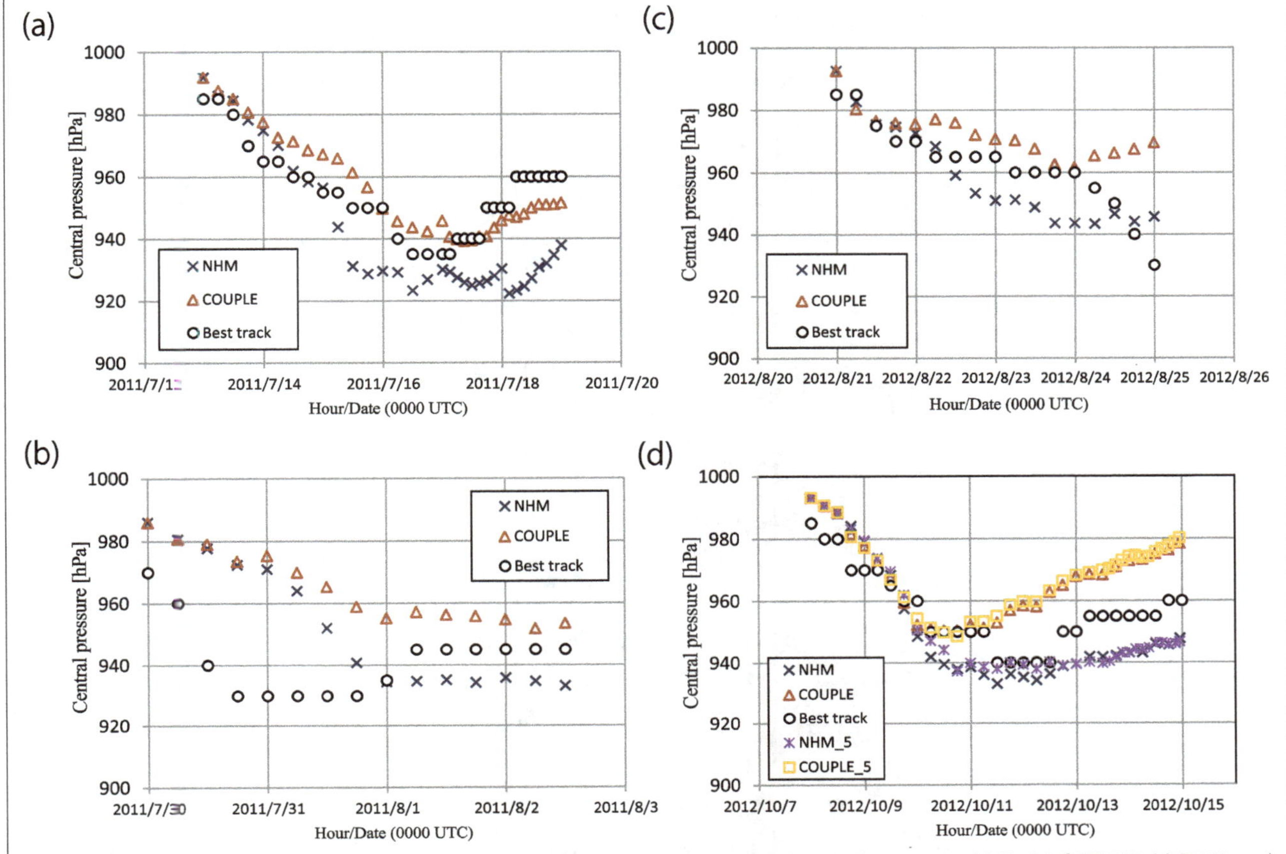

Figure 9 Time series of the RSMC-Tokyo best-track central pressures and simulated central pressures. (a) T1106, **(b)** T1109, **(c)** T1215, and **(d)** T1221L and T1221H. Open circles indicate RSMC-Tokyo best-track central pressures.

central pressure for T1215 (Figure 9c) simulated using the coupled model was lower than the RSMC-Tokyo best-track central pressure at around 0000 UTC on 25 August 2011 during the rapid intensification phase. This low central pressure was caused by an error in the track simulation (Figure 8c). During the intensification phase, in contrast, the central pressures simulated using the uncoupled model were much higher than the best-track central pressures for T1109 (Figure 9b). The central pressures simulated using the coupled model were higher than the central pressures simulated using the uncoupled model for all of the typhoons. This result, which shows that the simulated sea surface cooling induced by the typhoons decreased the magnitude of the TC intensification and the final (maximum) intensity, is consistent with the results of previous studies (Bender et al. 1993; Shade and Emanuel 1999; Wada 2009). This negative-feedback effect, however, did not improve the central pressure simulations, particularly in the rapid intensification phase, for the typhoons that were studied.

In addition, the impact of the horizontal resolution of the MOVE data on both the track (Figure 8d) and central pressure simulations for T1221 was small (T1221L and T1221H; Table 2 and Figure 9d). This is also consistent with previous findings (Wada et al. 2013). This suggests that the impact of the horizontal resolution difference in the oceanic analysis data on the typhoon simulation was negligible.

The horizontal sea surface temperature distributions used in the uncoupled model (Figure 10a,b,c,d,e) were compared with the distributions simulated using the coupled model (Figure 10f,g,h,i,j) for T1106 at 120 h, T1109 at 84 h, T1215 at 96 h, and T1221L and T1221H at 120 h. The sea surface temperature differences between the simulations using the uncoupled and coupled models are shown in Figure 11. The horizontal sea surface temperature distributions used in the uncoupled model did not change during the integration period. The results of the simulations using the coupled model revealed that sea surface cooling was induced by T1106 (Figures 10f and 11a), T1109 (Figures 10g and 11b), and T1215 (Figures 10h and 11c) on the right-hand side of the simulated track. The horizontal sea surface temperature distributions for T1221 were greatly changed by ocean coupling because of the simulated recurved irregular tracks with a slow translation. However, even using the

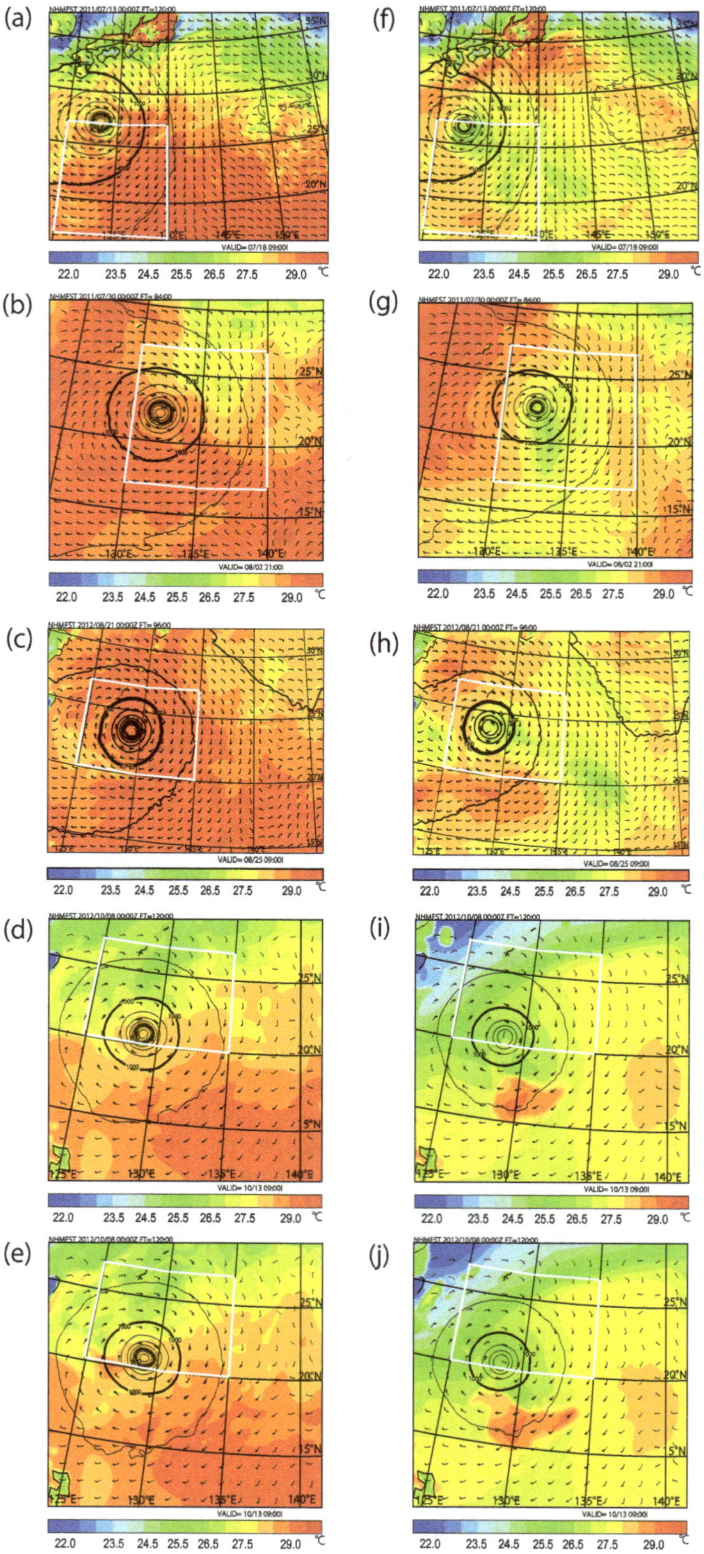

Figure 10 (See legend on next page.)

(See figure on previous page.)
Figure 10 Horizontal distributions of the sea surface temperature, sea-level pressure, and wind. Horizontal distributions of the sea surface temperature (color shading), sea-level pressure (contours; contour intervals are 8 hPa (thin line) and 40 hPa (thick line)), and wind at the 850 hPa level (vectors) simulated using **(a** to **e)** the uncoupled model and **(f** to **j)** the coupled model for (a and f) T1106, (b and g) T1109, (c and f) T1215, (d and i) T1221L, and (e and j) T1221H. White boxes indicate the domains wherein the three profiling floats were monitored (see Figure 1).

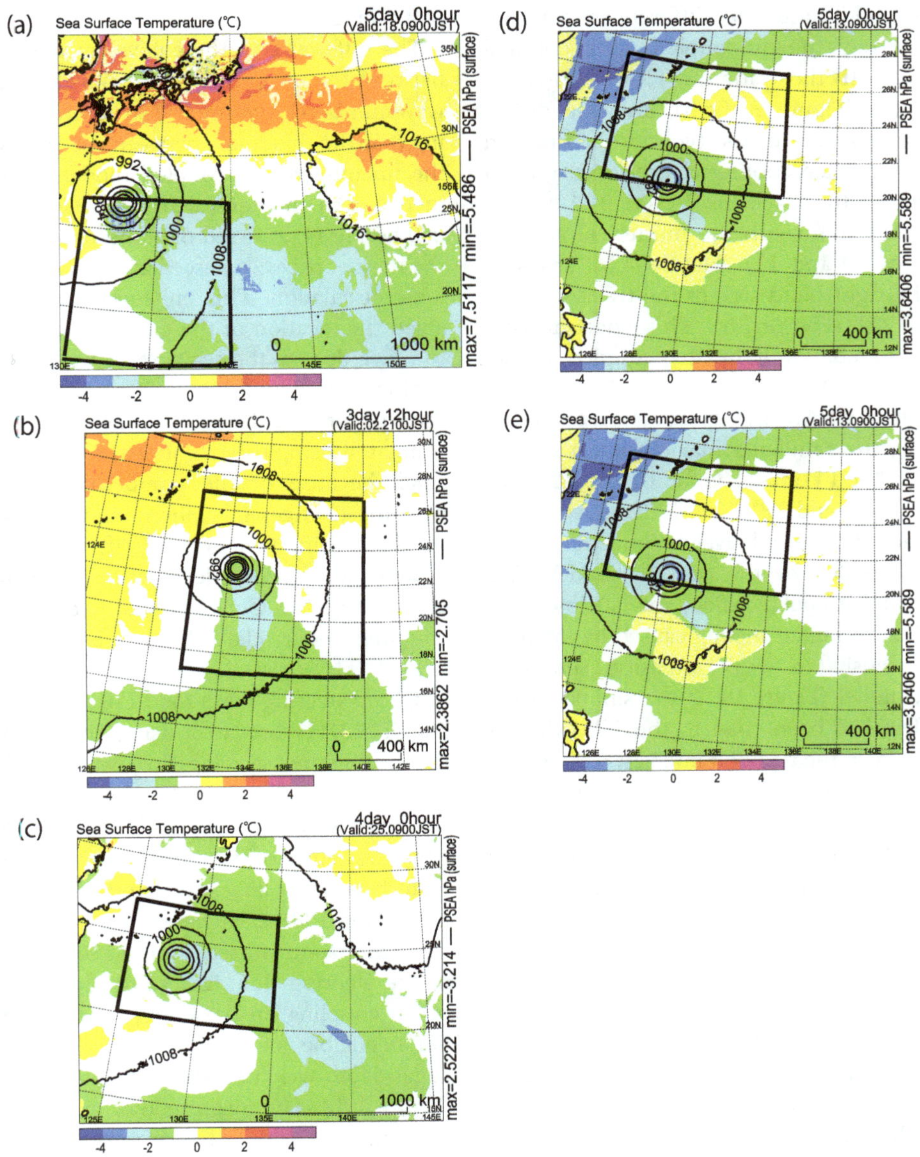

Figure 11 Horizontal distributions of the differences in sea surface temperatures simulated using the uncoupled model. Horizontal distributions of the differences in sea surface temperatures (color shading; the sea surface temperature simulated using the uncoupled model subtracted from the sea surface temperature simulated using the coupled model), and sea-level pressures simulated using the coupled model (contours; contour intervals are 8 hPa) **(a)** at 120 h for T1106, **(b)** at 84 h for T1109, **(c)** at 96 h for T1215, **(d)** at 120 h for T1221L, and **(e)** at 120 h for T1221H. Black boxes indicate the domains wherein the three profiling floats were monitored (see Figure 1).

coupled model (Figure 10i,j), the sea surface temperatures in the simulation (Table 3) were high compared with the observations (Table 1). The difference between the sea surface temperatures in T1221 that were observed in the simulations using the uncoupled and coupled models was relatively small compared with that for the other typhoons.

It is difficult to compare the simulated sea surface temperature with the TRMM/TMI sea surface temperature at a certain point/grid because of errors in the track and intensity simulations. Therefore, we compared the horizontal simulated sea surface temperature distributions (Figure 10f,g,h,i,j) and the differences between the simulations using the uncoupled and coupled models (Figure 11) with those for the TRMM/TMI sea surface temperatures (Figure 3e,f,g,h), and the differences between before and after the typhoon had passed (Figure 3i,j,k,l). Sea surface cooling clearly occurred on the right-hand side of the track for T1106, T1109, and T1215, which is consistent with the numerical simulation results (Figures 10 and 11). TRMM/TMI sea surface cooling on the right-hand side of the track was not remarkable for T1221 (Figure 3h). This characteristic was simulated reasonably well by the coupled model (Figures 10g,h, and 11d,e).

Compared with the TMI/AMSRE microwave satellite observations (Table 3), the simulated maximum sea surface cooling effects, 5.4°C for T1106, 2.7°C for T1109, 3.2°C for T1215, 2.9°C for T1221L, and 3.3°C for T1221H, were relatively low, indicating that a relatively high sea surface temperature was simulated by the coupled model. This relatively low sea surface cooling effect was attributed to the weakened vertical turbulent mixing caused by the high central pressures (causing a weak intensity) simulated using the coupled model (Figure 9). If the simulated sea surface cooling effect had been close to the TRMM/TMI sea surface cooling effect, the typhoon intensity would have been further weakened because of excessive cooling. This implies that the air-sea enthalpy fluxes within the inner core of a simulated typhoon were insufficient to simulate intensification and the maximum intensity of the typhoon. In other words, much higher enthalpy fluxes from the ocean to the atmosphere are needed, particularly within the inner core, to simulate the intensification phase and the maximum intensity.

Next, we explored the impact of sea surface cooling induced by T1109 on the difference between the tracks simulated using the uncoupled and coupled models. Figure 12 shows the deviations in equivalent potential temperatures at the 200 hPa level and wind velocities at the 500 hPa level that were simulated using the coupled and uncoupled models (the uncoupled results subtracted from the coupled results) for T1109. The typhoon-induced sea surface cooling effect led to a decrease in the equivalent potential temperature within the inner warm core of the typhoon at the 200 hPa level at 36 h (Figure 12a). This decrease caused anticyclonic circulation anomalies, which helped suppress cyclonic circulation, at the 500 hPa level in the center of the typhoon.

A decrease in the equivalent potential temperature was also apparent around the rainband on the western side of the typhoon center after 36 h (data not shown). In contrast, the equivalent potential temperature at the 200 hPa level at 42 h increased east of the area where the decrease occurred (Figure 12b). This warming was accompanied by enhanced cyclonic circulation at the 500 hPa level.

The difference in the equivalent potential temperatures between the uncoupled and coupled models at 60 h had a wavenumber-1 asymmetric structure (Figure 12c), indicating that the difference between the tracks simulated using the uncoupled and coupled models was significant. The increases in the equivalent potential temperatures at the 200 hPa level occurred on a mesovortex scale at 42 h, and the mesovortices coalesced at 60 h, strengthening the cyclonic circulation. This enhanced cyclonic circulation enabled the location of the center of the simulated typhoon to move despite the sea surface cooling that occurred beneath the typhoon.

In general, differences between the typhoon tracks simulated using the uncoupled and coupled models were small because of the competing processes associated with the changes in typhoon asymmetry and changes in the beta drift, which is a drift attributed to the background potential vorticity represented by the meridional gradient of the Coriolis parameter β, caused by ocean coupling under relatively weak background flow (Bender et al. 1993; Wu et al. 2005). Figure 13a shows the depth-weighted steering flows for T1106, T1109, T1215, T1221L, and T1221H simulated using the coupled model. The depth-weighted steering flow was calculated as a mass-weighted average every 3 h using the winds and geopotential heights at pressure levels between 300 and 850 hPa from the center of the typhoon to a radius of 400 km. The easterly steering flow was dominant in all of the typhoons except for T1109. However, the direction of the steering flow was southwesterly for T1109, and the magnitude of the steering flow was relatively small. The impact of ocean coupling on the depth-weighted steering flow was not negligible for T1109 because the magnitude of the zonal depth-weighted steering flow was less than 2 m s^{-1} (Figure 13b). The magnitude of the zonal depth-weighted steering flow for T1109 was quite different from that for T1215. Therefore, the weakening of both the typhoon intensity and the outer wind speeds caused the steering flow to change toward the northeast. In other words, more westward tracks simulated using the uncoupled model after 48 h may have been caused by northwestward beta

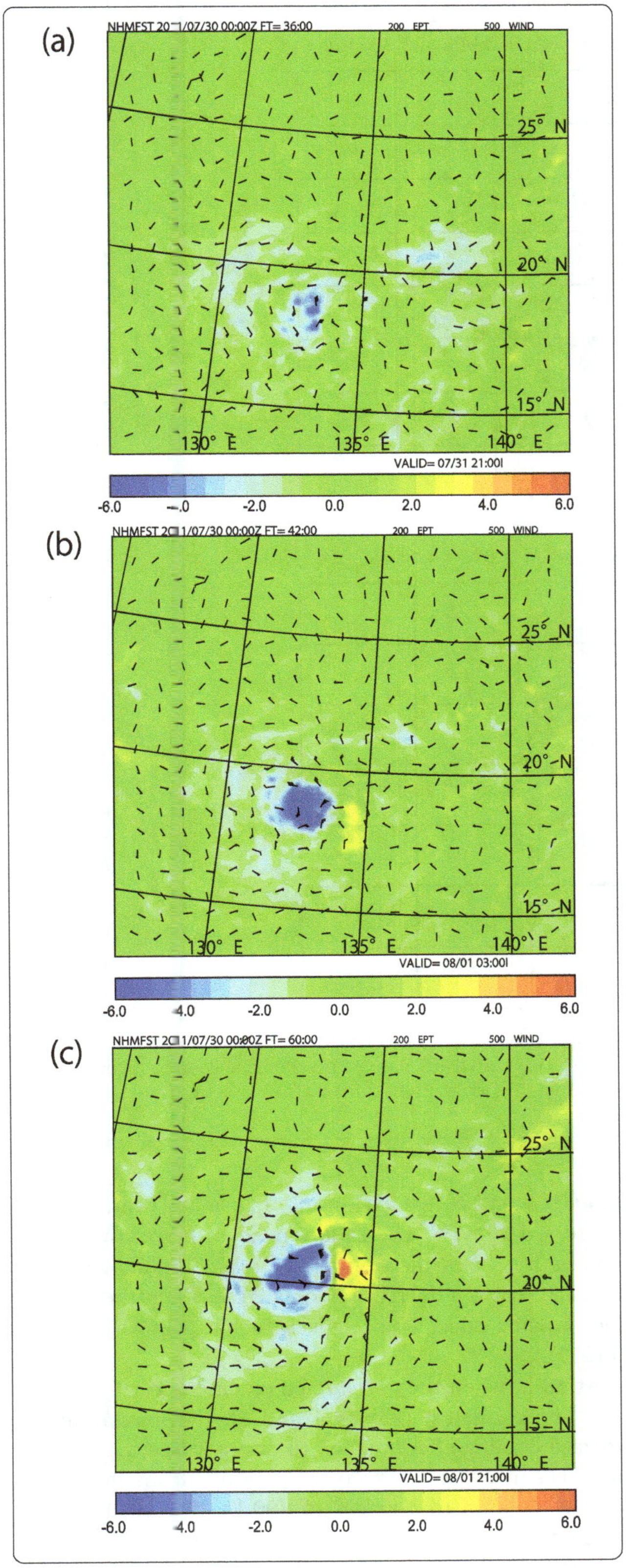

Figure 12 Differences in equivalent potential temperatures and wind simulated using uncoupled and coupled models. Differences in the 200 hPa equivalent potential temperatures (color shading) and 500 hPa wind (vectors) simulated using the uncoupled and coupled models (uncoupled results subtracted from the coupled results) for T1109 at **(a)** 36 h, **(b)** 42 h, and **(c)** 60 h.

drift under a relatively strong vortex and weak zonal basic flow. If the rapid intensification of T1109 had been successfully simulated the simulated track would have had a more northwestward bias. However, the larger amount of typhoon-induced sea surface cooling (see Table 3) would have changed the bias from northwestward to northeastward. This suggests that ocean coupling has the potential to improve track predictions under relatively weak basic flow.

Simulated sea surface cooling and latent heat fluxes

From the viewpoint of dynamics, sea surface cooling beneath a typhoon plays an essential role in delaying the merger of mesovortices within the inner core (Wada 2009). From the viewpoint of thermodynamics, typhoon-induced sea surface cooling leads to decreases in the latent heat fluxes from the ocean to the atmosphere. In this subsection, we will assess how typhoon-induced sea surface cooling affects the typhoon from the viewpoints of both dynamics and thermodynamics. Because latent heat fluxes are much higher than sensible heat fluxes over the ocean, we will focus on latent heat fluxes within the inner core of a typhoon, the inner core being defined as the region from the center of the typhoon to a radius of 200 km.

We will consider the axisymmetrical mean profiles of the sea surface temperature and the latent heat flux within the inner core at 24, 48, and 72 h, because the axisymmetric sea surface cooling effect during the intensification phase plays a decisive role in weakening the typhoon intensity, despite its relatively small magnitude compared with the asymmetric component of sea surface cooling after the passage of the typhoon (Wu et al. 2005). The three integration times we will use correspond to the intensification or mature phases of the typhoons we studied, except for T1221, which was transitioning from the mature phase to the decaying phase at 72 h. The bin sizes used when calculating the axisymmetrical mean values were 4 km for T1106, T1109, and T1215, and 6 km for T1221L and T1221H.

Tables 4 and 5 show the maximum values of the axisymmetrical mean latent heat fluxes calculated using the uncoupled and coupled models, and the maximum sea surface cooling derived from the deviations in the axisymmetrical mean sea surface temperatures between the uncoupled and coupled models at 24, 48, and 72 h for T1106, T1109, T1215, T1221L, and T1221H. The latent

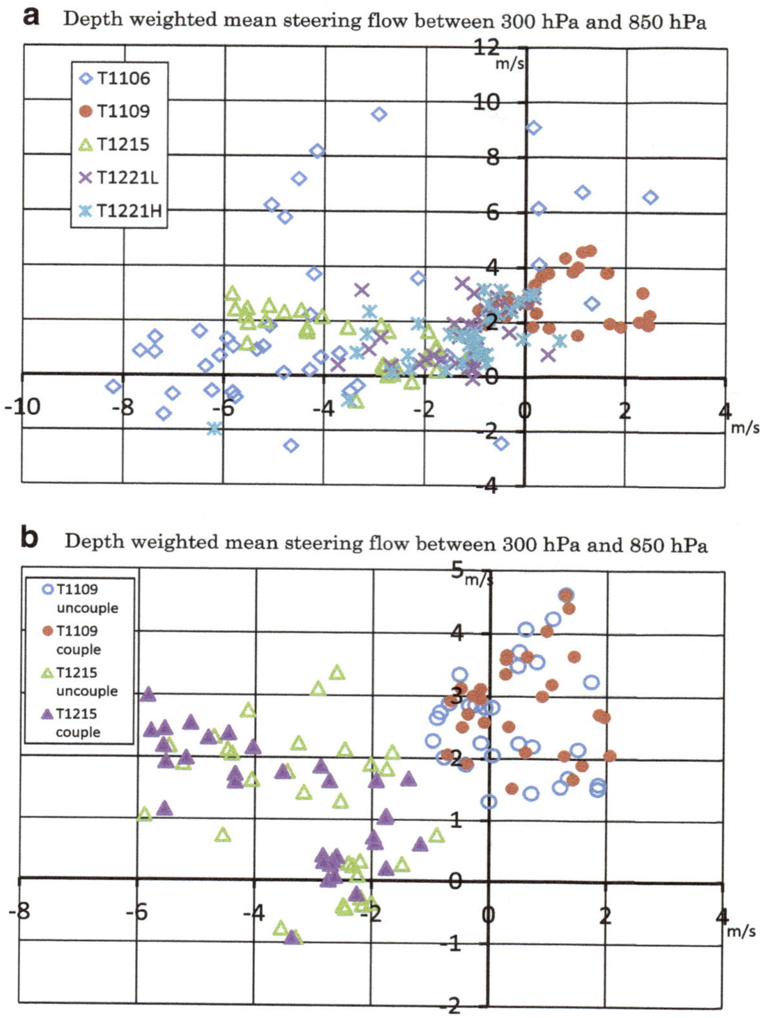

Figure 13 Scatter diagrams of the depth weighted mean steering flows. (a) Simulated using the coupled model for T1106 (open diamonds), T1109 (close circles), T1215 (open triangles), T1221L (crosses), and T1221H (asterisks). **(b)** Simulated using the uncoupled model (open marks) and coupled model (closed marks) for T1109 (circles) and T1215 (triangles). Positive values in the azimuthal axis indicate westerly steering flows, and positive values in the vertical axis indicate southerly steering flows.

heat fluxes calculated using the uncoupled model ranged from 275.1 W m^{-2} (for T1221L at 24 h) to 713.4 W m^{-2} (for T1215 at 48 h), whereas the latent heat fluxes calculated using the coupled model ranged from 209.4 W m^{-2} (for T1221L at 24 h) to 484.1 W m^{-2} (for T1215 at 72 h). The ratio of the latent heat flux calculated using the coupled model to the latent heat flux calculated using the uncoupled model ranged from 0.53 to 0.76. This implies that the latent heat fluxes were decreased by 24% to 47% due to ocean coupling, including the typhoon-induced sea surface cooling effect. Interestingly, the ratio decreased as the maximum sea surface cooling increased. For example, the highest maximum sea surface cooling effect was 1.8°C for T1109 at 48 h; however, the ratio for T1109 (0.53) at that time was the lowest of the typhoons that were studied (Tables 4 and 5), which means that latent heat flux was decreased by 47%. The ratio tended to be relatively high during the intensification phase (e.g., 0.76 at 24 h for T1221L that is a decrease of only 24%), when the sea surface cooling effect was relatively small (0.5°C at 24 h for T1221L).

At 48 h, when the sea surface cooling effect for T1109 was at its greatest (Tables 4 and 5), the maximum sea surface cooling effect tended to occur where the axisymmetrical mean sea surface temperature from the center of the typhoon to a radius of 200 km was at its lowest (Figure 14a). Moreover, the radial profile of the standard deviation of the axisymmetrical mean sea surface temperature indicated that the response of the sea surface temperature was asymmetrical outward along the

Table 4 Axisymmetrical mean latent heat fluxes and sea surface cooling effects for T1106 and T1109

Integration time	T1106			T1109		
	Latent heat flux (uncoupled; W m^{-2})	Latent heat flux (coupled; W m^{-2})	Sea surface cooling (°C)	Latent heat flux (uncoupled W m^{-2})	Latent heat flux (coupled; W m^{-2})	Sea surface cooling (°C)
24 h	451.9	324.2	1.0	331.1	227.5	1.0
(Ratio)		0.72			0.69	
48 h	459.0	266.8	1.5	569.9	303.3	1.8
(Ratio)		0.58			0.53	
72 h	566.4	388.3	1.2	492.9	260.6	1.6
(Ratio)		0.69			0.53	

radius of the maximum sea surface cooling effect. This is consistent with the asymmetry (rightward bias) seen in the oceanic response to typhoons in the Northern Hemisphere. In contrast, the radial profile of the standard deviation of the axisymmetrical mean latent heat flux hardly showed any asymmetry (Figure 14b) because the primary circulation of a typhoon is axisymmetric. The decrease in the wind speed within the inner core also played an essential role in decreasing the latent heat flux within the inner core.

We will now assess the impact of a decrease in the latent heat flux on the structure of the inner core of T1109, focusing on the lower troposphere, from the surface to 3,000 m. Changes in the latent heat flux within the inner core led to changes in the axisymmetrical mean equivalent potential temperature, the specific humidity, and the radial wind velocity. The vertical structure of the axisymmetrical mean equivalent potential temperature for T1109 showed a relatively high temperature around the center, which was a sharp horizontal temperature gradient around a radius of 50 km from the center above 1,000 m at 48 h (Figure 15a). The vertical structure of the axisymmetrical mean specific humidity showed that the specific humidity was highest around the center of the typhoon below 1,250 m, whereas it was highest at approximately 50 km from 1,250 to 3,000 m (Figure 15a). Below 1,250 m, the radial inflow was dominant from the surface to approximately 750 m, indicating the existence of an inflow boundary layer (Barnes and Powell 1995; Kepert and Wang 2001). The standard deviations were relatively low within the inflow boundary layer (Figure 15). This indicates that the primary circulation was axisymmetrically dominant in that area.

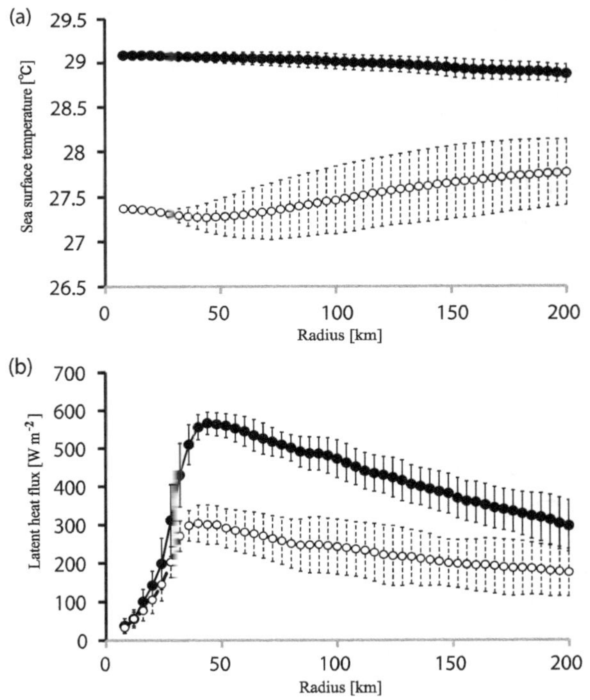

Figure 14 Axisymmetrical mean profiles. Sea surface temperature **(a)** and latent heat flux **(b)**, with standard deviations, for T1109 at 48 h. Closed circles with solid lines indicate the results simulated using the uncoupled model and open circles with dashed lines indicate the results simulated using the coupled model.

A decrease in the latent heat flux at 48 h led to a decrease in the axisymmetrical mean equivalent potential temperature around the center of the typhoon, weakening the horizontal gradient around a radius of 50 km from the center and slightly shifting outward the area wherein the standard deviation of the temperature was high (Figure 15c). The decrease caused by sea surface cooling led to a decrease in the specific humidity, particularly around a radius of 50 km in the inflow boundary layer (Figure 15d). In fact, the radial inflow simulated using the coupled model (Figure 15d) was much weaker than the radial inflow simulated using the uncoupled model (Figure 15b). This suggests that secondary circulation within the inner core of the simulated T1109 was weakened by a decrease in the latent heat flux caused by sea surface cooling induced by T1109.

Even though a sophisticated atmosphere-wave-ocean coupled model was used for the typhoon simulations, the track (Figure 8) and intensity (Figure 9) simulations

Table 5 Axisymmetrical mean latent heat fluxes and sea surface cooling effects for T1215, T1221L, and T1221H

Integration time	T1215			T1221L			T1221H		
	Latent heat flux (uncoupled; W m^{-2})	Latent heat flux (coupled; W m^{-2})	Sea surface cooling (°C)	Latent heat flux (uncoupled; W m^{-2})	Latent heat flux (couple; W m^{-2})	Sea surface cooling (°C)	Latent heat flux (uncoupled; W m^{-2})	Latent heat flux (coupled; W m^{-2})	Sea surface cooling (°C)
24 h	596.3	402.5	1.3	275.1	209.4	0.5	302.6	219.8	0.6
(Ratio)		0.67			0.76			0.73	
48 h	713.4	415.3	1.7	401.1	283.2	1.0	409.9	245.4	1.0
(Ratio)		0.57			0.71			0.50	
72 h	690.1	484.1	1.0	395.3	213.6	1.5	409.9	247.6	1.4
(Ratio)		0.67			0.54			0.60	

were not necessarily improved. In particular, rapid intensification in T1109 and T1215 was not successfully simulated even when the horizontal resolution was 2 km. Much higher enthalpy fluxes from the ocean to the atmosphere could be realized in the coupled model by introducing sea spray parameters (e.g., Bao et al. 2000), for example. Not only fine horizontal resolution but also improvements in the atmospheric physics used in the atmosphere model will be required to improve the typhoon predictions. In addition, uncertainties still remain in the initial and lateral boundary conditions in the atmosphere and ocean in the typhoon simulations.

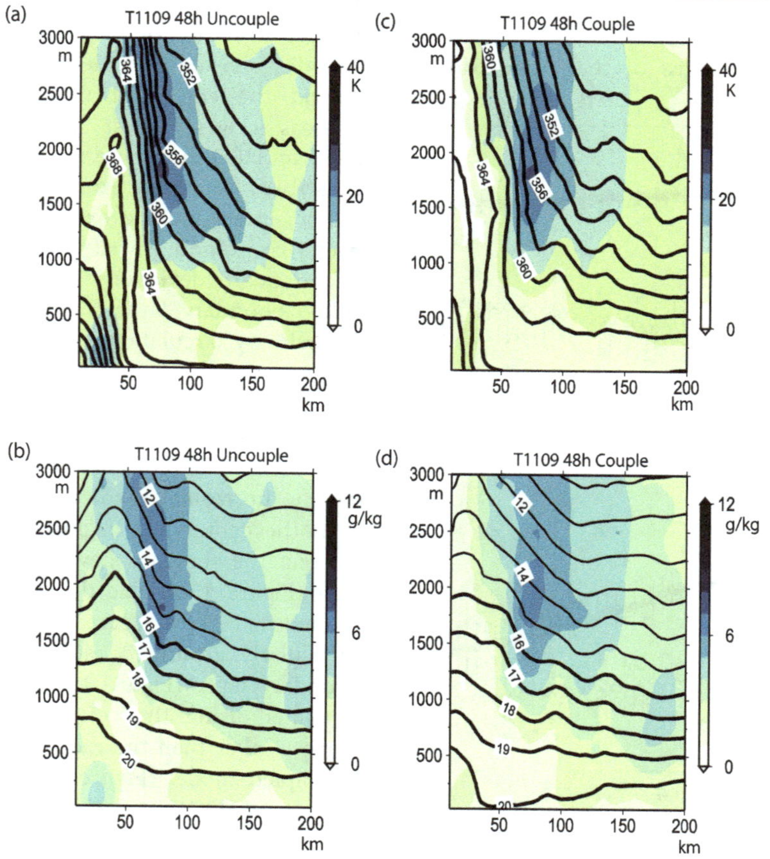

Figure 15 Axisymmetrical mean vertical profiles. (a) Equivalent potential temperature (solid contours) with the standard deviation (shaded), **(b)** specific humidity (solid contours) with the standard deviation (shaded) and radial inflow (dashed contours) simulated using the uncoupled model, **(c)** equivalent potential temperature (solid contours) with the standard deviation (shaded), and **(d)** specific humidity (solid contours) with the standard deviation (shaded) and radial inflow (dashed contours) simulated using the coupled model.

Solving these modeling problems will help clarify the mechanism involved in typhoon-induced sea surface cooling.

Conclusions

Daily observations using three profiling floats deployed around 20°N, 22°N, and 24°N along the 137°E meridian by the R/V Keifu Maru captured typhoon-induced sea surface cooling measurements during the 2011 and 2012 typhoon seasons. In this study, we investigated the characteristics of sea surface cooling effects from the viewpoint of different positions relative to the centers of the typhoons and different times through the passage of the typhoons, although sea surface cooling was partly attributed to pre-existing oceanic conditions and precipitation accompanying the typhoons. The impacts of sea surface cooling on typhoon simulations using the atmosphere-wave-ocean coupled model were assessed, focusing on the tracking and intensities of the typhoons, decreases in latent heat fluxes within the inner cores of simulated typhoons accompanying simulated sea surface cooling, and structural changes in the inner cores. The main conclusions are given below:

1. The oceanic response to a typhoon was examined according to the position of the float relative to the best-track typhoon center. Even though the oceanic response was affected by pre-existing oceanic conditions, precipitation accompanying the typhoon, the typhoon intensity, and the translation speed, the oceanic response clearly differed depending on the relative position studied. Upwelling was found to be dominant in the vicinity of the typhoon center, and the water temperature decreased from the surface to 200-m deep after the passage of a typhoon. Vertical turbulent mixing led to warming in the seasonal thermocline on the right-hand side of the typhoon track at a distance of 100 km or more from the typhoon center, whereas the oceanic response caused by vertical turbulent mixing was relatively weak on the left-hand side. Changes in the observed salinity profiles clearly differed from the changes in observed temperature profiles near the surface both before and after the passage of a typhoon in the vicinity of the center of the typhoon and on the right-hand side. This was found to be caused by stratification caused by the precipitation that accompanies typhoons and upwelled saline water in the seasonal thermocline. Deepening of the mixed layer caused by vertical turbulent mixing was clearly apparent in the temperature and salinity profiles during the passage of a typhoon.
2. Numerical simulations of typhoons Ma-on (T1106), Muifa (T1109), Bolaven (T1215), and Prapiroon (T1221) showed that typhoon-induced sea surface cooling significantly affected the extent to which simulated central pressures increased during the intensification and mature phases of the typhoons. Sea surface cooling also affected the track simulation of T1109 under weak basic flow because the weakened typhoon intensity altered the beta effect and changed the depth-weighted steering flows. Sea surface cooling beneath a typhoon led to axisymmetrical decreases in the latent heat fluxes of 24% (in the intensification phase, with sea surface cooling of less than 1°C) to 47% (in the mature phase, with sea surface cooling of more than 1°C). The decreases in the latent heat fluxes led to changes in the inner core structure of T1109, particularly in the inflow boundary layer and around the eye and eyewall, through changes in secondary circulation.

Abbreviations

AMSRE: Advanced Microwave Scanning Radiometer-Earth observing system; AVISO: Archiving: Validation and Interpretation of Satellite Oceanographic data; ID: Identification number; JMA: Japan Meteorological Agency; MOVE: Meteorological Research Institute Ocean Variational Estimation; MRI: Meteorological Research Institute; NHM: Nonhydrostatic atmosphere model; NOAA: National Oceanic and Atmospheric Administration; RSMC-Tokyo: Regional Specialized Meteorological Center-Tokyo; TC: Tropical cyclone; TMI: Tropical rainfall measuring mission microwave imager; WMO: World Meteorological Organization.

Competing interests

The authors declare that there are no conflicts of interests regarding the publication of this article.

Authors' contributions

AW proposed the topic and conceived and designed the study, including the numerical simulations using an atmosphere-wave-ocean coupled model. TU was in charge of the analysis of the profiling float data. SI conducted the oceanic reanalysis for the numerical simulations. All authors read and approved the final manuscript.

Authors' information

AW is a senior researcher in the Typhoon Research Department of the Meteorological Research Institute. TU and SI are technical officers in the Japan Meteorological Agency.

Acknowledgements

All authors would like to thank two anonymous reviewers for their thorough and helpful reviews. All authors are grateful to colleagues for observational and analysis support. This work was funded by KAKENHI Grant Number 25106708 from MEXT.

Author details

[1]Meteorological Research Institute, Japan Meteorological Agency, 1-1 Nagamine, Tsukuba, Ibaraki 305-0052, Japan. [2]Japan Meteorological Agency, 1-3-4 Otemachi, Chiyoda, Tokyo 100-8122, Japan.

References

Akima H (1970) A new method of interpolation and smooth curve fitting based on local procedures. J Assoc Comp Mach 17:589–602

Bao JW, Wilczak JM, Choi JK, Kantha LH (2000) Numerical simulations of air–sea interaction under high wind conditions using a coupled model: a study of hurricane development. Mon Weather Rev 128:2190–2210, doi: http://dx.doi.org/10.1175/1520-0493(2000)128<2190:NSOASI>2.0.CO;2

Baranowski DB, Flatau PJ, Malinowski SP (2011) Tropical cyclone turbulent mixing as observed by autonomous oceanic profilers with the high repetition rate. J Phys Conf Ser 318:072001, doi:10.1088/1742-6596/318/7/072001

Barnes GM, Powell MD (1995) Evolution of the inflow boundary layer of Hurricane Gilbert (1988). Mon Weather Rev 123:2348–2368, doi: http://dx.doi.org/10.1175/1520-0493(1995)123<2348:EOTIBL>2.0.CO;2

Bender MA, Ginis I, Kurihara Y (1993) Numerical simulations of the tropical cyclone-ocean interaction with a high-resolution coupled model. J Geophys Res 98:23245–23263, doi:10.1029/93JD02370

Bender MA, Ginis I, Tuleya R, Thomas B, Marchok T (2007) The operational GFDL coupled hurricane-ocean prediction system and a summary of its performance. Mon Weather Rev 135:3965–3989, doi: http://dx.doi.org/10.1175/2007MWR2032.1

Chang SW, Anthes RA (1979) The mutual response of the tropical cyclone and the ocean. J Phys Oceanogr 9:128–135, doi: http://dx.doi.org/10.1175/1520-0485(1979)009<0128:TMROTT>2.0.CO;2

Chen SS, Price JF, Zhao W, Donelan MA, Walsh EJ (2007) The CBLAST-Hurricane program and the next generation fully coupled atmosphere-wave-ocean models for hurricane research and prediction. Bull Am Meteorol Soc 88:311–317, doi:10.1175/BAMS-88-3-311

Chen SS, Zhao W, Donelan MA, Tolman HL (2013) Directional wind-wave coupling in fully coupled atmosphere-wave-ocean models: results from CBLAST-Hurricane. J Atmos Sci 70:3198–3215, doi:10.1175/JAS-D-12-0157.1

Cione JJ, Uhlhorn EW (2003) Sea surface temperature variability in hurricanes: implication with respect to intensity change. Mon Weather Rev 131:1783–1796, doi: http://dx.doi.org/10.1175//2562.1

D'Asaro E, Black P, Centurioni L, Harr P, Jayne S, Lin I-I, Lee C, Morzel J, Mrvaljevic R, Niiler P, Sanford T, Tang T (2011) Typhoon-ocean interaction in the western North Pacific: part 1. Oceanography 24:24–31, doi:10.5670/oceanog.2011.91

Deardorff JW (1980) Stratocumulus-capped mixed layers derived from a three-dimensional model. Boundary Layer Meteorol 18:495–527, doi:10.1007/BF00119502

Deardorff JW (1983) A multi-limit mixed-layer entrainment formulation. J Phys Oceanogr 13:988–1002, doi: 10.1175/1520-0485(1983)013<0988:AMLMLE>2.0.CO;2

Emanuel KA (1999) Thermodynamic control of hurricane intensity. Nature 401:665–669, doi:10.1038/44326

Emanuel KA, DesAutels C, Holloway C, Korty R (2004) Environmental control of tropical cyclone intensity. J Atmos Sci 61:843–858, doi: http://dx.doi.org/10.1175/1520-0469(2004)061<0843:ECOTCI>2.0.CO;2

Jeong YY, Moon I-J, Kim S-H (2013) A study on upper ocean response to typhoon Ewinar (0603) and its impact. Atmos Chem Phys 23:1–16

Joyce RJ, Janowiak JE, Arkin PA, Xie P (2004) CMORPH: a method that produces global precipitation estimates from passive microwave and infrared data at high spatial and temporal resolution. J Hydro 5:487–503

Kawai Y, Wada A (2007) Diurnal sea surface temperature variation and its impact on the atmosphere and ocean: a review. J Oceanogr 63:721–744, doi:10.1007/s10872-007-0063-0

Kepert J, Wang Y (2001) The dynamics of boundary layer jets within the tropical cyclone core. Part II: nonlinear enhancement. J Atmos Sci 58:2485–2501, doi: http://dx.doi.org/10.1175/1520-0469(2001)058<2485:TDOBLJ>2.0.CO;2

Klemp JB, Wilhelmson R (1978) The simulation of three-dimensional convective storm dynamics. J Atmos Sci 35:1070–1096, doi: 10.1175/1520-0469(1978)035<1070:TSOTDC>2.0.CO;2

Kondo J (1975) Air-sea bulk transfer coefficients in diabatic conditions. Bound-Lay Meteorol 9:91–112, doi:10.1007/BF00232256

Lin I-I (2012) Typhoon-induced phytoplankton blooms and primary productivity increase in the western North Pacific subtropical ocean. J Geophys Res 117: C03039, doi:10.1029/2011JC007626

Lin YH, Farley RD, Orville HD (1983) Bulk parameterization of the snow field in a cloud model. J Clim Appl Meteorol 22:1065–1092, doi: 10.1175/1520-0450(1983)022<1065:BPOTSF>2.0.CO;2

Lin I-I, Wu C-C, Emanuel KA, Lee I-H, Wu C-R, Pan I-F (2005) The interaction of supertyphoon Maemi (2003) with a warm ocean eddy. Mon Weather Rev 133:2635–2649, doi:10.1175/MWR3005.1

Lin I-I, Wu C-C, Pun I-F, Ko D-S (2008) Upper-ocean thermal structure and the western North Pacific category 5 typhoons. Part I: ocean features and the category 5 typhoons' intensification. Mon Weather Rev 136:3288–3306, doi:10.1175/2008MWR2277.1

Lin I-I, Pun I-F, Wu C-C (2009) Upper-ocean thermal structure and the western North Pacific category 5 typhoons. Part II: dependence on translation speed. Mon Weather Rev 137:3744–3757, doi:10.1175/2009MWR2713.1

Louis JF, Tiedtke M, Geleyn JF (1982) A short history of the operational PBL-parameterization at ECMWF. In: Proceedings of Workshop on Planetary Boundary Layer Parameterization. ECMWF, Reading, pp 59–79

Moon I-J, Kwon SJ (2012) Impact of upper-ocean thermal structure on the intensity of Korean peninsula landfall typhoons. Prog Oceanogr 105:61–66, doi: http://dx.doi.org/10.1016/j.pocean.2012.04.008

Nam S-H, Kim DJ, Moon WM (2012) Observed impact of mesoscale circulation on oceanic response to Typhoon Man-Yi (2007). Ocean Dyn 62:1–12, doi:10.1007/s10236-011-0490-8

Price JF (1981) Upper ocean response to a hurricane. J Phys Oceanogr 11:153–175, doi: 10.1175/1520-0485(1981)011<0153:UORTAH>2.0.CO;2

Pun IF, Chang Y-T, Lin I-I, Tang TY, Lien R-C (2011) Typhoon-ocean interaction in the western North Pacific: part 2. Oceanography 24:32–41, doi:10.5670/oceanog.2011.92

Qiu B, Chen S (2010) Interannual variability of the North Pacific Subtropical countercurrent and its associated mesoscale eddy field. J Phys Oceanogr 40:213–225, doi:10.1175/2009JPO4285.1

Saito K (2012) The Japan Meteorological Agency nonhydrostatic model and its application to operation and research. In: Yucel I (ed) Atmospheric model applications. InTech, Rijeka, Croatia, pp 85–110. doi:10.5772/35368

Sanford TB, Price JF, Girton JB, Webb DC (2007) Highly resolved observations and simulations of the ocean response to a hurricane. Geophys Res Lett 34: L13604, doi:10.1029/2007GL029679

Sanford TB, Price JF, Girton JB (2011) Upper ocean response to Hurricane Frances (2004) observed by EM-APEX floats. J Phys Oceanogr 41:1041–1056, doi:10.1175/2010JPO4313.1

Shade LR, Emanuel KA (1999) The ocean's effect on the intensity of tropical cyclones: results from a simple coupled atmosphere-ocean model. J Atmos Sci 56:642–651, doi: http://dx.doi.org/10.1175/1520-0469(1999)056<0642:TOSEOT>2.0.CO;2

Shay LK (2010) Air-sea interactions in tropical cyclones. In: Chan JCL, Kepert JD (eds) Global perspective on tropical cyclones from science to mitigation. World Scientific, Singapore, pp 93–132

Shay LK, Mariano AJ, Jacob SD, Ryan ED (1998) Mean and near-inertial ocean current response to hurricane Gilbert. J Phys Oceanogr 28:858–889, doi: http://dx.doi.org/10.1175/1520-0485(1998)028<0858:MANIOC>2.0.CO;2

Soloviev A, Lukas R (2006) The near-surface layer of the ocean: structure, dynamics and application. Atmo Ocean Sci Lib 31:572

Sugi M, Kuma K, Tada K, Tamiya K, Hasegawa N, Iwasaki T, Yamada S, Kitade T (1990) Description and performance of the JMA operational global spectral model (JMA-GSM88). Geophys Mag 43:105–130

Taylor PK, Yelland MJ (2001) The dependence of sea surface roughness on the height and steepness of the waves. J Phys Oceanogr 31:572–590, doi: 10.1175/1520-0485(2001)031<0572:TDOSSR>2.0.CO;2

Usui N, Ishizaki S, Fujii Y, Tsujino H, Yasuda T, Kamachi M (2006) Meteorological Research Institute multivariate ocean variational estimation (MOVE) system: some early results. J Adv Space Res 37:806–822, doi:10.1016/j.asr.2005.09.022

Wada A (2002) The processes of SST cooling by typhoon passage and case study of Typhoon Rex with a mixed layer ocean model. Pap Met Geophys 52:31–66, doi:10.2467/mripapers.52.31

Wada A (2009) Idealized numerical experiments associated with the intensity and rapid intensification of stationary tropical-cyclone-like vortex and its relation to initial sea-surface temperature and vortex-induced sea-surface cooling. J Geophys Res 114:D18111, doi:10.1029/2009JD011993

Wada A (2012) Oceanic influences for a large eye of Typhoon Talas in 2011. CAS/JSC WGNE Res Activ Atmos Oceanic Model 42:9.09–9.10

Wada A, Usui N (2007) Importance of tropical cyclone heat potential for tropical cyclone intensity and intensification in the Western North Pacific. J Oceanogr 63:427–447, doi:10.1007/s10872-007-0039-0

Wada A, Niino H, Nakano H (2009a) Roles of vertical turbulent mixing in the ocean response to Typhoon Rex (1998). J Oceanogr 65:373–396, doi:10.1007/s10872-009-0034-8

Wada A, Usui N, Sato K, Kawai Y (2009b) Comment on "Importance of pre-existing oceanic conditions to upper ocean response induced by Super Typhoon Hai-Tang" by Z.-W. Zheng, C.-R. Ho and N.-J. Kuo. Geophys Res Lett 36:L09603

Wada A, Kohno N, Kawai Y (2010) Impact of wave-ocean interaction on Typhoon Hai-Tang in 2005. SOLA 6A:13–16, doi:10.2151/sola.6A-004

Wada A, Cronin MF, Sutton AJ, Kawai Y, Ishii M (2013) Numerical simulations of oceanic pCO_2 variations and interactions between Typhoon Choiwan (0914) and the ocean. J Geophys Res 118:1–18, doi:10.1002/jgrc.20203.2013

Walker ND, Leben RR, Balasubramanian S (2005) Hurricane-forced upwelling and chlorophyll a enhancement within cold-core cyclones in the Gulf of Mexico. Geophys Res Lett 32:L18610, doi:10.1029/2005GL023716

Warner JC, Armstrong B, He R, Zambon JB (2010) Development of a coupled ocean-atmosphere-wave-sediment transport (COAWST) modeling system. Ocean Model 35:230–244, doi:10.1016/j.ocemod.2010.07.10

Wu L, Wang B, Braun SA (2005) Impacts of air-sea interaction on tropical cyclone track and intensity. Mon Weather Rev 133:3299–3313, doi: http://dx.doi.org/10.1175/MWR3030.1

Wu C-C, Lee CY, Lin I-I (2007) The effect of the ocean eddy on tropical cyclone intensity. J Atmos Sci 64:3562–3578, doi:10.1175/JAS4051.1

Zhu H, Ulrich W, Smith RK (2004) Ocean effects on tropical cyclone intensification and inner-core asymmetries. J Atmos Sci 61(Zhu H, Ulrich W, Smith RK): 1245–1257, doi: http://dx.doi.org/10.1175/1520-0469(2004)061<1245:OEOTCI>2.0.CO;2

2

Role of soil moisture-atmosphere feedback during high temperature events in 2002 over Northeast Eurasia

Enkhbat Erdenebat* and Tomonori Sato

Abstract

In summer 2002, abnormally high-temperature events and associated low soil moisture conditions were observed in Northeast Eurasia. In this study, two regional climate model experiments, one with and the other without soil moisture–atmosphere interaction, were conducted to investigate the role of soil moisture in surface air temperature and precipitation in Northeast Eurasia utilizing the Weather Research and Forecasting (WRF) model. In the experiment without interaction, satellite-based soil moisture was prescribed. Under the persistent mid-tropospheric ridge pattern with prolonged clear skies, both experiments captured the magnitude of the extreme surface air temperature events. A comparison of the model experiments demonstrated that the mid-tropospheric ridge pattern was intensified by dry soil moisture conditions that further increase the surface air temperature. The temporal variability of surface air temperature for the experiments with and without soil moisture–atmosphere interaction reveals the strong coupling of soil moisture to surface air temperature in June–August, raising the possibility that extreme hot temperature events in Northeast Eurasia were strengthened by the soil moisture anomaly. These results also indicate that the impact of soil moisture–atmosphere coupling on surface air temperature varies spatially and temporally, having higher importance for the predictability of extreme high-temperature events in Northeast Eurasia during mid-summer. This also suggests that the performance of the land surface model in simulating appropriate land–atmosphere coupling intensity is a key to evaluating the impact of climate change on extreme heat events in this region.

Keywords: Extreme temperature, Soil moisture, Land–atmosphere interaction, Northeast Eurasia

Introduction

The global mean surface air temperature (SAT) has been increasing steadily since the middle of the 1950s (Hartmann et al. 2013). It is widely known that the spatiotemporal variation in mean and extreme temperature increase is not uniform. Thus, several hot spots where warming rate is greater than the surrounding areas have been identified across the world (Alexander et al. 2006; Della-Marta et al. 2007; Seneviratne et al. 2012). In America, Europe, Australia, and Asia, the number of extreme-temperature-related weather events has increased as mean SAT has increased since 1950 (Seneviratne et al. 2012). Record-breaking extreme high-temperature events that occurred recently in Europe and North America have focused attention on the interannual variation in extreme high-temperature events and the physical mechanisms responsible for them, such as atmospheric circulation and the possible contribution of the land surface (Black et al. 2004; Meehl and Tebaldi 2004; Schär et al. 2004; Ferranti and Viterbo 2006; Fischer et al. 2007; Lau and Nath 2012).

A persistent anticyclonic circulation and blocking are the important atmospheric forcing conditions that cause hot weather events such as heat waves (HWs) in the mid-latitude regions (Dole et al. 2011; Schubert et al. 2014). In the 2000s, the mid-tropospheric ridge pattern was dominant during June–July–August (JJA) in Northeast Eurasia, providing favorable conditions for HWs (Erdenebat and Sato 2016). Soil moisture (SM) in JJA had declined in this region due to the lack of precipitation, and this possibly intensified the land–atmosphere interaction (Erdenebat and Sato 2016).

* Correspondence: enkhbat@ees.hokudai.ac.jp
Graduate School of Environmental Science, Hokkaido University, Kita-10, Nishi-5, Sapporo 060-0810, Japan

Water availability in the soil regulates evapotranspiration that alters air temperature and humidity in the lower atmosphere. Pre-existing dry soil conditions enhance surface sensible heat flux and limit evapotranspiration, leading to a rapid increase of SAT (Zaitchik et al. 2006). Observations and reanalysis data suggested a possible contribution of SM to the recent increasing number of HWs (Beniston 2004; Weisheimer et al. 2011; Erdenebat and Sato 2016). However, analysis of observations and reanalysis data has limitations for examining the role of SM in HWs because it is difficult to distinguish SM-to-HW and HW-to-SM contributions as both processes occur simultaneously during the hot events.

Several numerical studies have been conducted to determine the importance of SM conditions on precipitation variability (Koster et al. 2004; Su and Dickinson 2017) and HWs (Fischer et al. 2007; Seneviratne et al. 2006). The SM deficit was found to have a role in intensifying the European HW in 2003 (e.g., Fischer et al. 2007). Fischer et al. (2007) concluded that the warm summer temperature anomalies in 2003 would have been reduced by 40% if the SM had been at the level of the climatological mean. In North America, the SM anomaly in April tends to persist until July as a result of local positive feedback processes between SM and the atmosphere, as the surface heat flux anomaly in July induces a wave pattern in the upper troposphere (Koster et al. 2014). This suggests the influence of SM–atmosphere interaction on the predictability of hot events. Berg et al. (2015) found that the co-variability of temperature and precipitation is controlled by SM through surface heat flux partitioning. Thus, a reliable hydrological cycle in the model may reduce uncertainty in future projections. Although the effect of global SM–atmosphere interactions on the intensity of severe HWs has been studied, HW events in Northeast Eurasia have not been studied intensively, especially from the viewpoint of SM–atmosphere interaction during a hot summer.

Summer 2002 was the hottest summer in Northeast Eurasia over the period 1981–2010. In Mongolia, the accumulated summer duration of the strong mid-tropospheric ridge in 2002 was the longest over the period 1981–2010 (Erdenebat and Sato 2016). In addition, the SM averaged over 42° N–52° N, 90° E–120° E (black box in Fig. 1a) derived from ERA-Interim at surface layer (0–7 cm depth) was 0.213, 0.215, and 0.205 $m^3 m^{-3}$ for June, July, and August 2002, respectively, lower than the JJA mean SM for 1981–2010 (0.223 $m^3 m^{-3}$). The significantly low SM in 2002 may have played a role in further intensifying HWs. In Mongolia, the SM affects the predictability of interannual variation in precipitation (Sato and Xue 2013). However, the contribution of SM to HW in Northeast Eurasia is not known. Therefore, this study aims to investigate the role of SM–atmosphere interactions during extreme HW events in summer 2002 over Northeast Eurasia.

Methods/Experimental

Data

ERA-Interim (Dee et al. 2011) reanalysis data at 0.75° × 0.75° resolution from the European Centre for Medium-range Weather Forecasts (ECMWF) were used to examine atmospheric circulation, geopotential height, SM, and temperature variations from June to August 2002. The ERA-Interim has been shown to be the most reliable reanalysis dataset in terms of extreme temperature variation (Gross et al. 2017). ERA-Interim data were used as initial and boundary conditions for the regional model experiments and to validate the model results over a large area. For SM validation and analysis, the shallowest level (0–7 cm depth) of ERA-Interim data was used at time intervals of 1 day. Daily maximum SAT and precipitation at 70 meteorological stations in Mongolia provided by the National Agency for Meteorology and Environment Monitoring of Mongolia were also used. Daily gridded precipitation data from Asian Precipitation Highly Resolved Observational Data Integration Towards Evaluation (APHRODITE; Yatagai et al. 2012) were used to evaluate precipitation variation. Anomalies in observations and reanalysis data are calculated with respect to the corresponding 30-year (1981–2010) mean. We used the European Space Agency-Climate Change Initiative (ESA-CCI; Liu et al. 2011) version 2.2 multi-satellite SM product as an initial SM condition for model experiments and to constrain the SM in the model. This SM product has a spatial resolution of 0.25° × 0.25° and represents the volumetric soil water content ($m^3 m^{-3}$) in the uppermost 10 cm of the soil layer.

Model setup

To evaluate the role of SM in HW events, numerical experiments were conducted with the non-hydrostatic Weather Research and Forecasting (WRF) model version 3.6 (Skamarock et al. 2005) using the Advanced Research WRF (ARW) solver. The physics schemes are mostly as in Sato and Xue (2013), who assessed the model performance over Northeast Eurasia: the Eta (Ferrier) microphysics scheme (Rogers et al. 2001), the Kain–Fritsch cumulus parameterization scheme (Kain 2004), the rapid radiative transfer model longwave radiation scheme (Mlawer et al. 1997), the Dudhia shortwave radiation scheme (Dudhia 1989), and the Unified Noah land-surface model (LSM; Ek et al. 2003). The Noah LSM considers four layers in the ground (10, 30, 60, and 100 cm depth below the ground surface). The model domain has 340 × 260 grid points with a resolution of 20 km covering most of Northeast Eurasia centered around Mongolia (Fig. 1a). Initial and boundary conditions are obtained from ERA-Interim data (Dee et al. 2011) at 6-hour intervals. To consider the nonlinear atmospheric response to SM, we conducted five ensemble experiments with slightly different initial dates starting at 0000 UTC on 13, 14, 15, 16, and 17 June 2002.

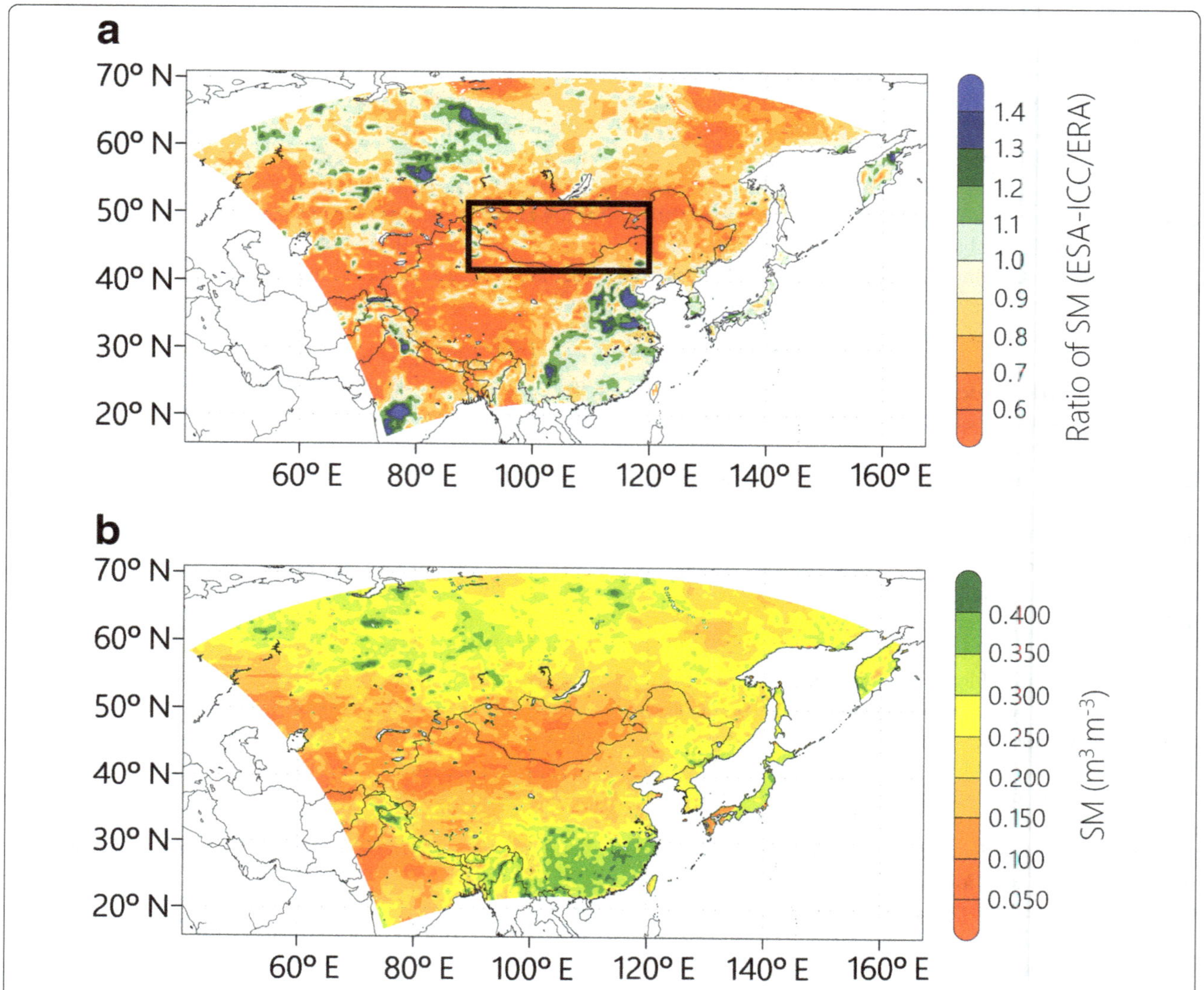

Fig. 1 a Domain of the numerical experiment. Color shading represents the SM ratio between ESA-CCI and ERA-Interim (ESA-CCI divided by ERA-Interim) for the surface layer during 11–20 June 2002. The black box presents the sub-domain used for area averaging. **b** Initial SM condition (m^3 m^{-3}) at 0–10 cm for the WRF experiment initialized at 0000 UTC 13 June 2002

All experiments ended at 0000 UTC on 1 September 2002. The ensemble-mean model output between 20 June and 31 August 2002 (referred to as JJA in the model analysis) is used in the analysis unless indicated otherwise.

Experimental design

To evaluate the impact of SM on HW, two numerical experiments were conducted, one with and one without SM–atmosphere interaction. In the experiment with SM–atmosphere interaction, SM varies according to the hydrological cycle simulated by two-way coupling between land and atmospheric processes (hereinafter, the CSM experiment). In the experiment without SM–atmosphere interaction (hereinafter, the PSM experiment), SM was prescribed using the original SM dataset as described below and thus varies independently of the atmospheric conditions.

To obtain a realistic spatial distribution of initial SM for both experiments in 2002, we created a new SM dataset based on ERA-Interim after applying a bias correction using the ESA-CCI satellite-based product. Firstly, the climatological mean (1981–2010) field of SM was created from the ESA-CCI for the period 11–20 June to use as a reference field to fill missing grid points. Then, pentad interval SM was created using ESA-CCI during 11–20 June 2002. To interpolate the missing grid points, we applied a ratio of pentad climatological SM between the missing grid point and available neighboring grid points onto the pentad SM field in 2002. Secondly, pentad interval SM data were created using the ERA-Interim in the same period for each layer. We calculated the ratio of ESA-CCI (0–10 cm depth) to ERA-Interim (shallowest level, 0–7 cm) SM defined for each pentad during 11–20 June 2002 (Fig. 1a). The ratio below 1.0 indicates that ERA-Interim SM is wetter than ESA-CCI SM. Although there is a small difference in the thickness of the layer in each case, we consider that both datasets reflect surface SM conditions. Then, the ratio was multiplied by the ERA-Interim SM time series in four layers (0–7, 7–28, 28–100, 100–255 cm), assuming that the ratio is the same for all layers. Finally, SM in the four soil layers (0–10, 10–40, 40–100, 100–200 cm) of the Noah LSM was obtained by linear interpolation (Fig. 1b). In the following analysis, we used the SM in the 0–10 cm depth layer. In the PSM experiment, the SM of all ground layers was prescribed at hourly intervals. To create hourly SM data, we prepared pentad SM fields from 11 June to 1 September 2002 as described above for the SM fields for initial conditions. Then, the pentad SM was linearly interpolated to the hourly interval. The same initial condition of SM was used for the CSM and PSM experiments if their initialization dates are the same. Each ensemble member uses different initial SM since they are initialized at different date.

In the PSM experiment, the variation in SM is independent of the precipitation predicted in the WRF model. By comparing the CSM and PSM experiments, we will discuss the impact of SM–atmosphere interaction on HW events in summer 2002.

Results and discussion

Atmospheric circulation and observed temperature in JJA 2002

In this study, the occurrence of HW was judged for each day and for each station in summer 2002 based on the definition of Erdenebat and Sato (2016) which is summarized as follows. A HW was assigned if the daily maximum temperature exceeded the 30-year average (1981–2010) of daily maximum temperature by 5 °C, and remained at this level for at least four

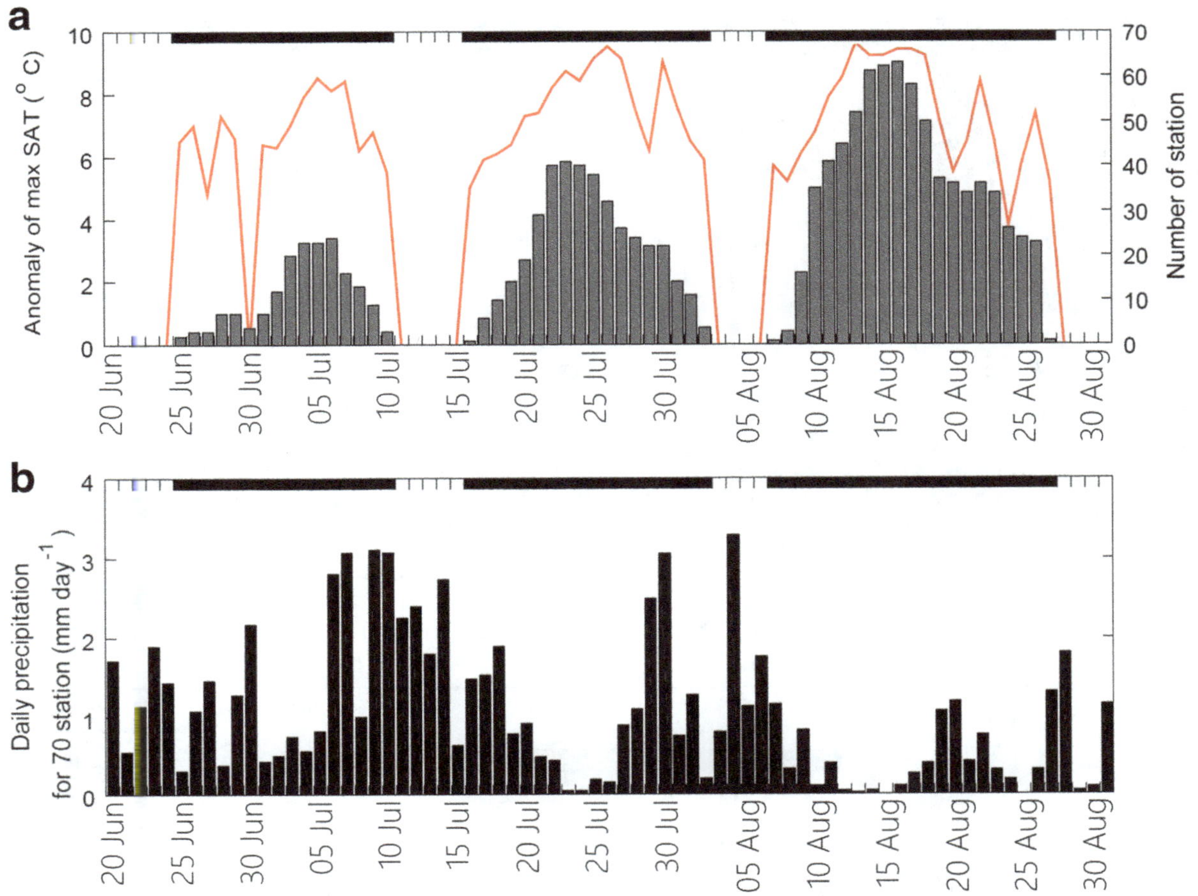

Fig. 2 a SAT anomalies at stations observing HW in Mongolia with respect to their daily climatology. The number of stations observing HW is indicated by gray bars. **b** Daily precipitation averaged for 70 stations in Mongolia. The black bars on the top of **a** and **b** indicate the duration of HW consecutive days. Events with a 1-day interruption were also regarded as HW.

In Mongolia, there were three distinct HW events in summer 2002. The average daily maximum SAT anomaly for stations with HW relative to the 30-year average of each calendar day remained high (Fig. 2a). In 2002, the number of stations experiencing a HW increased toward the end of summer, suggesting that the HW spread over a wider area (Fig. 2a). Summer of 2002 (JJA) was the driest summer during 1981–2010 in Mongolia (Erdenebat and Sato 2016). The daily precipitation in JJA 2002 has decreased gradually toward the end of the summer in Mongolia (Fig. 2b). In general, an anomalous planetary wave is a key factor in triggering abnormal SAT events in the mid-latitudes (e.g., Black et al. 2004; Meehl and Tebaldi 2004). The mid-tropospheric atmospheric circulation was investigated during JJA of 2002 using reanalysis datasets. Figure 3a shows the SAT and 500 hPa wind and geopotential height anomalies in JJA of 2002 relative to the climatology. Note that SAT is examined at 0600 UTC (1400 LT) and all the other parameters are daily means. In the JJA mean wind field at 500 hPa geopotential height, a weak southwesterly over Northeast Europe and Northeast Asia and a northwesterly over the west Siberian plain were apparent, corresponding to positive and negative geopotential height anomalies, respectively. There was a pair of positive anomalies in geopotential height over Northeast Europe and Asia, corresponding to regions with warm SAT. The anomalous geopotential height in JJA 2002 was very similar to the decadal mean for 2001–2010 when many intense HWs occurred (Erdenebat and Sato 2016). This suggests that the summer of 2002 is a characteristic period representing recent hot summers around the study area.

There occurred three massive HW events during summer 2002: from 25 June to 10 July, from 16 July to 02 August, and from 07 August to 27 August (hereafter, the first, second, and third HW events, respectively). The first

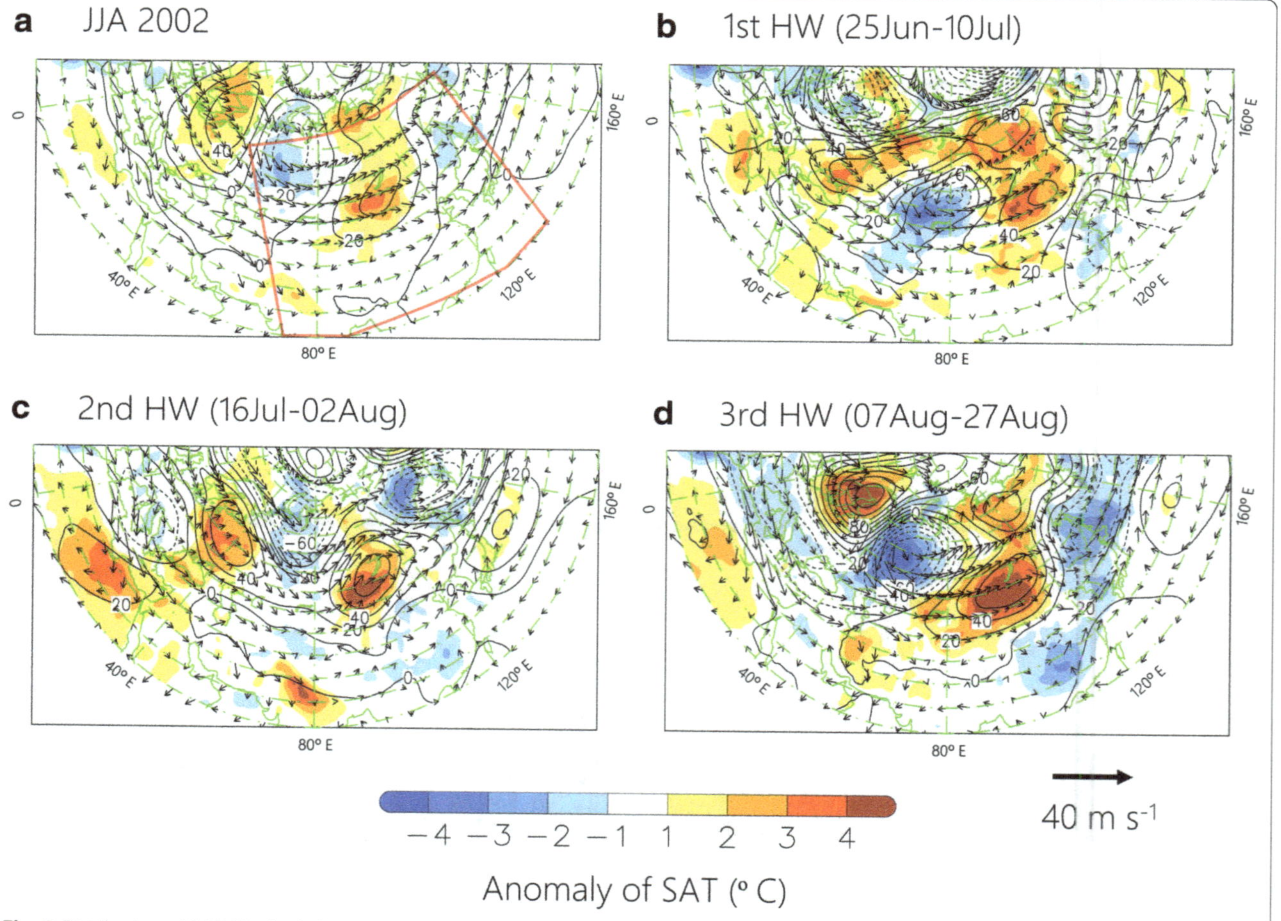

Fig. 3 Distribution of SAT (°C, shaded), geopotential height (m, contours), and wind (m s^{-1}, arrows) at 500 hPa, in **a** JJA 2002, **b** first HW (25 June to 10 July), **c** second HW (16 July to 2 August), and **d** third HW (7 August to 27 August). Variables in **b–d** are expressed as anomalies with respect to the 30-year average (1981–2010) of the corresponding period, but as the mean absolute value for wind at the 500 hPa level. The red line in **a** delineates the study domain by model. Negative contours are shown by dashed lines

HW event in late June (Fig. 3b) had a different wave propagation pattern from the subsequent two events (Fig. 3c, d). Figure 3b shows a positive anomaly of 500 hPa geopotential height over Northeast Eurasia associated with a negative anomaly over Scandinavia and the west Siberian plain. In the second HW, the height anomaly during mid-July to early August was highest (+ 80 m) over Mongolia. Similarly, high positive anomalies (+ 60 m) were observed for the first and third HW in Mongolia. The third HW event (7–27 August) had the largest spatial extent. The persistent positive anomalous field at 500 hPa geopotential height suggests that the prevailing ridge pattern in Northeast Eurasia was associated with the atmospheric circulation in regions upstream of the dominant westerly winds. For aforementioned three events, the mid-latitude wave trains seem to be a main trigger of HWs over Northeast Eurasia although each event has slightly different wave propagation patterns.

To clarify the local intensification of HWs and the origin of the warm air, a Hovmöller diagram of SAT is depicted for the meridional average between 42° N and 52° N (Fig. 4a). There are two regions of anomalously high SAT: Europe around 40° E–60° E and East Asia around 90° E–120° E. The warm SAT anomaly relative to the 30-year daily mean (1981–2010) propagating from Europe to Asia was weak in early summer (1 June to 15 July 2002). This weak SAT anomaly before the first HW event is due to its different wave propagation pathway at 500 hPa (Fig. 3b) which is not captured by the Hovmöller diagram. There was a cold anomaly of daily SAT (– 2.2 °C, averaged over 1 June to 15 July 2002) between 60° E and 80° E (Fig. 4b), but this tended to weaken toward the end of summer. There was eastward propagation of SAT anomalies from Europe to Asia as indicated by the mean warm anomaly of 0.4 °C over 60° E–80° E from 16 July to 31 August 2002 (Fig. 4b). Eastward propagation of the warm SAT anomaly is intensified around Northeast Eurasia between 90° E and 120° E.

The temporal variations in local SAT and SM were investigated over Mongolia (Fig. 4c). There is a pronounced negative relationship between SM and SAT. The correlation coefficients between daily mean SM and SAT from ERA-Interim (at 1400 LT) averaged around Mongolia (black box in Fig. 1a) were – 0.632, – 0.634, and – 0.625 for June, July, and August, respectively. The correlation coefficients are enhanced during HW periods, with values of – 0.872, – 0.885, and – 0.836 for the

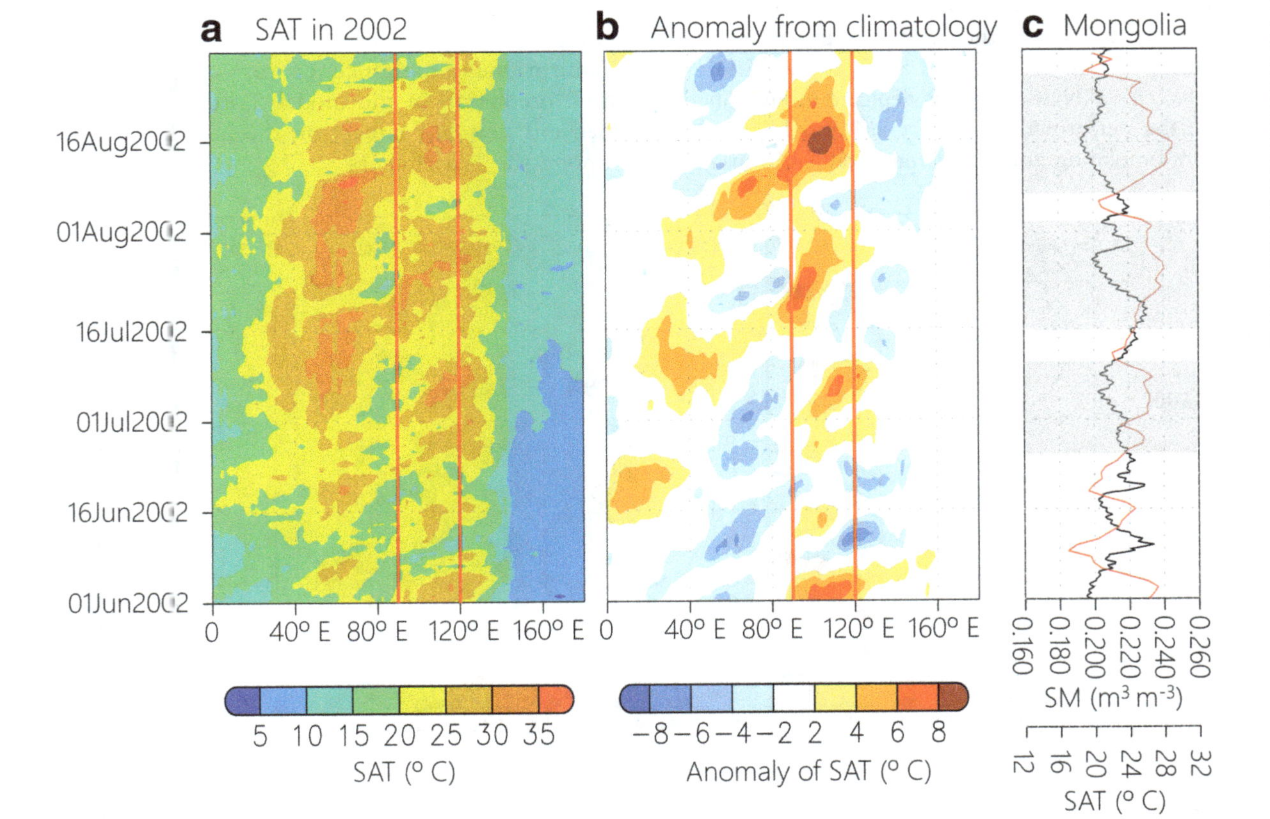

Fig. 4 The Hovmöller diagram (averaged between 42° N and 52° N) of **a** SAT in 2002 summer and **b** its anomaly from the 30-year mean daily SAT in ERA-Interim. **c** Time series of SAT (red line) and SM (black line) in ERA-Interim averaged over Mongolia (black box in Fig. 1a). Red straight lines in **a** and **b** represent Mongolia. The gray shadings in **c** represent the durations of HW events

first (25 June to 10 July), second (16 July to 2 August), and third (7 to 27 August) HW events, respectively. The high negative correlation (95% significance) during HW periods strongly suggests that SM–SAT interaction was crucial during the HW events. In the following section, the relationship between SM and HW events is investigated by numerical experiment to elucidate the role of land–atmosphere interaction.

Experiment on SM–atmosphere interaction

WRF experiments were conducted to examine the role of SM–atmosphere interaction during the HW event in 2002. The ensemble mean SM (0–10 cm depth) variations averaged around Mongolia (black box in Fig. 1a) in the CSM and PSM experiments are shown in Fig. 5a. There is a gap in SM between the CSM and PSM throughout the summer. Although the CSM and PSM have the same initial SM, the CSM gets wetter during the spin up duration. As indicated by the wetter SM in ERA-interim than the satellite SM (Fig. 1a), it is reasonable to speculate that hydrological cycle in ERA-interim may be stronger than the real atmosphere, and thereby, our WRF experiments have weak biases in precipitation and SM (Table 1), allowing the SM gap between two experiments throughout the summer (Fig. 5a). The fluctuations of SM are larger in the CSM than in the PSM. The Hovmöller diagram of simulated SAT (at 1400 LT) averaged between 42° N and 52° N is depicted in Figs. 5b, d to evaluate the performance of the CSM and PSM experiments in simulating zonal and temporal SAT variations. Most of the cold and warm biases appear between 90° E and 120° E. Figure 5c, e shows the area-averaged deviation of SAT in the CSM and PSM with respect to ERA-Interim, respectively. The JJA SAT variation in the PSM experiment is statistically different ($p = 0.10$) from ERA-Interim, whereas that in the CSM experiment is not. On a subseasonal timescale, the CSM and PSM experiments underestimate SAT compared with ERA-Interim, especially during late July and August. The CSM experiment underestimates the JJA mean SAT by 0.6 °C, with a root mean square error (RMSE) of daily SAT for JJA of 1.9 °C. In the PSM experiment, the mean bias and RMSE are − 0.7 and 2.0 °C for JJA, respectively. In both experiments, in comparison with ERA-Interim, the RMSE of area-averaged SAT around Mongolia was low for the first and second HWs, i.e., 1.4 and 1.2 °C in the CSM and 1.5 and 1.6 °C in the PSM, respectively. In the first HW, SATs of the CSM and PSM experiments are not significantly different from ERA-Interim, which indicates that the model agrees well with reanalysis data. During the second HW, temporal- and area-averaged SAT in the CSM is close to ERA-Interim, whereas PSM underestimates SAT by 0.93 °C ($p = 0.10$). The RMSE of SAT is highest in the third HW (7–27 August), with 1.9 °C in the CSM ($p = 0.05$) and 2.4 °C in PSM ($p = 0.01$). Those worsening of the SAT estimations in the PSM experiment confirm the importance of correct SM variation to SAT prediction in the model. Further analysis is carried out how SM impact on the heat flux partitioning among the experiments.

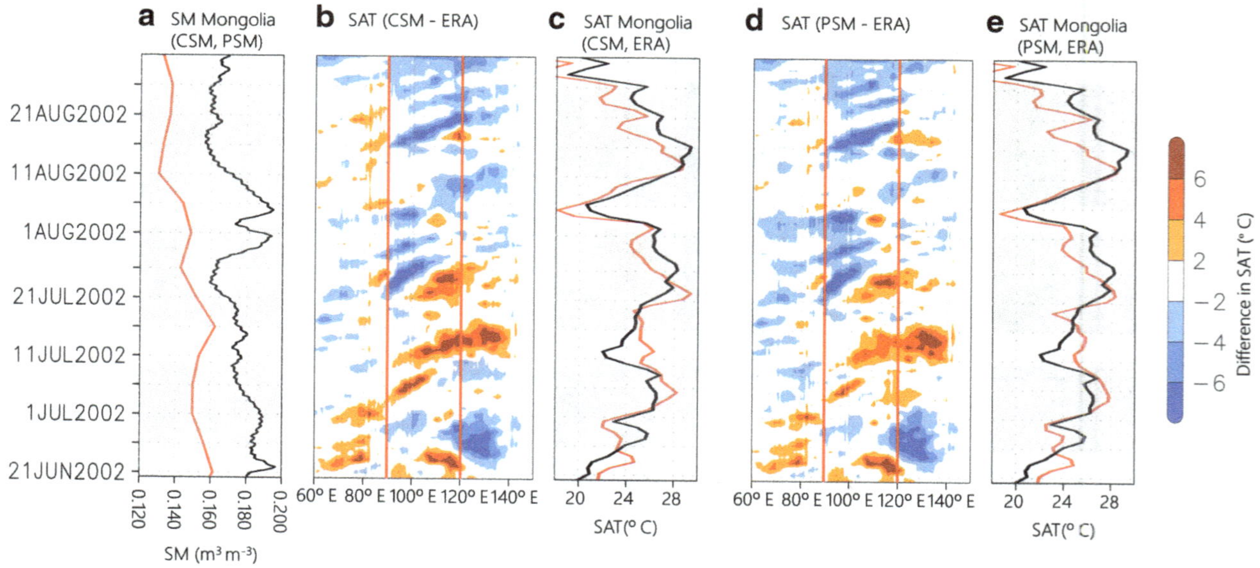

Fig. 5 a Time series of ensemble mean SM over Mongolia (black box in Fig. 1a) in the CSM (black line) and PSM (red line) experiment. **b** As in Fig. 4b but for the deviation of SAT in the CSM experiment relative to ERA-Interim. **d** As in **b** but for PSM. **c, e** Time series of area-averaged (black box in Fig. 1a) SAT in the CSM and PSM (red lines), respectively, referred to ERA-Interim (black lines). The gray shadings in **a,c** and **e** indicate the duration of HWs

Table 1 lists the area-averaged (box in Fig. 1a) surface variables for the CSM, PSM, and reference datasets of ERA-Interim. Note that heat fluxes, SAT, and radiation are examined at 0600 UTC (1400 LT). As mentioned above, the SATs in the two experiments are similar to ERA-Interim, with the highest SAT during the second HW. Although the event-to-event changes of surface variables in WRF experiments are similar with ERA-Interim, some differences are present. The JJA-mean SM is pronouncedly wetter in ERA-Interim than the WRF model. Thus, in the WRF runs, less latent heat and more sensible heat fluxes are found in comparison to ERA-Interim. In particular, sensible heat flux in the WRF runs is much larger than that in ERA-Interim. Since simulated SM in the CSM is more tightly constrained by ERA-Interim fields than in PSM, the CSM shows greater precipitation and SM than the PSM. As a result, the latent (sensible) heat flux in PSM is smaller (larger) than in the CSM for the JJA-mean and also for each HW. The downward shortwave radiation during first and second HWs is far exceeding the JJA average, indicating less cloud cover, while it is relatively low in third HW because of the seasonal cycle of solar radiation. The downward radiations in CSM and PSM are comparable which suggests SM's influence on cloud cover is negligible for the JJA average although CSM tends to have larger downward radiation during HW events. Since the analysis was limited to the particular region and for three HW events, spatial distribution of the WRF's skill and its sensitivity to SM for simulating sub-seasonal SAT and precipitation variations were investigated as follows.

To investigate the impact of SM–atmosphere interaction on the geographical distribution of subseasonal variation, the temporal correlations of pentad mean time series between simulated and observed variables (SAT and precipitation) are shown in Fig. 6. In both the CSM and PSM experiments, a high correlation of SAT is obtained near the northern and western lateral boundaries where the atmosphere is less influenced by land surface conditions (Fig. 6a, c). The correlation of SAT deteriorates toward the center of the study domain, with the lowest correlation around southern China and the Far Eastern region of Russia. The southeastward decreasing tendency of the correlation coefficient of SAT suggests that the dependence of SAT on land surface conditions increases southeastward. The CSM outperforms the PSM for most of the center of the study domain (Fig. 6e), which means land–atmosphere interaction improves the temporal variation in SAT for continental regions where land surface processes play a crucial role in influencing SAT. On the other hand, the correlation was higher in the PSM than the CSM around the east coast of China. This suggests that having the correct SM in the PSM is more important than having SM–atmosphere interaction for the subseasonal variation in SAT in this area.

Table 1 Area-averaged (see Fig. 1a) mean surface variables in the CSM and PSM experiments and ERA-interim for corresponded periods. Note that heat fluxes, surface air temperature, and radiation are examined at 0600 UTC (1400 LT)

		JJA	1st HW	2nd HW	3rd HW
Surface air temperature (°C)	CSM	24.4	25.2	26.1	24.5
	PSM	24.2	25.3	25.5	24.2
	ERA-interim	25.0	25.3	26.4	26.2
Soil moisture (m^3 m^{-3})	CSM	0.173	0.180	0.172	0.164
	PSM	0.146	0.152	0.150	0.135
	ERA-interim	0.210	0.211	0.215	0.201
Latent heat (W m^{-2})	CSM	128.6	140.3	140.3	117.0
	PSM	126.7	131.9	138.8	119.8
	ERA-interim	141.2	152.6	162.9	128.0
Sensible heat (W m^{-2})	CSM	245.7	257.7	248.1	230.3
	PSM	255.0	270.5	252.7	236.0
	ERA-interim	193.5	200.6	188.1	189.6
Precipitation (mm day^{-1})	CSM	1.4	1.6	1.6	0.7
	PSM	1.3	1.4	1.4	0.8
	APHRODITE	1.1	1.3	1.1	0.5
Downward shortwave radiation (W m^{-2})	CSM	823.7	871.6	846.9	783.8
	PSM	823.8	867.5	837.3	782.1
	ERA-interim	723.8	751.3	741.1	720.8

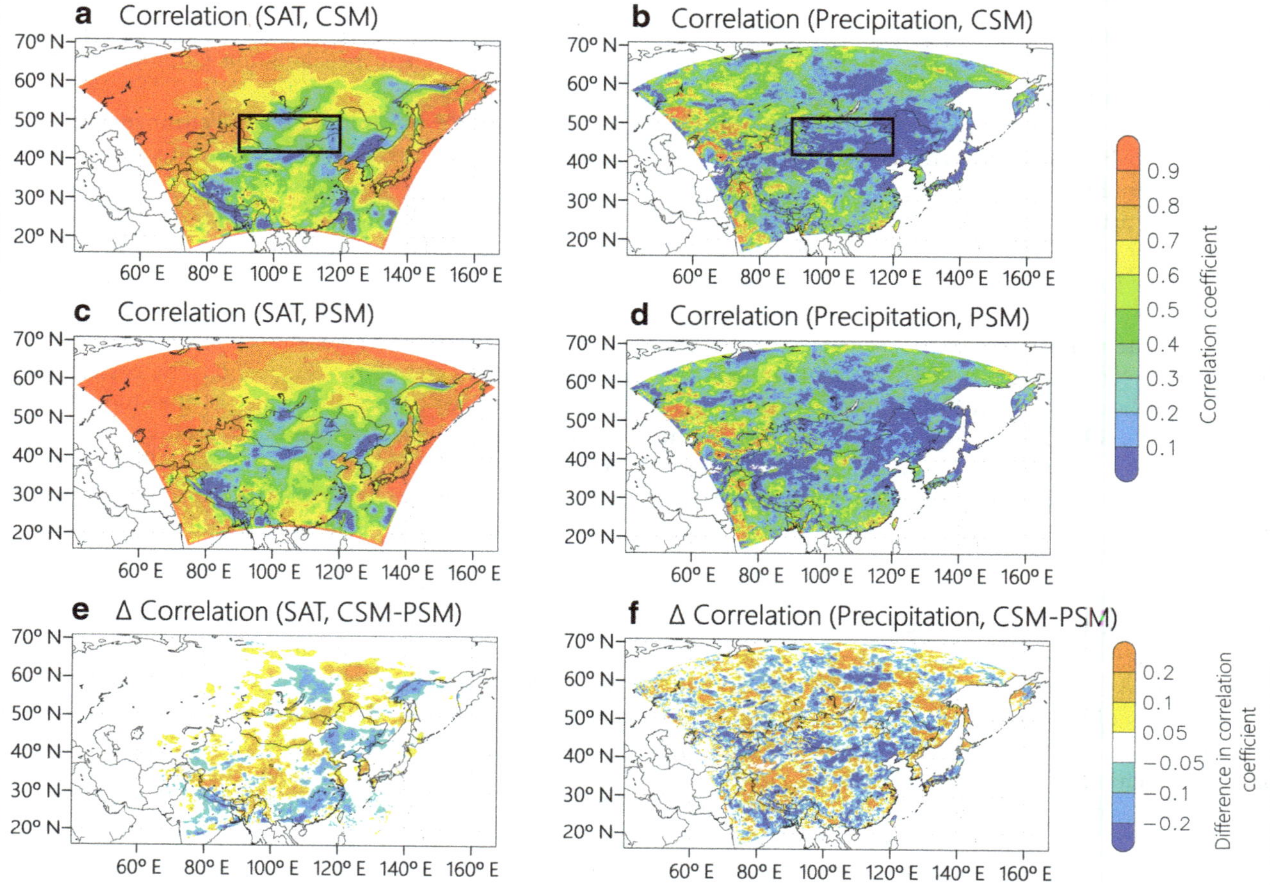

Fig. 6 Spatial distribution of temporal correlation for pentad-mean SAT (**a**, **c** and **e**) and precipitation (**b**, **d** and **f**) during JJA between the WRF experiment and reference dataset. ERA-Interim and APHRODITE were used as reference datasets for SAT and precipitation, respectively. (top) CSM, (middle) PSM, and (bottom) the difference in correlation (CSM minus PSM). The black box in **a** and **b** presents the sub-domain used for area averaging

The correlation of precipitation is also analyzed at each grid point in Fig. 6b, d, and f. In both experiments, however, the correlation for pentad mean precipitation time series between APHRODITE and WRF is low (R_{CSM} = 0.24 and R_{PSM} = 0.17 over the study domain). There is no significant difference in area-averaged (black box in Fig. 6b) daily precipitation between CSM and PSM during JJA (Student's t test, 95% confidence level). This unclear response of precipitation to SM indicates that the influence of SM on precipitation is complicated in space and time or that the expected impact is weak. Since the current study focused on high-temperature events in which the atmosphere is stable under high pressure and hence precipitation is low, additional analysis of years with more precipitation events is needed to understand the possible influences of SM on precipitation.

We have examined the performance of the CSM and PSM experiments against observations and reanalysis. In the following part of this study, further analysis was used to clarify the SM–atmosphere interaction, focusing on the SM difference between the CSM and PSM experiments.

The difference in SM between the CSM and PSM is inhomogeneously distributed (Fig. 7). It also varies on a subseasonal timescale, especially during the first and third HWs (Fig. 7b, d) over Mongolia by about 0.03 $m^3\ m^{-3}$. In contrast, the area-averaged (black box in Fig. 7a) SM difference is small (0.02 $m^3\ m^{-3}$) for the second HW (Fig. 7c). The horizontal distribution of anomalous latent and sensible heat flux is consistent with that of the anomalous SM (not shown), with more latent heat (less sensible heat) flux as SM increases. We will later discuss the SM's impact on middle tropospheric circulation through a comparison with SM patterns in Fig. 7.

Figure 8 depicts the difference in 500 hPa geopotential height fields between the CSM and the PSM. The mean JJA geopotential field at 500 hPa has a weak ridge over Mongolia. The strength of the ridge varies for each HW, and the strongest ridge occurred in the second HW. In the JJA mean field, the difference in geopotential height between the CSM and PSM was small (Fig. 8a). The greater differences occurred for the first (Fig. 8b) HW events in regions with lower SM in the CSM experiment

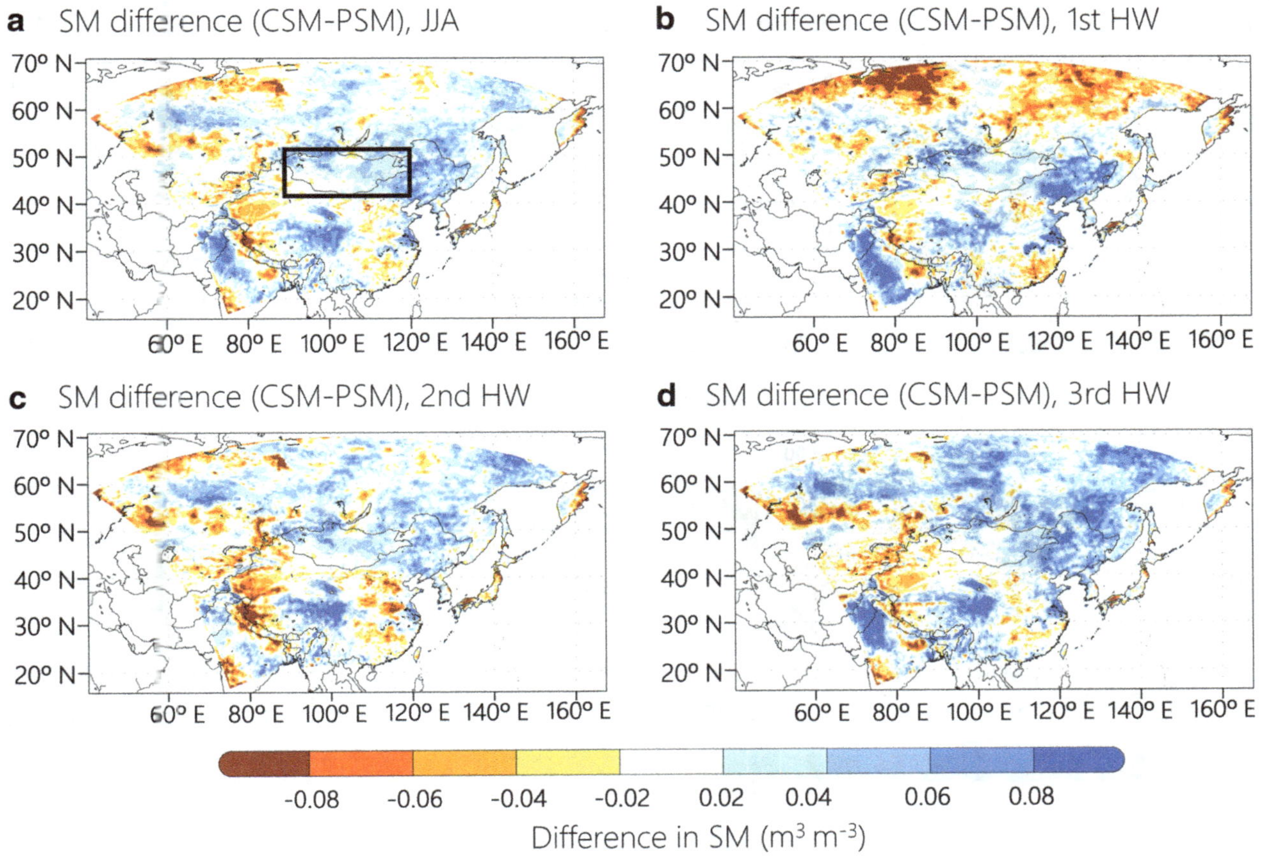

Fig. 7 Horizontal distribution of SM difference between the CSM and PSM experiments for **a** JJA, **b** first HW (25 June to 10 July), **c** second HW (16 July to 2 August), and **d** third HW (7 August to 27 August). The black box in **a** presents the sub-domain used for area averaging

(Fig. 7b). The atmospheric response to SM is strong in the first and second HW events, in which the 500 hPa geopotential height increased by more than 25 m (Fig. 8b) and 20 m (Fig. 8c) in the PSM relative to the CSM experiment. On the other hand, the influence of SM to the middle troposphere tends to be weak in the third HW (Fig. 8d) when SM is much wetter in the CSM (Fig. 7d). This suggests that the impact of SM on the mid-tropospheric circulation differs among three HW events. Kanae et al. (2001) suggested that the impact of land surface anomaly is likely to present during the period when prevailing wind weakens. To investigate the relationship between the environmental winds and impact of SM anomaly, upper-level wind speeds were compared among HW events. The areal mean (black box in Fig. 8a) daily wind speeds in the CSM at 500 hPa were 6.2, 7.9, and 8.1 m s^{-1} for the first, second, and third HWs, respectively. Under the relatively strong westerly wind during the third HW event, the impact of the SM anomaly might propagate weakly upward. It is speculated that the mid-troposphere is more sensitive to a SM anomaly when Northeast Eurasia is covered by a strong ridge and weak westerly wind, and less sensitive when strong westerly wind dominates. Further studies that consider different environmental conditions will be needed to verify this speculation.

Here, we discuss the strength of the coupling between land and atmosphere. To investigate the spatial distribution of temperature and SM interaction in the model experiments, a coupling strength is estimated by using the output from the CSM and PSM experiments. We followed the methodology proposed by Seneviratne et al. (2006), which considers the standard deviation of temporal variability (hereinafter, coupling strength) of particular variables derived from the CSM and PSM experiments. In this study, coupling strength is estimated by $(\sigma_{CSM}^2 - \sigma_{PSM}^2)/(\sigma_{CSM}^2)$ where σ_{CSM} and σ_{PSM} represent the standard deviations of daily SAT (1400 LT) or precipitation during JJA in the CSM and PSM outputs, respectively. The coupling strength for SAT is highest in Siberia and east China (Fig. 9a). Positive coupling strength implies that SAT variability in the CSM experiment is larger than in the PSM experiment. The spatiotemporal distribution of the coupling strength for all JJA is not uniform. A high coupling strength is widely distributed over Siberia during the first

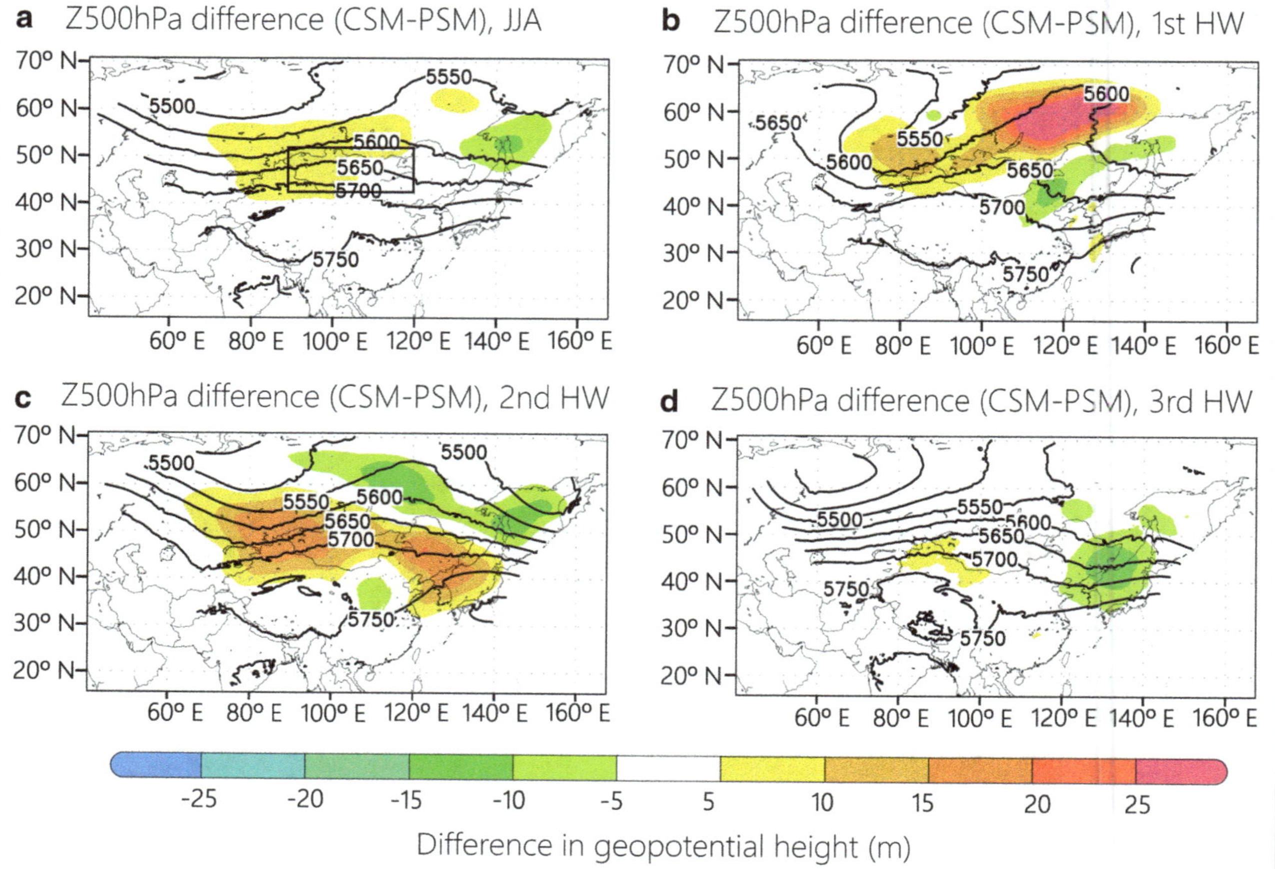

Fig. 8 a–d Same as Fig. 7, but for 500 hPa geopotential height (Z500hPa) difference (shading). Contours represent 500 hPa geopotential height in the CSM experiment

HW event (Fig. 9b), which remains in JJA average (Fig. 9a). Since low SM (Fig. 7b) and high geopotential (Fig. 8b) are co-locating over Siberia in the CSM run, it is speculated that positive feedback in local land–atmosphere interaction is intensified during the first HW event. The area-averaged (black box in Fig. 9a) monthly based coupling strengths are 0.02, 0.17, and 0.04 for June, July, and August, respectively. The coupling strength during the first HW is high in the western Mongolia, which is likely to be caused by temperature advection rather than local SM–SAT coupling. A weaker surface westerly in the CSM results in weaker cold advection in the western Mongolia during the decaying period of first HW, leading to higher SAT variance in the CSM (Fig. 5b). The highest coupling strength appears in central China around the Yellow River basin during the third HW (0.68), suggesting that the impact of SM on SAT is strongest in this region (Fig. 9d). A negative coupling strength, namely higher variance of SAT in the PSM than the CSM, occurs when the availability of SM in the PSM is lower than in the CSM. In general, the coupling strength between SM and SAT was found to be low over the Mongolia. Here, we raise two possible reasons. Firstly, the sign of SM difference between CSM and PSM varies spatiotemporally. This allows the SAT difference being both positive and negative signs, and consequently, the difference in SAT variance between two runs becomes small, leading to low coupling strength. Secondly, the study domain covers various hydrological regimes from dry to humid regions, which may result in diverse atmospheric response against SM anomaly as indicated by Guo and Dirmeyer (2013).

The coupling strength for precipitation does not show meaningful signals in the studied area (Fig. 10). This result is consistent with other studies (Koster et al. 2004; Seneviratne et al. 2010; Zhang et al. 2011), which found weak coupling strength for SM and precipitation in multiple GCM experiments in this region. Guo and Dirmeyer (2013) found strong SM-to-precipitation coupling dominates over Northeast Asia in dry years. Since the targeting year was dry in our studied area, we expected a strong coupling in the current study. Although positive coupling strength in eastern Mongolia (Fig. 10a) agrees with Guo and Dirmeyer (2013), the coupling strength over the surrounding region seems to be randomly distributed. This

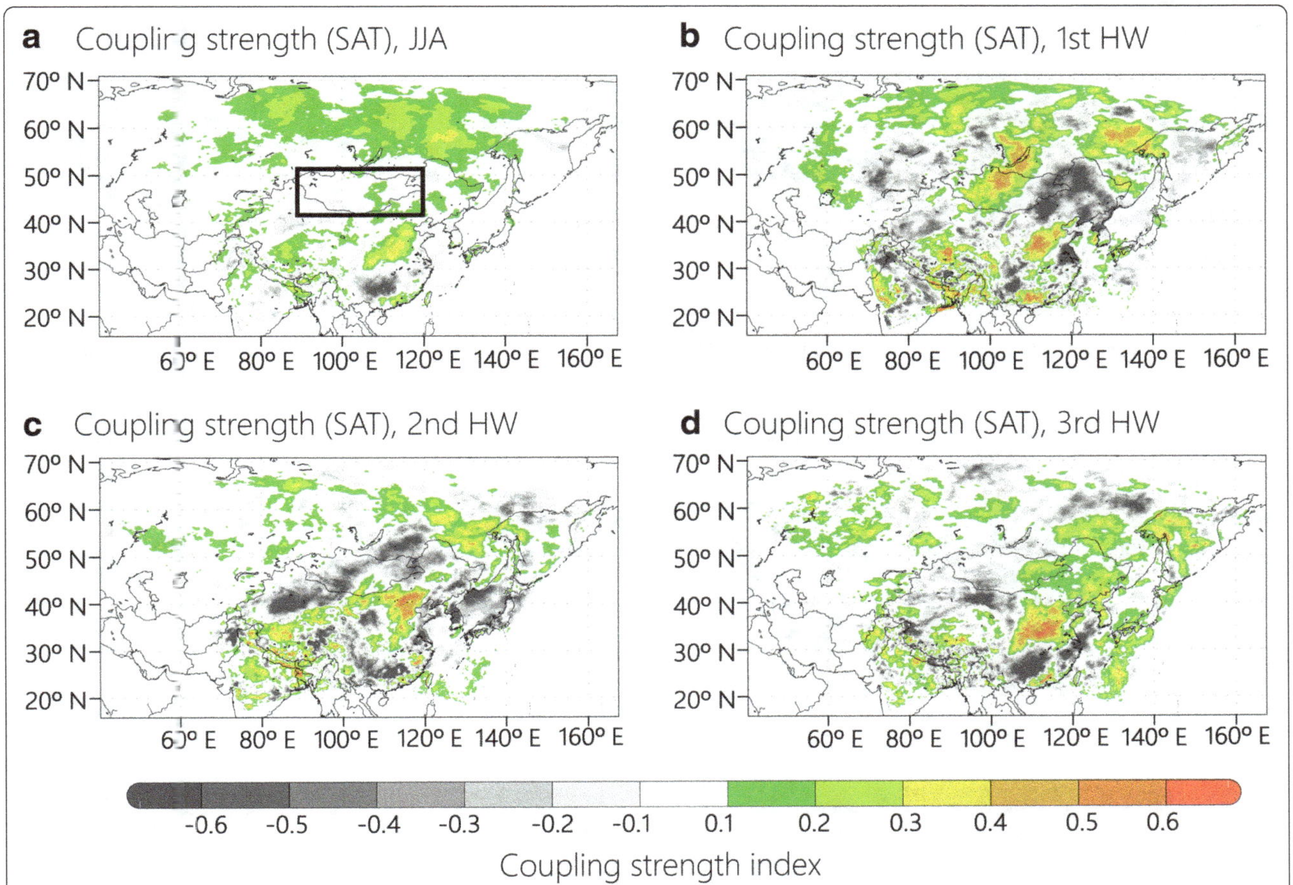

Fig. 9 Horizontal distribution of coupling strength for SAT for **a** JJA, **b** first HW (25 June to 10 July), **c** second HW (16 July to 2 August), and **d** third HW (7 August to 27 August)

may be due to different analysis period and different physics schemes in the model. Since this study is limited for typical hot summer in 2002, further investigation is needed by taking account of large number of cases for the analysis of SM-to-precipitation coupling.

Conclusions

In this study, the role of SM in the intensification of HW events in Northeast Eurasia has been investigated with observational datasets and numerical experiments for the extremely hot summer of 2002. A mid-tropospheric circulation analysis using the ERA-Interim reanalysis shows that HWs over Northeast Eurasia were triggered by mid-latitude wave trains. The Hovmöller diagram for SAT (Fig. 4b) demonstrates that HWs are concentrated around Mongolia associated with a low SM anomaly that can potentially enhance sensible heat flux and the mid-tropospheric ridge. Numerical experiments with (CSM) and without (PSM) SM–atmosphere interaction were conducted to study the importance of land–atmosphere interaction during the strong HW events in 2002. The correlations of pentad SAT and precipitation are high in Northeast Eurasia and become lower toward the southeast of the study domain, suggesting that the contribution of SM to SAT is stronger in Northeast Eurasia. Dry (wet) SM conditions in the PSM experiment increase (decrease) SAT by enhancing (limiting) the sensible heat flux in comparison with the CSM experiment. The dry (wet) SM in the CSM experiment can intensify (weaken) the positive anomaly at 500 hPa geopotential height. The coupling strength index reveals that there is higher coupling strength in Northeast Eurasia in this study domain and that land–atmosphere interaction varies on a sub-seasonal timescale. The CSM experiment resembles the variation in SAT seen in the reanalysis more closely than does the PSM, suggesting that land–atmosphere interaction is important in this region. On the other hand, in the PSM experiment, the model performance in simulating the temporal variation in SAT is better than the CSM over the east coast of China. Thus, correction of SM variation in the model using the satellite-based SM dataset has the potential to improve the sub-seasonal variability in SAT by constraining the internal variability in atmospheric circulation through land–atmosphere interaction. The correct

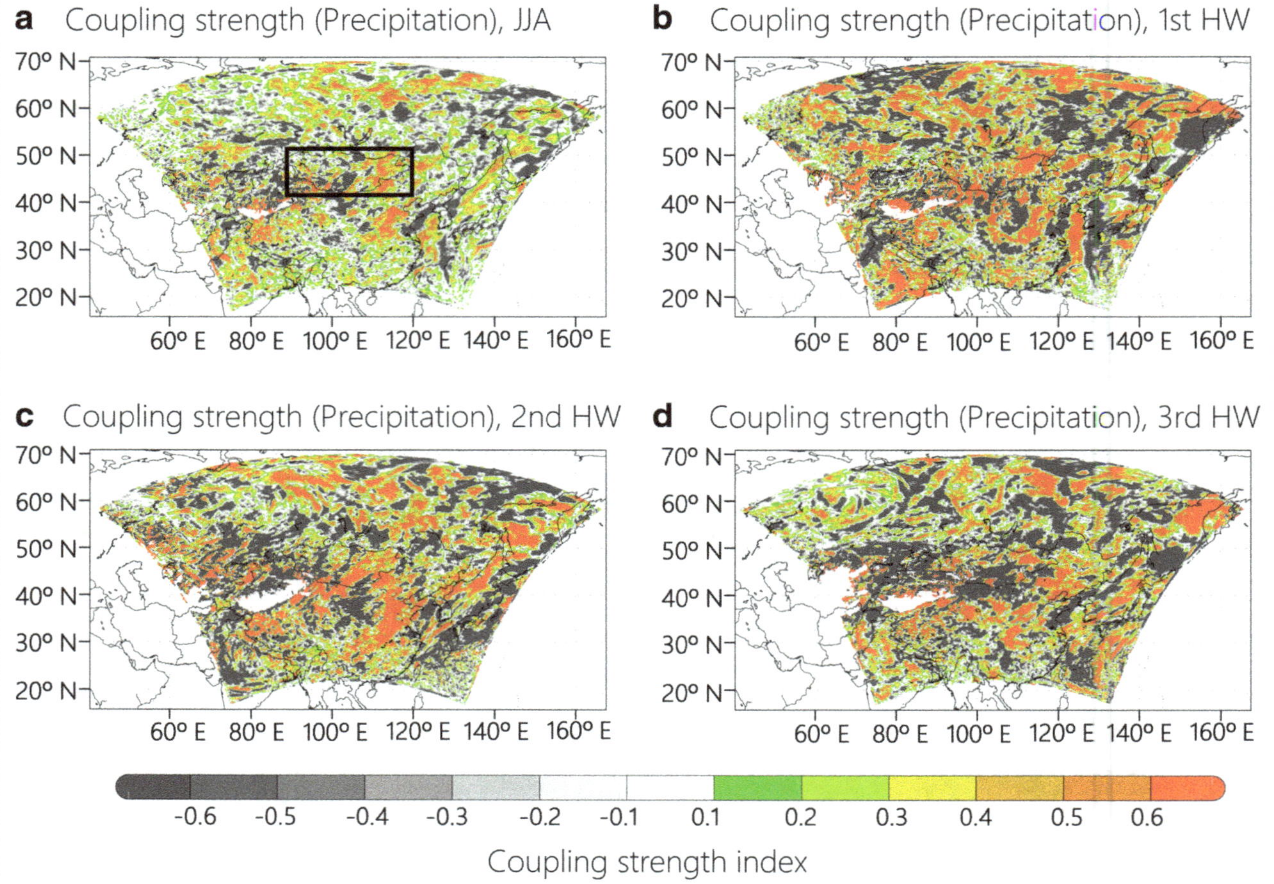

Fig. 10 a–d Same as Fig. 9 but for precipitation

land–atmosphere coupling in a model is expected to regulate the relationship between SM and SAT. The intensity of SM–atmosphere interaction varies spatially and temporally, i.e., there is weak interaction under strong westerly wind (during the third HW) and strong interaction under weak westerlies in the mid-troposphere, as occurred in summer 2002 over Northeast Eurasia. The CSM and PSM experiments suggest the importance of SM in regulating SM–atmosphere coupling that should eventually improve the representation of subseasonal variability in SAT during HW events.

Finally, we must point out the limitations of this study. In comparison to SM–SAT coupling, the SM–precipitation coupling strength is heterogeneously distributed. Although this might represent a complicated nonlinear interaction between SM and precipitation in mid-latitude, five ensemble experiments in this study may be too few to detect SM's impact on precipitation. A comprehensive analysis by using a large number of ensemble experiment is demanded in the future study. The SM dataset for the prescribed SM experiment also requires improvements. Recently, satellite-based and model-based SM products become available for such a purpose. However, there is a gap in seasonal mean SM values between the SM products and SM derived from atmospheric reanalysis. In such a case, simulated hydrological cycle in a regional climate model is likely to be biased that may hinder a proper evaluation of SM–precipitation interaction. With respect to future climate change, summer precipitation change around the studied region yet contains large uncertainty (Collins et al. 2013). Since SM–atmosphere interaction plays a key role in attributing the future change in summer precipitation (Sato et al. 2007), further studies are needed to enhance our understanding in land–atmosphere interactions over Northeast Eurasia.

Abbreviations

APHRODITE: Asian Precipitation Highly Resolved Observational Data Integration Towards Evaluation; ARW: Advanced Research WRF; CSM: Coupled soil moisture; ECMWF: European Centre for Medium-range Weather Forecasts; ERA: ECMWF reanalysis; ESA-CCI: European Space Agency-Climate Change Initiative; HW: Heat wave; JJA: June–July–August; LSM: Land-surface model; LT: Local time; PSM: Prescribed soil moisture; RMSE: Root mean square error; SAT: Surface air temperature; SM: Soil moisture; UTC: Coordinated Universal Time; WRF: Weather Research and Forecasting; Z500hPa: Geopotential height at 500 hectopascal

Acknowledgements

The authors acknowledge the Arctic Challenge for Sustainability Project and the Social Implementation Program on Climate Change Adaptation Technology (SI-CAT) project funded by the Ministry of Education, Culture, Sports, Science and Technology (MEXT) of Japan, and Grant-in-Aid for Young Scientists A (Grant No. 15H05464) funded by the Japan Society for the Promotion of Sciences (JSPS). The authors also would like to thank the two anonymous reviewers for their comments and suggestions to improve the quality of the paper.

Funding

This study was funded by the Arctic Challenge for Sustainability (ArCS) Project and Social Implementation Program on Climate Change Adaptation Technology (SI-CAT) by MEXT, Japan, and also the Grant-in-Aid for Young Scientists A (Grant No. 15H05464) by the Japan Society for the Promotion of Sciences (JSPS).

Authors' contributions

EE and TS proposed the topic. EE conceived the study and analyzed the data. TS designed the experiment and interpreted the results of the experiment. EE wrote the first draft, and TS revised the manuscript. Both authors read and approved the final manuscript.

Competing interests

The authors declare that they have no competing interests.

References

Alexander LV, Zhang X, Peterson TC, Caesar J, Gleason B, Klein Tank AM, Haylock M, Collins D, Trewin B, Rahimzadeh F, Tagipour A (2006) Global observed changes in daily climate extremes of temperature and precipitation. J Geophys Res 111:05109. https://doi.org/10.1029/2005JD006290

Beniston M (2004) The 2003 heat wave in Europe: a shape of things to come? An analysis based on Swiss climatological data and model simulations. Geophys Res Lett 31:02202. https://doi.org/10.1029/2003GL018857

Berg A, Lintner BR, Findell K, Seneviratne SI, van den Hurk B, Ducharne A, Chéruy F, Hagemann S, Lawrence DM, Malyshev S, Meier A (2015) Interannual coupling between summertime surface temperature and precipitation over land: processes and implications for climate change. J Clim:1308–1328. https://doi.org/10.1175/JCLI-D-14-00324.1

Black E, Blackburn M, Harrison G, Hoskins B, Methven J (2004) Factors contributing to the summer 2003 European heatwave. Weather 59:217–223. https://doi.org/10.1256/wea.74.04

Collins M, Knutti R, Arblaster J, Dufresne JL, Fichefet T, Friedlingstein P, Gao X, Gutowski WJ, Johns T, Krinner G, Shongwe M, Tebaldi C, Weaver AJ, Wehner M (2013) Long-term climate change: projections, commitments and irreversibility. In: Stocker TF, Qin D, Plattner G-K, Tignor M, Allen SK, Boschung J, Nauels A, Xia Y, Bex V, Midgley PM (eds) Climate Change 2013: The Physical Science Basis. Contribution of Working Group I to the Fifth Assessment Report of the Intergovernmental Panel on Climate Change. Cambridge University Press, Cambridge, United Kingdom and New York, NY, USA

Dee DP, Uppala SM, Simmons AJ, Berrisford P, Poli P, Kobayashi S, Andrae U, Balmaseda MA, Balsamo G, Bauer P, Bechtold P (2011) The ERA-interim reanalysis: configuration and performance of the data assimilation system. Q J R Meteorol Soc 137:553–597. https://doi.org/10.1002/qj.828

Della-Marta PM, Haylock MR, Luterbacher J, Wanner H (2007) Doubled length of western European summer heat waves since 1880. J Geophys Res 112:15103. https://doi.org/10.1029/2007JD008510

Dole R, Hoerling M, Perlwitz J, Eischeid J, Pegion P, Zhang T, Quan XW, Xu T, Murray D (2011) Was there a basis for anticipating the 2010 Russian heat wave? Geophys Res Lett 38:06702. https://doi.org/10.1029/2010GL046582

Dudhia J (1989) Numerical study of convection observed during the winter monsoon experiment using a mesoscale two-dimensional model. J Atmos Sci 46:3077–3107. doi: https://doi.org/10.1175/1520-0469(1989)046<3077:NSOCOD>2.0.CO;2

Ek MB, Mitchell KE, Lin Y, Rogers E, Grunmann P, Koren V, Gayno G, Tarpley JD (2003) Implementation of Noah land surface model advances in the National Centers for Environmental Prediction operational mesoscale Eta model. J Geophys Res 108:8851. https://doi.org/10.1029/2002JD003296

Erdenebat E, Sato T (2016) Recent increase in heat wave frequency around Mongolia: role of atmospheric forcing and possible influence of soil moisture deficit. Atmos Sci Lett 17:135–140. https://doi.org/10.1002/asl.616

Ferranti L, Viterbo P (2006) The European summer of 2003: sensitivity to soil water initial conditions. J Clim 19:3659–3680. https://doi.org/10.1175/JCLI3810.1

Fischer EM, Seneviratne SI, Vidale PL, Lüthi D, Schär C (2007) Soil moisture–atmosphere interactions during the 2003 European summer heat wave. J Clim 20:5081–5099. https://doi.org/10.1175/JCLI4288.1

Gross MH, Donat MG, Alexander LV, Sisson SA (2017) The sensitivity of daily temperature variability and extremes to dataset choice. J Clim 31:1337–1359. https://doi.org/10.1175/JCLI-D-17-0243.1

Guo Z, Dirmeyer PA (2013) Interannual variability of land–atmosphere coupling strength. J Hydrometeorol 14:1636–1646. https://doi.org/10.1175/JHM-D-12-0171.1

Hartmann DL, Tank AM, Rusticucci M, Alexander LV, Brönnimann S, Charabi YA, Dentener FJ, Dlugokencky EJ, Easterling DR, Kaplan A, Soden BJ (2013) Observations: atmosphere and surface. In: Climate change 2013 the physical science basis: Working Group I Contribution to the Fifth Assessment Report of the Intergovernmental Panel on Climate Change 2013 Jan 1. Cambridge University Press. https://doi.org/10.1017/CBO9781107415324.008

Kain JS (2004) The Kain–Fritsch convective parameterization: an update. J Appl Meteorol 43:170–181. https://doi.org/10.1175/1520-0450(2004)043<0170:TKCPAU>2.0.CO;2

Kanae S, Oki T, Musiake K (2001) Impact of deforestation on regional precipitation over the Indochina Peninsula. J Hydrometeorol 2:51–70. https://doi.org/10.1175/1525-7541(2001)002<0051:IODORP>2.0.CO;2

Koster RD, Chang Y, Schubert SD (2014) A mechanism for land–atmosphere feedback involving planetary wave structures. J Clim 27:9290–9301. https://doi.org/10.1175/JCLI-D-14-00315.1

Koster RD, Dirmeyer PA, Guo Z, Bonan G, Chan E, Cox P, Gordon CT, Kanae S, Kowalczyk E, Lawrence D, Liu P (2004) Regions of strong coupling between soil moisture and precipitation. Science 305:1138–1140. https://doi.org/10.1126/science.1100217

Lau N, Nath MJ (2012) A model study of heat waves over North America: meteorological aspects and projections for the twenty-first century. J Clim 25:4761–4784. https://doi.org/10.1175/JCLI-D-11-00575.1

Liu YY, Parinussa RM, Dorigo WA, De Jeu RA, Wagner W, Van Dijk AI, McCabe MF, Evans JP (2011) Developing an improved soil moisture dataset by blending passive and active microwave satellite-based retrievals. Hydrol Earth Syst Sci 15:425. https://doi.org/10.5194/hess-15-425-2011

Meehl GA, Tebaldi C (2004) More intense, more frequent, and longer lasting heat waves in the 21st century. Science 305:994–997. https://doi.org/10.1126/science.1098704

Mlawer EJ, Taubman SJ, Brown PD, Iacono MJ, Clough SA (1997) Radiative transfer for inhomogeneous atmospheres: RRTM, a validated correlated-k model for the longwave. J Geophys Res 102:16663–16682. https://doi.org/10.1029/97JD00237

Rogers E, Black T, Ferrier B, Lin Y, Parrish D, DiMego G (2001) Changes to the NCEP Meso Eta Analysis and Forecast System: increase in resolution, new cloud microphysics, modified precipitation assimilation, modified 3DVAR analysis. NWS Technical Procedures Bulletin 488:15

Sato T, Kimura F, Kitoh A (2007) Projection of global warming onto regional precipitation over Mongolia using a regional climate model. J Hydrol 333:144–154. https://doi.org/10.1016/j.jhydrol.2006.07.023

Sato T, Xue Y (2013) Validating a regional climate model's downscaling ability for East Asian summer monsoonal interannual variability. Clim Dyn 41:2411–2426. https://doi.org/10.1007/s00382-012-1616-5

Schär C, Vidale PL, Lüthi D, Frei C, Häberli C, Liniger MA, Appenzeller C (2004) The role of increasing temperature variability in European summer heatwaves. Nature 427:332–336. https://doi.org/10.1038/nature02300

Schubert SD, Wang H, Koster RD, Suarez MJ, Groisman PY (2014) Northern Eurasian heat waves and droughts. J Clim 27:3169–3207. https://doi.org/10.1175/JCLI-D-13-00360.1

Seneviratne SI, Corti T, Davin EL, Hirschi M, Jaeger EB, Lehner I, Orlowsky B, Teuling AJ (2010) Investigating soil moisture–climate interactions in a changing climate: a review. Earth Sci Rev 99:125–161. https://doi.org/10.1016/j.earscirev.2010.02.004

Seneviratne SI, Lüthi D, Litschi M, Schär C (2006) Land–atmosphere coupling and climate change in Europe. Nature 443:205–209. https://doi.org/10.1038/nature05095

Seneviratne SI, Nicholls N, Easterling D, Goodess CM, Kanae S, Kossin J, Luo Y, Marengo J, McInnes K, Rahimi M, Reichstein M, Sorteberg A, Vera C, Zhang X

(2012) Changes in climate extremes and their impacts on the natural physical environment. In: Field CB, Barros V, Stocker TF, Qin D, Dokken DJ, Ebi KL, Mastrandrea MD, Mach KJ, Plattner GK, Allen SK, Tignor M, Midgley PM (eds) Managing the risks of extreme events and disasters to advance climate change adaptation, A Special Report of Working Groups I and II of the Intergovernmental Panel on Climate Change (IPCC). Cambridge University Press, Cambridge, UK, and New York, NY, USA, pp 109–230

Skamarock WC, Klemp JB, Dudhia J, Gill DO, Barker DM, Wang W, Powers JG (2005) A description of the advanced research WRF version 2. NCAR Tech. Note NCAR/TN-468+STR 88.

Su H, Dickinson RE (2017) On the spatial gradient of soil moisture–precipitation feedback strength in the April 2011 drought in the southern Great Plains. J Clim 30:829–848. https://doi.org/10.1175/JCLI–D–13–00185.1

Weisheimer A, Doblas-Reyes FJ, Jung T, Palmer TN (2011) On the predictability of the extreme summer 2003 over Europe. Geophys Res Lett 38:05704. https://doi.org/10.1029/2010GL046455

Yatagai A, Kamiguchi K, Arakawa O, Hamada A, Yasutomi N, Kitoh A (2012) APHRODITE: constructing a long-term daily gridded precipitation dataset for Asia based on a dense network of rain gauges. Bull Amer Meteor Soc 93: 1401–1415. https://doi.org/10.1175/BAMS–D–11–00122.1

Zaitchik BF, Macalady AK, Bonneau LR, Smith RB (2006) Europe's 2003 heat wave: a satellite view of impacts and land–atmosphere feedbacks. Int J Climatol 26: 743–769. https://doi.org/10.1002/joc.1280

Zhang J, Wu L, Dong W (2011) Land-atmosphere coupling and summer climate variability over East Asia. J Geophys Res 116:05117. https://doi.org/10.1029/2010JD014714

3

MF and HF radar techniques for investigating the dynamics and structure of the 50 to 110 km height region

Iain Murray Reid[1,2]

Abstract

The application of medium-frequency (MF) and high-frequency (HF) partial reflection radar to investigate the neutral upper atmosphere is one of the oldest such techniques still regularly in use. The techniques have been continuously improved and remain a robust and reliable method of obtaining wind velocities, turbulence intensities, electron densities, and measurements of atmospheric structure in the mesosphere lower thermosphere (MLT) region (50 to 110 km). In this paper, we review recent developments, discuss the strengths and weaknesses of the technique, and consider possible improvements.

Keywords: Medium-frequency radar techniques; High-frequency radar techniques; Partial reflection radar; Spaced antenna technique; Full correlation analysis; Mesosphere lower thermosphere region; D region; MLT region dynamics; Meteor radar; Imaging Doppler interferometer

Review

Introduction

This review is about the investigation of the upper neutral atmosphere in the 50 to 110 km height region by radars operating in the medium-frequency (MF, 0.3–3 MHz) and high-frequency (HF, 3–30 MHz) bands. In the MF band, the focus is on the range from about 2 to 3 MHz and in the HF band on the frequency range up to around 3.2 MHz. These radars are sometimes referred to as partial reflection (PR) radars and make use of the ionized component of the atmosphere as a tracer for the neutral motions in the ionosphere between heights of around 50 to 110 km. We do not consider the case of total reflection from the ionosphere, except where it complicates partial reflection radar measurements of the neutral atmosphere. Of course, we do consider other results where appropriate.

The 50 to 110 km height range is generally referred to as the mesosphere and lower thermosphere (MLT). The mesosphere is that region between the temperature maximum near 50 km (the stratopause) and the temperature minimum near 85 km (the mesopause). The thermosphere lies above the mesosphere. It is perhaps worth noting that the actual boundary heights change with location and season, and the mesopause height may be nearer 100 km at mid-latitudes in winter (e.g., Fan et al. 2014).

The state of the MLT is determined by dynamical coupling from below via waves and by photoionization from above by the sun. The basic thermal structure is set by the balance between absorption of solar UV and infrared radiation to space, but it is significantly modified by momentum deposition by atmospheric waves, notably internal atmospheric gravity (buoyancy) waves, that drives the atmosphere away from radiative equilibrium. For example, this causes the summer polar MLT to be up to 100 K colder than radiative equilibrium acting alone and the winter polar MLT to be correspondingly warmer (e.g., Andrews et al. 1987).

The investigation of the MLT region relies largely on remote sensing techniques. The direct techniques available generally rely on rocket delivered instruments, and so observations are necessarily episodic in nature. Indirect techniques include passive (airglow) and active (lidar) optical and radar techniques. A form of radar, the focus of this paper, was used to confirm the existence of the ionosphere (Appleton and Barnett 1925; Breit and Tuve 1925,

Correspondence: iain.reid@adelaide.edu.au
[1]Department of Physics, School of Physical Sciences, The University of Adelaide, Adelaide 5005, Australia
[2]ATRAD Pty Ltd, 20 Phillips Street, Thebarton 5031, Australia

1926), and radar has since been extensively used to investigate its properties.

There are two main mechanisms believed to result in coherent returns of radio waves from the MLT region. These are turbulent (or Bragg) scatter and Fresnel Reflection or Fresnel Scatter. Bragg scatter results from radar returns from turbulent irregularities in the electron density with scales of half the projection of the radar wavelength onto the wave vector. Fresnel reflection (also known as partial reflection) results from the presence of irregularities in refractive index transverse to the beam direction which are thin compared to the radar wavelength. Fresnel scatter results when the scattering medium is coherent in the two dimensions transverse to the probing wave and random in the direction parallel to the wave vector. This might correspond to a sharply bounded region of turbulence, for example. Generally at the MF/HF frequencies that we are considering here, the radar returns are produced by Fresnel reflection or Fresnel scatter. However, it is noteworthy in this context that the detailed nature of the backscattering and partially reflecting irregularities in the MLT is still not well understood.

The MLT height region corresponds to the ionospheric D region and the lower part of the ionospheric E region. The E region extends from heights of about 90 to 140 km and reaches its peak ionization levels near 110 km near noon. The actual peak height and height extent of the E region is dynamic, and it varies with location, time of day, season, solar cycle, and solar activity. The D region is that part of the ionosphere lying below the E region, and its lowest extent is about 50 km. Again, this lower bound is dynamic and depends strongly on location, time of day, and solar activity. Likewise, its upper extent is dynamic, and it varies with the changes in the extent of the E region. During the day, ionization rates in the D region are determined by solar Lyman-α (Ly-α; ~65–80 km), solar X-rays (≥80 km) and galactic cosmic ray fluxes (≤65 km), and at night by solar Ly-α and Lyman-β (Ly-β) scattered from the geocorona (~80–90 km) and by galactic cosmic ray fluxes (e.g., Nicolet and Aikin 1960; Thomson et al. 2007; Hedin et al. 2008; Pavlov 2014).

It is sometimes suggested that the D region can be considered to be formed by two different "layers": a "D layer" formed by the solar photoionization of NO by Ly-α and peaking near 90 km, and a "C layer" near 64 to 68 km formed by cosmic ray ionization (e.g., Deeks 1966; Bain and May 1967; Abdu et al. 1973). The idea of the C layer was first introduced to understand very low frequency (VLF, 3–30 kHz) radar propagation observations. In more recent work, Bertoni et al. (2013) briefly review some of these and suggest that a C layer of enhanced electron density does exist near 64 to 68 km for a few hours after dawn, but that its formation is associated with sunrise rather than with cosmic ray ionization. However, in the 50 to 110 km height region, the concept of a clear division between "E" and "D" regions and single "D" and "C" layers is slightly misleading, as the 50 to 100 km height region is stratified (e.g., Reid 1990), and most often consists of multiple regions of enhanced backscatter, with a tendency for this to occur at certain "preferred" heights for radar returns (e.g., Gregory 1961; Titheridge 1962; Reid 1990). This strong tendency to form layer-like regions of enhanced backscatter is overlaid with a strong diurnal variation in the heights of occurrence of radar returns. Near 85–90 km, a layer persists both day and night, and during the day, partial reflections extend down to heights near 50 km. This behavior might be a more logical justification for splitting the D region into D and C layers.

Whatever the actual height extent of the D region, or its subdivisions, in this review, we are focused on partial reflection of radar returns from the ionosphere in the region above 50 km, and below the total reflection height of the E region, for the purposes of investigating the neutral atmosphere.

Brief historical overview

Partial reflection from the 50 to 100 km height region
Echoes that appeared to be from mesospheric heights were reported as early as 1930 (Appleton 1930), and in the years that followed, a number of reports of echoes, mainly not only from the 50–100 km height interval but also some of uncertain origin from the 15 to 65 km height range (see e.g., Watson-Watt et al. 1937) were made. These early observations were all obtained from low frequencies (LF, 30–300 kHz) to HF and are summarized in Ellyett (1947) and Ellyett and Watts (1959). The first unequivocal indications of radar that returns from heights near but below the E region appear to be those presented by Appleton and Piddington (1938). They found echoes below the E region (~110 km) and which extended down to heights of about 80 km. Their results were obtained at MF using an ionosonde and were followed by similar results at MF and HF obtained by Dieminger (1952), Dieminger and Hofmann-Heyden (1952), and by Gnanalingham and Weekes (1952).

The work of Dieminger represents the first systematic investigation of the characteristics of these echoes and is not inconsistent with present knowledge. The first results of an experiment specifically designed to investigate the lower ionosphere below the E region were presented by Gardner and Pawsey (1953). The ordinary (O) and extraordinary (E or X) magnetoionic components each suffer different amplitude and phase variations with height due to the height variations in electron density (N_e) and collision frequency (ν) (e.g., Thrane and Piggott 1966; Friedrich and Torkar 1983; Holdsworth et al. 2002; Vuthaluru et al. 2002), and this was exploited by Gardner and Pawsey to measure the electron densities. The technique they used was the foundation of the differential absorption experiment (DAE) and differential phase experiment (DPE), amongst the few techniques capable of providing regular ground-based measurements of electron density in the mesosphere (for reviews, see Belrose 1970; Manson and Meek 1984). They also found that the ionization had a tendency to form in strata near 70 and 90 km.

As we have noted above, the evidence that there is a tendency for layers observed by radar to occur at certain preferred heights regardless of location has a long history. For example, Dieminger (1952), Gregory (1956, 1961), Titheridge (1962), Reid (1990), Jones et al. (2004), and Hall et al. (2006) discuss this for MF and HF, Bailey et al. (1955) and Pineo (1956) for very high frequency (VHF; 30 to 300 MHz) forward scatter results, Bowles et al. (1964) for both vertical and forward scatter VHF results, and Flock and Balsley (1967) for vertical incidence VHF results. At MF/HF, such preferred heights are subject to seasonal and annual variations, but persistent echoes occur from mean heights of about 65–68, 74–75, 83–85, and 92 km. The origin of the layers below 80 km is not well understood and could be aeronomic, dynamical, or a combination of both in origin. Layers do appear near sunrise and disappear near sunset, and so photo detachment is suggested.

The authors noted above, and many others, have found that one of the most persistent regions of echo occurrence is that near 70–75 km. This particular preferred height seems to be related to an equally persistent temperature inversion in the same height range (Schmidlin 1976; Champion 1987; Offermann et al. 1987; Hauchecorne et al. 1987), and Hauchecorne et al. (1987) suggest a similar height variation in the inversion to that found in VHF echo occurrence and that an enhancement in gravity wave breaking occurs in the region of increased stability in and above the temperature minimum. This idea has not been explored in a systematic manner using a modern MF/HF radar collocated with other observing instruments, the ideal instrument being a LIDAR. This, and the underlying formation mechanisms for the other preferred heights, is a topic certainly worth for future investigation.

Investigation of dynamics in the 50 to 100 km height region

In the MLT region, the first attempts to measure the motion of the ionosphere used spaced receiving antennas to measure the apparent motion of the ground diffraction pattern across the observing region (e.g., Ratcliffe and Pawsey 1933; Mitra 1949; Krautkrämer 1950). These early results were for total reflections from the ionosphere, and spaced antenna (SA) measurements ultimately proved very difficult to interpret when applied to total reflection from the ionosphere if the object was to measure neutral atmospheric dynamics, and they are not suitable for this purpose. Spaced antenna techniques are still used for investigating ionospheric dynamics in the form of ionosondes and dynasondes, and Reinisch et al. (1998) and Rietveld et al. (2008) give contemporary accounts of this application.

Gardner and Pawsey's (1953) work laid the foundations for an entire field of study for measuring electron densities and investigating the structure of the D region MLT region, and weak D region partial reflections and backscatter were subsequently exploited using a SA technique to investigate the dynamics of the neutral atmosphere (Fraser 1965). Both Gardner and Pawsey's and Fraser's techniques are still in use in essentially the same basic experimental form as they were more than 60 and 50 years ago, respectively.

The evolution of MF/HF SA techniques since the 1950s has followed the development of radar hardware and computing. In 1965, Fraser (1965) used the spaced antenna technique to determine partial reflection winds at Birdling's Flat near Christchurch in New Zealand using a simple binary correlation analysis. A more complete description is included in Fraser (1965). Stubbs (1973) used the Buckland Park medium frequency (BPMF) array to make SA measurements and analyzed them using the full correlation analysis (FCA) to determine D region winds. The FCA is a more sophisticated correlation analysis for the SA experiment which we will discuss below. Stubbs also validated the MF SA FCA winds against meteor radar winds from a nearby location.

Originally, data analysis generally took place offline and offsite, and most observing sites were limited to operation on a campaign basis. The advent of cheap personal computers allowed the introduction of the first real-time analysis of the SA experiment with a simplified FCA, the so-called poor man's FCA (Meek 1980). The original system hardware used in the SA experiment often only measured the amplitude of the returned radio waves, and phase information was not generally available. An example of the amplitude-only variation of radio waves received at one antenna on the BPMF radar in the 1970s is shown in Additional file 1: Video 1. This variation in amplitude was originally referred to as "fading", and the mean fading time is still one of the output parameters of a typical SA analysis (see Table 5).

Later, once coherent receiver systems were introduced, phase information became available, and Doppler-based and interferometric techniques became more common at MF (e.g., Vincent and Reid 1983; Adams et al. 1986). The most sophisticated ultimate implementation of this was perhaps the development of the BPMF radar (Reid et al. 1995) which included 30 independent phase-controlled transmitters and 16 independent receivers, allowing great flexibility in experimental configurations (e.g., Holdsworth and Reid 2004a). Currently, the Saura HF large array radar is the most capable in the world (Singer et al. 2008).

In the sections below, we explore the application of MF/HF radar techniques to investigate the MLT region, consider their weaknesses and strengths, and suggest possible future developments.

MF and HF radar techniques for the MLT region

There are a variety of radar techniques available for investigating the MLT region. These split naturally into those that use the ionization produced by solar radiation as a target and those that use ionized trails produced by meteors as targets. A number of these techniques are summarized in Table 1. Each technique has its own strengths and weaknesses, and it has become more common to utilize clusters of instruments to investigate the MLT region in order to maximize strengths and minimize weaknesses. For completeness, we have included two techniques in this table that are no longer used. These are the LF ionospheric drift experiment (see Sprenger and Schminder 1969; Jacobi 2014) which used transmitters of opportunity, in the form of commercial broadcast radio transmitters, and the "partial reflection experiment" (PRE) radar for measuring electron densities which was established near Tromsø, Norway in 1975 (see Brekke et al. 1985). This radar was the last of the large dedicated instruments for measuring electron densities in the lower ionosphere (see Belrose 1970 for a summary of instruments existing at that time). The Tromsø radar now operates in modified form as an MF SA FCA radar (e.g., Hall 2000). As we note briefly below, the ability to measure electron densities has been incorporated into newer MF/HF SA and Doppler beam swing (DBS) radars (see e.g., Holdsworth et al. 2002; Holdsworth and Reid 2004a; Singer et al. 2011).

In the last decade, this reapplication of older techniques and the validation of existing techniques for measuring neutral winds in the MLT region have emerged. In addition to measuring electron densities, these include the use of dynasondes for measuring neutral winds in the MLT (Jones et al. 2003) and the use of super dual auroral radar network (SuperDARN) ionospheric radars as meteor radars (e.g., Kleinknecht et al. 2014).

While the application of radars operating in the MF and HF bands to investigate the neutral upper atmosphere is one of the oldest such techniques still regularly in use, the techniques have been continuously improved, largely through the availability of better hardware (e.g., Reid et al. 1995; Holdsworth and Reid 2004b; Singer et al. 2008; Li et al. 2012) and readily available and economical but powerful computers, and provide a robust and reliable method of obtaining wind velocities (e.g., Stubbs 1973; Pancheva et al. 2002), turbulence intensities (e.g., Hocking 1983; Holdsworth et al. 2001), electron densities and collision frequencies (e.g., Thrane and Piggott 1966; von Biel 1977; Friedrich and Torkar 1983; Holdsworth et al. 2002; Singer et al. 2011), measurements of atmospheric structure (e.g., Gregory 1956; Hall 2000), temperatures (e.g., Tsutsumi et al. 1999; Holdsworth et al. 2006), and measurements of energy and momentum transfer (e.g., Reid and Vincent 1987; Murphy and Vincent 1993; Placke et al. 2015) in the MLT region.

There have been a number of reviews of the techniques. Briggs (1980) carefully describes the underlying principles of SA and DBS measurements of the neutral atmosphere, and Whitehead et al. (1983) provides a complementary review, which extends the discussion to ionospheric measurements. Briggs (1992, 1994) provides

Table 1 Radar techniques for investigating the MLT region

Radar technique	Typical frequency (wavelength)	Frequency range	Typical height coverage	Height resolution	Time resolution	Measured quantities	Status
Meteor radar	30 MHz (10 m)	MF to UHF	75–110 km	1–5 km	30–60 min	u, v, T, ρ, radiants, meteoroid velocities	Current
LF ionospheric spaced antenna drift	200 kHz (1500 m)	200 kHz (1500 m)	90–100 km (night)	1 height uncertainty ~2 km	1 min	u, v	Retired
Dedicated DAE/DPE PR radar	3 MHz (100 m)	3 MHz (100 m)	60–90 km (day) 85–100 km (night)	1–3 km	1–5 min	Ne, υ	Retired
MF/HF PR SA radar	3 MHz (100 m)	1.94 to 3.18 MHz	60–100 km (day) 80–100 km (night)	1–3 km	1–5 min	u, v, Ne, υ	Current
MST radar	50 MHz (6 m)	44.75 to 64 MHz	0.5–20 km 60–80 km (day)	150–600 m	5 min	u, v, w	Current
SuperDARN radar	12 MHz (25 m)	8–20 MHz	As for meteor radar	As for meteor radar	As for meteor radar	As for meteor radar	Current

The symbols represent the following parameters: u is the zonal wind magnitude, v is the meridional wind magnitude, w is the vertical wind magnitude, T is the temperature, ρ is the neutral density, N_e the electron number density, and υ is the collision frequency a general overview and history of MF/HF radar techniques of direct relevance to this review that can be applied to the MLT region. Reid (1996) reviewed MF techniques in terms of their ability to make measurements of mean winds, atmospheric tides, and gravity waves, and Hocking (1997) considers the strengths and limitations of mesosphere stratosphere troposphere (MST) radar measurements of middle atmosphere winds with a focus directly related to MF and HF radar techniques. Reid et al. (1995) and Holdsworth and Reid (2004a, 2004b) provide an overview of the capabilities of the BPMF radar which serves as a good description of the capabilities of MF/HF radars generally, Holdsworth et al. (2004a) provide a good description of a modern all-sky meteor radar, Fraser (2005) provides a history of the development of scientific radar in New Zealand very relevant to this review, and Hocking (2011) provides a general review of radar techniques for investigating the MLT region.

As we noted above, in this review, we concentrate on MF/HF SA radar techniques for measuring the neutral atmosphere using returns from the ionized atmosphere, but include meteor radars where relevant. The most common meteor radar operating frequencies are currently in the lower VHF band, between 30 and 55 MHz, but there are advantages to operating at lower frequencies because the meteor count rate N goes as (McKinley 1951)

$$N \propto \frac{{P_T}^{1/2}\ G\lambda^{3/2}}{{P_R}^{1/2}}$$

where λ is the radar wavelength, P_T is the transmitted power, G is the system gain, and P_R is the received power. Meteor radars historically operated close to, and often below, 30 MHz. The peak of the meteor detection distribution is also higher with a lower frequency (e.g., Younger et al. 2014). This makes SuperDARN radars (8 to 20 MHz) potentially very powerful meteor radars for measuring winds at heights in the data poor region between above 100 and 120 km.

MF radars have been successfully used as meteor radars, particularly, but not exclusively, at higher latitudes (e.g., Tsutsumi et al. 1999; Holdsworth and Reid 2004a; Tsutsumi and Aso 2005), and for measuring winds up to 120 km. However, a disadvantage here is that as the frequency is lowered, meteor detections are harder to discriminate from ionospheric echoes. In addition, above about 110 km, the Earth's magnetic field increasingly influences the motion of the ionized component of the atmosphere, (e.g., Elford and Elford 2001), and so both the SA and meteor techniques may not measure the neutral dynamics above this height.

The radar techniques applied at MF and at HF to investigate the MLT height region are basically the same techniques that can be applied at other radar frequencies. They differ in that the wavelengths in the MF and HF bands are relatively long (1000 to 100 m in the MF band; 100 to 10 m in the HF band), in that ionospheric radar returns are generally present above 80 km both day and night at these frequencies as we shall see below, and in that the irregularities responsible for the backscatter and partial reflection are very aspect sensitive at these frequencies (see e.g., Pawsey 1935; Holdsworth and Reid 2004a). These differences give rise to a number of unique considerations which we discuss below.

Spaced antenna techniques

The most common technique applied at MF and lower HF is the SA technique, and MF/HF radars currently operating using the technique are summarized in Table 2 along with some of their operating characteristics. Radars of this type that have been retired since the review of Reid (1996) are summarized in the same way in Table 3. Some of these radars were only ever intended to provide shorter term observations, but for others, ongoing long-term funding has proved difficult to obtain. Figure 1 shows the location of the radars summarized in Tables 2 and 3.

In SA techniques, a minimum of three non-collinear receiving antennas are used to receive partial reflections from the D region. The transmitting antenna used is usually located very close to the receiving antennas in a bi-static configuration. More commonly, the same antennas are used for both transmission and reception. A typical antenna arrangement is shown in Fig. 2. In this particular case, which is for the Poker Flat MF radar (see e.g., Kawamura et al. 2007), all four antennas are used for both transmission and reception, and each antenna is connected to an independent receiver. Another arrangement is shown in Fig. 3, which is for the Saura HF radar (Singer et al. 2011). The long arms allow this radar to be used in DBS mode, and the antennas shown in green allow it to be used as a SA radar. In DBS, narrow beams are formed in order to measure the radial velocities in three or more beam directions, and the horizontal and vertical winds are calculated from the radial velocities.

The D region ionosphere is birefringent, and both ordinary (O) and extraordinary (E or X) polarizations are used. The E mode is strongly absorbed during the day, so historically, for wind measuring radars, the E mode was used at night and the O mode during the day. For this reason, the antennas shown in Fig. 2 are crossed dipoles. This allows both modes to be transmitted and received. It is now common to transmit O and E modes on either alternate radar pulses, or on alternate data records, and then apply the DAE or DPE to measure the electron density (see e.g., Holdsworth et al. 2002; Singer et al. 2003; Singer et al. 2011).

Table 2 A summary of current MF/HF SA radars

Facility	Lat./long	Mode	Freq. (MHz)	Power (kW)
Adelaide, Australia	35 S, 138 E	SA, IDI, DBS	1.98	100
Andenes, Norway	69 N, 16 E	SA	1.98	50
Bear Lake, USA	42 N, 111 W	IDI	2.2 (day), 3.8 (night)	10
Christchurch, NZ	44 S, 172 E	SA	2.40	100
Davis Station, Antarctica	69 S, 78 E	SA	1.98	25
Kolhapur, India	16.8 N, 74.2 E	SA	1.94	25
Juliusruh, Germany	55 N, 13 E	SA, IDI, DBS	3.18	128
Langfang, China	39.41 N, 116.71 E	SA	1.99	64
Pameungpeuk, Indonesia	7.7 S, 107.7 E	SA	2.008	30
Poker Flat, USA	65 N, 148 W	SA	2.43	50
Pontianak, Indonesia	0 S, 109 E	SA	1.98	25
Qujing, China	25.6 N, 103.8 E	SA	2.138	64
Saskatoon, Canada	52 N, 107 W	SA	2.22	50
Saura, Norway	69 N, 16 E	SA, IDI, DBS	3.17	116
Scott Base, Antarctica	78 S, 167 E	SA	2.90	60
Syowa Base, Antarctica	69 S, 39.6 E	SA	2.4	50
Tirunelveli, India	8.7 N, 77.8 E	SA	1.98	25
Tromso, Norway	70 N, 19 E	SA	2.78	50
Wakkanai, Japan	45.4 N, 141.7 E	SA	1.955	50

The radar at Bear Lake is a dynasondes that can operate in an IDI mode. The most recent MF radar installations have been in China, as part of the Chinese Meridian project. The Indian radars and the Saura 3.18 MHz radar (IAP, Germany) have been recently upgraded

Figure 4 shows one example from Singer et al. (2003) which shows the electron densities derived from the DAE and DPE, as well as the O and X mode signal powers obtained using interleaved operation.

Vertical incidence pulsed transmission is used for all current MF/HF SA systems. A unique frequency modulated continuous wave chirp (FMCW-chirp) 3.18 MHz system was operated at Juliusruh in Germany for a number of years. This system commenced operation in 1982

Table 3 MF/HF radars that have been retired since the review of Reid (1996)

Facility	Lat./long	Mode	Freq. (MHz)	Power (kW)	Remarks
Bribie Island, Australia	28 S, 153 E	SA, DBS	1.98	25	Long term
Christmas Island	2 N, 157 W	SA	1.98	25	Long term
Halley Base, Antarctica	76S, 26 W	IDI	2.75	10	Long term
Hawaii, USA	22 N, 156 W	SA	1.98	25	Long term
London, Canada	43 N, 81 W	SA	2.22	25	Long term
McMurdo, Antarctica	78 S,166 E	SA	1.98	50	Long term
Nanjing, China	32 N, 119 E	SA	2.00	25	Long term
Platteville, USA	40 N, 105 W	SA	2.219	25	Long term
Puerto Rico	19 N, 67 W	SA	1.95	25	Long term
Rarotonga, Cook Islands	22 S, 160 W	SA	1.98	25	Long term
Robsart, Canada	49 N, 109 W	SA	2.22	25	Campaign
Sylvan Lake, Canada	52 N, 114 W	SA	2.22	25	Campaign
Tindal, Australia	15 S, 132 E	SA	1.98	25	Campaign
Urbana, USA	40 N, 88 W	SA	2.66	25	Long term
Yamagawa, Japan	31 N, 131 E	SA	1.95	50	Long term

The Halley dynasonde operated successfully in IDI mode until it was relocated in 2008

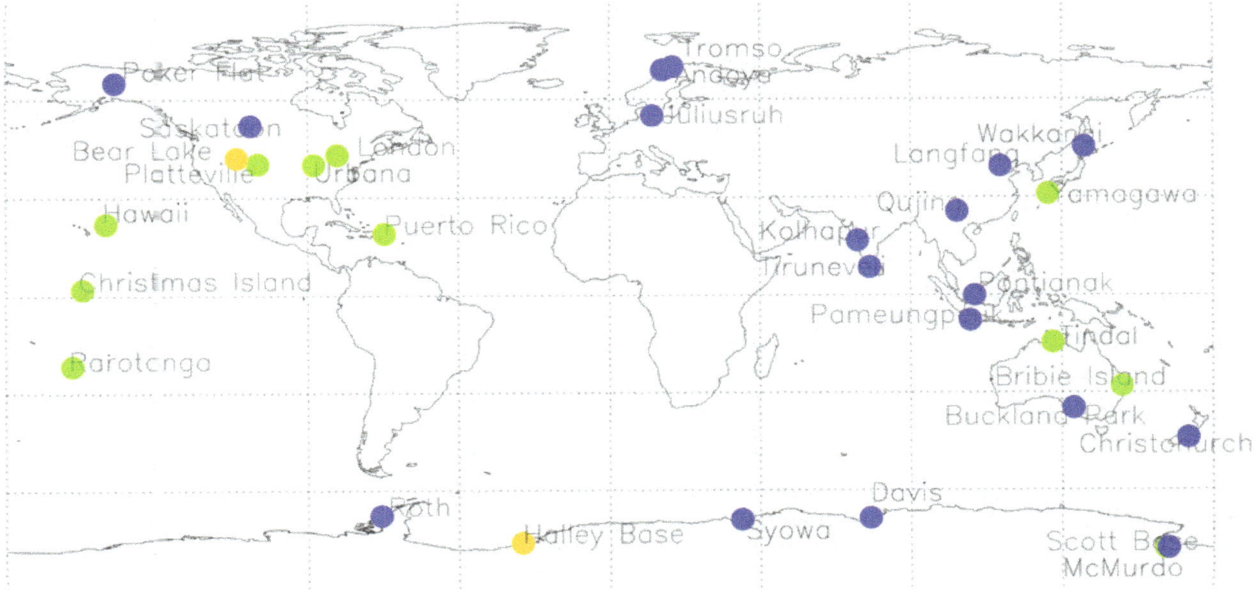

Fig. 1 MF and HF radars used to measure the MLT region. *Green* indicates retired radars, *blue* current radars, and *yellow* dynasondes that have operated as IDI radars. See text and Tables 2 and 3 for further details

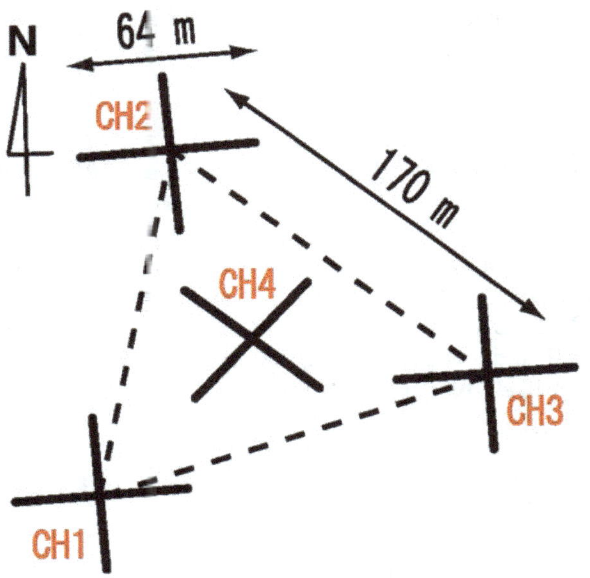

Fig. 2 A typical antenna layout for the spaced antenna technique at 2.43 MHz. This arrangement is for the Poker Flat MF radar (see Kawamura et al. 2007). All antennas are used for both transmission and reception using circular polarization. This optimizes system sensitivity, as transmitter and receiver polar diagrams are matched. The antenna orientation is designed to reduce coupling, and the overall arrangement facilitates the use of the FCA, FSA, SCA, IDI, and meteor analyses. Generally, the antenna spacing is selected so that cross-correlation functions calculated between antennas on the long sides of the triangle takes peak values of about 0.5. Because the fading times vary with height (see Fig. 8), the outer triangle spacing is generally optimized for a particular height. Having an additional antenna at the center of the triangle provides the potential for better cross-correlation values (~0.5) over a greater range of heights

to measure electron densities (Priese and Singer 1984; Singer et al. 1988) and was modified to be capable of making spaced antenna measurements in 1990 (e.g., Hoffmann et al. 1990). It was replaced with a pulsed system in 2005 (e.g., Hoffmann et al. 2011).

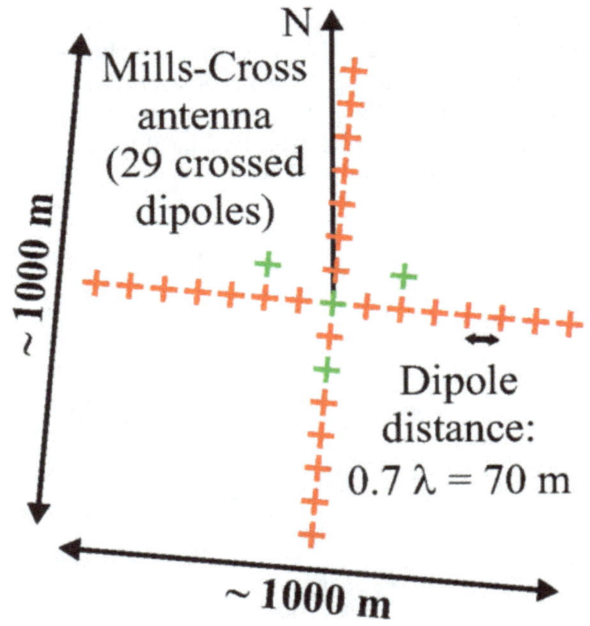

Fig. 3 The antenna arrangement for the Saura HF radar (after Placke 2013). All antennas of the main arms (the "Mills Cross") are used for transmission and reception. The crossed dipoles shown in *green* are used for SA applications

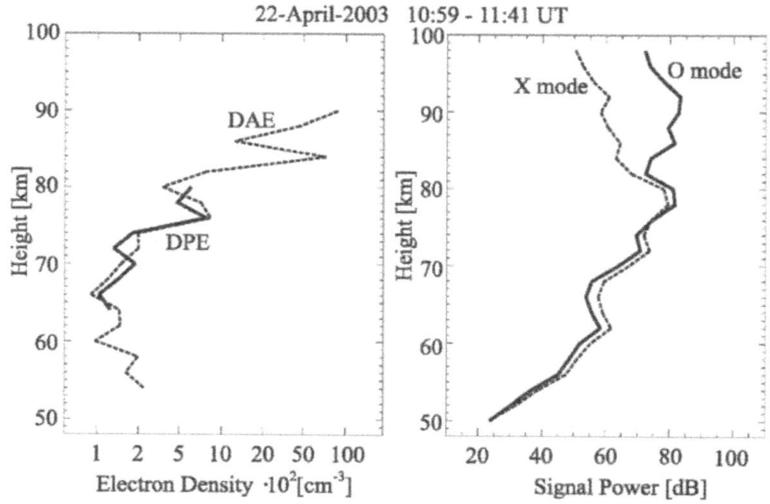

Fig. 4 Electron densities derived from the DAE and DPE using the Saura HF radar (*left panel*) and the corresponding signal powers for the ordinary (O mode) and X modes obtained using interleaved observations (*right panel*) (after Singer et al. 2003)

With a typical 3 dB pulse width of 26.6 μs for a system operating at 2 MHz, the height resolution is about 4 km, which is commonly oversampled to yield a 2 km range bin. However, as we discuss below, because of the presence of the strong returns from the E region, it would be preferable to use a shorter pulse. At 2 MHz, the wavelength is 150 m, so a 4 km resolution at 2 MHz is comparable to a 160 m height resolution for a 50 MHz (6 m wavelength) atmospheric radar. A more typical best range resolution at 50 MHz would be 100 m, which would correspond to a 2.5 km pulse width at 2 MHz. A 1 km pulse width (which would be 1.6 km at 2 MHz) is used successfully on the 3.17 MHz Saura radar (e.g., Singer et al. 2008, 2011).

The maximum usable pulse repetition frequency (PRF) is determined by the total reflection height. An illustrative range time diagram is shown in Fig. 5. This is taken from Czechowsky et al. (1983) and shows two pulse repetition periods (PRP = τ), of 1 and 4 ms, corresponding to PRFs of 1000 and 250 Hz, respectively. In simple range aliased terms, the maximum unaliased range $r_{\max}$ is given by

$$r_{\max} = c\tau/2,$$

where c is the speed of light, and so these PRFs would correspond to maximum usable ranges of 150 and 600 km, respectively. However, as the figure shows, returns from ranges 1, 2, and 3 in the E region appear at lower ranges at later times as a consequence of multiple reflections from the E region. At E region heights, and at MF and lower HF during the day, these returns would most likely be from total reflections. Consequently, we

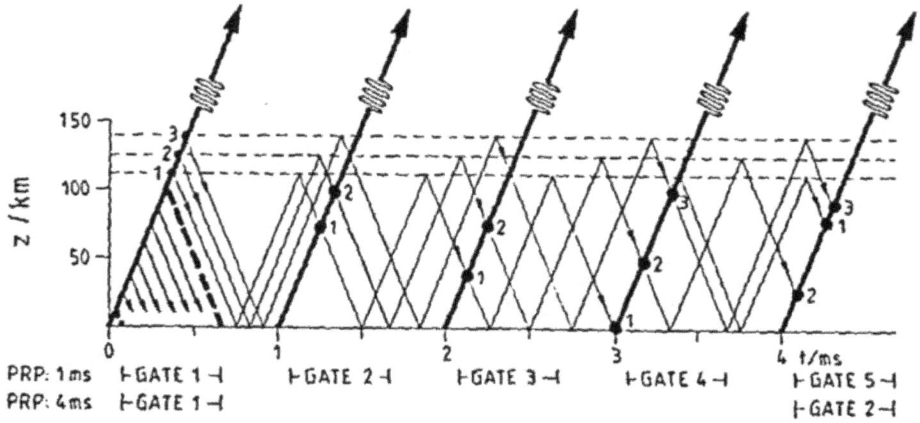

Fig. 5 Range-time diagram showing the influence of multiple reflections or scattering at E region heights on the received signals for two different pulse repetition periods (after Czechowsky et al. 1983). For the case of total reflection from the E region, the returns would be attenuated by each ground reflection and by any absorption through the D region. This is significant during daytime

expect these returns to be very strong and to potentially dominate the weaker partial reflections (typically 10^{-3} to 10^{-5} that of the E region) from the D region. Of course, the strength of these returns would be reduced by each ground reflection and by any D region absorption present during each transit, but the returns could still be strong enough to contaminate the weak D region returns even after multiple reflections. D region absorption is very strong during the day. As a consequence of this, with typical E region peak heights during the day of 110 km, and typical F region peak heights of above 210 km during the night, PRF values from 80 to 120 Hz during the day and from 20 to 40 Hz at night are used.

These restrictions on PRF mean that the duty cycles of PR radars are low, and in this, the radars could be considered to be underutilized. The use of pseudo-random pulse codes and of pulse stuttering on transmission are avenues for investigation that could lead to better use of such radar's capabilities.

Choice of operating frequency

Most MF/HF SA radar techniques for investigating the neutral atmosphere do not routinely make measurements above about 100 km. This is because a significant issue with radars operating at less than the critical frequency for the E region (between 2 and 4 MHz depending on latitude, time of day, season, and solar cycle; see e.g., Jursa et al. 1985) is that the total reflection height varies between around 95 and 110 km. This represents an upper operating height during the day for radars operating in the MF and lower HF bands. In addition, close to, but below the total reflection height, group retardation means the reported heights (the virtual heights) are greater than the actual backscatter/partial reflection heights. This results in different ranges (that is, group delays) corresponding to very similar heights. This also means that the maximum useable height is not fixed for a particular frequency radar or location, and the actual maximum useable height may not be readily identifiable for a particular location unless some care is taken (see e.g., Namboothiri et al. 1993). However, it is possible that with care, the actual height can be determined, and there is also good evidence that winds above 100 km can be recovered at some times of the day and at some locations using MF/HF SA techniques (Namboothiri et al. 1994) and MF meteor methods (e.g., Tsutsumi and Aso 2005).

Given the issues around in the influence of the height of the total reflection, the selection of the operating frequency is very important. This is more so the case because of the proximity of the operating frequencies to those of the AM broadcast band, MF and HF amateur bands, radiolocation bands, and maritime communication and mobile communications bands. In most instances, the radar frequency is determined by the available spectrum and licensing. The best operating frequency (f) is a balance between the competing requirements of licensing, good backscattered power, which goes as $1/f^2$, and favors a lower frequency, and the reflection height, which goes as f, and favors a higher frequency. As a consequence of these limitations, current PR radars operate in the range from 1.94 to 3.18 MHz as shown in Table 2.

In addition to forming an upper limit to the upper measurement height of MF and lower HF radar techniques, strong E region returns may have another impact on winds above about 92 km during the middle of the day. Because of the long tail on the Gaussian transmitter pulses used, and the steep gradient in returned power in height approaching the E region, height gates well away from the height of the total reflection layer may be contaminated by returns from the E region, possibly making the derived winds unrepresentative, and potentially unusable. This was first suggested by Hocking (1997) in the context of finite receiver bandwidths and who suggested that this was most likely to be an issue around mid-day. He suggested that all data over a measured height of 92 km should be discarded and repeated this in a subsequent review (Hocking 2011). This conservative approach is appealing in its simplicity, and a safe approach, but may not be entirely necessary when with some effort, suspect data could be excluded from further consideration after initial processing.

As noted above, the influence of the E region represents an upper limit to the technique during daytime. The majority of investigators have taken the view that below about 100 km at mid-latitudes any group retardation effects are negligible, that the real and virtual heights are equivalent below 100 km, and E region contamination was minimal. As we have noted above, this is too lax. However, the quality control criteria applied within most analysis techniques mean that the quality of the data is high and that some results contaminated by total reflections are most likely omitted. This is one reason for the relatively high rejection rates above 90 km during the day (greater than 50 % in ordinary mode (O-mode) and greater than 75 % in E mode) as we shall see below. The current quality control parameters (or equivalently the rejection criteria) for the BPMF SA FCA are summarized in Table 6 by way of example.

However, there is no doubt that great care must be taken to avoid contamination from total reflection, and the addition of other quality control measures to address this is recommended. In addition, an active consideration of the likely total reflection height, the consequent restriction of data to unaffected height ranges, an active consideration of group retardation and a correction made for it, and the use of smaller range resolutions to increase the number of unaffected range gates at least above 90 km should be implemented. At 2 MHz, a 2 km pulse oversampled to 1 km is suggested.

Character of the ground diffraction pattern

An example of a snapshot in time of a D region diffraction pattern from a weak partial reflection from a height of 80 km is shown in Fig. 6. This was obtained using the BPMF radar using a technique described by Felgate and Golley (1971), and further descriptions relevant to this are provided by Briggs and Holmes (1973), Holmes (1975), and Briggs (1993). It is the motion of this ground diffraction pattern that is used to infer the motion of the atmosphere at the partial reflection or backscatter height. In an attempt to do this by sampling the ground diffraction pattern at a limited number of locations, the various spaced antenna techniques that we discuss here were used. Generally, the number of locations sampled is limited by cost. If all of the pattern was available as it is in Additional file 2: Video 2, then in a modern radar system, image processing techniques could be applied to determine the motion of the pattern as is done for cloud images in both the lower and upper atmosphere (e.g., Rong et al. 2015). This does assume that all of the motion of the ground diffraction pattern is due to the horizontal motion of the atmospheric irregularities producing it, and this does not always hold true. For example, differential vertical motion in the radar pulse volume can induce an apparent horizontal motion of the pattern (e.g., Røyrvik 1983). We will return to this below. Additional file 2: Video 2 shows the sequence from which Fig. 6 was obtained. The fading shown in Additional file 1: Video 1 corresponds to the variation in the strength of the electric field at a particular position on the ground as the diffraction pattern moves over the receiving antennas.

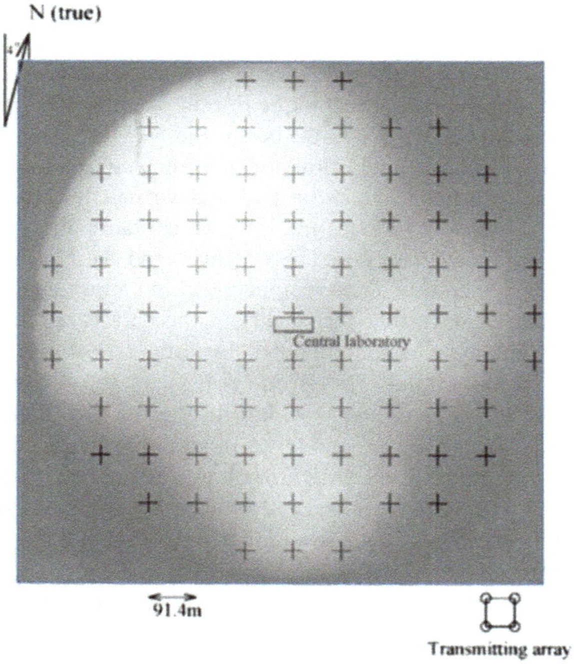

Fig. 6 The instantaneous ground diffraction pattern at 2 MHz for a partial reflection height of 80 km. The scale of the diffraction pattern is indicated by the overlay of the BPMF large array. The basic antenna spacing is 91.4 m. For this work, the small transmitting array at the bottom right of the figure was used for transmission. The image of the diffraction pattern is taken from movies produced by Briggs and Holmes (1973) and Holmes (1975) and described by Briggs (1993). Videos of the movies are included as additional material with this paper

A more contemporary example of a SA data record is shown in Fig. 7. This shows the returned signal measured during a 120 s period every 0.4 s (after coherent integration) over a range of heights for a single receiver for the BPMF radar, along with the corresponding mean power profile. This figure shows the variation in power with height and the tendency to layering. The corresponding figure for this record showing the variation of the amplitude of the signal from each range gate (the fading) is shown in Fig. 8. In the simplest spaced antenna analysis, the signals detected at three spaced antennas corresponding to the fading shown in Fig. 7 are analyzed using a correlation analysis to determine the motion of the diffraction pattern of the partially reflected signal from the D region over the ground. The detection may utilize coherent receivers, but only the amplitude information is used in the analysis.

A schematic representation of the SA experiment is shown in Fig. 9 (MacKinnon et al. 2002). The ground diffraction pattern is sampled at three ground locations. In this case, which corresponds to the arrangement used at VHF for the boundary layer version of the SA experiment, groups of nine antennas are used for both reception and transmission. Groups of three or four antennas are similarly used at each received location with the BPMF radar with SA mode (see e.g., Holdsworth and Reid 2004a), and there is a consequent increase in gain and a high frequency filtering of the ground diffraction pattern in both cases.

Figure 10 shows a sequence of a few hours of the signal to noise ratio (SNR) for one receiver for the BPMF radar again showing the tendency for layering. In this example, sunrise occurred at 21:05 UT (6.35 am LT). The layer between 60 and 65 km that forms around sunrise is consistent with VLF observations of the C layer described by Bertoni et al. (2013) and mentioned above. The layer near 75 km is also commonly observed and, as discussed above, is often associated with a "preferred height".

Figure 11 shows the height profile of the SNR for the Davis Station MF SA radar for one receiver for 1 year of observations for O mode (McIntosh et al. 2010). A distinction is often made between the nature of the scattering processes above and below about 80 km. In this context,

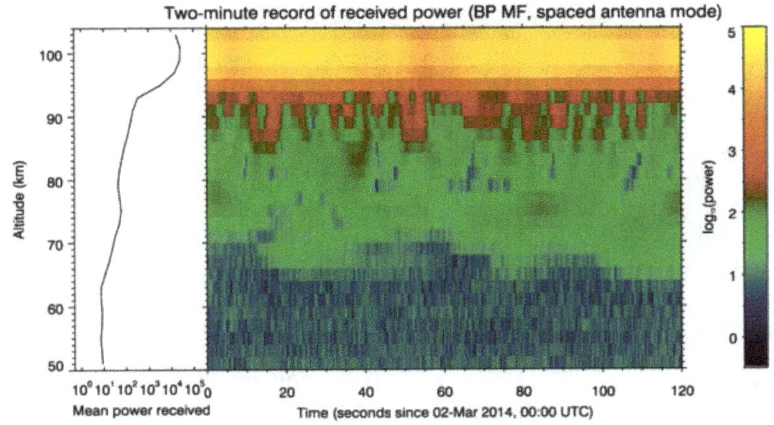

Fig. 7 The returned power on one receiver for one randomly selected 120 s observation at 00:00 UTC (LT = UTC + 9.5 h) from the BPMF radar. Note the returned power spans more than 3 orders of magnitude in this height range and the layered structure evident around 75 km

we note the rapid increase in SNR between 82 and 84 km, which corresponds to a gradient of about 8 dB/km. This rapid increase in returned power is the rationale for using two sampling regimes and two gain settings on some radars. Another example of a height profile of returned power for a brief period of observation is given in Fig. 4, in this case for the Saura HF radar in northern Norway. This example shows height profiles of both the O and X mode returned powers, which range over 65 dB.

Generally, slow fading times are associated with more aspect sensitive backscatter, and so with larger scales of the ground diffraction pattern. The aspect sensitivity is parameterized as the half width of the backscatter polar diagram and given by θ_S (see e.g., Holdsworth and Reid 2004a). A review of the methods for investigating thin sheets of temperature gradients in the lower atmosphere by Luce et al. (2001) provides a useful review of techniques for measuring aspect sensitivity and of the relationship between atmospheric structure and aspect sensitivity.

In addition to the rapid increase in power, there is a change in character of the backscatter at heights of around 80 km, from more to less aspect sensitive (Figs. 7 and 11). This suggests that one challenge of spaced antenna techniques is to adequately characterize the ground diffraction pattern through all sampling heights. This may not be possible when the same sampling scheme is used for all heights.

Hocking (1981) estimated that minimum data lengths of (80 – 100) × $T_{0.5}$, where $T_{0.5}$ is the fading time, are required to produce the true spectrum of velocities in the backscattering region. Inspection of Fig. 8, which is typical, suggests that data lengths of 160–500 s may be required

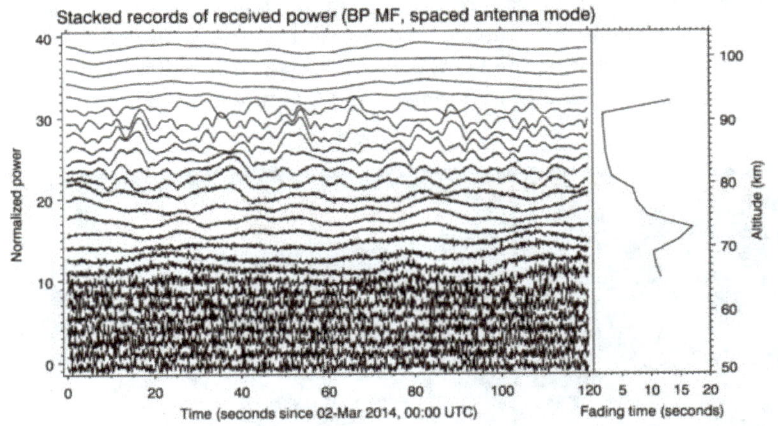

Fig. 8 The time series of the normalized voltage corresponding to Fig. 7. The amplitude variation with time is the "fading" analyzed in the original SA technique. The mean fading time for the record is shown in the *panel* on the *right hand side*. In any cross-correlation analysis, data records need to be of sufficient length to characterize the correlation functions. Because of this, the QC parameters applied to these types of analysis (see, e.g., Table 6) reject records for "slow fading", and indicators of poor correlation function form. Ideally, the record length should be selected based on the fading characteristics at a particular height. This would be simple to implement on current systems

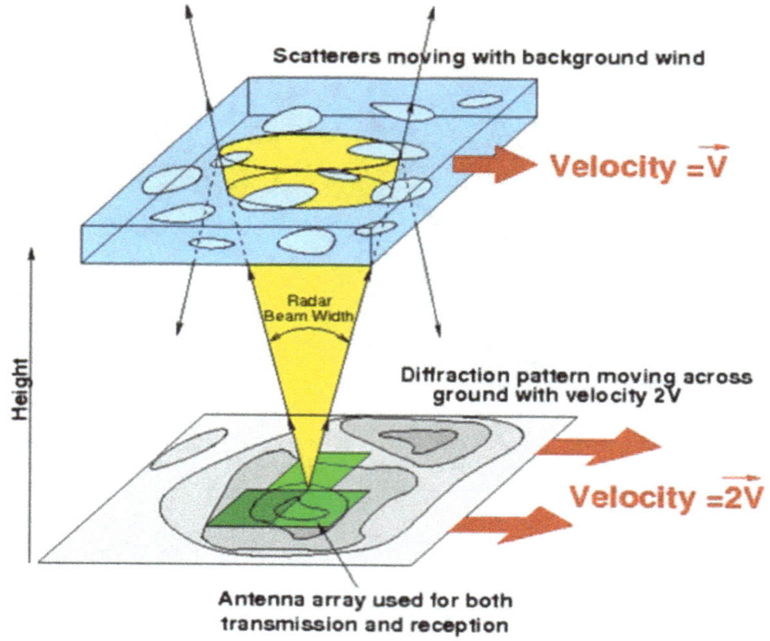

Fig. 9 Schematic representation of the SA experimental arrangement (after MacKinnon et al. 2002). In this figure, the scattering irregularities are moving at velocity v at a particular height, and the ground diffraction pattern is produced by the radar return from these irregularities. The ground diffraction pattern from a particular height moves across the ground with a velocity $2v$, and this is sampled at three locations indicated by the *green squares*. In this example, which is for a VHF boundary layer radar, three groups of nine antennas are used for reception. This increases the gain and also introduces a high frequency filter on the diffraction pattern

between 80 and 90 km and much longer lengths required above and below this height range. Typical data lengths used at Saskatoon were about 5 min (Meek and Manson 2001), but at Adelaide and many of the other radars in Tables 2 and 3, typical data lengths used for SA analyses were around 2 min, a length originally determined by computer limitations at the field site. Two minute record lengths are still typical, although now unnecessarily so, and the use of longer, fading time-dependent data record lengths is recommended. We note here that implicit in many of the SA analysis techniques, including the FCA, is the assumption of volume scatter. If discrete scattering targets exist, then interferometric techniques could measure radial velocities in shorter time intervals than indicated by Hocking's analysis. We will return to this point below and discuss the formation of the ground diffraction pattern when we discuss DBS techniques.

Spaced antenna analysis techniques

Similar fades The oldest and simplest analysis technique for spaced antenna data is the "method of similar fades" (see e.g., Mitra 1949; Krautkrämer 1950; Sprenger and Schminder 1969). This approach involves measuring

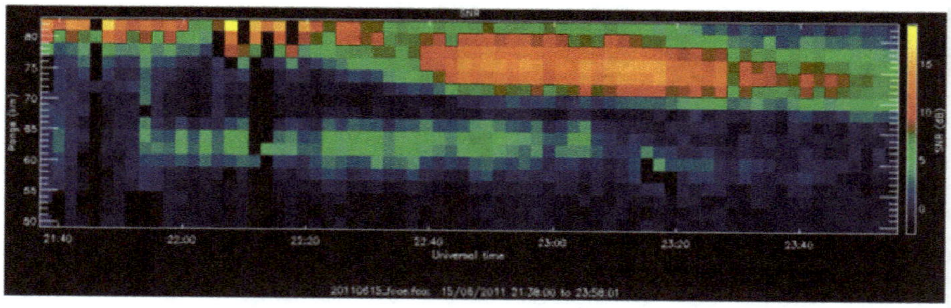

Fig. 10 A few hours of SNR for E mode randomly selected for one receiver of the BPMF radar showing layering typical of this type of observation. LT = UT + 9.5 h, and sunrise occurred at 21:05 UT (6.35 am LT). The layer between 60 and 65 km that forms around sunrise is consistent with VLF observations of the "C-layer" described by Bertoni et al. (2013) and mentioned in the text. The layer near 75 km is also commonly observed and is often associated with a "preferred height"

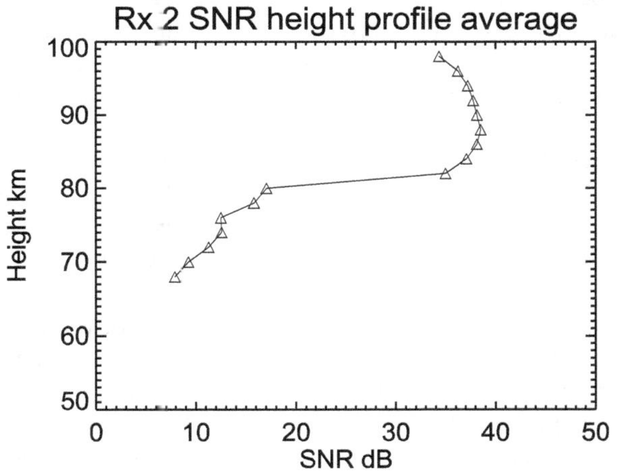

Fig. 11 The average height profile of returned signal to noise (SNR) for 1 year of observations for one receiver channel for O-mode on the Davis Station MF radar (after McIntosh 2010). Note the rapid increase in SNR between 80 and 82 km, which corresponds to a gradient of 18 dB in 2 km. This rapid increase in returned power with height was the rationale for using two sampling regimes and two gain settings on some older radars

and recording the fading of a radio signal returned from the ionosphere at three points on the ground at the corners of a triangle. The receiving locations are separated by distances of the order of one wavelength or less. Originally, the triangle was right-angled, but this was later shown to introduce a bias into the derived velocity, and subsequently, equilateral triangles were generally used (Stubbs 1973). The velocity of the ground diffraction pattern is then determined from the time displacements of the fading measured at each receiving location. A simple correction is made for the point source effect (the derived velocity is divided by two; see Felgate 1970; Briggs 1980). A series of experiments in the late 1960s and early 1970s determined that the motion of the ground diffraction pattern could be well characterized using only three non-collinear receiving locations (Briggs 1980; 1994). Varieties of this spaced antenna analysis all attempt to determine the motion of the atmosphere from the same basic information provided by the ground diffraction pattern. We will return to this below.

It was realized very soon after the introduction of the simple "similar fades" analysis that a complete evaluation of such measurements needed to take into account the effects of any random changes and any anisotropy of the ground diffraction pattern as it moves across the receiving antennas. An analysis to do this was developed by Briggs et al. (1950), Phillips and Spencer (1955), Ratcliffe (1956), and others and called the FCA method. A more contemporary description is given by Briggs (1984) and Holdsworth and Reid (1995a). However, as explained by Sprenger and Schminder 1969, at that time the FCA was relatively laborious and expensive, and so the simple similar-fade method was often preferred. They also describe another approach used with similar fades. This was to average the "drift" vectors measured between the spaced receiving locations to determine the velocity of the ground diffraction pattern.

When a simple correlation analysis (effectively a similar fades analysis) is used to determine the movement of the ground diffraction pattern across three non-collinear antennas, the resulting velocity does not take into account random changes in the pattern as it moves, or any anisotropy in the pattern. This results in a velocity that overestimates the magnitude of the true velocity. For this reason, this velocity is called the "apparent" velocity in the FCA. This is illustrated in Fig. 12 which is taken from Briggs (1993). Part (a) of this figure illustrates a changing pattern $E(x, t)$ moving past two antennas separated by distance ξ_0 with velocity V to produce records $E_1(t)$ and $E_2(t)$. Part (b) shows the cross-correlation function ρ_{12} that takes a maximum value a lag τ_m, smaller than the "correct" value $\frac{\xi_0}{V}$ because of these

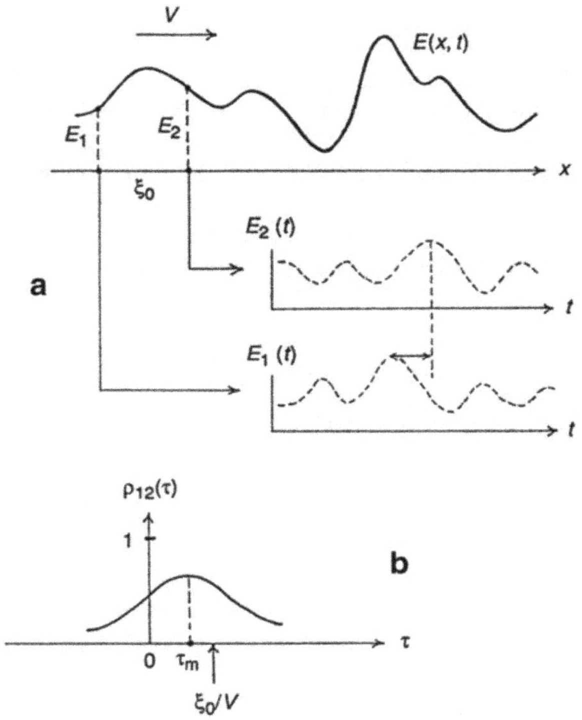

Fig. 12 (After Briggs 1993) Part **a** of this figure illustrates a changing pattern $E(x, t)$ moving past two antennas separated by distance ξ_0 with velocity V to produce records $E_1(t)$ and $E_2(t)$. Part **b** shows the cross-correlation function ρ_{12} calculated from $E_1(t)$ and $E_2(t)$ that takes a maximum value a lag τ_m, smaller than the "correct" value $\frac{\xi_0}{V}$ because of these changes, and consequently, an overestimation of the correct velocity V, the "apparent" velocity V_a is produced

changes, and consequently an overestimation of V, the apparent velocity. It is easy to understand how this comes about if we consider the limiting case of $V = 0$ and only random motions. In this case, the maximum cross correlation will be at $\tau_m = 0$.

Sprenger and Schminder (1969) argued that the simple similar fades approach produces a result very similar to the "apparent velocity" derived in the FCA method, whereas the velocity obtained from averaging the drift vectors produces a result in a term more similar to the "true velocity" in the FCA method. In this, the technique is similar to the simplest imaging Doppler interferometer (IDI) techniques, which we discuss below, and which produce a term similar to the apparent velocity.

The similar fades spaced antenna analysis using the drift vector approach was applied at LF and oblique incidence using commercial radio transmitters about 165 km distant from the receiving site in Germany from 1959 until 2007, with height information being available from 1982 onwards. This approach is effectively a single height measurement, with the mean reflection height being typically 90 to 95 km, and the measurements were restricted to nighttime because of strong daytime absorption. A description of the experiment, the analysis and results is given by Schminder and Kürschner (1994).

Full correlation analysis

Like the method of similar fades, the FCA utilizes a minimum of three spaced antennas to sample the ground diffraction pattern produced by the backscattering irregularities. Table 4 shows the typical operating parameters for the BPMF radar operating in SA mode. Cross correlation techniques are used to determine its velocity of motion, and the analysis takes account of random changes in the pattern, and also any anisotropy of the pattern, so that both horizontal components of the wind field are measured at a particular height in a single region directly above the radar. This analysis yields the horizontal wind velocity, as well as a number of parameters related to the ground diffraction pattern itself. These are summarized in Table 5. The analysis was originally applied to amplitude only time series data, but current systems all use coherent data acquisition systems. However, the FCA itself does not make full use of the phase information available. The apparent vertical velocity, angle of arrival (AOA), and corrected vertical velocity in this table are calculated outside of the FCA.

Table 4 Experimental parameters used for routine Spaced Antenna FCA for the Buckland Park MF radar

Parameter	Day value	Night value
Start range, km	50	70
Range resolution	26.6 μs (4 km)	26.6 μs (4 km)
Sampling range resolution, km	2	2
Number of ranges	25	15
Polarization	Linear	Linear
PRF, Hz	100	40
Coherent integrations	40	16
Effective sampling time, s	0.4	0.4
Number of complex samples	280	280
Record length, s	112	112

A typical 112 s data record is shown in Fig. 7. In common with most modern radar techniques, it is suggested that coherent integration should be avoided. Also, as discussed in the text, it is suggested that the sampling times and record lengths be better matched to the characteristics of the fading at each sampling height

The FCA is the most common analysis routinely applied to spaced antenna radars. It has been routinely applied at MF to investigate the atmosphere in the 60–100 km height region since the early 1970s on a campaign basis (e.g., Stubbs 1973), then as a continuous and routine real-time measurement of the MLT region (Gregory et al. 1979; Meek 1980). This development continued and in the early 1990s, commercial three or four receiving antenna MF/HF SA radar systems became available for the first time, and a de facto antenna arrangement standard as shown in Fig. 2 for small partial reflection MF/HF radars using the FCA was established.

However, a variety of transmitting and receiving antenna arrangements were and are still common, particularly with older larger systems established for the DAE or as large general purpose antenna arrays. For example, at Saskatoon, a vertically pointing 4 × 4 array of folded half-wave dipoles with a half-power beam width of about 15° operating at 2.219 MHz was used for transmission. Either four half-wave receiving antennas spaced at 1.16λ and arranged in a "Y" arrangement or three half-wave dipoles arranged in a triangle with a spacing of 1.35λ were available for reception for routine observations (see e.g., Meek and Manson 1987). At Adelaide, a square array of four half-wave dipoles was used until the early 1990s for transmission, and three groups of four half-wave dipoles were used for reception for routine observations (see e.g., Stubbs 1973).

Such arrangements affect the sampling of the ground diffraction pattern by performing a spatial average on it by removing finer scale structure. This is equivalent to restricting the angles of arrival of the backscatter detected by using narrower polar diagrams on reception. This means that the filtering of the ground diffraction pattern is different on different antenna systems, and this may potentially influence the results of the analysis.

The height coverage of MF/HF PR radars is amongst the best available from typical upper atmospheric radar systems. By way of example, typical acceptance results for 2 min records for both O and E modes, and for day

Table 5 Parameters derived from the SA FCA

Symbol	Parameter	Description
V_a	Apparent velocity	The velocity of the ground diffraction pattern over the spaced antennas calculated from the time delays of the cross-correlation function determined between pairs of antennas.
V_t	True velocity	The velocity of the ground diffraction pattern over the spaced antennas calculated from the time delays of the cross-correlation function determined from pairs of antennas corrected for random changes in the pattern in time and for any pattern anisotropy. Additional quality control parameters are also applied to obtain this parameter (see Table 6).
w_a	Apparent vertical velocity	Vertical velocity measured from the phase of the complex autocorrelation function measured on each antenna uncorrected for angle of arrival
w_{cor}	Corrected vertical velocity	Vertical velocity corrected for effects of off-zenith mean angle of arrival
AOA	Mean angle of arrival	Mean angle of arrival
$\tau_{0.5}$	Fading time	Mean auto-correlation function half-width
$T_{0.5}$	Pattern lifetime	Fading time in the frame of the ground diffraction pattern
$S_{0.5}$	Pattern scale	Major axis of the characteristic ellipse
R_{ax}	Axial ratio	Ratio of the major to minor axes of the characteristic ellipse
θ_{ax}	Axial rotation	Direction of major axis of characteristic ellipse
PTD	Percentage time discrepancy	Ratio of sum of cross-correlation function maximum delay times to sum of absolute cross-correlation function delay times
SNR	Signal to noise ratio	The signal to noise ratios for each receiver channel
P	Received power	The received power for each receiver channel

and night, are shown in Figs. 13 and 14 for routine FCA operation of the BPMF radar. These results correspond to a 3 year period of observation, and the acceptance criteria applied are those shown in Table 6. Daytime O mode acceptance rates peak at around 70 % near 82 km and acceptance rates for X mode peak at about 40 % near 71 km. During nighttime, O mode rates peak at around 40 % near 88 km, and X mode around 50 % near 84 km.

The FCA has also been applied at other frequencies. It was first applied at VHF to measure wind velocities in the 2–20 km height region in the mid-1970s (Vincent and Röttger 1980). Since the mid-1990s, it has been applied at VHF to mesospheric data using MST radars (e.g., Morris et al. 2006) and to boundary layer data using 55 MHz radars (0.3 to 8 km) (e.g., Dolman and Reid 2014), and at ultrahigh frequency (UHF) (Dolman and Reid, private communication, 2015).

It has proven to be a reliable and robust analysis but does underestimate the magnitude of the wind velocity. In general, any noise in the FCA analysis will lead to an underestimate of the derived wind velocity. This has been investigated extensively using a numerical model by Holdsworth and Reid (1995a, 1997). In addition, there is a tendency for the derived velocity to be too small if antennas are too closely spaced, the so-called triangle size effect (TSE), if the sampling rate is too low, if there is receiver saturation, if there are digitizer quantization errors, or if there are channel differences associated with antenna matching or receiver phase characteristics (see e.g., Holdsworth 1999).

When all of these effects are taken into account, there is still an underestimation of the wind magnitude. Holdsworth and Reid's modeling and their work using the BPMF radar suggested a 10 % magnitude underestimation at MF/HF. For some small MF/HF systems, the underestimation is found to be up to 15 to 40 % and strongly height dependent. For comparison, it is typically up to 5 to 10 % at VHF and only weakly height dependent, if at all (e.g., Reid et al. 2005; Dolman and Reid 2014). It is likely that antenna system mismatches play some role in the larger underestimation at MF/HF for modern systems, and this is an important aspect of maintenance for such systems, particularly after some years of operation. However, antenna mismatches cannot explain the height dependence of the bias. We return to this below.

An example of a wind intercomparison between a small MF/HF SA FCA radar and a meteor radar is shown in Fig. 15. These data represent 1 year of observations from a 2 MHz MF PR SA radar which has an arrangement similar to that shown in Fig. 2 and the 33 MHz "all-sky" meteor radar at Davis Station in Antarctica (McIntosh et al. 2010). This is for a height of 88 km, which is near the peak meteor count rate, and also generally corresponds to the highest acceptance rates for the MF winds at this location. The red and blue lines indicate least-squares fits of MF on meteor, and meteor on MF winds respectively, and so are the bounds of the fits assuming errors in only one technique. The dashed line indicates $y = x$, and the green line indicates

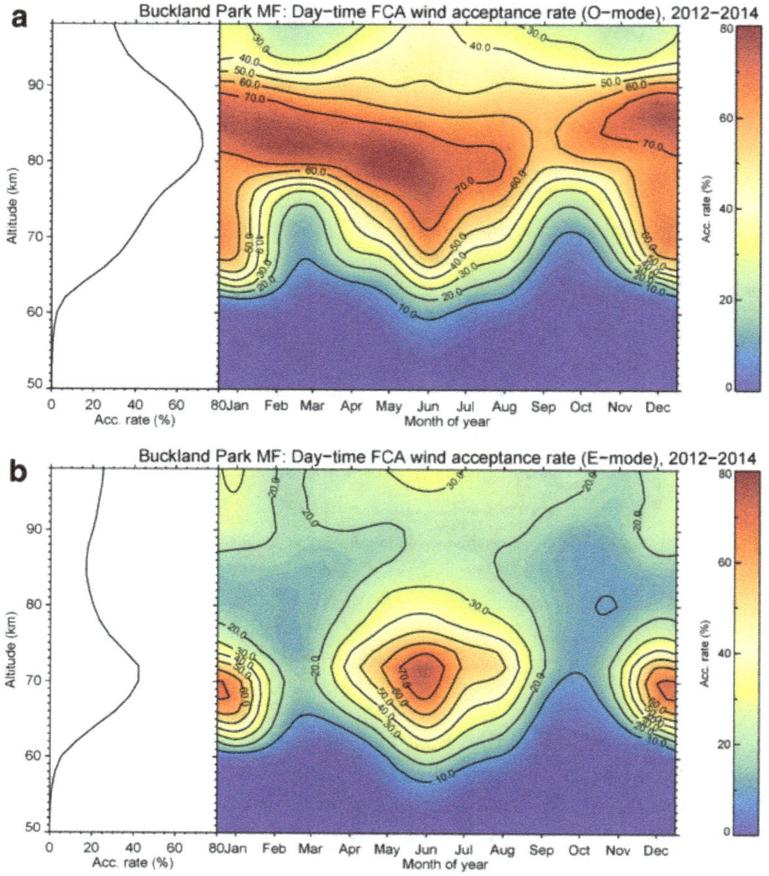

Fig. 13 The daytime SA FCA acceptance rates for the BPMF radar for O mode (*top*) and E mode (*bottom*) measured over a 3 year period. These figures show the acceptance rates for 2 min data records for O and E mode for daytime. The current quality control parameters (or equivalently the rejection criteria) for the BPMF SA FCA are summarized Table 6 by way of example

the slope of fit after estimating the actual errors in each technique. The purple line represents the slope derived using error values obtained for the meteor radar from a similar comparison of the 55 and 33 MHz meteor radars at the same site. This analysis indicates the MF winds underestimate of the meteor winds in the zonal component, with a value of 0.82, with the corresponding meridional value being 0.77.

In this context, it is interesting to note that Dolman and Reid (2014) have described an empirical correction for the bias on operational boundary layer tropospheric (BLT) VHF SA FCA wind profilers. The form of their correction was determined by extensive intercomparison with radiosonde winds and appears to be very robust. With a suitable technique on which to base an intercomparison, it should be possible to also correct for the magnitude bias evident in the MF/HF SA FCA winds. However, it should be noted that unlike the FCA correction for VHF SA radars, experience shows that the MF/HF FCA correction is likely to be strongly height dependent and that there are few options for intercomparison techniques below about 80 km. Needless to say, bias correction will not remove the effects of E region contamination.

Just why the underestimation is height dependent is a question that still needs to be addressed. There are a number of possibilities, including inadequate sampling of the ground diffraction pattern. The ideal antenna spacing for the FCA is that for which the cross-correlation values are typically one half. We have seen how slower fading is associated with smaller values of the aspect sensitivity parameter (θ_S) and with a larger pattern scale. We have also seen how the fading times change with height and the sharp demarcation between the character of the fading and signal powers at heights near 80 km during the day. This means that the antenna spacing cannot be matched to the diffraction pattern scale for every height. The use of the arrangement shown in Fig. 2 does help because there is more than one basic spacing, but this does seem to be a fundamental issue when using the FCA. Furthermore, when a fixed sampling rate and record length are used for all heights, it is probable that this is better suited to some heights than

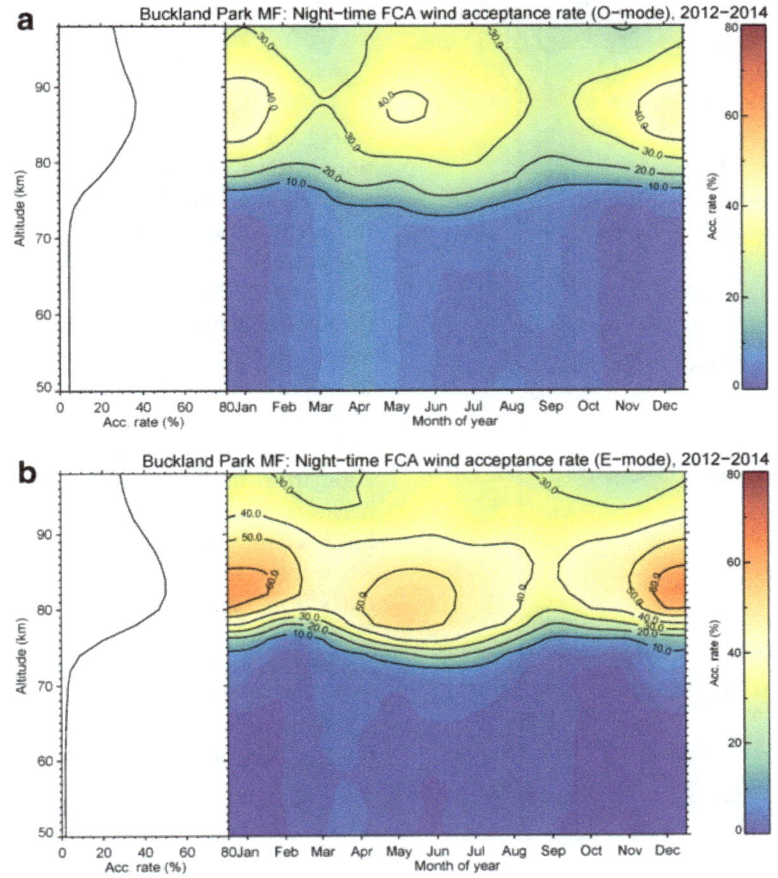

Fig. 14 The nighttime SA FCA acceptance rates for the BPMF radar for O mode (*top*) and E mode (*top*) measured over a 3 year period. These figures show the acceptance rates for 2 min data records for O and E modes for nighttime

others. Dolman and Reid (2014) did note a weak dependence on wind speed, again suggesting a sampling issue. In the MLT region, large vertical shears in magnitude are common, and so, the quality of the characterization of the ground diffraction pattern may also be height dependent. Modern systems have the ability to vary their sampling schemes with sampling height, and this should be investigated.

Dolman and Reid (2014) also describe a quality control (QC) approach for the winds derived from the FCA analysis. QC at the spectral level is not an option for SA radars running the FCA technique as it is for DBS radars, and outliers must be identified in the wind profiles themselves. Dolman and Reid (2014) follow Weber and Wuertz (1991) in developing a successful QC technique for SA FCA-derived winds. This raises the question of whether MF/HF SA FCA winds should be quality controlled. Of course, this is the norm for most observing systems (e.g., satellite-based instruments), and since most radars are now operated more as turnkey wind measuring systems, rather than as pure research radars, this seems appropriate.

Additional information that can be recovered from the FCA analysis relates to the turbulent velocity and the aspect sensitivity and general form of the scattering irregularities.

Turbulent velocity

The basic theory of turbulence velocity estimation using the FCA is described by Briggs (1980), and examples of its application are provided by Holdsworth et al. (2001) and Holdsworth and Reid (2004a). Holdsworth et al. (2001) demonstrated that wider transmit beam widths lead to an overestimation of the turbulent velocity. This confirmed the suggestions made by previous authors (e.g., Briggs 1980; Vandepeer and Hocking 1993; Hall et al. 1998). They concluded that using relatively narrow transmitter beams (in their case, the BPMF radar operated with transmit beam half-power widths of 9°) reduced the effects producing

Table 6 Quality control criteria for the FCA of the SA experiment

Code	Meaning
0	No error—analysis result ok.
1	Time series unsuitable for analysis (signal amplitude too low or interference present)
2	Signal to noise ratio < -3 dB
3	Fading time $>6_S$
4	At least one cross-correlation function maxima lies outside end points
5	Percentage time discrepancy >35 %
6	$V_c^2 < 0$
7	Ground diffraction pattern analysis breaks down
8	Apparent velocity magnitude $V_a > 400$ ms^{-1}
	True velocity magnitude $V_t > 200$ ms^{-1}
9	Difference between apparent V_a and true V_t direction exceeds 40°.
10	$V_a > 10V_t$
11	$V_t > 1.5V_a$
12	Poor fit to cross-correlation function (CCF) maximum
13	At least *one CCF maximum* <0.2
14	Poor zero-lag auto-correlation function (ACF) interpolation
15	Insufficient number of lags on main peak of mean auto-correlation function (MACF) to estimate fading time
16	Poor fit to MACF
17	Secondary maxima in MACF >0.5 (oscillatory)
18	Mean ACF >0.5 at end lags
19	One or more CCF maxima exceeds 1.0

overestimation of turbulent velocities and resulted in a usable result. Holdsworth and Reid (2004a) later showed that turbulent velocities derived using the IDI technique were equivalent to those from the FCA.

The FCA estimate of the turbulent velocity is given by

$$v_{\text{FCA}} = \frac{\lambda\sqrt{2ln2}}{4\pi T_{0.5}},$$

where λ is the wavelength and $T_{0.5}$ is the ground diffraction pattern lifetime in the reference frame of the pattern (e.g., Holdsworth et al. 2001). Figure 16 shows the turbulent root mean square velocities obtained by Holdsworth and Reid (2004a) for a period of 7 years using the BPMF radar. Turbulent velocities are observed to increase with height and solsticial maxima and equinoctal minima are observed below 80 km. Inspection of this figure indicates that there is a maximum extending from 93 km in summer to above 98 km in winter. Holdsworth and Reid (2004a) attribute this to contamination from the E region total reflection because routine harmonic analysis of the turbulent velocities at these heights and times yields values which peak at mid-day when the E region contamination is expected to be most likely (e.g., Holdsworth et al. 2001)

Aspect sensitivity

The FCA also allows an estimate of the aspect sensitivity θ_S of the backscattering irregularities to be estimated as (Lesicar et al. 1994; Holdsworth and Reid 1995a)

$$\sin^{-2}\theta_S = \sin^{-2}\theta_{SO} - \sin^{-2}\theta_0, \text{ with } \theta_{SO} = \frac{15.2\lambda\sqrt{R_{\text{ax}}}}{S_{0.5}}$$

and where $S_{0.5}$ is the pattern scale, R_{ax} is the axial ratio, θ_{SO} is the $1/e$ half width of the effective polar diagram of the backscatter including the antenna polar diagram. Figure 17 is from Holdsworth and Reid (2004a) and shows the mean aspect sensitivity calculated for a 7 year

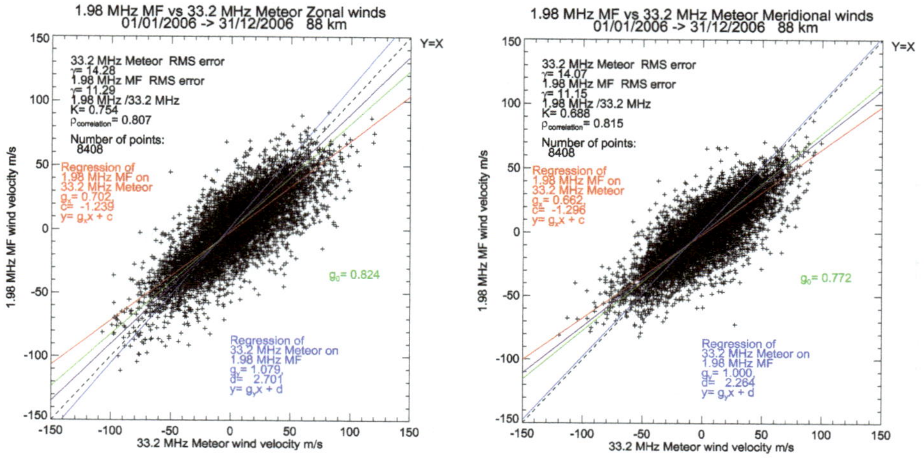

Fig. 15 Intercomparison of meteor and MFSA winds from Davis Station for 1 year of observations showing the zonal component (*left*) and the meridional component (*right*) (after McIntosh 2010). See text for details

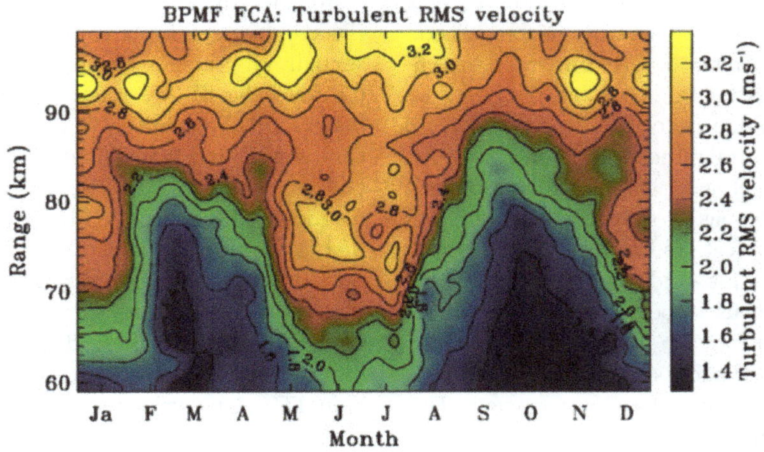

Fig. 16 The turbulent root mean square velocity measured at Buckland Park for a period of 7 years using the spaced antenna technique and the full correlation analysis (after Holdsworth and Reid 2004a)

period at 2 MHz for the BPMF radar derived using this technique. Inspection of this figure illustrates a general increase in the aspect sensitivity with decreasing height, and a seasonal variation, with the least anisotropic scatter occurring above 80 km in winter. The backscatter is different in character above and below about 80 km, with scatter from a range of angles up to about 10° to 15° centered on the zenith above 80 km, indicating a mixture of quasi-isotropic and specular scatter, while below, scatter is from a more restricted range of angles, typically less than 5°, indicating a scattering process rather more specular in character.

Imaging Doppler interferometry

There are a number of interferometric analyses available for use with spaced antenna systems, and Chau (2000) provides a good review. The most commonly applied at MF/HF is IDI. The raw data acquired by the IDI and other SA techniques are essentially identical. However, as in all interferometry, multiple antenna spacings are desirable in order to reduce the possibility of ambiguity in the phase corresponding to the angle of arrival of the returned signal, when the spacing is larger than a wavelength (by also using a spacing less than a wavelength), while reducing the uncertainty in the phase (by using a larger spacing). This requirement perhaps gives some insight into the limitations imposed on the sampling of the ground diffraction pattern by the fixed antenna spacing used by current small MF/HF FCA systems. The IDI and non-interferometric SA techniques also differ in the assumptions they make about the backscattering medium, and consequently, in the subsequent analysis, and in that,

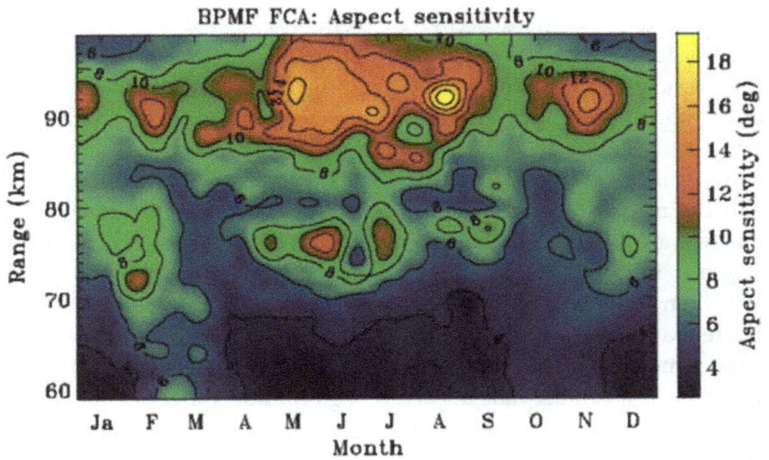

Fig. 17 The aspect sensitivity θ_S for 2 MHz for Buckland Park for a period of 7 years measured using the spaced antenna technique and the full correlation analysis (after Holdsworth and Reid 2004a)

IDI makes use of the available phase information. IDI-like techniques assume that discrete scattering regions exist in the atmosphere. The other non-interferometric SA techniques assume volume scatter.

The aim of IDI technique is to track individual radio wave scatterers in the neutral atmosphere, and it was first applied at MF (Adams et al. 1986). The experimental setup utilizes spaced antennas to sample the radio waves partially reflected or backscattered from the atmosphere. The analysis then looks for evidence of down-coming plane wave radiation and uses spectral techniques to locate a scattering position for each frequency bin. A radial velocity is associated with each scattering position, and the entire wind field can then be reconstructed. This approach has some similarity to the analysis of all sky meteor radar data but tends to require more scattering centers to derive a stable result in practice.

The fundamental assumption of IDI is that the phase information at each Doppler frequency results from a single discrete scattering location, but this original underlying theory of its operation is suspect (see Briggs 1995; Holdsworth and Reid 2004b).

The simple IDI analysis was extended by Franke et al. (1990) to remove the "single scatterer" criterion and the associated limitations regarding antenna configuration. They also introduced rejection criteria to quality control the data. These included limiting the acceptable range of zenith angles for angle of arrival data and setting limits on acceptable radial velocities. Without these, the IDI-derived velocities appeared to be the FCA apparent velocities (or equivalently, the similar fades velocities), a result predicted by Vandepeer and Reid (1995a).

In this context, it is noteworthy that Pfister (1971) anticipated this result. He applied the SA experiment arrangement to E region radar returns in what seems to be the first attempt to make use of the phase information in the returned signals using this technique. He found that the phase of the cross-spectral functions measured between the spaced receivers was a linear function of the associated Doppler velocities of the reflecting surface, and he used the slopes of these phases to estimate the horizontal velocity of the surface. In doing so, he found that the interferometric wind velocities were in good agreement with the velocities obtained with the method of similar fades. While the Pfister (1971) results are for total ionospheric reflections, Bennett and Dyson (1993) later showed that this result generally applied when fixed ground sites were used to observe an ionosphere that moved without change of shape. This means that the same conclusions that apply to simple IDI in the MLT, namely that similar-fade apparent velocities correspond to the case for which changes in the backscatter, or equivalently, in the ground diffraction pattern over time, are not adequately accounted for, apply to the case of total reflections from the ionosphere.

Holdsworth and Reid (2004a) further investigated the IDI technique using a 3 year data set. They applied essentially the same analysis as that of Frank et al. (1990), but used the BP MF radar, which has a much narrower beam than the standard SA arrangement. In this special case, they found that IDI and FCA winds are comparable and concluded that IDI is a useful technique with three distinct and significant advantages over the FCA. The first is that it is significantly simpler to apply than FCA, requiring fewer criteria to reject bad data. The second is that it yields considerably higher 2 min acceptance rate percentages. The third and most important is that they found that the IDI velocities were around 10 % larger than FCA true velocities and in good agreement with spatial correlation function (SCA; see below) and hybrid Doppler interferometer (HDI; see below) velocities, which are better agreement with the actual wind velocity. That is, the magnitude bias evident in the FCA was not evident in the IDI technique in this special case.

One apparent disadvantage to IDI when compared to other non-interferometric SA techniques is that accurate phase calibration is required. However, it should be noted that the phase calibration of SA radars running the FCA analysis is also essential as discussed by Holdsworth and Reid (1997). They found in simulations using the Holdsworth and Reid (1995b) model that even small phase differences between the receiver channels resulted in an underestimation of the actual wind magnitude. The conclusion here then is that phase calibration is as important for SA radars running the FCA analysis as it is for those running the IDI analysis, and Holdsworth and Reid (2004a) describe how phase calibration is conducted periodically for the BPMF radar.

Briefly, Holdsworth and Reid (2004a) followed the approach of Golley and Rossiter (1970) and used a commercially available eight-way splitter to equally divide the signal received by a single antenna into multiple receivers. They did this manually at regular intervals as part of routine maintenance of their radar. They then calculated the cross-correlation functions (CCFs) of the resulting signals, and after interpolating across zero-lag to remove the effects of noise correlated between receiver channels (e.g., Briggs 1984), the zero-lag CCF magnitudes were estimated to measure the statistical similarity of the receiving channels. The zero-lag CCF phases were estimated to measure the phase difference between channels.

Holdsworth and Reid (2004a) do note that the technique they describe does not account for complex receiver gain variations that may occur over smaller time scales (e.g., diurnal variations as found by Vandepeer and Reid 1995b in older radar equipment). It also does not account for antenna differences. It would be possible to supplement the Holdsworth and Reid (2004a) approach to apply an active automated self-calibration technique for the entire receiver chain from antenna to digitizers as is done for meteor radar interferometers (e.g., Holdsworth et al. 2004b; Younger et al. 2013), and this approach is recommended using either meteor echoes or attenuated returns from the E region.

In summary, in spite of the issues around the actual form of the backscatter from the MLT region, IDI has been successfully applied by Franke et al. (1990) and Holdsworth and Reid (2004b) on SA MF/HF radars and also on MF/HF dynasonde radars to produce usable wind velocities (Jones et al. 1997; Jones et al. 2003). It makes better use of the phase information available than the FCA and is not susceptible to the underestimation bias of the FCA. With a radar with a suitably narrow transmit beam (~9° half-power full-width) and good phase calibration, it would be preferable to the standard FCA analysis.

Other SA analyses techniques

There are a number of other analysis approaches which are not routinely applied at MF and lower HF, but which are worth noting.

Full spectral analysis (FSA)

The full spectral analysis (FSA) is a variation of the FCA that differs from it in that the analysis takes place in the frequency, rather than time domain (e.g., Briggs and Vincent 1992; Holdsworth and Reid 1995b). This has some advantages, but the simple implementation of the technique at MF does not appear to be as robust as the FCA (Holdsworth and Reid 1997). Further development of the technique is described by Holloway et al. (1997) and by Venkatesh and Frasier (2013).

Spatial correlation analysis (SCA)

The SCA originated with Briggs et al. (1969). It determines the displacement of the ground diffraction pattern using the spatio-temporal autocorrelation function. When more than three receiving sites are used, it has some advantages over the FCA technique (Holdsworth and Reid 1995b, 1997). In particular, it is less susceptible to the triangle size effect and does not suffer from the magnitude bias evident in the FCA winds. It is noteworthy that the antenna arrangement shown in Fig. 2 does support analysis using the FCA, FSA, IDI, SCA, time domain interferometry (TDI), and meteor techniques, as well as the DAE and DPE experiments.

Time domain interferometry (TDI) and hybrid Doppler interferometry (HDI)

A problem when using the broad beams associated with the SA technique is that measurements of the vertical velocities are compromised because the mean angle of arrival of the backscatter is not typically from the vertical and so has large uncertainties. In TDI (Vandepeer and Reid 1995b), complex time series from spaced receivers are analyzed using the zero lag cross-correlation phase measured between multiple antenna pairs to determine a single angle of arrival of the backscatter from each data record. The average radial velocity measured from the autocorrelation functions of the time series sampled from each of the antennas can be associated with the mean angle of arrival to yield the average motion along the line of sight.

In the case of vertically directed beams, this can be used to calculate the vertical velocity and in the case of off-vertical beams, to determine the effective beam direction corresponding to the measured radial velocity. Because of their small relative magnitudes, vertical velocity measurements are always challenging. Furthermore, the apparent bias towards downwards vertical velocities observed in atmospheric radars with vertically directed Doppler beams has led to the results being treated with some suspicion.

Consequently, even when TDI is applied, vertical velocities obtained with MF/HF SA radars have been little used. Insight into the vertical bias, at least in the convective boundary layer, has been provided by Muschinski and Sullivan (2013), who modeled covariances of turbulently fluctuating clear-air radar reflectivity and the turbulently fluctuating radial wind velocity. They found that such covariances can lead to Doppler velocity biases similar to those observed with vertically pointing wind profiling radars operating in this region. It is likely that this is a viable mechanism for the vertical velocity biases generally observed by atmospheric radars throughout the atmosphere.

An extension of the TDI analysis for wider beam DBS radars to correct for uncertainties in the actual beam direction, called HDI, was applied by Holdsworth and Reid (2004a) at MF and by Reid et al. (2005) at VHF. In HDI, the transmitter beams are formed and directed as for normal DBS radars, but the effective pointing angle of the receiver beam is formed and measured interferometrically. Post statistics beam steering (PSS) (e.g., Kudeki and Woodman 1990) is used to form a receive beam which is steered through a grid of beam directions. The power for each beam direction is determined, and the effective beam position (EBP) is found by applying a 2-D Gaussian to estimate the direction of maximum power. The receive beam is then re-steered in the direction of the EBP, and the standard DBS parameters are estimated. The term "hybrid Doppler interferometry" is used as the technique is a hybrid of Doppler and interferometric techniques. Holdsworth and Reid (2004a) also used in HDI preference to time domain interferometry, since the analysis can be applied in either the time or frequency domain.

The use of PSS allows the EBP to be determined more accurately (and to lower SNRs) than using the mean angle of arrival (MAOA), since it uses combined receiver outputs, while the MAOA is estimated by cross-correlating single receiver outputs. The technique can also be usefully applied to wider vertically directed transmitter beams to determine the mean angle of arrival. The ability to determine the actual beam form is clearly an advantage, and the technique appears to work well.

With sufficient spread in the angles of arrival of the measured radial velocities in the TDI technique, it is also possible to measure the horizontal wind components. Thorsen et al. (1997) provided a more extensive theoretical formulation for the TDI approach and considered ordinary and total least squares fitting to TDI radial velocities to derive the mean and fluctuating components of the wind field. In so doing, they extended TDI to measurements of the three-dimensional mean winds and of the Reynolds stress tensor for small MF/HF PR SA radars. In this, they anticipated a similar approach applied by Hocking (2005) to derive the same parameters from all-sky meteor radars. Thorsen et al. provided promising results from the Urbana MF radar, but their insightful approach has not been further investigated.

However, preliminary investigation of the application of HDI to the BPMF radar indicates good results and suggests that it should perhaps be applied in preference to hardware beam forming when using this radar because of the effects of aspect sensitivity on the relatively wide polar diagram of the antenna array (Spargo et al. 2015). How well-horizontal velocities can be measured in the wide vertically directed beams of small MF/HF PR SA radars depends on the aspect sensitivity of the backscatter, and as we have seen, this is very high at heights below about 80 km. This may restrict the application of the technique to height ranges where more isotropic backscatter is typical.

By way of example, Fig. 18 shows the mean angles of arrival determined using the HDI technique for a 13 day observational period using the BPMF radar in 1998. For these observations, the radar transmitter beam was successively steered between five beam positions. These were 12° off-zenith towards the east, west, north, south, and vertical, and so a cycle took 10 min to complete. The transmitter beam 3 dB contours are shown in the figure. Reception was on 16 receivers. Inspection of this diagram indicates the effect of the aspect sensitivity on the effective beam position for each 2 min observation and the great variability in the actual angle of arrival over time.

Figure 19 shows an example of the mean density normalized upward flux of zonal (< $u'w'$ >) and meridional (< $v'w'$ >) momentum measured over a 6 day period using the same configuration as shown in Fig. 18, but

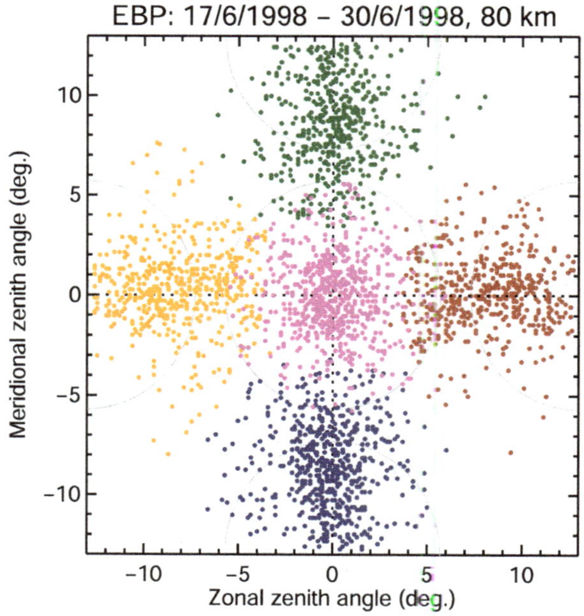

Fig. 18 Shows a typical example of the effective receiver beam positions for the BPMF radar operating in an east (*red*), west (*yellow*), north (*green*), south (*blue*), and vertical (*pink*) transmitter beam sequence. Each point corresponds to a 2 min determination. The 3 dB transmitter beam is shown as a *gray line* for this range (80 km). The effective beam positions (EBPs) are determined using hybrid Doppler interferometry on reception using 16 receivers (after Spargo et al. 2015)

with an off-zenith angle of 13°. To calculate these terms, three approaches were used. In the first, the mean aspect sensitivity was calculated and used to correct the apparent beam direction and the density normalized Reynolds stress terms $\left(\overline{u'^2}, \overline{v'^2}, \overline{w'^2}, \overline{u'w'}, \overline{v'w'}, \overline{u'v'}\right)$ were calculated using the approach of Vincent and Reid (1983) by fitting to all five beams. This results for this approach for the $\overline{u'w'}$ and $\overline{v'w'}$ terms are shown as the red line. The second approach calculated the density normalized Reynolds stress tensor by taking the radial velocities corresponding to the angle of arrivals and performing a least squares fit to these to obtain the six Reynolds stress terms. The results are shown as the black line. The last analysis used this same approach but only used the angles of arrival from the vertically directed beam. This resulted in this observation being made over only 20 % of the observation time for the red and black profiles. This is shown as the gray line in Fig. 19.

Inspection of this diagram indicates good agreement in form between the red and black height profiles. While noisier, the gray profile looks similar in form. Note that this last analysis is a version of that suggested by Thorsen et al. (1997) and for this case, with a relatively narrow transmitter beam (~20° for these observations), appears to deliver a tolerable result for this difficult to measure

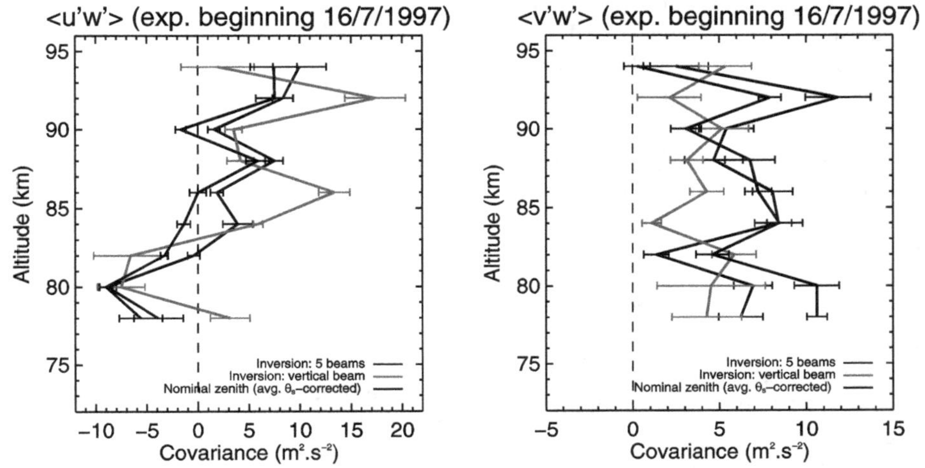

Fig. 19 The mean upward flux of horizontal momentum for the zonal (*left*) and meridional (*right*) directions for a 5 day period in July 1997 using the five beam arrangement shown in Fig. 18. The *red line* indicates these terms calculated by using the Vincent and Reid (1983) approach with the five nominal beam directions after correcting them for the aspect sensitivity of the atmosphere. The *black line* uses the radial velocities corresponding to each of the calculated 2 min effective beam positions and calculates the Reynolds stress tensor from these. The results indicated by the *gray line* are calculated in the same way as the *black line* but using only the vertically directed beam. These then correspond to the Thorsen et al. (1997) approach described in the text (after Spargo et al. 2015). For further details, see the text

parameter. This is a preliminary result but suggests some hope of measuring these terms with small MF/HF PR radar systems.

Lataitis slope technique

Other analysis approaches to SA data are described by Lataitis et al. (1995), Doviak et al. (1996), and Holloway et al. (1997). Lataitis et al. (1995) showed that the slope of the cross-correlation amplitude at zero lag, normalized by the level of the cross-correlation amplitude at zero lag, is directly proportional to the component of the wind velocity along the antenna pair baseline. Using this approach and three non-collinear antenna locations, they were able to determine the wind velocity. They found that their "slope technique" was sensitive to the degree of vertical anisotropy of the scattering medium and suggested that it was best suited to the situation of scattering from isotropic to moderately anisotropic turbulence. This would correspond to wind profiling in the atmospheric boundary layer. Doviak et al. (1996) and Holloway et al. (1997) further considered the application of spaced antenna technique in situations of more isotropic turbulence, a condition more likely to hold only as generally true in the troposphere when using upper VHF, UHF, or higher frequencies. For example, in a typical study, Tsuda et al. (1997) found significant aspect sensitivity at lower VHF in the troposphere and lower stratosphere when using the MU radar in Japan.

More recent work has considered the application of the technique to UHF boundary layer radar (Cohn et al. 2001), and to weather radar (e.g., Zhang and Doviak 2007; Venkatesh and Frasier 2013), and has revisited the theoretical formulation of the analysis. Venkatesh and Frasier (2013) provide a good brief contemporary review of SA analysis techniques.

Doppler beam swinging

Traditional Doppler beam steering radar, in which transmitter and receiving beams are formed and used to measure radial velocities in a number of look directions, is rare at MF and lower HF because of the physically large array size needed to obtain the relatively narrow beams needed for the technique at these wavelengths. In DBS, beams may be formed in hardware or software, but we use the DBS term to apply where there is no further interferometric correction to determine the actual angle of arrival of the returned signal. There are only three MF/HF radars currently capable of operating in this mode. These are the BPMF large array (Briggs et al. 1969; Reid 1987; Reid 1988; Reid et al. 1995; Holdsworth and Reid 2004a), the large Saura HF crossed array radar (Singer et al. 2003, 2008, 2011), and the smaller Juliusruh crossed array HF radar. The BPMF uses a filled array as shown in Fig. 6, while the Saura HF radar uses a cross type antenna array as shown in Fig. 3. A large MF/HF cross type antenna array on Bribie Island in Australia used a sophisticated raster style of operation to image the ionosphere at three different frequencies (see e.g., Lingard 1996a,b) until retired in the 1990s, and is particularly noteworthy as a precursor to the Saura HF radar (Singer et al. 2008), and to crossed type partially filled radars generally (Hocking and Thayapaxan 1997).

A very important point here is that the complex data received on any one receiver has the same form for FCA, FSA, SCA, IDI, HDI, TDI, or Doppler beam steering, but the analysis applied differs. The Doppler beam swinging and spaced antenna methods use the same information. The DBS technique exploits the change in the mean Doppler shift with azimuth angle and samples small spatially separated volumes. The SA correlation methods make use of the fact that the variation in the Doppler shift with angle is such that the resultant field moves over the ground as a random pattern with a velocity twice that of the horizontal wind at the backscatter height. Briggs (1980) provides a complete discussion.

We illustrate the idealized situation in Fig. 20 which is adapted from Briggs (1980). In this figure, the atmosphere at height z is moving with a horizontal velocity u, so that the electric fields from paths 1 and 2, E_1 and E_2, are given by

$$E_1 = E_0 \exp i\left\{\omega_0\left(1+\frac{2u}{c}\sin\theta\right)t-\frac{2\pi x\sin\theta}{\lambda}+\frac{2\pi z\cos\theta}{\lambda}\right\},$$

and

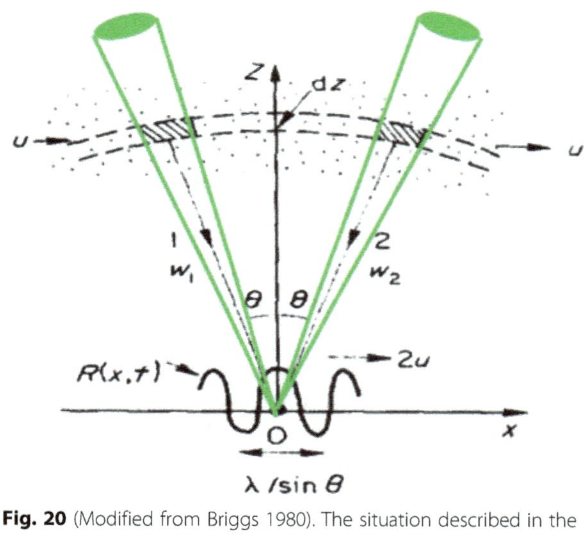

Fig. 20 (Modified from Briggs 1980). The situation described in the text outlining the basis of the formation of the ground diffraction pattern in the SA experiment. Here, the electric fields produced by volume scatter from a moving atmosphere at angles of $\pm\theta$ sum to produce an electric field on the ground, the real part of which is a sinusoid of wavelength $\lambda/\sin\theta$ that moves with twice the velocity of the atmosphere at the backscatter height. When a broad range of angles is illuminated, the sinusoids from each pair of angles sum to produce the moving ground diffraction pattern. DBS techniques use a narrower angular range of angles (such as those outlined in *green*) to measure the line of sight Doppler shift. When these are obtained from a number of angles and statistical homogeneity of the wind field is assumed, the three-dimensional wind field can be reconstructed. Interferometric techniques use Doppler shifts associated with randomly distributed scattering centers to determine a mean velocity

$$E_2 = E_0 \exp i\left\{\omega_0\left(1-\frac{2u}{c}\sin\theta\right)t+\frac{2\pi x\sin\theta}{\lambda}+\frac{2\pi z\cos\theta}{\lambda}\right\},$$

at the ground. Here ω_0, is the radar angular frequency, and ω_1 and ω_2 are the Doppler shifted radar frequencies given by

$$\omega_1 = \omega_0 + 2\omega_0\left(\frac{u}{c}\right)\sin\theta,$$

and

$$\omega_2 = \omega_0 - 2\omega_0\left(\frac{u}{c}\right)\sin\theta,$$

and the other terms have their usual meanings. The real part of the sum of the electric fields at the ground produces a sinusoidal variation in electric field strength on the ground with a wavelength $\lambda/\sin\theta$ that moves with a velocity $2u$ and is given by

$$R(x) = 2E_0\cos\left(\frac{2\pi\sin\theta}{\lambda}-\frac{2\omega_0 ut\sin\theta}{c}\right).$$

Each pair of angles $\pm\theta$ produces a particular sinusoid, and these all sum to produce the moving ground diffraction pattern. Smaller spatial scales in the diffraction pattern are produced by returns from larger off-zenith angles, and so the effect of using groups of antennas on reception is to filter out higher frequency spatial scales. Of course, this just corresponds to using a narrower polar diagram on reception. In the SA experiment, the antenna spacing is selected so that the cross-correlation functions between antenna pairs have values of around 0.5 on average. This varies with height (see Fig. 8), and so in this sense, the experiment is optimized for a particular height or range of heights. This can be thought of as being somewhat analogous to selecting a particular off-zenith angle in the DBS experiment.

Note that a differential vertical velocity, that is, one which is not the same in beams 1 and 2 will produce an apparent motion of the ground diffraction pattern. A similar effect occurs in the DBS technique, and as we discuss in the section on meteor radars below, this may limit the minimum time resolution available for measuring winds in all of these techniques. Generally, an averaging time of about 1 h is considered reliable.

DBS techniques use a narrower angular range to measure the line of sight Doppler shift, and when these are obtained from a number of angles, and statistical homogeneity of the wind field is assumed, the three-dimensional wind field can be reconstructed (e.g., Reid 1987). Interferometric techniques use Doppler shifts associated with randomly distributed scattering centers to determine a mean velocity. In this, they are similar to all-sky meteor radar techniques (Holdsworth et al. 2004a). The hybrid Doppler interferometer combines relatively narrow transmitter beams with interferometric reception, and so, we have described it in the SA section above.

Great care must be taken when using DBS at MF and HF because of the high aspect sensitivity of the MLT region at these frequencies (see e.g., Reid 1988) and the relatively wide beams available. This results in the effective beam pointing angle being less than the apparent beam angle (e.g., Reid 1988; Vandepeer and Reid 1995b; Holdsworth and Reid 2004a). This can be corrected by using HDI as noted above, but this is a further limitation for MF and lower HF radars even when it is possible to build arrays large enough for DBS.

MF/HF radar: strengths and weaknesses

As we have seen, the strengths of the MF/HF SA technique include moderate range and good time resolution, with values of 2 to 4 km height resolution and 2 to 5 min time resolution for wind measurements being typical. The height coverage is good, with typical values from 60 to 98 km during the day and 80 to 98 km during the night (see Figs. 13 and 14). The upper usable height depends on the total reflection height and also on the ability of the data acquisition system to cope with the very large dynamic range in echo strength in the 60 to 100 km height range. We will return to this point below. MF radars operate with high peak powers, but very low average powers and duty cycles (typically less than 0.5 %). Power consumption is low, and reliable continuous long-term operation has been demonstrated consistently at a number of locations; for example, the Saskatoon MF SA radar has operated continuously since 1979, and the Adelaide MF SA radar continuously since 1984.

The limitations of the techniques include a degradation in height resolution through range smearing if the angular scatter is wide (with a half-power half-width greater than about 10°). This is because the actual beam pattern is the product of the antenna polar diagram and the polar diagram of the backscattering irregularities. The simple antennas used in the spaced antenna technique have wide angular polar diagrams, and so, this product also produces a wide effective beam when the backscatter angular polar diagram is wide.

The large increase in the returned echo power as illustrated in Figs. 4 and 11 requires a large dynamic range within the receiving and data acquisition system, and a variety of approaches have been utilized. These have included a three-step variable gain system with height used with the Saskatoon MF radar (Meek 1979) and alternate low- and high-mode gains for the 60 to 80 and 80 to 100 km height ranges, respectively, and an adaptive gain system (e.g., Vandepeer and Reid 1995b). This was a particular issue for older systems with limited digitization. Most current systems now use 12-bit or larger digitization schemes and so have corresponding larger dynamic ranges.

The rapid increase in power with height typical at MF and lower HF for the MLT region has restricted the application of pulse coding techniques at these frequencies. However, this could have considerable utility in the 40 to 80 km height region with 4 bit and 8 bit codes and 1 km bit lengths.

Czechowsky et al. (1983) do report on an intriguing radar backscatter experiment at 2.75 MHz using a peak power of 1.2 MW and a maximum duty cycle of 4 %. Their experiment was conducted at Tromsø using the heating facility transmitter in winter and used "limited pulse coding". With this very high peak power, they reported finding echoes from 15 to 35 and from 52 to 100 km during their brief experiment. Belrose (1970) does report the ability to measure electron densities in the 40 to 80 km height region during times of abnormally high ionization, and Holdsworth and Reid (1997) report measurements from 52 km upwards, so there is some additional evidence to support their observations.

The Czechowsky et al. result for echoes below 40 km does not appear to have been replicated since, and there are potential complications because of range aliasing, but the results are interesting enough to justify repeating the experiment. This could be done in the region above 30 km using a (typically available) 64 kW transmitter, large antenna array, and an 8 bit code with 1 km bit length. This would lead to a power aperture (PA) product smaller than that of Czechowsky et al.'s (1983) experiment but would allow much longer observation times and a more systematic investigation of the possible appearance of stratospheric echoes at MF. However, it is noteworthy that stratospheric echoes have not been reported by the Saura HF radar at a similar latitude and location to Czechowsky et al.'s observations.

Simple 2 bit pulse coding may also be useful in discriminating out interfering signals, which is sometimes an issue at MF and HF, and this has been implemented on the Saura radar to reduce interference from HF communications (Mayo 2015).

Finally, relatively large land areas are required for the simple antennas used in the SA technique, although it is possible to share the land with cropping or other agricultural activities as is done with the Indian MF radars.

Investigation of the IDI and FCA analysis applied to MF and HF radar techniques

The MF/HF SA technique has been somewhat controversial as a technique when applied to measuring MLT winds. This has resulted from its evolution from its original application to total reflection from the ionosphere, and the elementary analyses applied to the original experiments, largely because of computational limitations.

These original measurements are certainly problematic. This was further compounded by the evolution of the acceptance criteria that were applied to constrain the data, the uneven adoption of both these quality control measures and the more sophisticated analyses such as the FCA, which took account of the real nature of the ground diffraction pattern, and a lack of care in excluding backscatter contaminated by total reflections above about 92 to 94 km at some times and locations. The current FCA includes 18 QC measures (see Table 6) that have often, but not exclusively, evolved based on experience in applying the analysis, rather than from derivations from first principles.

In the 1990s, considerable useful investigation of MF/HF SA radar techniques occurred. This was because of a number of factors, which we now discuss and which are still relevant and instructive, because some of the results were over generalized and not necessarily critically assessed subsequently.

AIDA 89 campaign

The first of these was the AIDA 89 campaign during which very mixed quality results were obtained above 80 km when an HF radar using the simple IDI analysis was compared to other techniques (Hines et al. 1993). Hines argued that this was a result of gravity wave contamination of the measured IDI winds resulting in the phase velocities of the gravity waves being measured instead of the actual background wind and generalized this argument to all MF/HF SA radars. This particular argument had previously been used for MF/HF SA radars using total reflections from the ionosphere and the similar fades technique, where it could be valid (see e.g., Hines and Rhagava Rao 1968). However, it is not valid in the case of partial reflections (see e.g., Hocking et al. 1989), and in any case, gravity wave effects, if any, would be the same as those experienced by all atmospheric radars operating at frequencies up into the VHF band. Hines subsequently agreed that this would be the case (private communication, 1994), but it is not observed at MF, or any other frequency.

The Hines et al. (1993) results motivated experimental (Vandepeer and Reid 1995a), theoretical (Briggs 1995), and modeling (Holdsworth and Reid 1995a,b) investigations of the IDI analysis. These found that the simplest IDI analysis produced results similar to the apparent velocity of the FCA, and that the presence of (apparent) discrete scattering centers was inevitable even in the case of volume scattering. The Holdsworth and Reid (1995a) model used point scatterers, which would correspond to a volume scatter situation, but included aspect sensitivity in the radar scattering volume, so that the anisotropy of the real atmosphere was accounted for.

The Hines et al. (1993) results also motivated modeling studies by Kudeki et al. (1993) and Surucu et al. (1995). These studies suggested that MF/HF SA wind measurements were suitable for measurement periods longer than about 1 h. Hocking (1987) noted that wind measurements made over shorter time intervals were potentially usable, but some care needed to be exercised to recognize the potential contaminating effects of vertical velocities associated with short period gravity waves on the results. In this, the SA techniques are no different to DBS (e.g., Reid 1987) or all-sky meteor radar techniques (see below).

Turek et al. (1995) revisited the Hines et al. (1993) AIDA IDI intercomparisons, included many more observations, and re-analyzed the IDI data using the Saskatoon and Adelaide implementations of the FCA. They found much better agreement between the various techniques, but they did find that the relationship between the true and apparent FCA velocities, and the interferometric velocities remained to be clarified. We have discussed this above, and as we noted there, subsequent detailed investigation by Holdsworth and Reid (2004b) found the IDI technique valid when relatively narrow beams were used and careful selection criteria were applied. They also concluded that with these improvements, the technique had some real advantages over the FCA. The numerical modeling work of Holdsworth and Reid (1995b, 1997) failed to find any evidence of the gravity wave contamination effects proposed by Hines et al. (1993) in the IDI. Holdsworth and Reid (1997) also investigated the FCA analysis in some detail using their model and confirmed that the gravity wave arguments of Hines were not valid for this analysis either.

From these investigations, it was clear that the AIDA 89 IDI results reported by (Hines et al. 1993) were not at all typical of modern MF/HF SA radars in general or of analyses that were more sophisticated than the similar fades analysis that was current in the 1960s. The subsequent investigations showed that although the FCA technique did underestimate the wind magnitude, it was a valid method of measuring winds in the MLT region, particularly for periods longer than about 1 h. The investigation of the IDI and FCA analyses is discussed in some detail by Reid (1996) and Holdsworth and Reid (2004b).

Interferometric techniques

The second reason for the close scrutiny of MF/HF SA techniques was the development of a new radar interferometric techniques and their application at MF (e.g., Meek et al. 1986a, 1986b). This was facilitated by the use of better hardware using coherent receiver systems, but more importantly, as we have noted above, more powerful affordable computers. This development in itself was an exciting evolution, but the technique we have discussed at some length, IDI, was offered as an MF/HF SA radar technique analysis not subject to the limitations or doubts of the FCA analysis.

However, as noted above, the simplest IDI analysis measures the apparent velocity. More sophisticated IDI analyses do not necessarily have an advantage over the FCA when applied to the standard SA configuration, which has relatively wide beams, and additional care must be taken with system phase calibration. However, again as noted above, it does have advantages in the case of narrower beam radars (Holdsworth and Reid 2004b), and importantly, it does allow some other MF/HF radar types to measure MLT winds. For example, the dynasonde (see e.g., Rietveld et al. 2008) for a description of a modern dynasonde at Halley was configured to operate in an IDI mode (Jones et al. 1997) and was successfully used to measure MLT winds prior to being relocated in 2008. Other dynasondes, such as that at Bear Lake in Utah, have since been successfully used in similar ways (Jones et al. 2003). These radars are included in Fig. 1 and Table 3.

Satellite intercomparisons

The third reason for intercomparisons was that satellite-derived MLT winds became generally available for the first time with the launch of the Upper Atmosphere Research Satellite (UARS) satellite and the high resolution Doppler interferometer (HRDI) and Wind Imaging Doppler Interferometer (WINDII) instruments it carried (e.g., Reber et al. 1993). These instruments provided a powerful global view of the wind field of the planet. In addition to studying MLT region atmospheric dynamics, they also provided the means and motivation for the intercomparison of MLT winds derived from different techniques for the purposes of ground proofing the satellite measurements (e.g., Palo et al. 1997), and provided a better understanding of the limitations of the various measurement techniques.

Intercomparisons between HRDI and a number of MF/HF SA FCA radars found that mean and tidal component winds in the satellite instrument were larger than the corresponding radar winds (e.g., Khattatov et al. 1996), and Burrage et al. (1996) concluded that HRDI measured larger winds than MF radars, but that the size of the difference varied from radar to radar. Burrage et al. (1996) concluded that the HDRI results were more consistent with falling sphere measurements and with WINDII measurements and so by inference that MF wind magnitudes were too small. It is worth noting that the HRDI measurements represented averages over several hundred kilometers in the meridional direction. This was certainly recognized in the HRDI results referenced above, where the importance of differences in time and space averaging, the short-term variability of tides, mean winds, and other motions, and the presence of non-migrating tides and other longitudinal irregularities on the intercomparison were noted. However, as we have seen elsewhere in this review, the most likely, and probably the dominant reason for the discrepancy between the satellite and MF radar results, was the underestimation of the wind magnitude in the SA FCA analysis.

Rebirth of the meteor radar technique for measuring MLT winds

The last reason for these intercomparisons was the re-development of meteor radars for measuring MLT region winds and the consequent comparison with the MF/HF SA FCA radar winds. This started with the application of narrow beam VHF stratospheric tropospheric (ST) radars to meteor studies (e.g., Valentic et al. 1996; Elford 2001) and followed with the development of purpose-built, compact, all-sky VHF meteor radars (e.g., Hocking et al. 2001; Holdsworth et al. 2004a). For example, in the first of these intercomparisons, Cervera and Reid (1995) found an underestimation of the MF/HF SA FCA winds in the 80 to 100 km height region when the Adelaide BPMF radar winds were compared to meteor winds measured using the collocated Adelaide narrow beam VHF ST radar operating in meteor mode. The underestimation varied with height, but FCA wind magnitudes were typically between 0.6 and 0.8 of the meteor wind magnitudes. This and the many comparisons that followed (e.g., Thayaparan and Hocking 2002) demonstrated that MF/HF SA FCA winds were an underestimate of the actual wind magnitude, but that their directions were correct.

As we have noted elsewhere in this review, other MF/HF SA analysis techniques, including IDI and the SCA do not appear to be as susceptible to this underestimation, although larger antenna arrays would be required to fully realize this advantage. MF meteor radars are also not subject to this underestimation but are limited in other ways.

MF/HF meteor radars

Meteor radars operating near 30 MHz in the HF and lower VHF bands are not subject to an upper useable height governed by total reflection but rather from the upper detectable height for meteor trails (e.g., Greenhow and Hall 1960). At these frequencies, the number of meteor detections tends to zero near 110 km. Generally, the lower the operating frequency, the greater the maximum useable height, so that for example at 2 MHz, meteor trails are detected to heights near 120 km (e.g., Steel and Elford 1991; Tsutsumi et al. 1999). However, as we have noted above, the drift and diffusion of meteor trails is increasingly affected by the Earth's magnetic field above 110 km and measuring parameters related to the neutral atmosphere requires some care. Nevertheless, Holdsworth and Reid (2004a) report nighttime winds to 120 km with the BPMF, and Tsutsumi and Aso (2005) report regular wind measurements to 120 km during geomagnetically quiet conditions at high latitudes using the Syowa MF radar (69°S). However, the routine use of MF radars as meteor radars is unlikely to be practical, and this application is most likely restricted to special observations at suitable locations.

One issue with all-sky meteor radars relates to the distribution of the radial velocities measured across their wide observing areas. This is never uniform, and varies significantly with time of day, through the year, and with latitude. Using a simple model, Reid (1987) demonstrated the limitations of Doppler radars with different beam geometries in measuring atmospheric parameters. Kudeki et al. (1993) also considered the impact of inhomogeneous winds fields on SA, DBS, and IDI radar measurements using theoretical arguments and concluded that such inhomogeneity could led to significant biases in such measurements. Surucu et al. (1995) extended the Kudeki et al. (1993) model to include both isotropic and non-isotropic scatterers and concluded that the SA radars produced the horizontal wind as long as any waves present had horizontal scales larger than the horizontal scale of the SA radar pulse volume. Hocking (1987) concluded that these effects would not be of concern for MF/HF SA wind measurements of periods longer than an hour or so.

If meteor radars are considered as Doppler radars with randomly varying beam directions, then it would not be too surprising to find that they are also subject to limitations in just what parameters that they can measure with reasonable uncertainties. More sophisticated modeling using gravity wave fields of meteor radar observations by Spargo et al. (2015) has shown similar limitation for meteor radars to those found for DBS radars by Reid (1987).

Nevertheless, modern meteor radars are powerful instruments, capable of measuring the horizontal winds for periods longer than about an hour, the radiants of meteor showers (Jones and Jones, 2006), the speed of the meteoroids (Holdsworth et al. 2007), the density of the neutral atmosphere (Younger et al. 2015), and its temperature in some instances (Holdsworth et al. 2006; Younger et al. 2014). The extension of these developments to meteor radars operating in the HF band would be significant, and Chisham and Freeman (2013) review the potential of SuperDARN radars as meteor radars. Generally, SuperDARN radars lack the optimized interferometric capability of all-sky meteor radars (e.g., Holdsworth et al. 2004a), and this limits their capability somewhat. However, consideration of the expected count rate using McKinley's (1951) relation suggests meteor counts with a typical SuperDARN configuration would be around 40,000 per day at the typical SuperDARN operating frequency of 12 MHz. This would be a significant extension of the capability of these HF band radars and a significant addition to our understanding of the dynamics of the MLT region.

Differential absorption experiment/differential phase experiment

The DAE proposed by Gardner and Pawsey (1953) estimates the electron density using the amplitude ratio of the extraordinary to ordinary mode signals returned from the ionospheric D region. The application of this technique was an active area of research until the early 1980s (see e.g., Belrose 1970; von Biel 1977; Manson and Meek 1984). The technique provides a coarse measure of the electron density, but relative changes in electron density can be closely monitored.

After not being utilized for some years, the DAE and DPE were applied to new MF/HF radars that began being produced in the early 2000s onwards (see, e.g., Holdsworth et al. 2002; Vuthaluru et al. 2002; Singer et al. 2003). They are now routinely applied on many MF/HF radars, with the most sophisticated results perhaps being obtained on the Saura radar in northern Norway (e.g., Singer et al. 2008, 2011). Figure 4 shows power profiles for both O and X modes as well as the electron densities calculated using both the DAE and DPE. Figure 21 shows an intercomparison from the Saura radar operating in DAE/DPE mode with results from a rocket borne Faraday rotation and absorption experiment (Singer et al. 2011). These authors also show a number of other intercomparisons that demonstrate the utility of the technique. The DAE and DPE techniques also allow the collision frequencies to be derived (Vuthaluru et al. 2002), and these are directly related to pressure, suggesting that this parameter might also be able to be recovered from the experiment.

Conclusions

MF/HF partial reflection radars remain one of the few techniques available to routinely measure winds in the 55 to 80 km height region. They also offer the ability to measure winds in the 80 to 92 km height region, as well as turbulent velocities, and D region electron densities. However, the detailed nature of the irregularities in the 50 to 90 km height region is still not understood nor is the underlying reason for the occurrence of preferred heights of echo occurrence. This lack of understanding limits the ability to clearly interpret some aspects of the measurements from these radars. The clustering of instruments, ideally including high powered LIDARs, offers the opportunity to resolve some of these uncertainties.

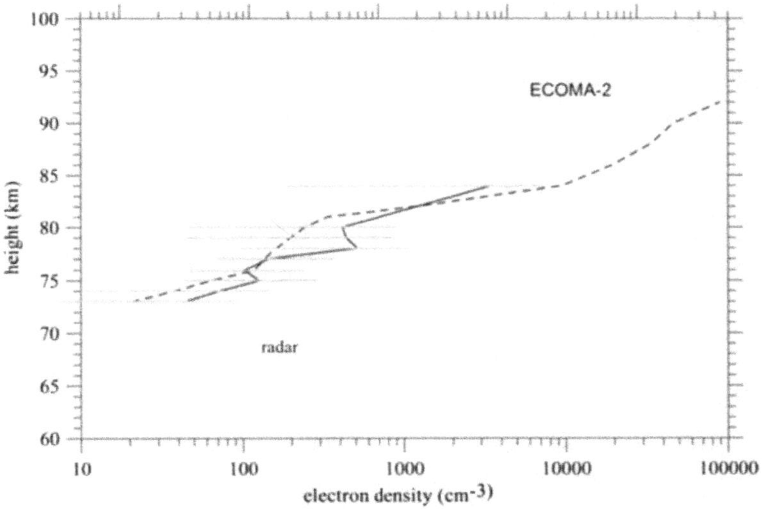

Fig. 21 Electron density profiles by DAE/DPE measurements (full lines, 20:50–21:20 UT, solar zenith angle 106°) and electron density profiles from the in situ radio wave propagation measurements (*dashed lines*) of the ECOMA-2 flight on September 17, 2006, 21:06 UT. The *error bars* represent the quartile limits (after Singer et al. 2011)

Uncertainty about the reliability of the technique and the availability of all sky meteor radars has led to some decline in its popularity. However, the technique is robust and offers a variety of analysis possibilities. The FCA is the most commonly applied analysis, its major deficiency being an underestimation of 15 to 40 % of the wind magnitude at MF/HF, but not its direction, for an ideally configured small system. This is unacceptably high for studies of dynamics. For comparison, typical values for small VHF SA systems may be closer to between 5 and 10 % and not obviously height dependent. For MF/HF SA radars running the FCA, there is a strong height dependence in the bias. The reason for this remains to be investigated, but inadequate sampling of the ground diffraction pattern may be an explanation. The use of sampling schemes optimized for the characteristics of the fading time series is recommended. Such a scheme would ideally be dynamic, so that it would potentially be different for each sampled height and vary with time.

The SCA, IDI, TDI, and HDI techniques offer additional possibilities for measuring MLT winds when appropriate acceptance criteria are applied to the data, and in the case of IDI, TDI, and HDI, when narrower transmitter beams are available, and when this is done, the underestimation bias characteristic of the FCA is not evident. Importantly, all SA systems should correct for receiver channel phase differences.

Care needs to be taken with winds measured above about 92 km with all MF/HF radar systems particularly near mid-day because the dominance of the E region total reflection and the potential for group retardation. We recommend using the best possible practical height resolution and utilizing the calculated total reflection heights to omit contaminated data. We also recommend the implementation of additional QC measures to detect and omit data contaminated by total reflections.

Based on experience at VHF, correction of the known bias in wind magnitude in the FCA should be possible and, if so, should be applied. One issue in this context is a general lack of techniques for intercomparison below about 80 km. The question of whether MF/HF SA wind fields should be quality controlled is still open, but we would recommend its adoption. The application of more sophisticated experimental techniques such as pulse coding to improve the height coverage at lower heights, improve range resolution, improve duty cycles, and reduce interference is suggested, as are the introduction of range imaging (RIM)-like techniques (e.g., Yu and Palmer 2001), and further exploitation of MF meteor echoes. These developments will continue to extend the technique, as will hardware improvements, which are continuing generally with atmospheric radar.

Long MF/HF SA FCA data sets exist, and in spite of the known underestimation in wind magnitude, investigation of long-term changes in the MLT wind field is providing promising results (e.g., Hoffmann et al. 2011). With very similar MF/HF SA and meteor systems now operating, the idea of a "ground based satellite" and the development of virtual observatories has great appeal. This is being pursued through a number of projects, including the Madrigal database (http://madrigal.haystack.mit.edu/madrigal/) and IUGONET global observation network (http://www.iugonet.org/en/). The availability of relatively inexpensive radars with complementary attributes such as MF/HF partial reflection radars and all-sky meteor radars with global geographical spread, combined with a global assimilation system will greatly facilitate our understanding of the MLT, and this is the likely future of these systems. This is certainly the case for the East Asia Oceania region shown in Fig. 1.

Abbreviations

AIDA: Arecibo Initiative Dynamics of the Atmosphere; AM: amplitude modulation; BLT: boundary layer troposphere; BPMF: Buckland Park medium frequency radar; BPVHF: Buckland Park very high frequency radar; CCF: cross-correlation function; DAE: differential absorption experiment; DBS: Doppler beam swinging; DPE: differential phase experiment; E-Mode or X-mode: extraordinary mode; FCA: full correlation analysis; FSA: full spectral analysis; HDI: hybrid Doppler interferometry; HF: high frequency; HRDI: high-resolution Doppler interferometer; IDI: imaging Doppler interferometry; LIDAR: laser radar; MAOA: mean angle of arrival; MF: medium frequency; MLT: mesosphere lower thermosphere; MST: mesosphere stratosphere troposphere; O mode: ordinary mode; PR: partial reflection; PRF: pulse repetition frequency; PRP: pulse repetition period; RADAR: radio detection and ranging; RIM: range imaging; SCA: spatial correlation analysis; SNR: signal to noise ratio; ST: stratospheric tropospheric; SuperDARN: super dual auroral radar network; TDI: time domain interferometry; TSE: triangle size effect; UARS: Upper Atmosphere Research Satellite; UHF: ultrahigh frequency; VHF: very high frequency; VLF: very low frequency; WINDII: Wind Imaging Doppler Interferometer.

Competing interests

The author declares that he has no competing interests.

Author's information

IMR is a Professor of Physics at the University of Adelaide and the Executive Director of the ATRAD group of companies.

Acknowledgements

IMR gratefully acknowledges Andrew Spargo, Bronwyn Dolman, Bob Vincent, and Andrew MacKinnon for reading early drafts the manuscript and Andrew Spargo for assistance in preparation of some of the figures.

References

Abdu MA, Ananthakrishnan S, Krishnan BA, Massambani O (1973) Cosmic ray ionization in the D region at sunrise: evidence from VLF phase measurements. Radio Sci 8(8–9):733–6, http://dx.doi.org/10.1029/RS008i008p00733

Adams W, Brosnahan JW, Walden DC, Nerney SF (1986) Mesospheric observations using a 2.66-MHz radar as an imaging Doppler interferometer: description and first results. J Geophys Res 91:1671–83, http://dx.doi.org/10.1029/JA091iA02p01671

Andrews DG, Holton JR, Leovy CB (1987) Middle atmosphere dynamics. Academic, San Diego, California, p 489

Appleton EV (1930) On some measurements of the equivalent height of the atmospheric ionized layer. Proc Roy Soc A126:542–69, http://dx.doi.org/10.1098/rspa.1930.0025

Appleton EV, Barnett MAF (1925) Local reflection of wireless waves from the upper atmosphere. Nature 115:333–4, http://dx.doi.org/10.1038/115333a0

Appleton EV, Piddington JH (1938) The reflection coefficients of ionospheric regions. Proc Roy Soc A164:467–76, http://dx.doi.org/10.1098/rspa.1938.0031

Bailey DK, Bateman R, Kirby RC (1955) Radio transmission at VHF by scattering and other processes in the lower ionosphere. Proc IRE 43:1181–231, http://dx.doi.org/10.1109/JRPROC.1955.277933

Bain WC, May BR (1967) D-region electron density distributions from propagation data. Proc IEE 114:1593, http://dx.doi.org/10.1049/piee.1967.0306

Belrose JS (1970) Radio wave probing of the ionosphere by the partial reflection of radio waves (from heights below 100 km). J Atmos Terr Phys 32:567, http://dx.doi.org/10.1016/0021-9169(70)90209-6

Bennett JA, Dyson PL (1993) On Pfister's theorem and its generalization. J Atmos Terr Phys 55:1307–10, http://dx.doi.org/10.1016/0021-9169(93)90055-4

Bertoni FCP, Raulin J-P, Gavilán HR, Kaufmann P, Rodriguez R, Clilverd M, Cardenas JS, Fernandez G (2013) Lower ionosphere monitoring by the South America VLF network (SAVNET): C region occurrence and atmospheric temperature variability. J Geophys Res Space Physics 118:6686–93, http://dx.doi.org/10.1002/jgra.50559

Bowles, K. L (1964) Radiowave scattering in the ionosphere, in Advances in Electronics and Electron Physics, 19, ed. L. Marton, Academic Press; 55-176, http://dx.doi.org/10.1016/S0065-2539(08)60845-8

Breit G, Tuve MA (1925) A radio method of estimating the height of the conducting layer. Nature 116:357, http://dx.doi.org/10.1038/116357a0

Breit G, Tuve MA (1926) A test of the existence of the conducting layer. Phys Rev 28:554, http://dx.doi.org/10.1103/PhysRev.28.554

Brekke A, Holt O, Dickinson PHG, Friedrich M, Hansen T, Stauning P (1985) Development of D-region electron and ion densities under various auroral conditions during the Energy Budget Campaign (EBC). J Atmos Terr Phys 47:101, http://dx.doi.org/10.1016/0021-9169(85)90127-8

Briggs BH (1980) Radar observations of atmospheric winds and turbulence: a comparison of techniques. J Atmos Terr Phys 42:823–33 http://dx.doi.org/10.1016/0021-9169(80)90086-0

Briggs BH (1984) The analysis of spaced sensor records by correlation techniques, Handbook for MAP, 13, 166–186, SCOSTEP Secretariat. Univ. of Illinois, Urbana

Briggs BH (1993) Observations of atmospheric dynamics using radar techniques. Aust J Phys 46(1):127–48, http://dx.doi.org/10.1071/PH930127

Briggs BH (1994) The spaced-receiver technique. J Atmos Terr Phys 56:831–4, http://dx.doi.org/10.1016/0021-9169(94)90140-6

Briggs BH (1995) On radar interferometric techniques in the situation of volume scatter. Radio Sci 30(1):109–14, http://dx.doi.org/10.1029/94RS02415

Briggs BH, Holmes N (1973) Ionospheric observations using an ultrasonic image forming technique. Nature Phys Sci 243:111, http://dx.doi.org/10.1038/physci243111a0

Briggs BH, Vincent RA (1992) Spaced-antenna analysis in the frequency domain. Radio Sci 27(2):117–29, http://dx.doi.org/10.1029/91RS03051

Briggs BH, Elford WG, Felgate DG, Golley MG, Rossitter DE, Smith JW (1969) Buckland park aerial array. Nature 223:1321–5, http://dx.doi.org/doi:10.1038/2231321a0

Burrage MD, Skinner WR, Gell DA, Hays DA, Marshall AR, Ortland DA, et al. Vincent (1996) Validation of mesosphere and lower thermosphere winds from the high resolution Doppler imager on UARS. J Geophys Res 101(D6):10365–92, http://dx.doi.org/10.1029/95JD01700

Cervera MA, Reid IM (1995) Comparison of simultaneous wind measurements using collocated VHF meteor radar and MF spaced antenna radar systems. Radio Sci 30:1245–61, http://dx.doi.org/10.1029/95RS00644

Champion KSW (1987) Middle atmosphere models and comparison with shuttle re-entry density data. Adv Space Res 7(10):77–82, http://dx.doi.org/10.1016/0273-1177(87)90080-9

Chau JL (2000) A review of radar interferometric/imaging techniques used in MST radars, Proc. 9th Workshop on Technical and Scientific Aspects of MST radar, SCOSTEP. pp 25–34

Chisham G, Freeman MP (2013) A reassessment of SuperDARN meteor echoes from the upper mesosphere and lower thermosphere. J Atmos Solar Terr Physics 102:207–21, http://dx.doi.org/10.1016/j.jastp.2013.05.018

Cohn SA, Brown WOJ, Martin CL, Susedik MS, Maclean G, Parsons DB (2001) Clear air boundary layer spaced antenna wind measurement with the multiple antenna profiler (MAPR). Ann Geophys 19:845–54, http://dx.doi.org/10.5194/angeo-19-845-2001

Czechowsky P, Schmidt G, Kopka H (1983) Medium frequency radar observations in the middle atmosphere. J Atmos Terr Phys 45:729–32 http://dx.doi.org/10.1016/S0021-9169(83)80031-2

Deeks DG (1966) D-region electron density distributions in middle latitudes deduced from the reflection of long radio waves. Proc Roy Soc A 291:413, http://dx.doi.org/10.1098/rspa.1966.0103

Dieminger W (1952) Über die Ursache der excessiven absorption in der Ionosphärean Wintertagen. J Atmos Terr Phys 2:340–9, http://dx.doi.org/10.1016/0021-9169(52)90074-3

Dieminger W, Hofmann-Heyden AE (1952) Reflexionen von Kurzwellen aus Höhen unter 100 km. Naturwissenschaften 39:84–5, http://dx.doi.org/10.1007/BF00631093

Dolman, B.K. and I.M. Reid (2014) Bias correction and overall performance of a VHF spaced antenna boundary layer profiler for operational weather forecasting. J Atmos. Solar Terr Physics. http://dx.doi.org/10.1016/j.jastp.2014.02.009i

Doviak RJ, Lataitis RJ, Holloway CL (1996) Cross correlation and cross spectra for spaced antenna wind profilers: 1. Theoretical analysis. Radio Sci 31:157–80, http://dx.doi.org/10.1029/95RS02318

Elford, W.G (2001) Novel applications of MST radars in meteor studies. J Atmos Solar Terr. Physics, 63. http://dx.doi.org/10.1016/S1364-6826(00)00145-0

Elford WG, Elford MT (2001). The effective diffusion coefficient of meteor trails above 100 km. In: Proceedings of the Meteoroids 2001 Conference, Kiruna, Sweden. ESA Publications Division, ESA SP-495, Noordwijk, pp 357–9, ISBN 92-9092-805-0, 2001

Ellyett WL (1947) Echoes at D-heights with special reference to the Pacific Islands. Terr Magn Atmos Electr 52(1):1–13, http://dx.doi.org/10.1029/TE052i001p00001

Ellyett C, Watts JM (1959) Stratification in the lower ionosphere. J Res NBS 63D:117–34, http://dx.doi.org/10.6028/jres.063D.016

Fan X, Weng LB, Zhang JB, Xie YQ (2014) Investigation of the mesospheric temperature over Fort Collins region. Sci China Tech Sci 57:1562–7, http://dx.doi.org/10.1007/s11431-014-5593-2

Felgate DG (1970) On the point source effect in the measurement of ionospheric drifts. J Atmos Ter Phys 32:241–245, http://dx.doi.org/10.1016/0021-9169(70)90196-0

Felgate DG, Golley MG (1971) Ionospheric irregularities and movements observed with a large aerial array. J Atmos Terr Phys 33:1353, http://dx.doi.org/10.1016/0021-9169(71)90009-2

Flock WL, Balsley BB (1967) VHF radar returns from the D-region of the equatorial ionosphere. J Geophys Res 72:5537–41, http://dx.doi.org/10.1029/JZ072i021p05537

Fraser GJ (1965) The measurement of atmospheric winds at altitudes of 64–120 km using ground-based radio equipment. J Atmos Sci 22:217–8, http://dx.doi.org/10.1175/1520-0469(1965)022<0217:TMOAWA>2.0.CO;2

Fraser GJ (2005) The antecedents and subsequent development of scientific radar in New Zealand. J Atmos Solar Terr Physics 67:1411–8, http://dx.doi.org/10.1016/j.jastp.2005.07.010

Friedrich M, Torkar KM (1983) Collision frequencies in the high-latitude D-region. J Atmos Terr Phys 45(4):267, http://dx.doi.org/10.1016/S0021-9169(83)80048-8

Gardner FF, Pawsey JL (1953) Study of the ionospheric D-region using partial reflections. J Atmos Terr Phys 3(6):321–4, http://dx.doi.org/10.1016/0021-9169(53)90084-1

Gnanalingham S, Weekes K (1952) Weak echoes from the ionosphere with radiowaves of frequency 1.42 MHz. Nature 170:113–4, http://dx.doi.org/10.1038/170113b0

Golley MG, Rossiter DE (1970) Some tests of methods of analysis of ionospheric drift records using an array of 89 aerials. J Atmos Terr Phys 32:1215–33, http://dx.doi.org/10.1016/0021-9169(70)90053-X

Greenhow JS, Hall JE (1960) The importance of initial trail radius on the apparent height and number distributions of meteor echoes. MNRAS 121:183–96, http://dx.doi.org/10.1093/mnras/121.2.183

Gregory JB (1956) Ionospheric reflections from heights below the E-region. Aust J Phys 9:324, http://dx.doi.org/10.1071/PH560324

Gregory JB (1961) Radio wave reflections from the mesosphere: 1. Heights of occurrence. J Geophys Res 66(2):429–45, http://dx.doi.org/10.1029/JZ066i002p00429

Gregory JB, Meek CE, Manson AH, Stephenson DG (1979) Developments in the radiowave drifts technique for measurement of high-altitude winds. J Appl Meteorol 18:682–9, http://dx.doi.org/10.1175/1520-0450(1979)018<0682:DITRDT>2.0.CO;2

Hall C (2000) The Ramfjordmoen MF radar (69N, 19E): application development 1990-2000. J Atmos Terr Phys 63:171–9, http://dx.doi.org/10.1016/S1364-6826(00)00144-9

Hall CM, Manson AH, Meek CE (1998) Seasonal variation of the turbopause: one year of turbulence investigation at 69°N by the joint University of Tromsø/University of Saskatchewan MF radar. J Geophys Res 103(D22):28769–73, http://dx.doi.org/10.1029/1998JD200002

Hall CM, Manson AH, Meek CE, Nozawa S (2006) Isolated lower mesospheric echoes seen by medium frequency radar at 70° N, 19° E. Atmos Chem Phys 6:5307–14, http://dx.doi.org/10.5194/acp-6-5307-2006

Hauchecorne A, Chanin ML, Wilson R (1987) Mesospheric temperature inversion and gravity wave breaking. Geophys Res Lett 14:933–6, http://dx.doi.org/10.1029/GL014i009p00933

Hedin J, Gumbel J, Khaplanov M, Witt G, Stegman J (2008) Optical studies of noctilucent clouds in the extreme ultraviolet. Ann Geophys 26:1109–19, http://dx.doi.org/10.5194/angeo-26-1109-2008

Hines CO, Rhagava Rao R (1968) Validity of three station methods of determining ionospheric motions. J Atmos Terr Phys 30:979–93, http://dx.doi.org/10.1016/S0021-9169(68)80048-0

Hines CO, Adams GW, Brosnahan JW, Djuth FT, Sulzer MP, Tepley CA, Van Baelen JS (1993) Multi-instrument observations of mesospheric motions over Arecibo: comparisons and interpretations. J Atmos Sol Terr Phys 55(3):241–87, http://dx.doi.org/10.1016/0021-9169(93)90069-B

Hocking WK (1981) Investigation of the movement and structure of D-region ionospheric irregularities. PhD Thesis, Adelaide University, Australia

Hocking WK (1983) On the extraction of atmospheric turbulence parameters from radar backscatter Doppler spectra—I. Theory. J Atmos Terr Phys 45:89–102, http://dx.doi.org/10.1016/S0021-9169(83)80013-0

Hocking WK (1997) Strengths and limitations of MST radar measurements of middle-atmosphere winds. Ann Geophys 15:1111–22, http://dx.doi.org/10.1007/s00585-997-1111-1

Hocking WK (2005) A new approach to momentum flux determinations using SKiYMET meteor radars. ANGEO 23(7):2433–9, http://dx.doi.org/10.5194/angeo-23-2433-2005

Hocking WK (2011) A review of mesosphere–stratosphere–troposphere (MST) radar developments and studies, circa 1997–2008. J Atmos Sol Terr Phys 73:848–82, http://dx.doi.org/10.1016/j.jastp.2010.12.009

Hocking WK, Thayapaxan T (1997) Simultaneous and co-located observations of winds and tides by MF and meteor radars over London, Canada (43N, 81W) during 1994-1996. Radio Sci 32:835–67, http://dx.doi.org/10.1029/1999GL003618

Hocking WK, May PT, Röttger J (1989) Interpretation, reliability and accuracies of parameters deduced by the spaced antenna method in middle atmosphere applications. PureAppl Geophys 130:571–604, http://dx.doi.org/10.1007/BF00874475

Hocking WK, Fuller B, Vandepeer B (2001) Real-time determination of meteor-related parameters utilizing modern digital technology. J Atmos Sol Terr Phys 63:155–69, http://dx.doi.org/10.1016/S1364-6826(00)00138-3

Hoffmann P, Singer W, Kreuer D, Schminder R, Kürschner D (1990) Partial reflection drift measurements in the lower ionosphere over Juliusruh during winter and spring 1989 and comparison with other wind observations. Zeit Meteorol 40:405–12

Hoffmann P, Rapp M, Singer W, Keuer D (2011) Trends of mesospheric gravity waves at northern middle latitudes during summer. J Geophys Res 116:D00P08, http://dx.doi.org/10.1029/2011JD015717

Holdsworth DA (1999) Influence of instrumental effects upon the full correlation analysis. Radio Sci 34(3):643–55. doi:10.1029/1999RS900001

Holdsworth DA, Reid IM (1995a) A simple model of atmospheric radar backscatter: description and application to the full correlation analysis of spaced antenna data. Radio Sci 30(4):1263–80, http://dx.doi.org/10.1029/95RS00645

Holdsworth DA, Reid IM (1995b) Spaced antenna analysis of atmospheric radar backscatter model data. Radio Sci 30(5):1417–33, http://dx.doi.org/10.1029/95RS01888

Holdsworth DA, Reid IM (1997) An investigation of biases in the full correlation analysis technique. Adv Space Res 20(6):1281–4, http://dx.doi.org/10.1016/S0273-1177(97)00784-9

Holdsworth DA, Reid IM (2004a) The Buckland Park MF radar: routine observation scheme and velocity comparisons. Ann Geophys 22:3815–28, http://dx.doi.org/10.5194/angeo-22-3815-2004

Holdsworth DA, Reid IM (2004b) Comparisons of full correlation analysis (FCA) and imaging Doppler interferometry (IDI) winds using the Buckland Park MF radar. Ann Geophys 22(11):3829–42, http://dx.doi.org/10.5194/angeo-22-3829-2004

Holdsworth DA, Vincent RA, Reid IM (2001) Mesospheric turbulent velocity estimation using the Buckland Park MF radar. Ann Geophys 19(8):1007–17, http://dx.doi.org/10.5194/angeo-19-1007-2001

Holdsworth DA, Vuthaluru R, Reid IM, Vincent RA (2002) Differential absorption measurements of mesospheric and lower thermospheric electron densities using the Buckland Park MF radar. J Atmos Sol Terr Phys 64:2029–42, http://dx.doi.org/10.1016/S1364-6826(02)00232-8

Holdsworth DA, Reid IM, Cervera MA (2004a) The Buckland Park all-sky interferometric meteor radar—description and first results. Radio Sci 39:RS5009, http://dx.doi.org/10.1029/2003RS003014

Holdsworth DA, Tsutsumi M, Reid IM, Nakamura T, Tsuda T (2004b) Interferometric meteor phase calibration using meteor echoes, Radio Sci. RS5012, http://dx.doi.org/10.1029/2003RS003026

Holdsworth DA, Morris RJ, Murphy DJ, Reid IM, Burns GB, French WJR (2006) Antarctic mesospheric temperature estimation using the Davis VHF radar. J Geophys Res 111:D05108, http://dx.doi.org/10.1029/2005JD006589

Holdsworth DA, Elford WG, Vincent RA, Reid IM, Murphy Damian J, Singer W (2007) All-sky interferometric meteor radar meteoroid speed estimation using the Fresnel transform. ANGEO 25(2):385–98, http://dx.doi.org/10.5194/angeo-25-385-2007

Holloway CL, Doviak RJ, Cohn SA, Lataitis RJ, Van Baelen JS (1997) Cross correlations and cross spectra for spaced antenna wind profilers: 2. Algorithms to estimate wind and turbulence. Radio Sci 32:967–82, http://dx.doi.org/10.1029/96RS03885

Holmes NE (1975) An ultrasonic image-forming system for use with multi-element antenna arrays. Proc Inst Radio Electron Eng Aust 36(10):321–8

Jacobi C (2014) Long-term trends and decadal variability of upper mesosphere/lower thermosphere gravity waves at midlatitudes. J Atmos Solar Terr Phys 118:90–5, http://dx.doi.org/10.1016/j.jastp.2013.05.009

Jones J, Jones W (2006) Meteor radiant activity mapping using single-station radar observations. MNRAS 367:1050–6, http://dx.doi.org/10.1111/j.1365-2966.2006.10025.x

Jones GOL, Charles K, Jarvis MJ (1997) First mesospheric observations using an imaging Doppler interferometer adaption of the dynasonde at Halley, Antarctica. Radio Sci 32(6):2109–22, http://dx.doi.org/10.1029/97RS01825

Jones GOL, Berkey FT, Fish CS, Hocking WK, Taylor MJ (2003) Validation of imaging Doppler interferometer winds using meteor radar. Geophys Res Lett 30(14):1743, http://dx.doi.org/10.1029/2003GL017645

Jones GOL, Clilverd MA, Espy PJ, Chew S, Fritts DC, Riggin DM (2004) An alternative explanation of PMSE-like scatter in MF radar data. Ann Geophys 22:2715–22, http://dx.doi.org/10.5194/angeo-22-2715-2004

Jursa, AS (1985) Handbook of geophysics and the space environment, Hanscom Air Force Geophysics Laboratory. http://www.cnofs.org/Handbook_of_Geophysics_1985/Handbook.pdf

Kawamura S, M Tsutsumi, and Y Murayama (2007) Wind estimations with meteor observations by MF radars at Poker Flat, Alaska and Wakkanai, Japan. J National Institute of Information and Communications Technology. 54, Nos.1/2

Khattatov BV, Geller MA, Yudin VA, Hays PB, Skinner WR, Burrage MD, Franke SJ, Friits DC, Isler JR, Manson AH, Meek CE, McMurray R, Singer W, Hoffmann P, Vincent RA (1996) Dynamics of the mesosphere and lower thermosphere as seen by MF radars and by the high-resolution Doppler imager/UARS. J Geophys Res 101(D6):10393–404, http://dx.doi.org/10.1029/95JD01704

Kleinknecht NH, Espy PJ, Hibbins RE (2014) The climatology of zonal wave numbers 1 and 2 planetary wave structure in the MLT using a chain of Northern Hemisphere SuperDARN radars. J Geophys Res Atmos 119:1292–307, http://dx.doi.org/10.1002/2013JD019850

Krautkrämer J (1950) Windmessungen in der Ionosphäre. Arch elektr Übertragung 4:133–41

Kudeki E, Woodman RF (1990) A poststatistics steering technique for MST radar applications. Radio Sci 25(4):591–4, http://dx.doi.org/10.1029/RS025i004p00591

Kudeki E, Rastogi PK, Sürücü F (1993) Systematic errors in radar wind estimation: implications for comparative measurements. Radio Sci 28(2):169–79, http://dx.doi.org/10.1029/92RS01931

Lataitis RJ, Clifford SF, Holloway CL (1995) An alternate method for inferring winds from spaced-antenna radar measurements. Radio Sci 30:463–74, http://dx.doi.org/10.1029/94RS02414

Lesicar D, Hocking WK, Vincent RA (1994) Comparative studies of scatterers observed by MF radars in tbe southern hemisphere mesosphere. J Atmos Terr Phys 56:581–91, http://dx.doi.org/10.1016/0021-9169(94)90099-X

Li G, Ning B, Hu L, Chu Y-H, Reid IM, Dolman B (2012) A comparison of lower thermospheric winds derived from range spread and specular meteor trail echoes. J Geophys Res 117:A03310, http://dx.doi.org/10.1029/2011JA016847

Lingard DM (1996a) A deconvolution technique for measuring D-region radio wave backscatter. J Atmos Sol Terr Phys 58(11):1201–9, http://dx.doi.org/10.1016/0021-9169(95)00150-6

Lingard DM (1996b) Comparison of a deconvolution technique for measuring mesospheric winds with the spaced antenna technique. J Atmos Sol Terr Phys 58(11):1211–8, http://dx.doi.org/10.1016/0021-9169(95)00151-4

Luce H, Crochet M, Dalaudier F (2001) Temperature sheets and aspect sensitive radar echoes. Ann Geophys 19:899–920, http://dx.doi.org/10.5194/angeo-19-899-2001

MacKinnon AD, Vincent RA, Reid IM (2002) Boundary layer radar measurements during DAWEX, in International Symposium on Equatorial Processes Including Coupling (EPIC). Int. Union of Radio Sci, Kyoto, Japan, pp 478–80, 18–22 March

Manson AH, Meek CE (1984) Partial reflection D-region electron densities, Handbook for MAP, 13, ed. R.A. Vincent, SCOSTEP Secretariat, 113-123

Mayo R (2015) Pulse Coding, Characteristics and Implementation on ATRAD Radars, ATRAD Technical Report, Document No: 15-70052. ATRAD Pty Ltd, Adelaide, Australia

McIntosh DL (2010) Comparisons of VHF meteor radar observations in the middle atmosphere with multiple independent remote sensing techniques, PhD thesis, University of Adelaide. http://hdl.handle.net/2440/60068

McKinley DWR (1951) Variation of meteor echo rates with radar system parameters. Can J Phys 29:403–26

Meek CE (1980) An efficient method for analysing ionospheric drifts data. J Atmos Terr Phys 42:835–9, http://dx.doi.org/10.1016/0021-9169(80)90087-2

Meek CE, Manson AH (1987) Mesospheric motions observed by simultaneous medium-frequency interferometer and spaced antenna experiments. J Geophys Res 92(D5):5627–39, http://dx.doi.org/10.1029/JD092iD05p05627

Meek CE, Manson AH (2001) MF radar spaced antenna experiment: wind variance vs. record length. J Atmos Sol Terr Phys 63:181–91, http://dx.doi.org/10.1016/S1364-6826(00)00148-6

Meek CE, Reid IM, Manson AH (1986a) Comparison of medium frequency pulsed radar interferometer and correlation analysis winds: 1, Handbook for MAP, 20. SCOSTEP Secretariat, University of Illinois, Urbana, USA, pp 293–8

Meek CE, Reid IM, Manson AH (1986b) Comparison of medium frequency pulsed radar interferometer and correlation analysis winds: 2, Handbook for MAP, 20. SCOSTEP Secretariat, University of Illinois, Urbana, USA, pp 299–302

Mitra SN (1949) A radio method of measuring winds in the ionosphere. Proc Instn Elect Engrs 96:441–6, http://dx.doi.org/10.1049/pi-3.1949.0094

Morris RJ, Murphy DJ, Vincent RA, Holdsworth DA, Klekociuk AR, Reid IM (2006) Characteristics of the wind, temperature and PMSE field above Davis Antarctica. J Atmos Solar Terr Physics 68:418–35, http://dx.doi.org/10.1016/j.jastp.2005.04.011

Murphy DJ, Vincent RA (1993) Estimates of momentum flux in the mesosphere and lower thermosphere over Adelaide, Australia, from March 1985 to February 1986. J Geophys Res 98(D10):18617–38, http://dx.doi.org/10.1029/93JD01861

Muschinski A, Sullivan PP (2013) Using large-eddy simulation to investigate intermittency fluxes of clear-air radar reflectivity in the atmospheric boundary layer, 2013 IEEE Antennas and Propagation Society International Symposium (APSURSI). 2321 – 2322. http://dx.doi.org/10.1109/APS.2013.6711819

Namboothiri SP, Manson AH, Meek CE (1993) E region real heights and their implications for MF radar-derived wind and tidal climatologies. Radio Sci 28(2):187–202, http://dx.doi.org/10.1029/92RS02458

Namboothiri SP, Manson AH, Meek CE (1994) Extension of MF radar tidal measurements to E-region heights (95-125 km): Saskatoon (52°N, 107°W), Canada. Ann Geophys 12:333–41, http://dx.doi.org/10.1007/s00585-994-0333-8

Nicolet M, Aikin AC (1960) The formation of the D region of the ionosphere. J Geophys Res 65:1469–83, http://dx.doi.org/10.1029/JZ065i005p01469

Offermann D, Gerndt R, Küchler R, Baker K, Pendleton WR, Meyer W, von Zahn U, Philbrick CR, Schmidlin FJ (1987) Mean state and long term variations of temperature in the winter middle atmosphere above northern Scandinavia. J Atmos Terr Phys 49:655–74, http://dx.doi.org/10.1016/0021-9169(87)90011-0

Palo SE, Hagan ME, Meek CE, Vincent RA, Burrage MD, McLandress CS et al (1997) An intercomparison between the GSWM, UARS, and ground based radar observations: a case-study in January 1993, Ann Geophysicae 15, 1123-1141, http://dx.doi.org/10.1007/s00585-997-1123-x

Pancheva D, Mitchell NJ, Hagan ME, Manson AH, Meek CE, Yi L, Ch J, Kürschner D, Clark RR, Hocking WK, MacDougall J, Jones GOL, Vincent RA, Reid IM, Singer W, Igarashi K, Fraser GI, Nakamura T, Tsuda T, Yu P, Merzlyakov E, Fahrutdinova AN, Stepanov AM, Poole LMG, Malinga SB, Kashcheyev BL, Oleynikov AN, Riggin DM (2002) Global-scale tidal structure in the mesosphere and lower thermosphere during the PSMOS campaign of June-August 1999 and comparisons with the global-scale wave model. J Atmos Solar Terr Physics 64:1011–35, http://dx.doi.org/10.1016/S1364-6826(02)00054-8

Pavlov AV (2014) Photochemistry of ions at D-region altitudes of the ionosphere: a review. Surv Geophys 35:259–334, http://dx.doi.org/10.1007/s10712-013-9253-z

Pawsey JL (1935) Further investigations of the amplitude variations of downcoming wireless waves. Proc Camb Phil Soc 31:125, http://dx.doi.org/10.1017/S0305004100013025

Pfister W (1971) The wave-like nature of inhomogeneities in the E-region. J Atmos Terr Phys 33:999–1025, http://dx.doi.org/10.1016/0021-9169(71)90123-1

Phillips GJ, Spencer M (1955) The effects of anisometric amplitude patterns in the measurement of ionospheric drifts. Proc Phys Soc, B 68:481, http://dx.doi.org/10.1088/0370-1301/68/8/301

Pineo VC (1956) Oblique-incidence measurements of the heights at which ionospheric scattering of VHF radio waves occurs. J Geophys Res 61:165–9, http://dx.doi.org/10.1029/JZ061i002p00165

Placke, M (2013) Gravity waves and momentum fluxes in the mesosphere and lower thermosphere region, PhD Thesis, University of Rostock. 51 pp

Placke MP, Hoffmann R, Latteck M, Rapp (2015) Gravity wave momentum fluxes from MF and meteor radar measurements in the polar MLT region, J Geophys Res Space Physics. http://dx.doi.org/10.1002/2014JA020460

Priese J, Singer W (1984) Measurements of partial reflections at 3.18 MHz using the CW-radar technique. Handbook for Middle Atmosphere Program 10:39

Ratcliffe JA (1956) Some aspects of diffraction theory and their application to the ionosphere. Rep Prog Phys 19:188, http://dx.doi.org/10.1088/0034-4885/19/1/306

Ratcliffe JA, Pawsey JL (1933) A study of the intensity variations of downcoming wireless waves. Math Proc Camb Philos Soc 29:301–18, http://dx.doi.org/10.1017/S0305004100011117

Reber CA, Trevathan CE, McNeal RJ, Luther MR (1993) The Upper Atmosphere Research Satellite (UARS) mission. Geophys Res 98:10,643–10,647, http://dx.doi.org/10.1029/92JD02828

Reid IM (1987) Some aspects of Doppler radar measurements of the mean and fluctuating components of the wind field in the upper middle atmosphere. J Atmos Terr Phys 49:467–84, http://dx.doi.org/10.1016/0021-9169(87)90041-9

Reid IM (1988) MF Doppler and spaced antenna measurements of upper middle atmosphere winds. J Atmos Terr Phys 50:117–34, http://dx.doi.org/10.1016/0021-9169(88)90019-9

Reid IM (1990) Radar observations of stratified layers in the mesosphere and lower thermosphere (50–100 km). Adv Space Res 10(10):7–19, http://dx.doi.org/10.1016/0273-1177(90)90002-H

Reid IM (1996) On the measurement of gravity waves, tides and mean winds in the low and middle latitude mesosphere and thermosphere with MF radar. Adv Space Res 18:131–40, http://dx.doi.org/10.1016/0273-1177(95)00852-6

Reid IM, Vincent RA (1987) Measurements of mesospheric gravity wave momentum fluxes and mean flow accelerations at Adelaide, Australia. J Atmos Terr Phys 49:443–60, http://dx.doi.org/10.1016/0021-9169(87)90039-0

Reid IM, Vandepeer BGW, Dillon SD, Fuller BG (1995) The new Adelaide MF Doppler radar. Radio Sci 30:1177–89, http://dx.doi.org/10.1029/95RS00731

Reid IM, Holdsworth DA, Kovalam S, Vincent RA, Stickland J (2005) Mount Gambier (38°S, 141°E) prototype VHF wind profiler. Radio Sci 40:RS5007, http://dx.doi.org/10.1029/2004RS003055

Reinisch B, Scali JL, Haines DM (1998) Ionospheric drift measurements with ionosondes. Ann Geofis 41:695–702

Rietveld MT, Wright JW, Zabotin N, Pitteway MLV (2008) The Tromsø dynasonde. Polar Science 2(1):55–71, http://dx.doi.org/10.1016/j.polar.2008.02.001

Rong PP, Yue J, Russell III JM, Lumpe JD, Gong J, Wu DL, Randall CE (2015) Horizontal winds derived from AIM CIPS polar mesospheric cloud images, J Geophys Res Atmos. 120, http://dx.doi.org/10.1002/2014JD022813.

Røyrvik O (1983) Spaced antenna drift at Jicamarca, mesospheric measurements. Radio Sci 18(3):461–76, http://dx.doi.org/10.1029/RS018i003p00461

Schmidlin FJ (1976) Temperature inversions near 75 km. Geophy Res Lett 3:173–6, http://dx.doi.org/10.1029/GL003i003p00173

Schminder R, Kürschner D (1994) Permanent monitoring of the upper mesosphere and lower thermosphere wind fields (prevailing and semidiurnal tidal components) obtained from LF D1 measurements in 1991 at the Collm Geophysical Observatory. J Atmos Terr Phys 56(10):1263–9, http://dx.doi.org/10.1016/0021-9169(94)90064-7

Singer W, Priese J, Hoffmann P (1988) Experimental technique of lower ionosphere electron density measurements by means of partial reflections. Adv Space Res 8(4):49, http://dx.doi.org/10.1016/0273-1177(88)90203-7

Singer W, Latteck R, Holdsworth DA, Kristiansen T (2003). A new narrow beam MF radar at 3 MHz for studies of the high-latitude middle atmosphere: system description and first results, Proc. of the 10th Workhop on Tech., Sci. Aspects of MST Radar in Piura, Peru, http://jro.igp.gob.pe/mst10/CD/ExtAbs/Session5/I5_015.pdf

Singer W, Latteck R, Holdsworth DA (2008) A new narrow beam Doppler radar at 3 MHz for studies of the high-latitude middle atmosphere. Adv Space Res 41(9):1488–94, http://dx.doi.org/10.1016/j.asr.2007.10.006

Singer W, Latteck R, Friedrich M, Wakabayashi M, Rapp M (2011) Seasonal and solar activity variability of D-region electron density at 69°N. J Atmos Sol Terr Phys 73:925–35, http://dx.doi.org/10.1016/j.jastp.2010.09.012

Spargo AJ, Reid IM, MacKinnon AD (2015) Gravity wave momentum transport measurements in the mesosphere lower thermosphere/ionosphere region using partial reflection and meteor radar, paper IUGG-5011, M14 Middle Atmosphere Science, 26th IUGG General Assembly, Prague

Sprenger K, Schminder R (1969) On some relationships between correlation analysis and similar-fade analysis results of drift measurements in the lower ionosphere. J Atmos Terr Phys 31:1085–98, http://dx.doi.org/10.1016/0021-9169(69)90107-X

Steel DI, Elford WG (1991) The height distribution of radio meteors: comparison of observations at different frequencies on the basis of standard echo theory. J Atmos Terr Phys 53:409–17, http://dx.doi.org/10.1016/0021-9169(91)90035-6

Stubbs TJ (1973) The measurement of winds in the D-region of the ionosphere by the use of partially reflected radio waves. J Atmos Terr Phys 35:909–19, http://dx.doi.org/10.1016/0021-9169(73)90072-X

Surucu F, Franke SJ, Kudeki E (1995) On the influence of specular reflections in MF radar wind measurements. Radio Sci 30:1229–44, http://dx.doi.org/10.1029/94RS03166

Thayaparan T, Hocking WK (2002) A long-term comparison of winds and tides measured at London, Canada (43°N,81°W) by co-located MF and meteor radars during 1994–1999. J Atmos Sol Terr Phys 64:931–46, http://dx.doi.org/10.1016/S1364-6826(02)00048-2

Thomson NR, Clilverd MA, McRae WM (2007) Nighttime ionospheric D region parameters from VLF phase and amplitude. J Geophys Res 112:A07304, http://dx.doi.org/10.1029/2007JA012271

Thorsen D, Franke SJ, Kudeki E (1997) A new approach to MF radar interferometry for estimating mean winds and momentum flux. Radio Sci 32(2):707–26, http://dx.doi.org/10.1029/96RS03422

Thrane EV, Piggott WR (1966) The collision frequency in the D- and E-regions of the ionosphere. J Atmos Terr Phys 28:721, http://dx.doi.org/10.1016/0021-9169(66)90056-0

Titheridge JE (1962) The stratification of the lower ionosphere. J Atmos Terr Phys 24:283–96, http://dx.doi.org/10.1016/0021-9169(62)90098-3

Tsuda T, Gordon WE, Saito H (1997) Azimuth-angle variations of specular reflection echoes in the lower atmosphere observed with the MU radar. J Atmos Terr Phys 59:777–84, http://dx.doi.org/10.1016/S1364-6826(96)00058-2

Tsutsumi M, Aso T (2005) MF radar observations of meteors and meteor-derived winds at Syowa (69S, 39E), Antarctica: a comparison with simultaneous spaced antenna winds. J Geophys Res 110:D24111, http://dx.doi.org/10.1029/2005JD005849

Tsutsumi M, Holdsworth D, Nakamura T, Reid I (1999) Meteor observations with an MF radar. Earth Planets Space 51:691–9, http://dx.doi.org/10.1186/BF03353227

Turek RS, Miller KL, Roper RG, Brosnahan JW (1995) Mesospheric wind studies during AIDA Act '89: morphology and comparison of various techniques. J Atmos Terr Phys 57:1321–43, http://dx.doi.org/10.1016/0021-9169(94)00118-8

Valentic TA, Avery JP, Avery SK, Cervera MA, Elford WG, Vincent RA, Reid IM (1996) A comparison of meteor radar systems at Buckland Park. Radio Sci 31(6):1313–29, http://dx.doi.org/10.1029/96RS02028

Vandepeer BGW, Hocking WK (1993) A comparison of Doppler and spaced antenna techniques for the measurement of turbulent energy dissipation rates. Geophys Res Lett 20(1):17–20, http://dx.doi.org/10.1029/92GL01116

Vandepeer BGW, Reid IM (1995a) On the spaced antenna and imaging Doppler interferometer. Radio Sci 30:885–901, http://dx.doi.org/10.1029/95RS00994

Vandepeer BGW, Reid IM (1995b) Some preliminary results with the new Adelaide MF radar. Radio Sci 30:1191–203, http://dx.doi.org/10.1029/95RS00732

Venkatesh V, Frasier SJ (2013) Simulation of spaced antenna wind retrieval performance on an X-band active phased array weather radar. J Atmos Ocean Technol 30:1447–59, http://dx.doi.org/10.1175/jtech-d-11-00203.1

Vincent RA, Reid IM (1983) HF Doppler measurements of mesospheric gravity wave momentum fluxes. J Atmos Sci 40:1321–33, http://dx.doi.org/10.1175/1520-0469(1983)040<1321:HDMOMG>2.0.CO;2

von Biel HA (1977) The phase switched correlation polarimeter—a new approach to the partial reflection experiment. J Atmos Terr Phys 39:769, http://dx.doi.org/10.1016/0021-9169(77)90138-6

Vuthaluru R, Vincent RA, Holdsworth DA, Reid IM (2002) Collision frequencies in the mid-latitude D-region. J Atmos Sol Terr Phys 64:2043–54, http://dx.doi.org/10.1016/S1364-6826(02)00220-1

Watson-Watt RA, Wilkins AF, Bowen EG (1937) The return of radio waves from the middle atmosphere, 1. Proc Roy Soc A161:181–96, http://dx.doi.org/10.1098/rspa.1937.0140

Weber BL, Wuertz DB (1991), Quality control algorithm for profiler measurements of winds and temperatures. NOAA Tech. Memo. ERL WPL-212, 32 pp. [NTIS PB92124080XSP]

Whitehead JD, From WR, Jones KL, Monro PE (1983) Measurement of movements in the ionosphere using radio reflections. J Atmos Terr Phys 45:345–51, http://dx.doi.org/10.1016/S0021-9169(83)80039-7

Younger JP, Reid IM, Vincent RA (2013) Mutual coupling of antennas in a meteor radar interferometer. Radio Sci 48:118–21, http://dx.doi.org/10.1002/rds.20026

Younger JP, Lee CS, Reid IM, Vincent RA, Kim YH, Murphy DJ (2014) The effects of deionization processes on meteor radar diffusion coefficients below 90 km. J Geophys Res Atmos 119:10,027–10,043, http://dx.doi.org/10.1002/2014JD021787

Younger JP, Reid IM, Vincent RA, Murphy DJ (2015) A method for estimating the height of a mesospheric density level using meteor radar. Geophys. Res. Lett., 42, http://dx.doi.org/10.1002/2015GL065066

Yu T-Y, Palmer RD (2001) Atmospheric radar imaging using multiple-receiver and multiple-frequency techniques. Radio Sci 36:1493–503, http://dx.doi.org/10.1029/2000RS002622

Zhang G, Doviak RJ (2007) Spaced-antenna interferometry to measure crossbeam wind, shear, and turbulence: theory and formulation. J Atmos Ocean Technol 24:791, http://dx.doi.org/10.1175/JTECH2004.1

4

Model depiction of the atmospheric flows of radioactive cesium emitted from the Fukushima Daiichi Nuclear Power Station accident

Teruyuki Nakajima[1,2*], Shota Misawa[1], Yu Morino[3], Haruo Tsuruta[1,4], Daisuke Goto[3], Junya Uchida[1], Toshihiko Takemura[5], Toshimasa Ohara[3], Yasuji Oura[6], Mitsuru Ebihara[6] and Masaki Satoh[1]

Abstract

In this study, a new method is proposed for the depiction of the atmospheric transportation of the ^{137}Cs emitted from the Fukushima Daiichi Nuclear Power Station accident. This method employs a combination of the results of two aerosol model ensembles and the hourly observed atmospheric ^{137}Cs concentration at surface level during 14–23 March 2011 at 90 sites in the suspended particulate matter monitoring network. The new method elucidates accurate transport routes and the distribution of the surface-level atmospheric ^{137}Cs relevant to eight plume events that were previously identified. The model ensemble simulates the main features of the observed distribution of surface-level atmospheric ^{137}Cs. However, significant differences were found in some cases, and this suggests the need to improve the modeling of the emission scenario, plume height, wet deposition process, and plume propagation in the Abukuma Mountain region. The contributions of these error sources differ in the early and dissipating phases of each event, depending on the meteorological conditions.

Keywords: Fukushima Nuclear Power Station accident, Aerosols, Radioactive materials, ^{137}Cs, Chemical transport modeling, Ensemble models

Introduction

A wide area of northeastern Japan, the Tohoku and Kantou regions, was contaminated by the radioactive material emitted from the accident at the Fukushima Daiichi Nuclear Power Station (FDNPS) of the Tokyo Electric Power Company (TEPCO), as manifested by various environmental investigations (Nakajima et al. 2014). The accident was caused by the Great East Japan Earthquake, which struck at 14:46 Japan Standard Time (JST; Coordinated Universal Time, UTC+ 9 h) on 11 March 2011.

Takemura et al. (2011) show that the negative anomaly of a 500-hPa height over the Okhotsk Sea area along 145° E made the westerly jet stronger than the climatological mean during mid-March; consequently, 70 to 80% of the radioactive material from the FDNPS was driven to the Pacific Ocean and the rest of the globe (Takemura et al. 2011; Stohl et al. 2012; Mészáros et al. 2016). The remaining material spread over and deposited onto the land area of Japan, producing characteristic hot spot patterns (Yasunari et al. 2011; JAEA 2012; SCJ 2014). The total emission of ^{137}Cs into the atmosphere until the end of April was estimated to be 14.6 ± 3.5 PBq (SCJ 2014). The ratio of the total deposition over the Japanese land area to the total atmospheric emission was estimated as 20 ± 6%, according to the airborne monitoring conducted by the Ministry of Education, Culture, Sports, Science, and Technology, Japan (MEXT 2011), whereas the ratio was calculated as 27 ± 10% based on the multi-model intercomparison by the Science Council of Japan (SCJ 2014). To date, this inconsistency has not been fully understood, owing to the lack of observation data, which is attributable to instrumental damage and electric outages

* Correspondence: terry-nkj@nifty.com
[1]Atmosphere and Ocean Research Institute (AORI), The University of Tokyo, 5-1-5 Kashiwanoha, Kashiwa, Chiba 277-8568, Japan
[2]Earth Observation Research Center (EORC), Japan Aerospace Exploration Agency (JAXA), 2-1-1 Sengen, Tsukuba, Ibaraki 305-8505, Japan
Full list of author information is available at the end of the article

as well as modeling uncertainties. In addition, there is still great uncertainty in the emission time series of the radioactive material, as shown in Fig. 1. Yumimoto et al. (2016) conducted an inverse analysis to optimally estimate the emission rate using the time series of the deposition map, but the result is very different from that of Katata et al. (2015).

Recently, Tsuruta et al. (2014) developed a method to directly measure the hourly time series of the atmospheric ^{137}Cs concentration at surface level, from the aerosol sampling tapes of the national suspended particulate matter (SPM) network. The SPM network monitors air pollution by employing beta-ray attenuation counters. Four laboratories, namely, those of Tokyo Metropolitan University, the Nuclear Professional School of the University of Tokyo, the Japan Atomic Energy Agency, and the Japan Chemical Analysis Center, retrieved the atmospheric loading from the hourly aerosol spots on the SPM tape. This method offers the potential for studying the atmospheric transport of ^{137}Cs, although the data is from surface level, during the entire post-accident period; the SPM dataset has high temporal and spatial sampling, with observations every hour at 90 out of 400 sites (Fig. 2). In Fig. 2, it can be seen that the Nakadori region is a channel basin area between the Ou and Abukuma mountains, while the Hamadori region is a coastal region to the east of the Abukuma mountains. The FDNPS is located in the northern part of the Hamadori region. In this report, we compare the ensemble results of two aerosol transport models with SPM data. An important purpose of the comparison is to investigate the validity of the combined use of SPM data and multi-model simulations to depict the transportation of atmospheric ^{137}Cs over the Japan land area. Once validated, further analysis can be performed on a larger volume of SPM data, such as the most recent data from 99 SPM sites, which has recently been made available to the public (Oura et al. 2015). In addition, the results could be a useful input for our second model intercomparison, which is intended as a follow-up to the first comparison, which was made by the SCJ (SCJ 2014), and this can contribute to future discussions of the use of models in emergency protocols.

Tsuruta et al. (2014) identified nine plumes, as listed in Fig. 3, that transported particulates to the land area of Japan and in which the maximum atmospheric ^{137}Cs concentration exceeded 10 Bq m^{-3}, based on a synoptic analysis using a time series of the SPM data and the wind vector field. For purposes of comparison, we selected plumes P2 to P9 in the period 14–24 March 2011. In this period, there were two migrations of low pressure systems over Japan; these occurred on 15 and 20 March, according to the weather maps shown in Fig. 4.

Methods/Experimental

Model simulation of the atmospheric ^{137}Cs concentration at surface level

In this study, two aerosol transport models were employed to simulate the atmospheric transportation of ^{137}Cs. The first model uses an online dynamic core of the nonhydrostatic icosahedron atmospheric model (NICAM; Tomita and Satoh 2004; Satoh et al. 2008) coupled with the spectral radiation-transport model for aerosol species (SPRINTARS; Takemura et al. 2000; Dai et al. 2014); we will refer to this as the N-model. The NICAM model is implemented with an air-mass flux-type dynamic scheme of high mass conservation, and it can cover a global-to-regional simulation with its three grid systems, which include a quasihomogeneous global grid, a stretched grid, and a diamond regional grid (Uchida et al. 2015). We adopted the diamond grid system for a regional simulation with the following settings: 10 s time step, 5-km grid resolution, and 40 layers, with the lowest layer being 20 m thick. The cloud microphysics scheme was the NICAM single-moment scheme with six water categories (NSW6), which is a simplified version of Lin's scheme (Lin et al. 1983). The turbulent diffusion calculation adopted a partial condensation scheme (Mellor and Yamada 1982) and an eddy diffusion scheme with a level 2.0 turbulent scale adjustment scheme (Nakanishi and Niino 2004). The numerical domain was a rhombus area, with apexes at 26.5° N, 144.4° E; 35.8° N, 132.0° E; 50.6° N, 133.4° E; and 39.9° N, 147.7° E; it covered the Tohoku region (the northeastern part of the Japanese islands, including the Fukushima and Miyagi prefectures, as shown in Fig. 2) and the Kantou Plain region (the area including the Tokyo, Saitama, Chiba, Kanagawa, and Ibaraki prefectures). After release from the FDNPS, the atmospheric ^{137}Cs was treated as a sulfate aerosol particle

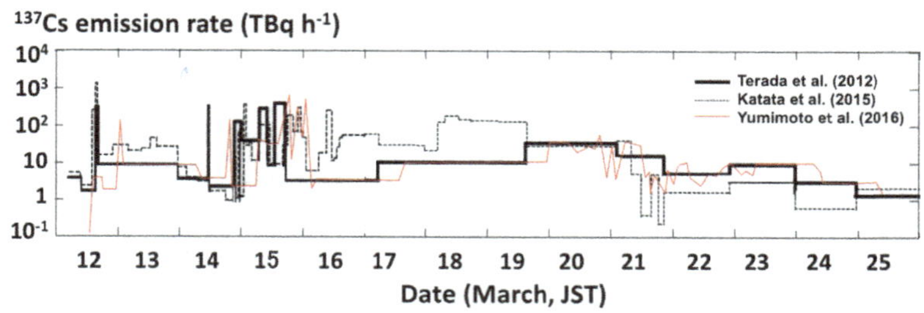

Fig. 1 Time series of the ^{137}Cs emission rate from the FDNPS, as estimated by Terada et al. (2012), Katata et al. (2015), and Yumimoto et al. (2016)

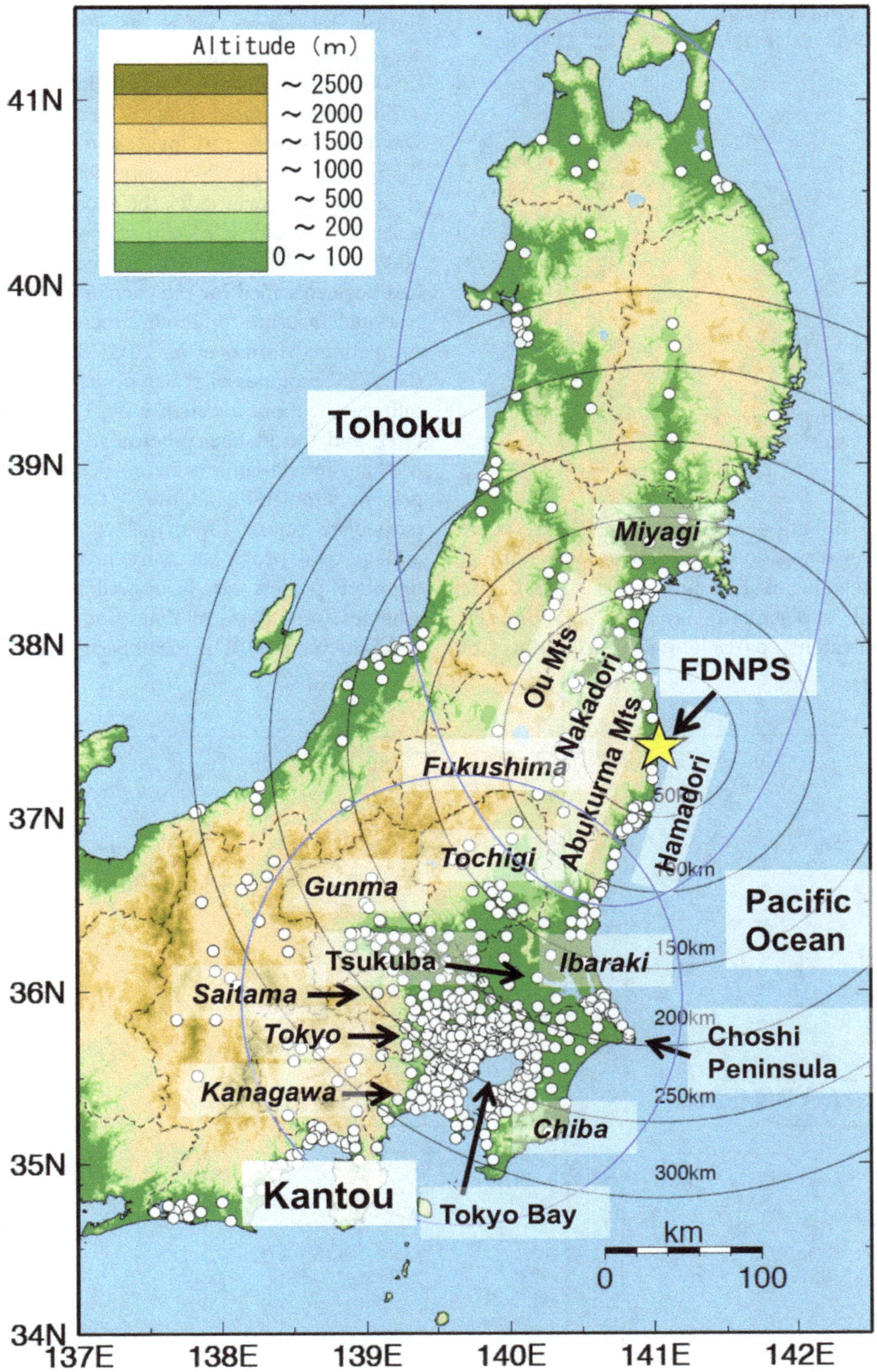

Fig. 2 Names of key regions and locations of SPM sites at the time of accident for the present study. The Tohoku region is the northeastern part of the Japanese islands and includes the Fukushima and Miyagi prefectures; the Kantou region is the area that includes the Tokyo, Saitama, Chiba, Kanagawa, and Ibaraki prefectures. The FDNPS is located in the northern part of the Hamadori region, a coastal area to the east of the Abukuma Mountains. The Nakadori region is a channel basin area between the Ou and the Abukuma mountains. *Open circles* are SPM monitoring sites managed and maintained by local governments in eastern Japan before the accident. The base map was modified by using the original map in Fig. 1 of Tsuruta et al. (2014)

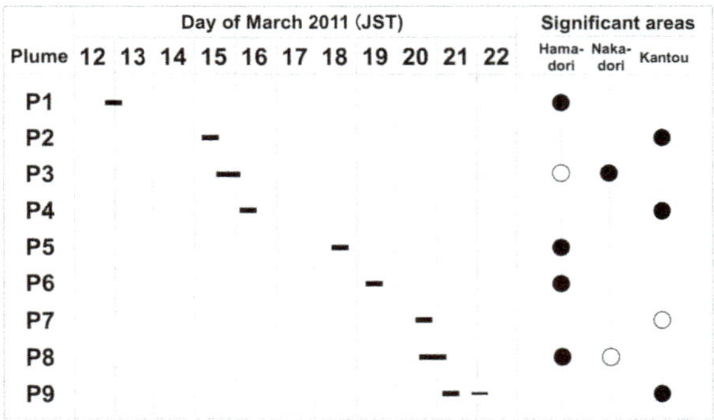

Fig. 3 Plumes identified by Tsuruta et al. (2014). *Horizontal bars* show the period with high ^{137}Cs concentrations (>10 Bq m^{-3}). *Closed and open circles* indicate areas in the Hamadori, Nakadori, and Kantou regions where the concentrations were larger or smaller, respectively, than 100 Bq m^{-2}

and assumed to undergo transportation, hygroscopic growth, dry and wet deposition, and gravitational settling, according to the scheme of Takemura et al. (2000). The present numerical experiment assumed a monodisperse sulfate-equivalent particle with a radius of 0.24 μm, taking into account the observed size distribution by Kaneyasu et al. (2012) and an in-cloud scavenging coefficient defined by Takemura et al. (2000) as 0.8 (Goto et al. 2015a).

The other model was a meteorological model, the weather research and forecast model (WRF) version 3.1 (Skamarock et al. 2008), coupled with a three-dimensional chemical transport model, the Models-3 Community Multiscale Air Quality (CMAQ) version 4.6 (Byun and Schere 2006); we will refer to this as the W-model. The CMAQ had been modified for radioactive material transport by the National Institute for Environmental Studies for a FDNPS simulation (Morino et al. 2011; SCJ 2014). We employed the WRF modules of scalar positive-definite advection, full diffusion, without the sixth-order horizontal-advection noise filter, and the planetary boundary parameterization of the Mellor-Yamada-Janjic scheme (Wang et al. 2012). In the present simulation, all the ^{137}Cs was assumed to be in particulate phase (Sportisse 2007), and the chemical and aerosol processes were not calculated because no detailed process has been well reported. The deposition schemes were those of Byun and Ching (1999) and Byun and Schere (2006). The mean particle radius was assumed to

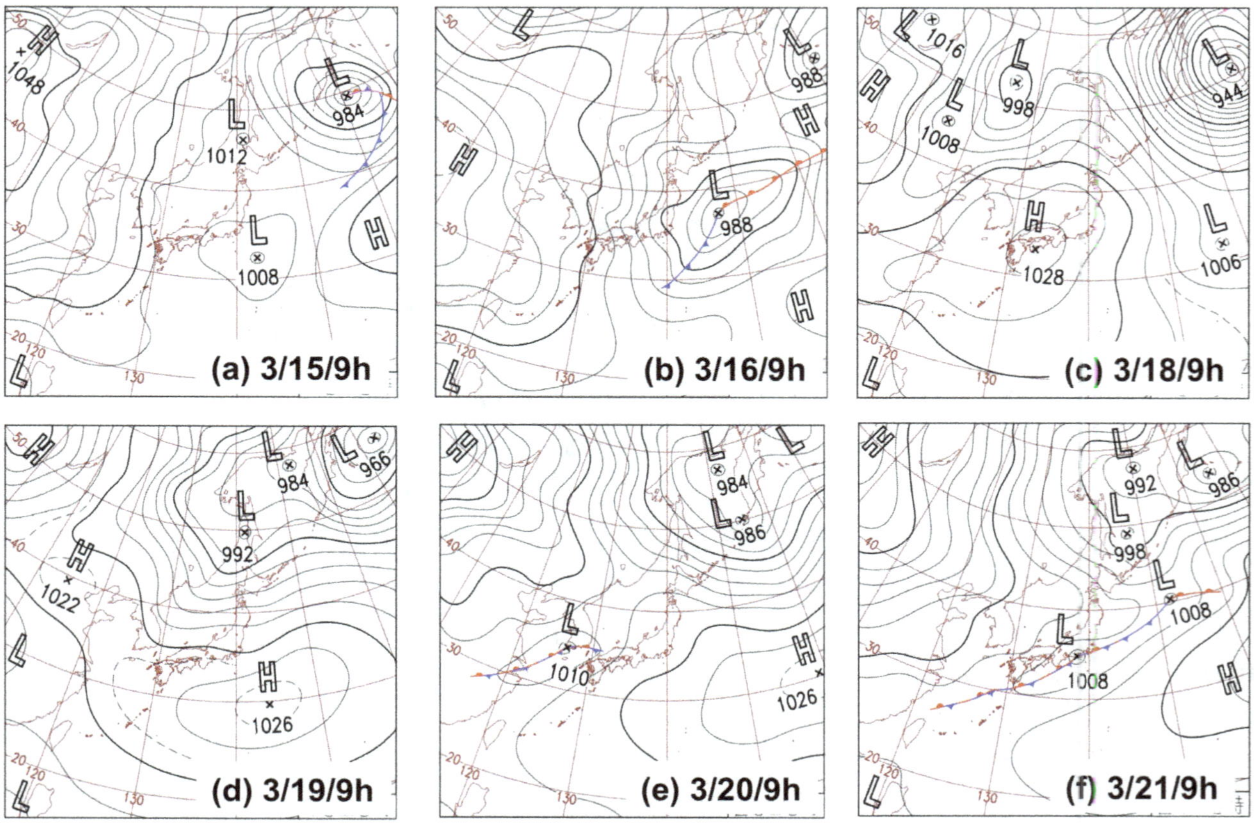

Fig. 4 a–**f** Weather maps based on JMA analysis at 9:00 JST in the analysis period

be 0.5 μm, and the geometrical dispersion σ_g was assumed to be 1.1 (Sparmacher et al. 1993; Sportisse 2007; Kaneyasu et al. 2012). The model domain was a 700 × 700 km^2 area, covering most of the Tohoku and Kantou regions, with a grid resolution of 3 km and 34 layers, with the lowest layer approximately 60 m thick. The simulation period was 14–24 March, and the analysis was performed by comparing the model ensemble results with the observed atmospheric ^{137}Cs concentration at surface level. The total deposition over Japanese land was simulated as 2.12 PBq by the N-model and 2.21 PBq by the W-model (Morino et al. 2013). These values belong to the small deposition group in the SCJ model comparison (SCJ 2014), which gave a mean value of total deposition of 2.92 PBq.

Both models were nudged with three-hourly 5 × 5 km mesoscale objective analysis data (MANAL) from the Japan Meteorological Agency (JMA). We selected the emission scenario of Terada et al. (2012), which has been used in many previous studies and is thus suitable for comparison with them. Figure 5 shows two examples of the simulated and observed atmospheric ^{137}Cs used to study the performance of both the N- and the W-model for specific cases, taken from the detailed time sequences as shown in Figs. 6 and 7 with precipitation maps of the JMA analysis and the N-model. These two examples were 15 March, 12:00 JST (3/15/12 h), and 21 March, 9:00 JST (3/21/9 h), when high-concentration plumes of more than 100 Bq m^{-3} were observed in the Kantou region. In Fig. 5, observed concentrations are indicated by color-coded circles, and the data below the detection limit of 0.1 Bq m^{-3} are indicated by triangles. On 15 March, a dry process governed the transportation, and there was no significant precipitation until late that evening, as indicated by the JMA analysis shown in Fig. 6. The distribution of the atmospheric ^{137}Cs was well simulated by the two models for a wide range of concentrations, from 0.1 to 100 Bq m^{-3}. It should be noted that the observed high-concentration area was confined to a narrow region, which was consistent with the simulated plume-wise distribution starting from the FDNPS for the Kantou region. The plumes simulated by the N-model were less diffused (narrower) than those simulated by the W-model. Judging by the correlation coefficients of the observed and simulated distributions, reality is assumed to fall between these two results; this is further discussed in the following paragraph.

For the March 21 event, the results of the N- and W-models differed significantly, as can be seen in Fig. 5c, d, respectively. Morning precipitation occurred in the Kantou region on 21 March, as will be shown in Fig. 7b. However, the simulated atmospheric concentration of ^{137}Cs was generally smaller than the observed values, suggesting that the simulated wet deposition was too strong for the W-model. Wet deposition during the total period over land in Japan was simulated as 1.68 and 2.12 PBq by the N- and W-model, respectively.

As indicated in Fig. 5, the accuracies of the simulated plume patterns of the two models differed from event to event. These differences were also reported by SCJ (2014). One way to reconcile the differences and construct statistically optimum distributions is to use a multi-model ensemble method to calculate the weighted mean concentration $\overline{C}_k$ for the kth grid point of the N-model at the jth time step, as given by

$$\overline{C}_{k,j} = \left(\frac{C_{\mathrm{N},k,j}}{\sigma_{\mathrm{N}}^2} + \frac{C_{\mathrm{W},k,j}}{\sigma_{\mathrm{W}}^2}\right) / \left(\frac{1}{\sigma_{\mathrm{N}}^2} + \frac{1}{\sigma_{\mathrm{W}}^2}\right), \quad (1)$$

where $C_{N,k,j}$ and $C_{W,k,j}$ are the values of the N- and W-model, respectively. The grid values of the W-model are linearly interpolated onto the N-model grid. The quantities σ_{N} and σ_{W} are the daily root mean square deviations of the respective model values from the observed values. Figure 8 shows a comparison of the observed values with the values obtained by the N-model alone and with the two-model ensemble means constructed by the proposed method. The daily correlation coefficients for specific days (15, 20, and 21 March) are listed in Table 1. The correlation coefficients were calculated using the daily means of the observed and simulated values at the surface level, as linearly interpolated at the site locations. The ensemble method constructs an atmospheric ^{137}Cs distribution that had a correlation coefficient better than that of either model alone. However, 20 March was an exception; the correlation coefficient of the two-model ensemble method was not as good as that of the N-model, although it was superior to that of the W-model. The 3-day average of the correlation coefficients was improved to 0.51 by the two-model ensemble method, from 0.48 obtained with the N-model and 0.46 with the W-model.

Results and discussion

Structure of atmospheric ^{137}Cs plumes

In this section, we examine the plume formation and dissipation during each of three periods: 15–16 March, 18–19 March, and 20–21 March 2011.

15–16 March

Land plume events in the Kantou and Tohoku regions lasted from the morning of 15 March until noon of the next day. Time sequences of the ^{137}Cs concentration of the two models and the observations are shown in Fig. 6a, b, along with precipitation maps of the JMA analysis and those of the N-model. Two-model ensemble means and the wind vectors of the JMA analysis are shown in Fig. 9 for several characteristic times. During the morning of 15 March, a low-pressure system with a weak trough structure was located in the Pacific region off the Kantou region (Fig. 4a). A northeasterly wind

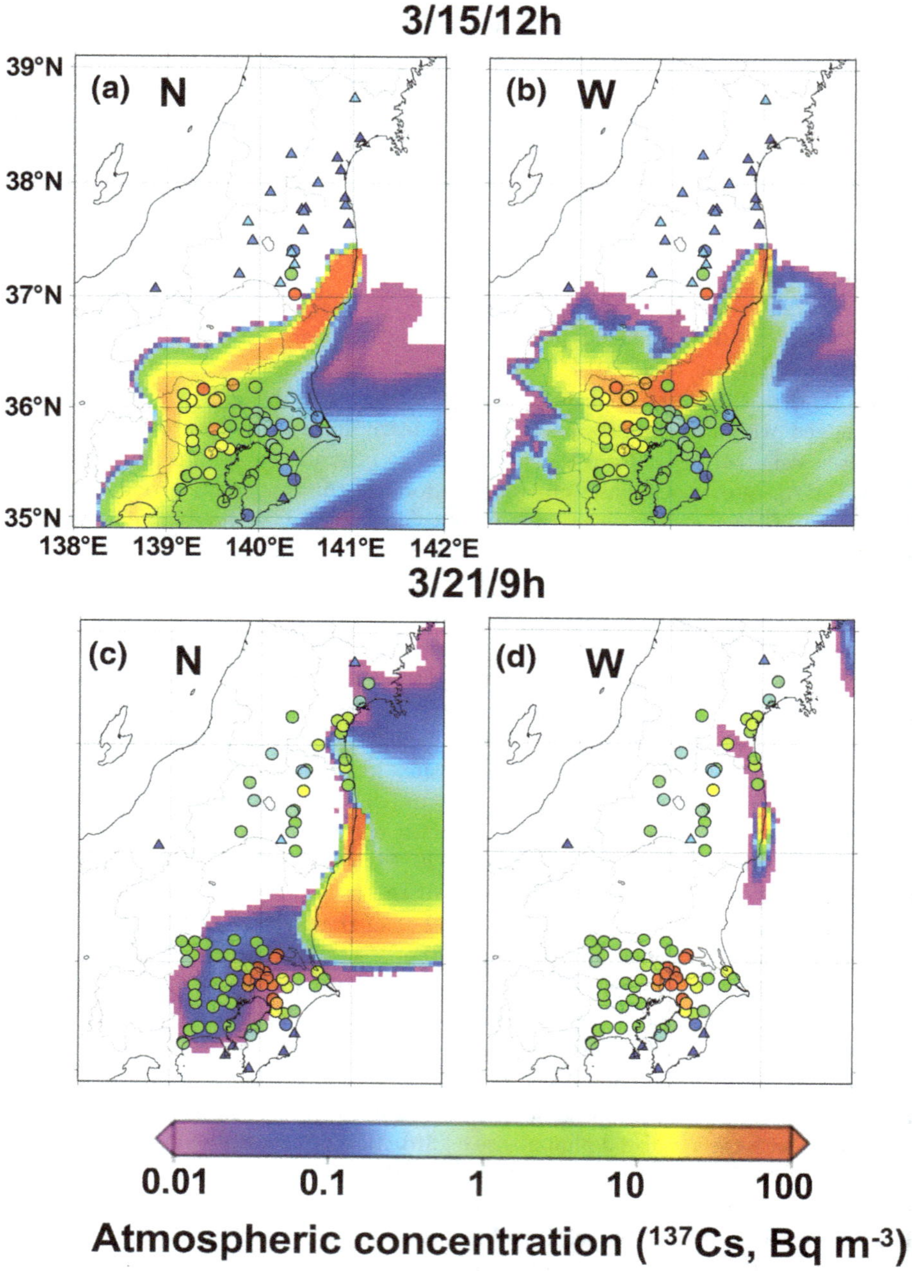

Fig. 5 Comparison of the atmospheric ^{137}Cs (Bq m^{-3}) for events on 15 March, 12:00 and 21 March, 9:00, as estimated by the NICAM-SPRINTARS model (*N*; **a**, **c**) and the WRF-CMAQ model (*W*; **b**, **d**). Observed values are indicated by *color-coded circles* (Bq m^{-3}); *triangles* indicate data below the detection limit (0.1 Bq m^{-3}) of SPM measurements

prevailed at the northern edge of the low-pressure system until 9:00, replacing the southwesterly wind associated with the high-pressure system of the previous day (not shown). This northeasterly wind transported radioactive material from the FDNPS southward to form plume P2, departing from the FDNPS at midnight and arriving at the Kantou region around 4:00 (Fig. 6a, panel 4:00), at which point the concentration exceeded several Bq m^{-3}, according to observations by the SPM network. Figure 6a, panels 0:00 to 4:00, indicates that the arrival time and location of the tongue of simulated plume P2 appear to be consistent with the observed concentrations shown by color-coded circles. The model simulation of plume transportation was relatively easy in this early phase during which there was no precipitation.

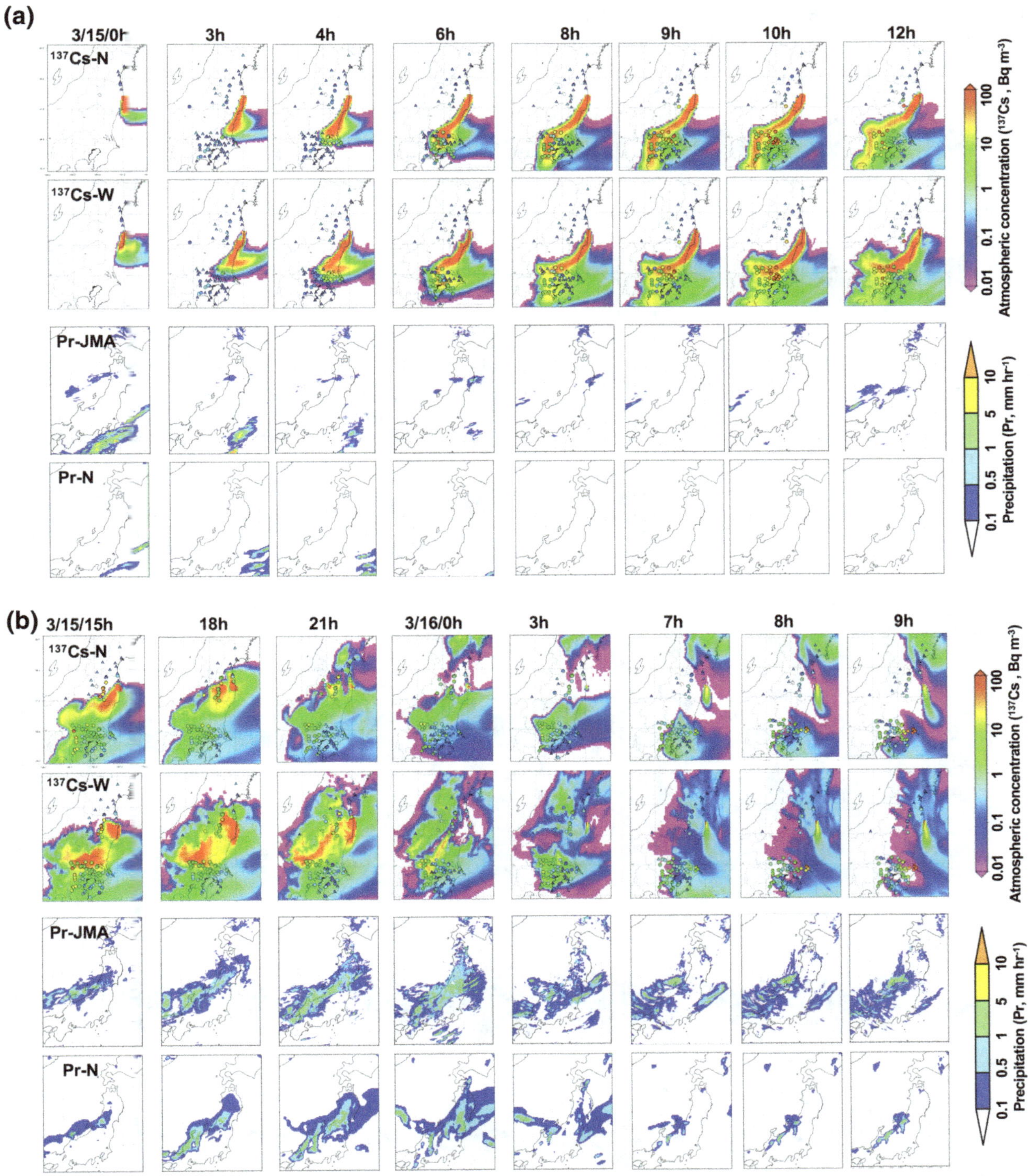

Fig. 6 Maps of the atmospheric ^{137}Cs concentration at surface level as simulated by the N- and W-models for the periods of **a** 3/15/0 h to 3/15/12 h and **b** 3/15/15 h to 3/16/9 h. The observed concentrations are indicated by *colored circles*. Precipitation maps are also shown for the JMA analysis and the N-model

Simulation results of the maximum stage of plume P2 in the Kantou region is, however, different from the SPM observations; a thick simulated plume as large as tens of Bq m^{-3} arrived around 6:00 at the Saitama, Tokyo, and Kanagawa prefectures (Fig. 6a, panels 6:00 to 9:00, and Fig. 9a, b), whereas the observed data did not

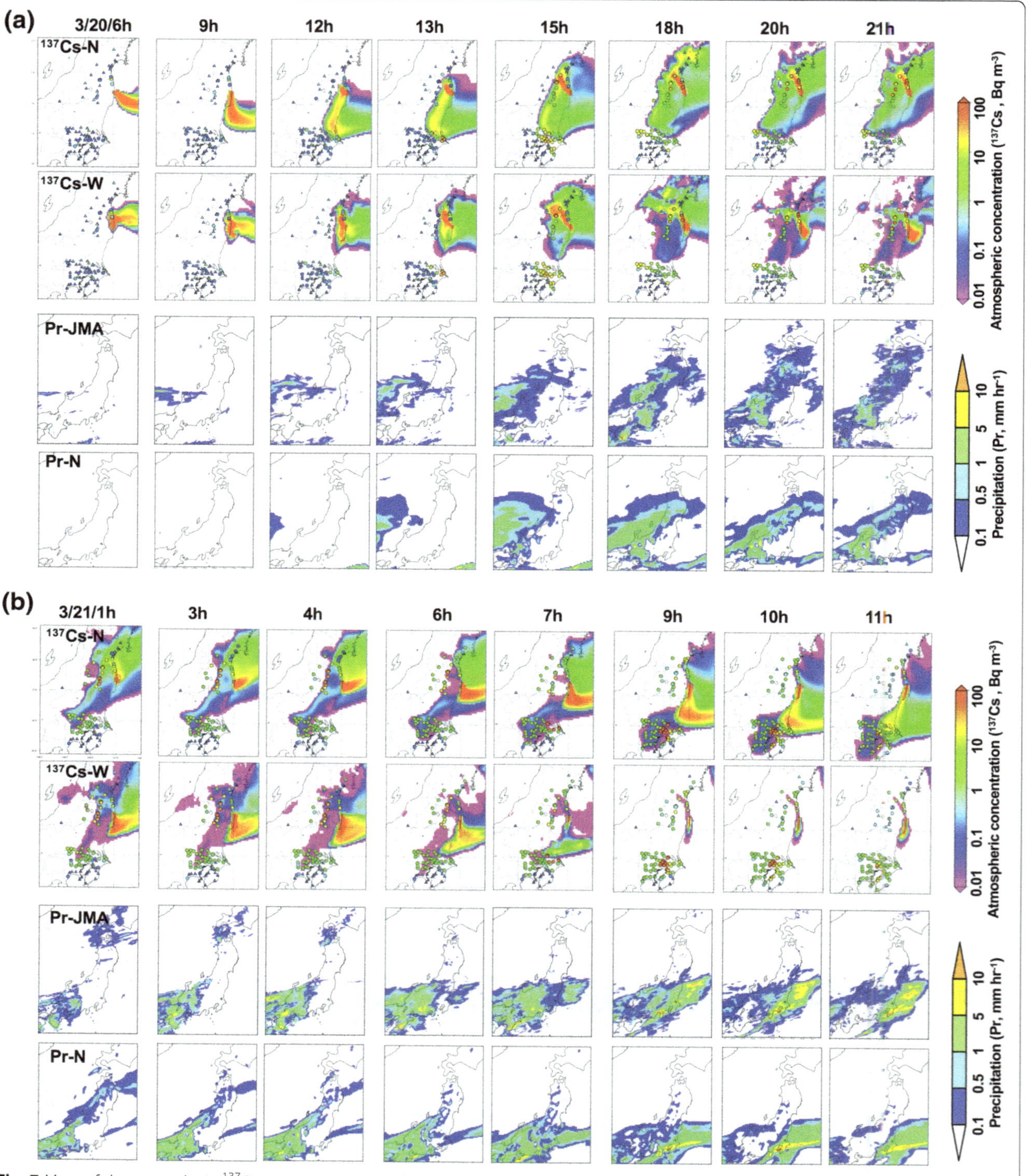

Fig. 7 Maps of the atmospheric ^{137}Cs concentration at surface level as simulated by the N- and W-models for the periods of **a** 3/20/6 h to 3/20/21 h and **b** 3/21/1 h to 3/21/11 h. The observed concentrations are indicated by *colored circles*. Precipitation maps are also shown for the JMA analysis and the N-model

show an increase to more than 10 Bq m^{-3} until 9:00, indicating a later arrival. Such a large difference in the arrival time of the plume cannot be explained by model errors, given that there were relatively simple meteorological conditions with a dry process, and the N- and W-models had similar results, as shown in Fig. 6a, panels 0:00 to 9:00.

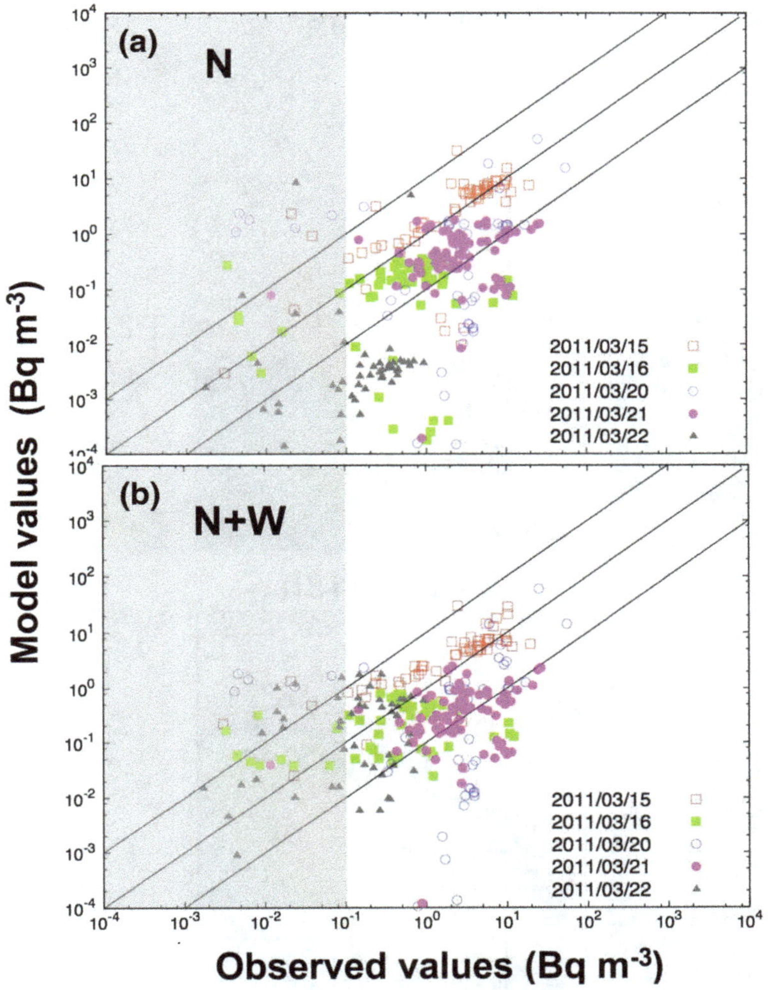

Fig. 8 Scatter plots of the observed atmospheric ^{137}Cs concentration (Bq m^{-3}) and the values obtained by the N-model and the two-model ensemble mean (**a**, **b**). The area in which the data was below the detection limit (0.1 Bq m^{-3}) is *shaded*

One possible cause of this is the large emission from FDNSP around 22:00 on 14 March, in the emission scenario of Terada et al. (2012). Figure 1 shows that the updated scenario of Katata et al. (2015), on the other hand, suggests a large emission occurred later, around 1:00 on 15 March. To investigate this possibility, Fig. 10 shows the simulation results of the N-model for the scenario of Terada et al. (2012) and that of Katata et al. (2015). It can be seen in the figure that with the scenario of Katata et al. (2015), the high-concentration area in the Kantou region, becomes smaller and is more consistent with the observations.

Table 1 Daily correlation coefficients between observed and simulated ^{137}Cs concentrations based on N-model, W-model, and two-model ensemble (N + W) results for ^{137}Cs

Day (JST)	N	W	N + W
2011/3/15	0.51	0.52	0.55
2011/3/20	0.57	0.50	0.55
2011/3/21	0.36	0.37	0.44
Average	0.48	0.46	0.51

The observed high-concentration area of plume P2 gradually spread from 9:00 to 12:00 in the Central Tokyo area, which is similar to simulations that show that the plume slightly shifted to the east, as shown in Fig. 6a, panels 9:00 to 12:00. Figure 9c shows that the ensemble model cannot account for this shift of the high-concentration area, but the emission scenario of Katata et al. (2015) gives a better result (Fig. 10). The SPM data suggest that dissipation of the P3 plume occurred during the evening of 15 March through the morning of 16 March, as shown in Fig. 6, panels 3/15/15:00 to 3/16/7:00, and Fig. 9d, e. The N-model adequately simulated the timing and distribution of the plume dissipation following a reduction in the transport of radioactive

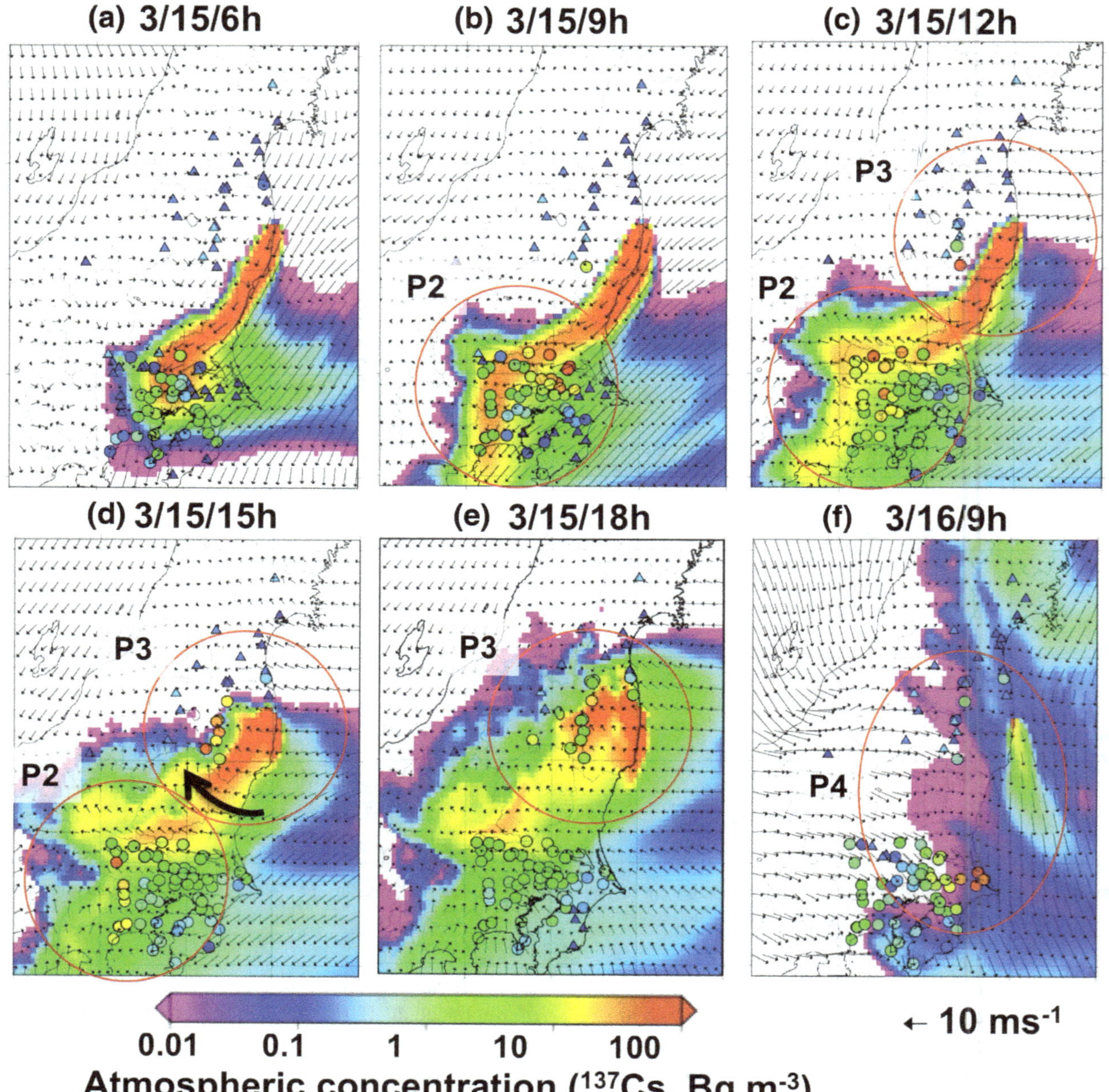

Fig. 9 a–f Maps of observations (*colored circles*) and model ensemble means of atmospheric ^{137}Cs concentration from 3/15/9 h to 3/16/9 h. Areas identified as plumes with concentration larger than 10 Bq m^{-3} are *encircled* and labeled with the name of the plume. JMA-MANAL wind vectors at 1000 hPa are superimposed. Movement of the P3 plume route is indicated by a *thick arrow* in panel **d**

material, due to a patch of precipitation between FDNSP and Kantou; however, the W-model simulation showed a slower dissipation.

The wind vector fields shown in Fig. 9c, d indicate that a weak cyclonic motion around the Central Fukushima area was generated until noon of 15 March; this was probably due to solar heating. It cause a gradual spread of plume P2 toward the western part of the Tokyo area, as indicated by a thick arrow in Fig. 9d. The wind field also transported plume P3 toward the Nakadori region in the afternoon of 15 March, when high concentrations of more than 100 Bq m^{-3} were observed consecutively from southern to northern sites (Fig. 6, panels 12:00 to 15:00, and Fig. 9c, d). In the late afternoon, a southeasterly wind became dominant at the FDNSP, and consequently, a part of the P3 plume took a northern detour to the edge of the Abukuma mountains and into the Nakadori region from north to south, as shown by

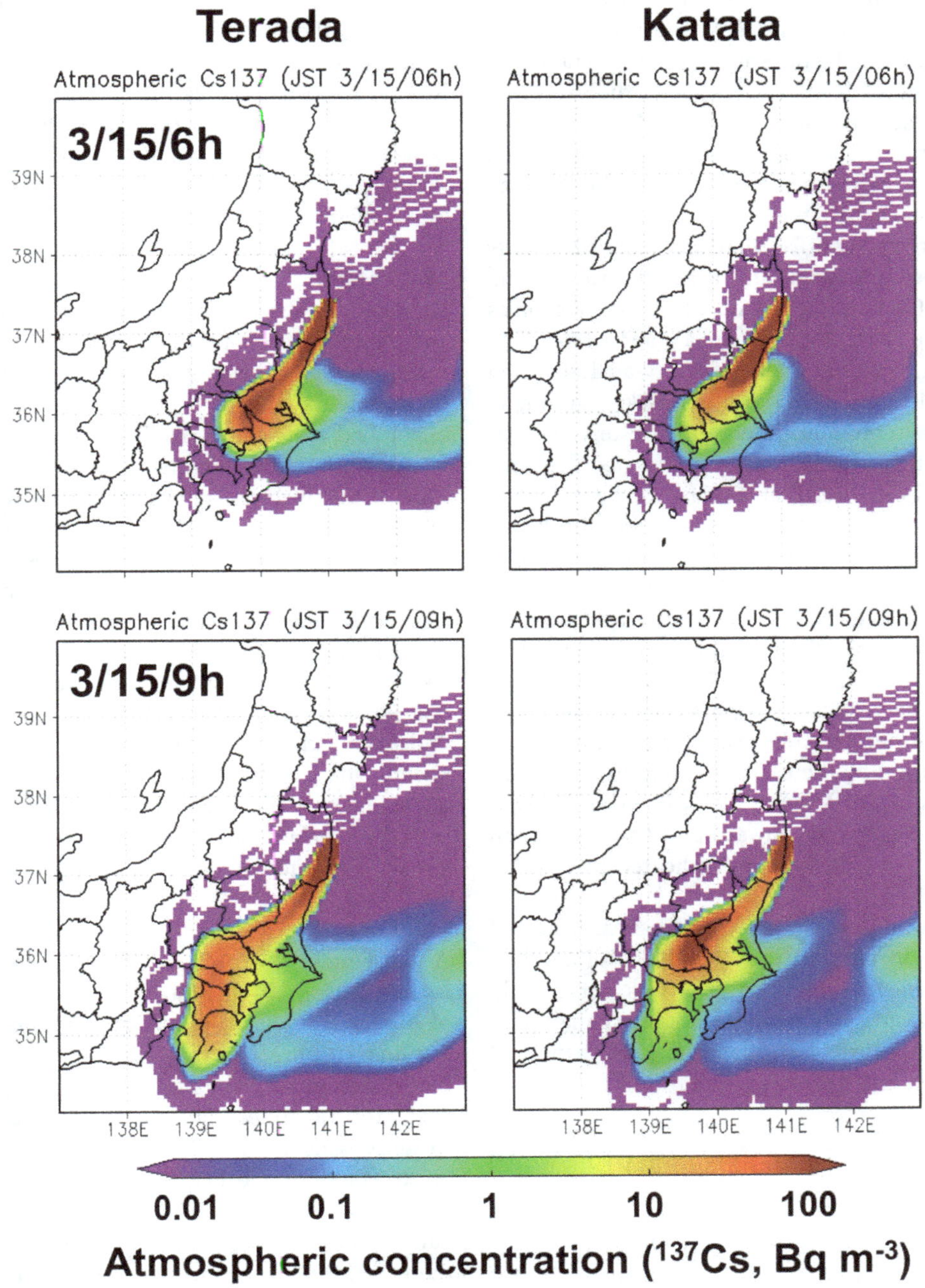

Fig. 10 Atmospheric ^{137}Cs distributions simulated by the N-model at 3/15/6 h and 3/15/9 h with emission scenarios by Terada et al. (2012) and Katata et al. (2015)

Fig. 6b, panels 15:00 to 18:00, and Fig. 9e. Tsuruta et al. (2014) also identified this northern branch. However, Figs. 6b and 9e show that without the plumes crossing the Abukuma mountains, the simulated plumes were too narrow to explain the elevated observed concentrations at the SPM sites in the Nakadori region. This problem is probably due to the spatial resolution (3–5 km), which is too coarse to simulate the fine-scale orographic horizontal advection of the plume traversing the Abukuma Mountains.

Subsequently, precipitation occurred over the Nakadori and Hamadori regions during the night of 15 March, lasting until the morning of 16 March and dissipating plume P3, as shown in Fig. 6b, panels 7:00 to 9:00, and Fig. 9f. These figures indicate that the next plume, P4, was transported by a northerly wind that became dominant during the morning of March 16, due to increased activity of the low in the Pacific (Fig. 4b). The SPM sites in the Choshi Peninsula observed a high concentration for several hours during the morning of 16 March, although the simulated plume failed to reach the target area (Fig. 9f). A possible explanation for the model failure is that the simulated aerosol height was higher than the actual height. According to the JMA analysis, there was considerable wind shear during the morning of 16 March, i.e., a northeasterly wind prevailed below 600 m, and a northwesterly wind prevailed above 600 m, as indicated by the vertical profiles of the atmosphere at the JMA Tsukuba station (36.06° N, 140.13° E) shown in Fig. 11. In a trajectory analysis using MANAL data, Miyasaka (Takafumi Miyasaka, University of Tokyo, personal communication, 2015) estimated that the aerosol height for the air mass that reached the Kantou region should be lower than 500 m. However, at the latitude 36.7° N, the simulated aerosols were distributed up to an altitude of 1500 m. Consequently, the simulated aerosols were transported to the Pacific area once they were uplifted to that level; this is indicated by the fact that the plume is not parallel to the 1000-hPa wind vector. There is a report that NICAM-SPRINTARS tends to overestimate the vertical transport of aerosols (Goto et al. 2015b). Another possible reason is overestimation of the wet deposition by the early morning precipitation along the plume on 16 March; the simulation showed this to be shifted slightly southward and was more than the MANAL objective analysis, as indicated by Fig. 6b, panel 3:00.

Another notable difference between observations and the model is that in plume P2, a medium-level concentration of between 1 and 10 Bq m^{-3} was persistently observed in a wide area of Kantou from the evening of 15 March until the morning of 16 March (Fig. 6b, panels 3/16/0:00 to 7:00; Fig. 9f), whereas the simulation results, in particular, those of the W-model, show a much smaller concentration in the southern part of the Hamadori and Kantou regions. Thus far, it is difficult for us to identify the cause of the model failures for plumes 2 and 4, but it is natural to assume that this is due to the difficulty of accurately simulating transport in a strongly sheared, thin atmosphere.

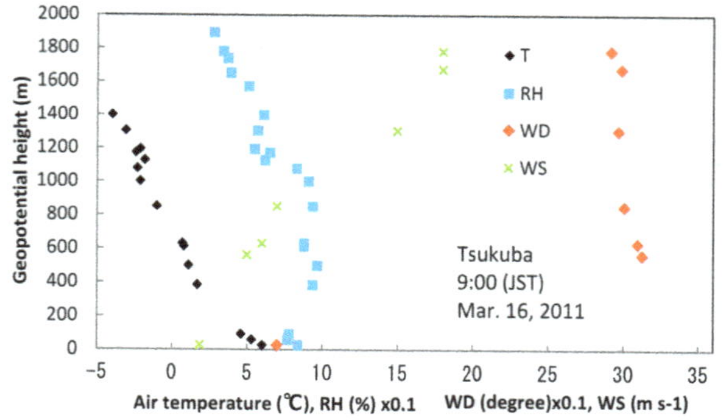

Fig. 11 Vertical profiles of temperature (*T*), relative humidity (*RH*), wind direction (*WD*), and wind speed (*WS*) versus geopotential height at 3/16/9 h for the JMA Tsukuba station. Wind directions and speeds are at the heights corresponding to the standard isobaric surfaces and characteristic points in the wind profile as determined by the JMA routine measurement protocol

18–19 March

Figure 12a, b shows the observed and simulated plumes P5 and P6 in the short-period events of 18 and 19 March. A migrating high-pressure system passed the Japanese islands on those days (Fig. 4c, d), and a weak wind condition was established around the FDNPS. The northwesterly component of the wind field turned southwesterly along the outskirts of the high-pressure zone at approximately 15:00 on 18 March (not shown). The P5 plume driven by the southwesterly wind was subsequently observed at sites north of the FDNPS. It should be noted that the wind vector at 18:00 was relatively perpendicular to the plume with this transport mechanism. A similar oblique relation between the wind vector and the plume axis was also simulated for plume P3 (Fig. 9d) and plumes P7 and P9 (shown below in Fig. 13b, f) under quick changes in the dominant wind direction. These phenomena indicate that the temporal change in wind direction must be accurately reproduced in order to successfully simulate the plume. Moreover, it is also necessary to have higher resolution of the land-sea breeze simulation.

20–21 March

A subsequent low-pressure system passed the Japanese islands during 20–21 March (Fig. 4e, f). Time sequences of observed and two-model atmospheric ^{137}Cs concentrations with precipitation maps are shown in Fig. 7, and model ensemble results are shown in Fig. 13 for several characteristic times. The figures indicate the dominant westerly wind on 19 March through the early morning of 20 March at the FDNPS blew the plume toward the Pacific, as shown in Fig. 13a (labeled A) and Fig. 7a, panel 6:00; however, during the morning of 20 March, there was an increasing easterly component due to a moving low-pressure system; this blew the plume back westward toward the Hamadori and established plume P7 (labeled B in Fig. 13a), which had a wide distribution covering the Hamadori and its offshore areas, as shown in Fig. 7a, panels 9:00 to 12:00, and Fig. 13a, b. This plume movement is indicated by a thick arrow in Fig. 13a. The simulated distribution of plume P7, however, was very different in the N- and W-models, resulting in a two-branch pattern of the ensemble mean (A and B in Fig. 13a).

The arrival of P7 at the Choshi Peninsula between 12:00 and 13:00, and its gradual westward spread until 15:00, as shown by the rotation of the plume indicated

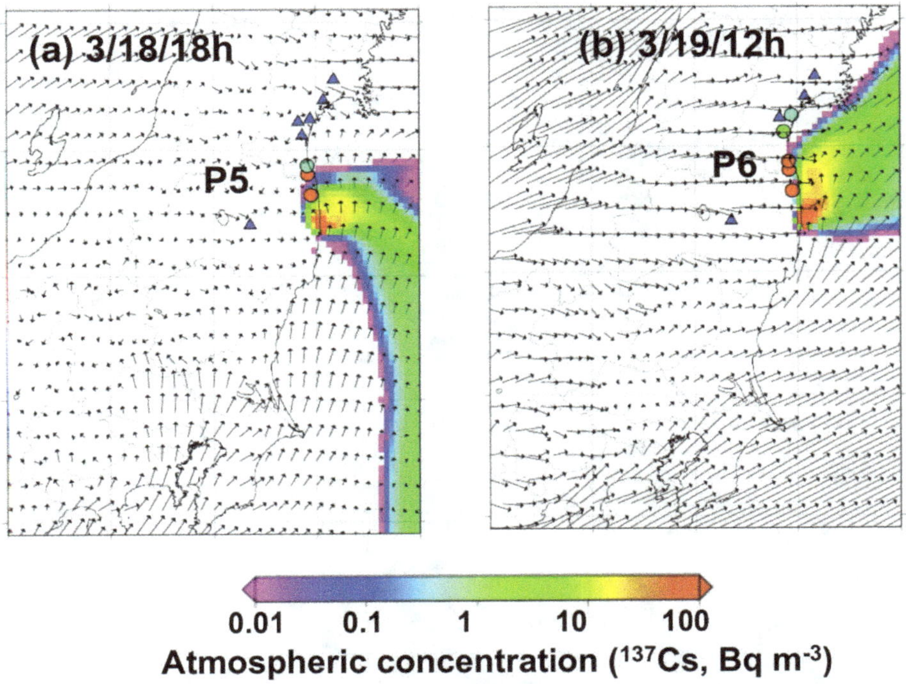

Fig. 12 Maps of observations (*colored circles*) and model ensemble means of atmospheric ^{137}Cs concentration in **a** 3/18/18 h and **b** 3/19/12 h. Areas identified as plumes with concentration larger than 10 Bq m^{-3} are *encircled* and labeled with the name of the plume. JMA-MANAL wind vectors at 1000 hPa are superimposed

by a thick arrow in Fig. 13a, was successfully simulated by the N-model, as shown in Fig. 7a, panels 12:00 to 15:00. On the other hand, the W-model failed to simulate this phenomenon. Since precipitation was not involved in the transport process on the morning of 20 March, this failure was considered to be a consequence of more transportation to the upper atmosphere near the emission source, as suggested by the shift to the north (relative to the results of the N-model) of the high concentration around the FDNPS (Fig. 7a, panels 12:00 to 13:00).

Northwestern transport started around 13:00 and lasted until that night, forming plume P8 in the Tohoku region (Fig. 7a, panels 15:00 to 21:00, and Fig. 13b, c). Each model satisfactorily simulates the transportation of plume P8 to the northern part of the Nakadori region, but they are both too coarse to simulate the successive increase in the concentration of ^{137}Cs observed from north to south along the Nakadori channel.

Plumes P7 and P8 started to dissipate at around 3:00 on March 21 (Fig. 7b, panels 1:00 to 7:00, and Fig. 13e). At this time, plume P9 started traveling to Kantou, due to a northeasterly wind ascribed to the combined effect of a migratory system and the Okhotsk low-pressure system (Fig. 4f). Plume P9 reached the Kantou region at around 9:00 on 21 March and collided with the weather front located at the southern part of the Kantou region, as shown by the wind vectors in Fig. 13f. The plume simulated by the N-model had a tongue shape and covered the land area of the northern Chiba prefecture; this was consistent with the distribution of the observed sites characterized by concentrations of more than 100 Bq m^{-3} around the Chiba prefecture and Tokyo Bay, as depicted in Fig. 7b, panels 9:00 to 10:00. However, the N-model plume was too short, and it shifted to the northern area without reaching the observed sites. In this event, the W-model also failed to transport ^{137}Cs to the Kantou region. Although not shown by a figure, we found that changing the emission scenario to that of Katata et al. (2015) did not improve the results. Another possibility is that the height of the aerosol layer was overestimated. Figure 14 shows the vertical profile of meteorological parameters at the JMA Tsukuba station at 9:00 on 21 March. The figure indicates that there was a strong temperature inversion at around 300 m geopotential height. Wind was northeasterly below the inversion and northwesterly above. Therefore, it can be assumed that the modeled aerosol layer was too high, and the aerosols were transported eastward by the northwesterly. This also might be due to the stronger precipitation that was simulated by the models in the Kantou region during the time period 3:00 to 7:00 (Fig. 13b, panels 3:00 to 7:00). These two reasons may

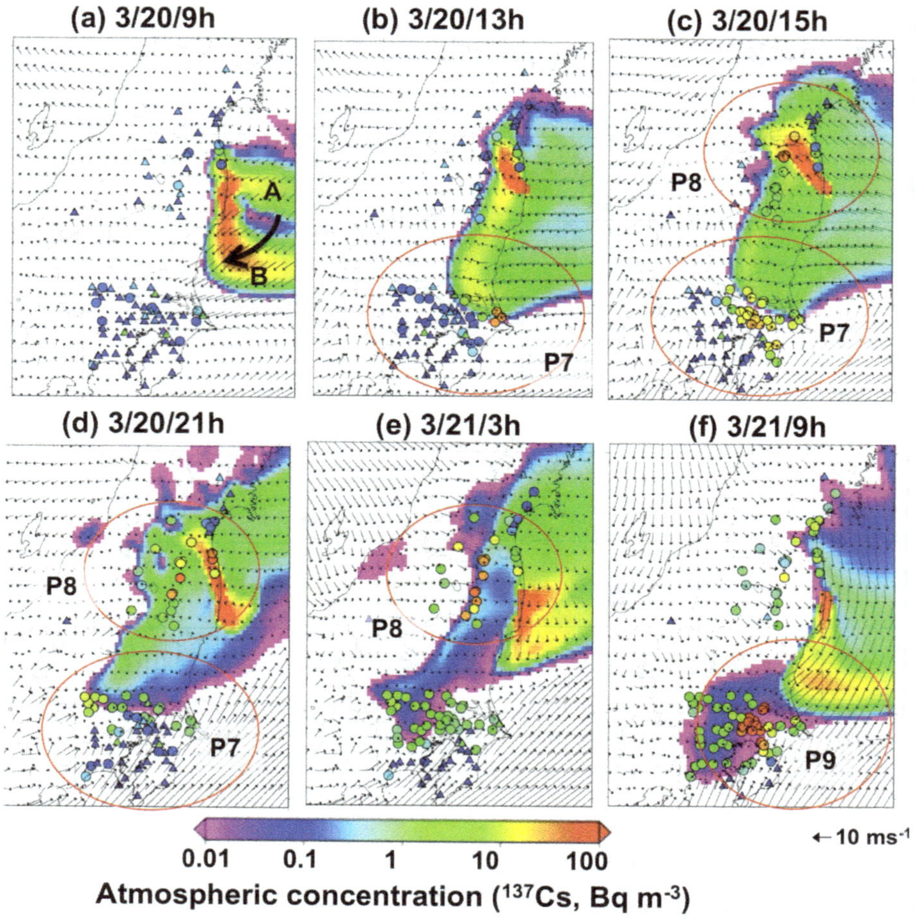

Fig. 13 a–f Maps of observations (*colored circles*) and model ensemble means of atmospheric ^{137}Cs concentration from 3/20/9 h to 3/21/9 h. Areas identified as plumes with concentration larger than 10 Bq m^{-3} are *encircled* and labeled with the name of the plume. JMA-MANAL wind vectors at 1000 hPa are superimposed. Movement of the P7 plume route is indicated by a *thick arrow* in panel **a**. See the main text for explanation of labels (**a**, **b**)

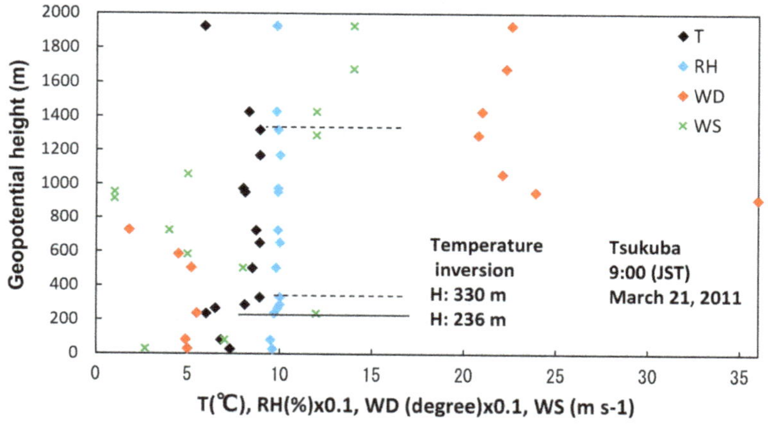

Fig. 14 Vertical profiles of temperature (*T*), relative humidity (*RH*), wind direction (*WD*), and wind speed (*WS*) versus geopotential height at 3/21/9 h for the JMA Tsukuba station. Wind directions and speeds are at the heights corresponding to the standard isobaric surfaces and characteristic points in the wind profile as determined by the JMA routine measurement protocol

explain the quick dissipation of simulated plume P7, in contrast to the observed persistent survival of the plume in the morning of 21 March (Fig. 7b, panels 3:00 to 9:00, and Fig. 13e, f).

Plume routes and time series

Figure 15 presents a summary of the plume routes analyzed in the preceding sections, and Fig. 16 shows the time series of the observations and the model ensemble mean ^{137}Cs concentrations at several sites. The relevant sites are B, C, E, and J in the Tohoku region and sites 9, 12, and 15 in the Kantou region, as defined by Tsuruta et al. (2014), as well as at the newly added sites K1, K2, and K3.

During a 6-hour period on 15 March, plume P2 spread over a large area of the Kantou region. The coverage was not uniform, and the plume was narrow, as indicated by the peak concentration at sites 9, K1, K2, and K3 in Fig. 16b. However, the simulated arrival time was too early, as indicated by the time series and as was already suggested by Fig. 9a.

Plume P3 transported the radioactive material toward the Nakadori region, crossing the Abukuma Mountains, as indicated by the successive peaks over time from the south to the north at sites E and C in the afternoon of 15 March (Fig. 16a); the models were too coarse to simulate the detailed progress of the plume along the Nakadori channel. The progress of plumes P2 and P3 is indicated by thick white arrows in Fig. 15.

As shown in Fig. 9f, plume P4 produced a high concentration of ^{137}Cs in the Choshi Peninsula, as shown by the peaks at sites 12 and 15, but the plume did not reach Kantou.

Short-duration plumes P5 and P6 were observed at site J on 18 and 19 March, and they were simulated by the models, but the simulated arrival time was earlier than the observed time. Moreover, the concentration was largely underestimated for the duration of the period.

In the early afternoon of 20 March, plume P7 was transported over a long distance, first to the ocean and then toward land by a clockwise rotation of the plume route, as indicated by the thick white arrow in Fig. 15; it covered a large part of the northern Kantou region, as shown in Fig. 16b.

Plume P8 was transported northward and was redirected southward along the Nakadori region at the northern edge of the Abukuma Mountains. Consequently, the high-concentration area moved from north to south, as illustrated by the time series of observations at sites B, C, and E. However, the models failed to indicate this phenomenon, underestimating the concentration at sites in the Nakadori region, as also shown in Fig. 7a, panels

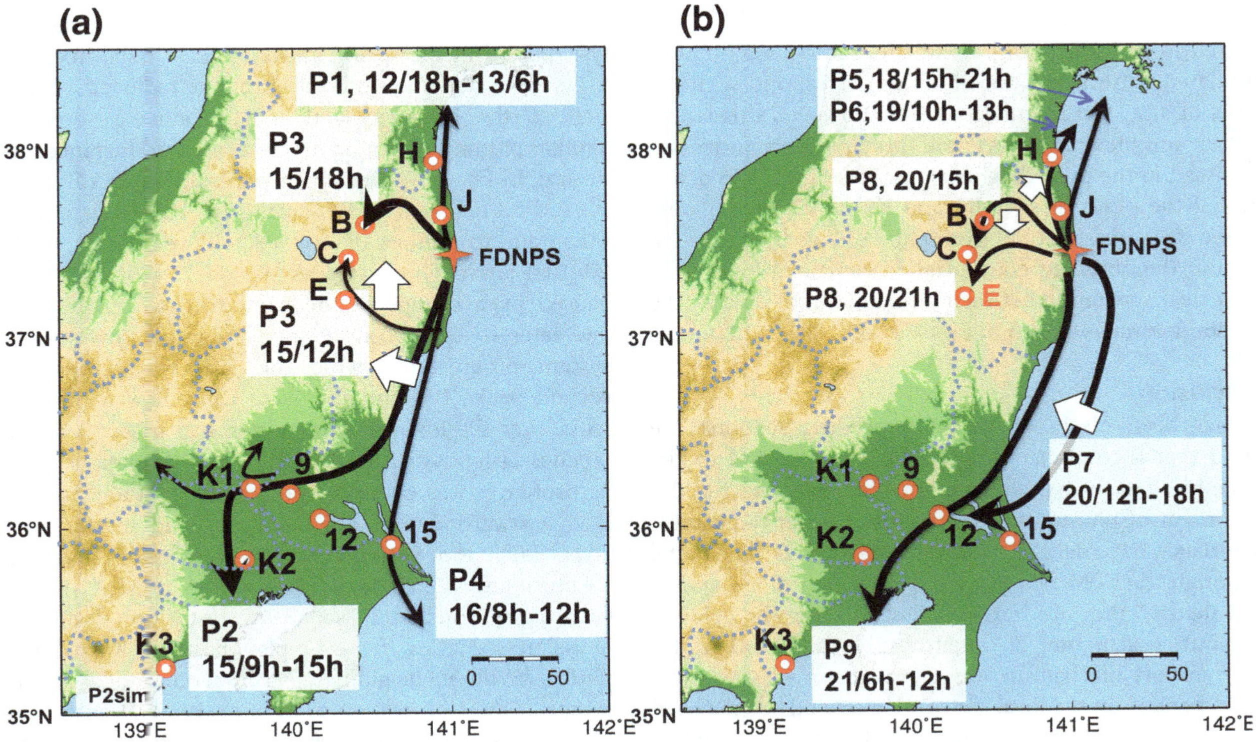

Fig. 15 a, b Schematic diagrams of transport routes analyzed in the present study for plumes P2, P3, P4, P7, P8, and P9. *Thick arrows* indicate the general trend of the movement of each plume

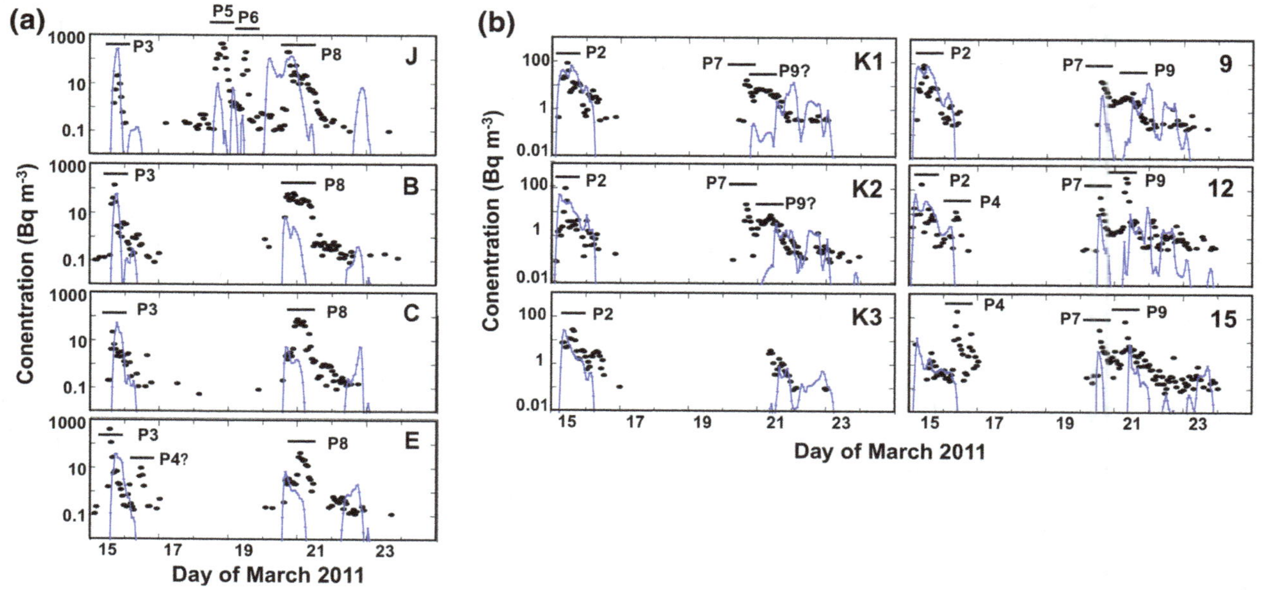

Fig. 16 Time series of observed (*dots*) and two-model ensemble mean (*lines*) atmospheric ^{137}Cs concentrations (Bq m^{-3}) from 14 to 24 March **a** in the Tohoku region at sites J, B, C, and E (see Fig. 15) and **b** in the Kantou region at sites 9, 12, 15, K1, K2, and K3 (see Fig. 15)

15:00 to 21:00. At the same time, the wind direction turned from northwest to north at the FDNPS, with the result that the northern part of P8 covered sites H and J, as illustrated by the thick northeastward arrow.

Plume P9 began on the morning of 21 March, taking a southern route until it collided with the weather front located in the southern part of the Kantou region. The simulated pattern was too short to correspond with the peaks at sites 9, 12, and 15 at 9:00; however, this is due to the complex transport and precipitation processes discussed in the preceding sections. It is difficult to determine if the peaks at sites K1 and K2 on 21 March were caused by a persistent tail from plume P7 or by plume P9, because the observed concentration remained high without a clear separation between plumes, as indicated by the distribution maps shown in Figs. 7b and 13d–f.

Conclusions

As has been discussed in the preceding sections, we found that a combined analysis of observed and model ensemble data is a useful method for analyzing the development of plumes and the distribution of radioactive materials, but neither approach alone is adequate. Although the SPM observational data are unique and of high density, they are not sufficient to show the detailed distribution structure of the atmospheric ^{137}Cs, because the transport mechanism was complex, varied over time, and depended on the local meteorological and geographical conditions. Although in some cases, the models failed to simulate the exact location and time of arrival of the plumes at the SPM sites, the spatial and temporal development of the plume structure was adequately simulated, making it possible to understand how the atmospheric ^{137}Cs was distributed.

The following statement is characteristic of the target area and period, and it is relevant to the atmospheric transportation of radioactive materials: during the analysis period in the spring season in East Asia, migratory pressure systems periodically brought radioactive materials to the Japanese land area, producing somewhat similar plume development patterns. For instance, two peaks can be seen in the time series: one due to plumes P2 and P3 followed by P4, during 15–16 March, and the other due to plumes P7 and P8 followed by P9, during 20–21 March. The first peak (P3 on 15–16 March) was caused by a change in the wind field to northeasterly and later to southeasterly, as a migratory low-pressure system progressed toward the Japanese islands. The second peaks (P4 and P9) were caused by a northerly wind after the low-pressure system had passed from the Japanese islands to the Pacific. When the height of the aerosol layer was overestimated, plumes P4 and P9 deviated eastward due to the westerly wind in the upper layers, such as at 900 hPa, as shown in Figs. 9f and 13f.

Future tasks include improving the present method, such as by conducting sensitivity tests for (1) different emission scenarios, for example, those by Katata et al. (2015); (2) plume height based on different model-layering setups; (3) wet deposition processes with different parameterizations; and (4) material transport across and along the northern edge of the Abukuma Mountains. Relevant to task 4, a report by Sekiyama et al. (2015) has claimed that there was no significant difference between 3-km and 500-m grid simulations of the JMA nonhydrostatic model for horizontal transport of radioactive material. However, in view of the differences in model parameterization for orographic waves, further analysis of this is necessary. It is also necessary to increase the spread of the model ensemble in order to obtain a more accurate reconstruction of the plume transportation. Another interesting and unresolved problem is the area of mid-level concentration, from 1 to 10 Bq m^{-3}, in the Kantou region that persisted from the evening of 20 March to the morning of 21 March. It is also important to carefully evaluate the performance of the model when simulating the vertical stratification of the atmosphere, which controls the dry deposition process. A future study should address aerosol survival under strong but stochastic precipitation conditions. On the other hand, SCJ (2014) suggested that all of the models evaluated tended to underestimate wet deposition in weak precipitation conditions. Consequently, it may be necessary to develop a nonlinear parameterization of the precipitation rate.

We have analyzed only 25% of the total SPM sampling tapes to date, so additional analysis is necessary. Further efforts are also needed to collect missing observational data, in particular, in the Hamadori region, in order to investigate the detailed atmospheric transportation processes that could not be addressed by the present study. Some of the SPM tapes were discarded by the network before we could retrieve them. Therefore, a special tape conservation effort would be required for large-scale disaster events. We expect the present study to be useful for future research.

Abbreviations
CMAQ: Community multiscale air quality; FDNPS: Fukushima Daiichi Nuclear Power Station; JAEA: Japan Atomic Energy Agency; JAXA: Japan Aerospace eXploration Agency; JMA: Japan Meteorological Agency; JST: Japan standard time; MANAL: Mesoscale objective-analysis data; MEXT: Ministry of Education, Culture, Sports, Science and Technology, Japan; NICAM: Nonhydrostatic icosahedral atmospheric model; NSW6: NICAM single-moment scheme with six water categories; SCJ: Science Council of Japan; SPM: Suspended particulate matter; SPRINTARS: Spectral radiation-transport model for aerosol species; TEPCO: Tokyo Electric Power Company; UNCEAR: United Nations Scientific Committee on the Effects of Atomic Radiation; UTC: Coordinated universal time; WRF: Weather Research and Forecast Model

Acknowledgements
We greatly appreciate the kind support of H. Tomita and H. Yashiro of RIKEN/AICS. We also thank all the local governments who offered the used SPM filter tapes at the request of the Ministry of the Environment, Japan. We have cited the weather maps of Japan Meteorological Agency. We used the meteorological data sets of the Japan Meteorological Agency.

Funding
Parts of this research were supported by funds from MOE/GOSAT2, JST/CREST/EMS/TEEDDA, JAXA/EarthCARE&GCOM-C, MEXT/KAKENHI/Innovative Areas 2409, MOEJ/ERTDF/S-12 and 5-1501, and the National Regulation Agency, Japan.

Authors' contributions
TN was the principal investigator of the modeling and SPM data analysis projects, the chair of the SCJ Model Comparison Committee, designed the combined analysis, and drafted the manuscript. SM, DG, JU, TT, and MS conducted the numerical simulations with NICAM-SPRINTARS. YM and TO conducted the numerical simulations with WRF-CMAQ. HT, TO, and ME conducted the SPM tape analysis. All co-authors participated in discussions about the results and commented on the original manuscript. All authors read and approved the final manuscript.

Authors' information
Correspondence and requests for materials should be addressed to TN (terry-nkj@nifty.com).

Competing interests
The authors declare that they have no competing interests.

Author details
[1]Atmosphere and Ocean Research Institute (AORI), The University of Tokyo, 5-1-5 Kashiwanoha, Kashiwa, Chiba 277-8568, Japan. [2]Earth Observation Research Center (EORC), Japan Aerospace Exploration Agency (JAXA), 2-1-1 Sengen, Tsukuba, Ibaraki 305-8505, Japan. [3]National Institute for Environmental Studies (NIES), 16-2 Onogawa, Tsukuba, Ibaraki 305-8506, Japan. [4]Remote Sensing Technology Center of Japan, 3-17-1 Toranomon, Minato-ku, Tokyo 105-0001, Japan. [5]Research Institute for Applied Mechanics, Kyushu University, Kasuga Park 6-1, Kasuga, Fukuoka 816-8580, Japan. [6]Graduate School of Science and Engineering, Tokyo Metropolitan University, 1-1 Minami-Osawa, Hachioji, Tokyo 192-0397, Japan.

References
Byun DW, Ching JKS (1999) Science algorithms of the EPA Models-3 Community Multi-Scale Air Quality (CMAQ) modeling system. Rep. USA EPA/600/R-99/030. US Environmental Protection Agency, Washington

Byun D, Schere KL (2006) Review of the governing equations, computational algorithms, and other components of the Models-3 Community Multiscale Air Quality (CMAQ) modeling system. Appl Mech Rev 59:51–77

Dai T, Goto D, Schutgens NAJ, Dong X, Shi G, Nakajima T (2014) Simulated aerosol key optical properties over global scale using an aerosol transport model coupled with a new type of dynamic core. Atmos Environ 82:71–82. doi:10.1016/j.atmosenv.2013.10.018

Goto D, Dai T, Satoh M, Tomita H, Uchida J, Misawa S, Inoue T, Tsuruta H, Ueda K, Ng CFS, Takami A, Sugimoto N, Shimizu A, Ohara T, Nakajima T (2015a) Application of a global nonhydrostatic model with a stretched-grid system to regional aerosol simulations around Japan. Geosci Model Dev 8:235–259. doi:10.5194/gmd-8-235-2015

Goto D, Nakajima T, Dai T, Takemura T, Kajino M, Matsui H, Takami A, Hatakeyama S, Sugimoto N, Shimizu A, Ohara T (2015b) An evaluation of simulated particulate sulfate over East Asia through global model intercomparison. J Geophys Res Atmos 120(12):6247–6270. doi:10.1002/2014JD021693

JAEA (2012) JAEA Open Workshop 'Reconstruction of the Emission and diffusion processes of materials released by the Fukushima Daiichi Atomic Power Plant accident', March 6, 2012, Tokyo, http://nsec.jaea.go.jp/ers/environment/envs/FukushimaWS/index.htm. Accessed 18 Jan 2017

Kaneyasu N, Ohashi H, Suzuki F, Okuda T, Ikemori F (2012) Sulfate aerosol as a potential transport medium of radiocesium from the Fukushima nuclear accident. Environ Sci Technol 46:5720–5726. doi:10.1021/es204667h

Katata G, Chino M, Kobayashi T, Terada H, Ota M, Nagai H, Kajino M, Draxler R, Hort MC, Malo A, Torii T, Sanada Y (2015) Detailed source term estimation of the atmospheric release for the Fukushima Daiichi Nuclear Power Station accident by coupling simulations of an atmospheric dispersion model with an improved deposition scheme and oceanic dispersion model. Atmos Chem Phys 15:1029–1070. doi:10.5194/acp-15-1029-2015

Lin YL, Farley RD, Orville HD (1983) Bulk parameterization of the snow field in a cloud model. J Clim Appl Meteorol 22:1065–1092

Mellor GL, Yamada T (1982) Development of a turbulence closure model for geophysical fluid problems. Rev Geophys Space Phys 20:851–875

Mészáros R, Leelőssy Á, Kovács T, Lagzi I (2016) Predictability of the dispersion of Fukushima-derived radionuclides and their homogenization in the atmosphere. Sci Rep 6:19915. doi:10.1038/srep19915

MEXT (2011) Airborne monitoring results after the first one conducted by the Ministry of Education, Culture, Sports, Science, and Technology in April 2011 are summarized by the Nuclear Regulation Authority. http://radioactivity.nsr.go.jp/ja/list/362/list-1.html. Accessed 6 May 2011

Morino Y, Ohara T, Nishizawa M (2011) Atmospheric behavior, deposition, and budget of radioactive materials from the Fukushima Daiichi Nuclear Power Plant in March 2011. Geophys Res Lett 38:L00G11. doi:10.1029/2011GL048689

Morino Y, Ohara T, Watanabe M, Hayashi S, Nishizawa M (2013) Episode analysis of deposition of radiocesium from the Fukushima Daiichi Nuclear Power Plant accident. Environ Sci Technol 47:2314–2322. doi:10.1021/es304620x

Nakajima T, Ohara T, Uematsu M, Onda Y (eds) (2014) Radioactive environmental pollution from the Fukushima Daiichi Nuclear Power Station accident: earth science perspectives. University of Tokyo Press, Tokyo (in Japanese)

Nakanishi M, Niino H (2004) An improved Mellor-Yamada level 3 model with condensation physics: its design and verification. Bound Layer Meteorol 112:1–31

Oura Y, Ebihara M, Tsuruta H, Nakajima T, Ohara T, Ishimoto M, Sawahata H, Katsumura Y, Nitta W (2015) A database of hourly atmospheric concentrations of radiocesium (^{134}Cs and ^{137}Cs) in suspended particulate matter collected in March 2011 at 99 air pollution monitoring stations in eastern Japan. J Nucl Radiochem Sci 15(2):1–12

Satoh M, Matsuno T, Tomita H, Miura H, Nasuno T, Iga S (2008) Nonhydrostatic Icosahedral Atmospheric Model (NICAM) for global cloud resolving simulations. J Comput Phys 227:3486–3514

SCJ (2014) A review of the model comparison of transportation and deposition of radioactive materials released to the environment as a result of the Tokyo Electric Power Company's Fukushima Daiichi Nuclear Power Plant accident, Report of Committee on Comprehensive Synthetic Engineering, Science Council of Japan, http://www.jpgu.org/scj/report/20140902scj_report_e.pdf. Accessed 18 Jan 2014

Sekiyama TT, Kunii M, Kajino M, Shimbori T (2015) Horizontal resolution dependence of atmospheric simulations of the Fukushima nuclear accident using 15-km, 3-km, and 500-m grid models. J Meteorol Soc Jpn 93:49–64

Skamarock WC, Klemp JB, Dudhia J, Gill DO, Barker DM, Duda MG, Huang XY, Wang W, Powers JG (2008) A description of the Advanced Research WRF Version 3. NCAR/TN.475 + STR. National Center for Atmospheric Research, Boulder

Sparmacher H, Fulber K, Bonka H (1993) Below-cloud scavenging of aerosol particles: particle-bound radionuclides—experimental. Atmos Environ 27(4): 605–618. doi:10.1016/0960-1686(93)90218-N

Sportisse B (2007) A review of parameterizations for modelling dry deposition and scavenging of radionuclides. Atmos Environ 41:2683–2698. doi:10.1016/j.atmosenv.2006.11.057

Stohl A, Seibert P, Wotawa G, Arnold D, Burkhart JF, Eckhardt S, Tapia C, Vargas A, Yasunari TJ (2012) Xenon-133 and cesium-137 releases into the atmosphere from the Fukushima Dai-ichi nuclear power plant: determination of the source term, atmospheric dispersion, and deposition. Atmos Chem Phys 12: 2313–2343. doi:10.5194/acp-12-2313-2012

Takemura T, Okamoto H, Maruyama Y, Numaguti A, Higurashi A, Nakajima T (2000) Global three-dimensional simulation of aerosol optical thickness distribution of various origins. J Geophys Res 105:17853–17873

Takemura T, Nakamura H, Takigawa M, Kondo H, Satomura T, Miyasaka T, Nakajima T (2011) A numerical simulation of global transport of atmospheric particles emitted from the Fukushima Daiichi Nuclear Power Plant. SOLA 7: 101–104. doi:10.2151/sola.2011-026

Terada H, Katata G, Chino M, Nagai H (2012) Atmospheric discharge and dispersion of radionuclides during the Fukushima Dai-ichi Nuclear Power Plant accident, part II: verification of the source term and analysis of regional-scale atmospheric dispersion. J Environ Radioact 112:141–154

Tomita H, Satoh M (2004) A new dynamical framework of nonhydrostatic global model using the icosahedral grid. Fluid Dyn Res 34:357–400

Tsuruta H, Oura Y, Ebihara M, Ohara T, Nakajima T (2014) First retrieval of hourly atmospheric radionuclides just after the Fukushima accident by analyzing filter-tapes of operational air pollution monitoring stations. Sci Rep 4:6717. doi:10.1038/srep06717

Uchida J, Mori M, Suzuki K, Satoh M, Nakamura H, Nakajima T (2015) Error and energy budget analysis of a non-hydrostatic stretched-grid global circulation model. Mon Weather Rev 144:1423–1447. doi:10.1175/MWR-D-15-0271.1

Wang W, Bruyère C, Duda M, Dudhia J, Gill D, Lin H-C, Michalakes J, Rizvi S, Zhang X, Beezley JD, Coen JL, Mandel J (2012) User's guide: Advanced Research WRF (ARW) Version 3.3 modeling system. http://www.mmm.ucar.edu/wrf/users/. Accessed 12 June 2016

Yasunari TJ, Stohl A, Hayano RS, Burkhart JF, Eckhardt S, Yasunari T (2011) Cesium-137 deposition and contamination of Japanese soils due to the Fukushima nuclear accident. PNAS 108:19530–19534. doi:10.1073/pnas.1112058108

Yumimoto K, Morino Y, Ohara T, Oura Y, Ebihara M, Tsuruta H, Nakajima T (2016) Inverse modeling of the 137Cs source term of the Fukushima Dai-ichi Nuclear Power Plant accident constrained by a deposition map monitored by aircraft. J Environ Radioact 164:1–12

5

Land-sea surface air temperature contrast on the western coast of Sumatra Island during an active phase of the Madden-Julian Oscillation

Peiming Wu[1*], Shuichi Mori[1] and Fadli Syamsudin[2]

Abstract

The land-sea surface temperature contrast on the western coast of Sumatra Island was examined using observation data obtained from the pre-Years of the Maritime Continent (YMC) field campaign from November to December 2015. Surface observations showed that, on most days, strong daytime solar radiation caused a pronounced diurnal cycle in surface air temperatures on the island, even during the local active phase of the Madden-Julian Oscillation (MJO). Sudden drops in surface air temperature occurred frequently on the island in the late afternoon and over the sea at nighttime, accompanied by precipitation. Temperatures on the island were higher than those over the sea during the daytime and lower in the night and early morning. Prior to the local active phase of the MJO, dual maxima in the land-sea surface air temperature contrast occurred in the evening and early morning. During the local active phase of the MJO, in spite of cloudy conditions, there were still large land-sea temperature contrasts during the daytime and in the early morning. In addition to the nighttime radiative cooling of the land surface, decreases in air temperature over the land due to precipitation cooling and the lower solar insolation in the MJO active phase caused the larger temperature differences in the morning. These results suggest that the decrease in air temperature caused by precipitation cooling had a substantial effect on the land-sea surface air temperature contrast on the western coast of Sumatra Island, particularly during an active phase of the MJO.

Keywords: Air temperature, Land-sea temperature contrast, Precipitation, Land and sea breeze, Sumatra Island, Diurnal cycle, Madden-Julian Oscillation (MJO)

Introduction

The Maritime Continent consists of many large and small islands with a complex topography, surrounded by warm and shallow oceans. The difference in the heat capacity between the sea and land can have a great impact on the weather due to the formation of sea and land breezes. The most fundamental mode of atmospheric variability over the Maritime Continent is the diurnal cycle, which is generated by the land-sea temperature contrast (Hara et al. 2009; Yamanaka 2016). The Madden-Julian Oscillation (MJO) is the largest modulator of atmospheric intraseasonal variability in the tropics. The MJO has been demonstrated to influence weather over the Maritime Continent at various spatial and temporal scales.

The MJO influences tropical convection by providing an envelope of enhanced or suppressed convection. When an active phase of the MJO propagates eastwards from the Indian Ocean to the Maritime Continent, the MJO's influence on the diurnal cycle of convection causes considerable variations in precipitation over the islands and their surroundings (e.g., Peatman et al. 2014; Birch et al. 2016, Wu et al. 2017). Both the amplitude and phase of the diurnal cycle of rainfall on the island are modulated by the category of the MJO (Rauniyar and Walsh 2011). Using satellite-derived brightness, temperature, and rainfall, Peatman et al. (2014) showed that MJO propagation throughout the Maritime

* Correspondence: pmwu@jamstec.go.jp
[1]Japan Agency for Marine-Earth Science and Technology, 2-15 Natsushima-cho, Yokosuka-City, Kanagawa 237-0061, Japan
Full list of author information is available at the end of the article

Continent is not smooth, with gaps in the MJO envelopes aligned with the islands. Based on numerical simulations, Birch et al. (2016) explained the reasons for the differences in the mean rainfall anomaly over the land and sea by scale interactions between the MJO-related large-scale environment and mesoscale circulation.

The mesoscale circulation, that is, land-sea breezes, is a result of the differences in air temperature between the land and the adjacent sea that are driven by the unequal heating and cooling of land and sea surfaces. To study the scale interaction between the large-scale circulation and the thermally induced local circulations, observation data over both the islands and the seas are necessary. However, most of the previous observations of the diurnal cycles of wind and cloud activity in the Maritime Continent region have been conducted only on the islands (e.g., Wu et al. 2003; Araki et al. 2006; Wu et al. 2008). There are insufficient observation data over the sea to verify the results of numerical simulations reported in previous studies. Exactly how the MJO interacts with local processes, such as sea and land breeze circulations, to influence precipitation over the region is not well understood.

To improve our understanding of convection over the Maritime Continent and its relation to the MJO and local atmospheric circulations, we performed intensive observations as part of the pre-Years of the Maritime Continent (YMC) field campaign on the western coast of Sumatra Island from November to December 2015. Land-based observations at Bengkulu on the western coast and ship-based observations over the sea off the western coast of the island were conducted simultaneously. A surface meteorological dataset with high temporal resolution was constructed for both the land and sea from these observations. The aims of the present study were therefore to examine the variations of air temperature over the western coast of Sumatra Island and to clarify the land-sea surface temperature contrast in an active phase of the MJO, using the observation data obtained from the field campaign.

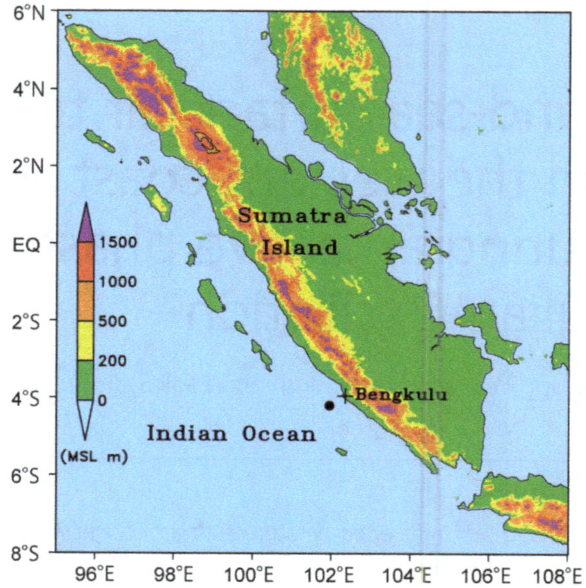

Fig. 1 Topography of Sumatra Island and its surroundings and the locations of Bengkulu Meteorological Observatory (cross) and the research vessel Mirai (closed circle). Shading denotes terrain elevation. MSL is meters above sea level

Methods/Experimental

Land-based observations were conducted at Bengkulu on the western coast of Sumatra Island, and ship-based observations were conducted over the sea 50 km off the western coast (Fig. 1). Surface meteorological observations, radar observations, and balloon soundings at intervals of 3 h were performed at Bengkulu Meteorological Observatory for 47 successive days from 9 November to 25 December 2015. The observation site is located on the coastal plain about 4 km from the shore, at an elevation of 16 m above sea level. Surface meteorological variables, including atmospheric pressure, temperature, relative humidity, wind direction and speed, rainfall, and global solar radiation, were measured at intervals of 1 min using an automatic weather station (MAWS 201, Vaisala, Finland). Rainfall was measured by a tipping bucket rain gauge, with a sensitivity of 1 tip per 0.2 mm.

Stationary ship observations were made by the research vessel Mira at a fixed point 4° 04′ 00″ S, 101° 54′ 00″ E from 23 November to 17 December 2015. A C-band meteorological radar was used, and 3-hourly radiosonde observations were conducted. Surface meteorological parameters were measured by the Mirai Surface Meteorological observation (SMet) system and the Shipboard Oceanographic and Atmospheric Radiation (SOAR) measurement system. Air temperature and relative humidity were measured by thermometer sensors (HMP155, Vaisala, Finland) installed in the compass deck at 21 m altitude from the sea surface. Rainfall rate was measured by a capacitive rain gauge (Model-50202, R. M. Young, USA) and an optical rain gauge (ORG-815DS, Osi, USA) on the compass deck at 19 m altitude from the sea surface. Short- and long-wave downward radiations were measured by a radiometer (MS-802, Eko Seiki, Japan) on the radar mast. Sea surface temperature was measured by a thermometer (RFN2-0, Koshin Denki, Japan) at an inlet of 5 m depth from the sea surface. Data were collected and processed by a KOAC-7800 weather data processor made by Koshin-Denki, Japan. Details of the ship-based observations can be found in the MIRAI MR15-04 Cruise report (JAMSTEC 2016). The present study used data for the sea surface and sea air temperatures, rainfall, and global solar radiation obtained from the observations.

Results and discussion

Diurnal variations of surface air temperature, rainfall, and solar radiation on the western coast of Sumatra Island

The oceanic and atmospheric features in December 2015 indicated strong and mature El Niño conditions (e.g., Deser and Wallace 1990). Sea surface temperatures were well above average across the central and eastern equatorial Pacific. Widespread enhanced convection developed across the central Pacific and suppressed convection over the Maritime Continent. Meanwhile, a positive Indian Ocean Dipole (IOD) event (Saji et al. 1999) occurred during the boreal summer and autumn in 2015 and then weakened rapidly. The Dipole Mode Index indicated a transition from the positive to negative IOD phase in December 2015. Sea surface temperatures were close to normal across the tropical Indian Ocean; seas to the west of Sumatra Island were warm, as usual.

During early December 2015, enhanced convection developed over the central Indian Ocean, while suppressed convection persisted over the Maritime Continent. Then, in mid-December, the MJO signal overcame the El Niño-Southern Oscillation background state over the Maritime Continent region. Enhanced convection developed and shifted east across the Maritime Continent and the west-central Pacific, indicating a steady eastward propagation of the robust MJO signal. The MJO index (Wheeler and Hendon 2004) showed that the enhanced convective phase was centered on the western Maritime Continent from 13 to 22 December 2015 (Wu et al. 2017).

The time series of hourly total rainfall, hourly averaged surface air temperature, and global solar radiation during the period from 23 November to 17 December 2015 are shown in Fig. 2. Rainfall occurred at Bengkulu on 17 days during the 25-day observation period, i.e., on roughly 2 out of every 3 days. Prior to the local active phase of the MJO until 12 December, most of the rainfall occurred over a short time period of 1 to 2 h during the late afternoon and evening. Meanwhile, a daily maximum hourly mean global solar radiation greater than 800 W m^{-2} was observed on most days, although on 8 December, daily maximum values of less than 400 W m^{-2} were recorded. On days with rain, strong solar radiation was still observed because rainfall occurred mostly during the late afternoon and evening.

The typical diurnal cycle of surface air temperature and precipitation at Bengkulu on the western coast of Sumatra Island is shown for 25 November and 1, 3, and 15 December 2015 at time intervals of 1 min in Fig. 3. The daily minimum air temperature was observed at about 0600 LT on these days, and the temperatures increased rapidly from about 0700 LT, i.e., soon after sunrise. The daily maximum air temperature occurred during the several hours before and after noon. The air temperature remained high during the daytime, with short time fluctuations. Then, a sudden drop in the air temperature occurred in the late afternoon, starting about 10 min before precipitation was observed at the observation site. The air temperature decreased rapidly from about 29 to 24 °C in under 30 min, and thereafter, the air temperature remained at around 24 °C during and after the precipitation. Daily maximum air temperatures above 30 °C and daily minimum air temperatures below 24 °C were observed during the observation period.

November and December are in the annual rainy season of the region (Hamada et al. 2002). In the mid-latitudes, diurnal variations in surface air temperature are usually weak during the rainy season because cloud cover prevents the land from heating rapidly in the daytime and cooling rapidly at night. However, the situation in the near-equatorial area of western Sumatra Island is unique. As described previously, before the active phase of the MJO, rainfall on the western coast of the island occurs mostly over a short time in the late afternoon and evening. The sky is almost cloudless over most areas of the island in the morning and early afternoon, even on days with heavy rain. Strong solar radiation was observed on most days, with the result that surface air temperature increased greatly during the morning hours. During the local active phase of the MJO, i.e., 13 to 17 December 2015 (except for 16 December), surface air temperatures at Bengkulu still displayed a clear diurnal cycle, with a maximum occurring during the several hours before and after noon.

Previous studies have shown that when the MJO main convective envelope is over the Indian Ocean, rainfall and its diurnal cycle over the land in the Maritime Continent reach their maxima due to an increase in the atmospheric instability caused by strong solar insolation in the clearer skies and moistening environment (e.g., Fujita et al. 2011; Peatman et al. 2014; Birch et al. 2016). During the daytime, the upslope winds driven by diurnal heating at the slope surface of the mountains on Sumatra Island reinforce the sea breeze, with the result that a notable thermally induced local circulation develops on the western coast on a daily basis, frequently resulting in convection over the island at similar times in the afternoon (Wu et al. 2003; Wu et al. 2009). Wu et al. (2017) examined the occurrence and movement of convection over the island prior to and during an active phase of the MJO in December 2015, using radar observation data obtained from the pre-YMC field campaign. Their results showed that prior to the active phase of the MJO in early December, convection with heavy rain occurred regularly over Sumatra Island in the afternoon and migrated westward to the sea west of the island in the evening. Results from surface meteorological observations

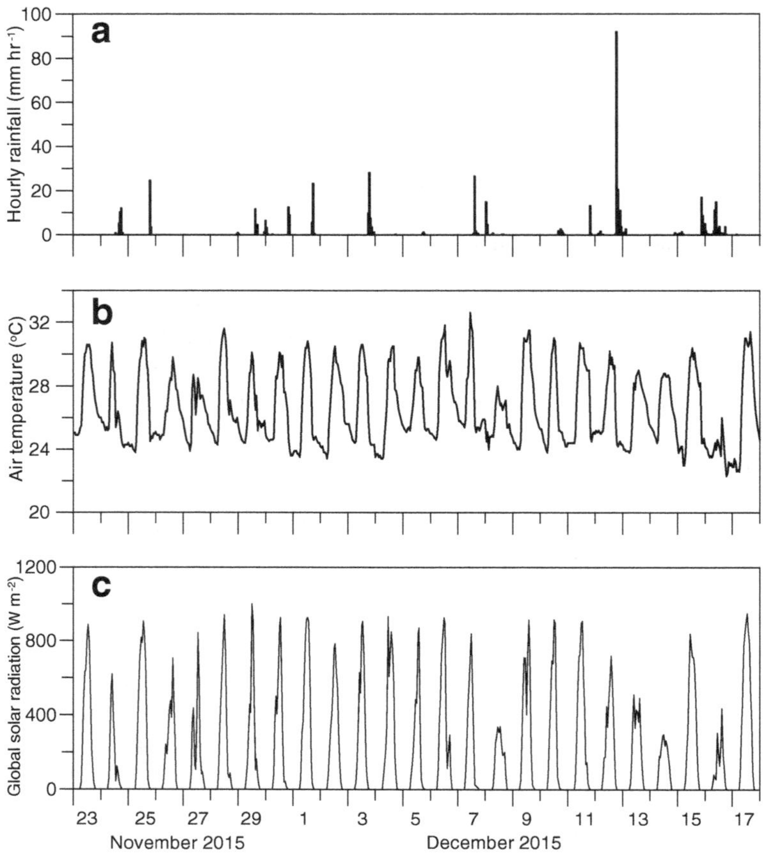

Fig. 2 Time series of **a** hourly total rainfall, **b** hourly averaged surface air temperature, and **c** hourly averaged global solar radiation observed at Bengkulu Meteorological Observatory (3°51′30″ S 102°20′11″ E) on Sumatra Island from 23 November to 17 December 2015. The tick marks in the abscissa indicate 00 LT for each of the days within the time period shown

in the present study showed strong solar radiation and the regular occurrence of precipitation on the island in the late afternoon prior to the beginning of the local active phase of the MJO until 12 December (except for 8 December), which is consistent with the results reported in previous studies. After 13 December, during the passage of the active phase of the MJO, convection frequently shifted eastward from the Indian Ocean to the island, whereas afternoon convection over the island was suppressed because of strong prevailing winds and lower solar insolation. The daily average amount of rainfall observed at Bengkulu during the local active phase of the MJO was less than that before its arrival.

Summarizing the results of this section, during the study period, strong solar radiation caused a clear diurnal cycle in surface air temperatures on most days on the western coast of Sumatra Island, even during an active phase of the MJO. The daily maximum air temperature occurred during the several hours before and after noon, with the minimum occurring in the early morning, before sunrise.

Diurnal variations of sea surface temperature, surface air temperature, rainfall, and solar radiation over the sea off the western coast of Sumatra Island

The time series of hourly total rainfall, hourly averaged sea surface temperature (SST), surface air temperature, and global solar radiation observed over the sea onboard the research vessel Mirai from 23 November to 17 December 2015 are shown in Fig. 4. Hourly mean sea surface temperatures greater than 29 °C were observed throughout the observation period, with an average SST of 29.7 °C over the whole observation period. Before the active phase of the MJO, i.e., until 12 December, there were continuous fluctuations in the SST, and an hourly SST greater than 30 °C was frequently recorded. Subsequently, in the local active phase of the MJO, the SST decreased gradually from 29.6 °C on 13 December to 29.3 °C on 16 December. The decrease in sea surface temperature was attributed to the strong prevailing winds and lower solar insolation during this period. During the passage of the active MJO over the Maritime Continent from 13 to 22 December, westerly wind bursts

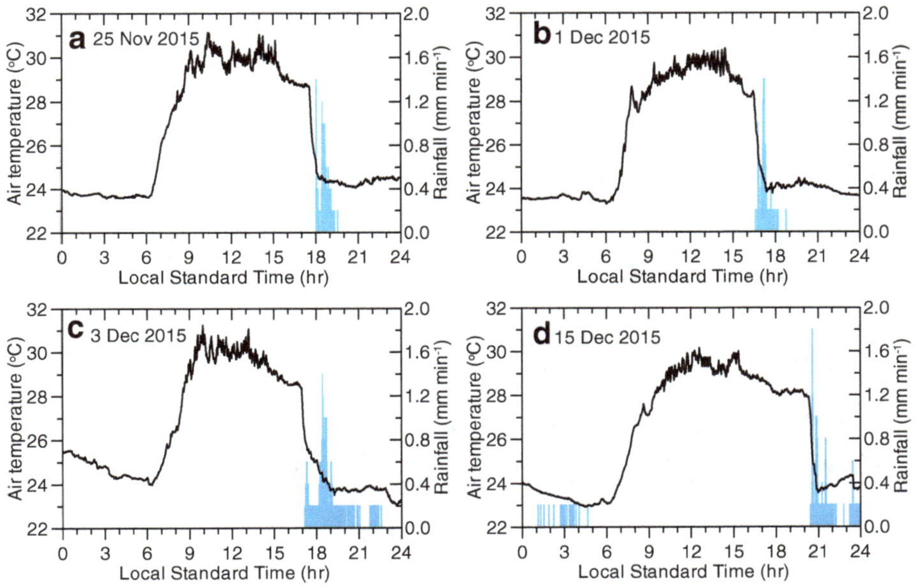

Fig. 3 Time series of surface air temperature and precipitation observed at Bengkulu Meteorological Observatory on Sumatra Island for **a**) 25 November, **b**) 1, **c**) 3 and **d**) 15 December 2015 at intervals of 1 min

were observed in the region (Wu et al. 2017). These strong winds increased evaporation from the sea surface, which in turn led to a decrease in the SST. Meanwhile, a lower than normal daily maximum hourly mean global solar radiation was recorded from 13 to 16 December (except for 15 December), due to the cloudy conditions in the active phase of the MJO. On 17 December, the SST increased during the daytime due to the strong solar radiation and weaker near-surface wind. On 15 December, solar radiation was relatively strong while SST remained low, which suggests that strong winds may contribute greatly, although detailed evaluation on the relative roles of solar radiation and wind is necessary for further verification.

Rainfall was observed onboard the research vessel Mirai on 21 days during the observation period of 25 days. Prior to the local active phase of the MJO, most of the rainfall occurred near midnight over a period of 1 to 2 h. The daily maximum rainfall was observed around 2300 LT on average, about 4 h later than at Bengkulu on the western coast of Sumatra Island. Strong solar radiation was observed on most days, even those with heavy rain. On days with rain, strong solar radiation was still observed because rainfall was concentrated mostly after sunset. On these days, there was a daily maximum hourly mean air temperature of around 29 °C, about 1.0 °C lower than that over the land.

The temporal variations of surface air temperature and precipitation observed on the research vessel Mirai off the western coast of Sumatra Island for four consecutive days, from 10 to 13 December 2015, are shown in Fig. 5. From Fig. 5, it can be seen that a sudden drop in air temperature occurred from the late afternoon to nighttime prior to the local active phase of the MJO, accompanied by the occurrence of precipitation at the observation site. After the rain stopped, air temperatures increased gradually. Then, the air temperatures dropped again when the next rainfall event occurred. If the drop in air temperatures was caused by radiative cooling of the sea surface, the decrease in air temperatures should have continued until the following morning, because radiative cooling would continue throughout the nighttime. The increase in air temperature was due to the air being heated by the warm water, because during the nighttime, there was no solar irradiance to heat the sea surface. On 13 December 2015 when the active phase of the MJO passed over the Maritime Continent, a drop in surface air temperature occurred during the daytime, corresponding to the occurrence of precipitation. Therefore, it is reasonable to conclude that the variations of surface air temperature over the sea were caused by a drop in temperature associated with precipitation, and the subsequent increase in temperature resulted from heating by the warm water.

Figure 6 shows a comparison of the diurnal changes in the surface air temperature observed at Bengkulu on the western coast and over the sea onboard the research vessel Mirai, before and during the local active phase of the MJO. Surface air temperatures on the island displayed a clear diurnal cycle both before and during the active phase of the MJO. In contrast, air temperatures over the sea varied little within a day, with the lowest temperatures

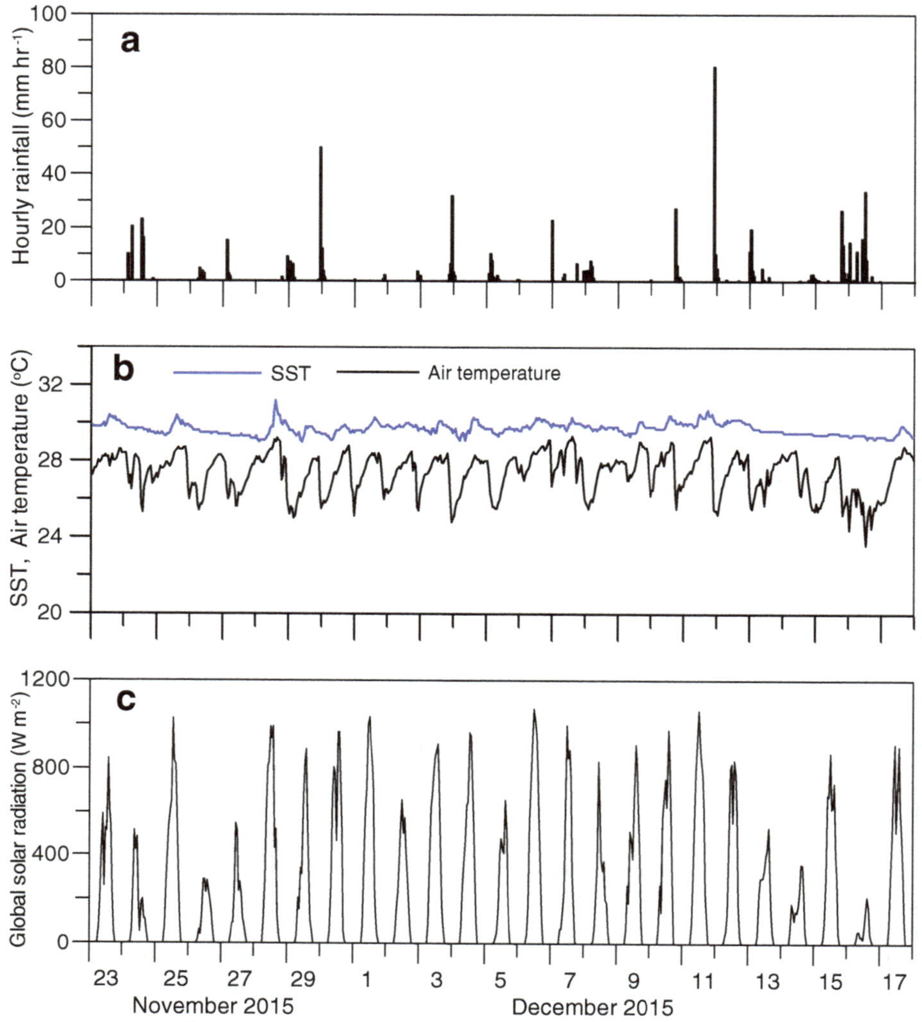

Fig. 4 Time series of **a** hourly total rainfall, **b** hourly averaged sea surface and sea air temperature (depth = 5 m), and **c** hourly averaged global solar radiation observed by the research vessel Mirai. Stationary ship observations were carried out at 4°04′00″ S 101°54′00″ E, 50 km off the western coast of Sumatra Island from 23 November to 17 December 2015. The tick marks in the abscissa indicate 00 LT for each of the days within the time period shown

occurring near midnight. Daily mean air temperatures decreased during the local active phase of the MJO, both over the land and sea. However, over the land, air temperatures in the evening did not decrease during the local active phase of the MJO.

The decrease in air temperature over the sea in the afternoon and evening during the active MJO corresponds to the increased rainfall in the afternoon and evening. Wu et al. (2008) investigated the diurnal cycle of convection over western Kalimantan (Borneo) Island and the adjacent seas. Their results suggest that the evaporative cooling from convection associated with rain, which frequently occurs in the afternoon on the island, causes a rapid drop in air temperature in the lower atmosphere in the evening. This produces a strong temperature contrast between the land and the sea, with lower temperatures occurring over the island. As mentioned previously, before the onset of the active phase of the MJO, rainfall on Sumatra Island frequently occurred over a short period of time in the late afternoon and evening. Similar decreases in air temperature in the evening caused by evaporative cooling associated with rain occurred over the western coast of Sumatra Island (Wu et al. 2009). Therefore, the rapid drop in air temperature over the sea at nighttime can be attributed to precipitation cooling of the lower atmosphere.

To summarize, the air temperature decreased over both the land and sea during an active phase of the MJO. However, air temperatures during the early morning hours over the sea varied little before and during the active MJO. A sudden drop in air temperature frequently

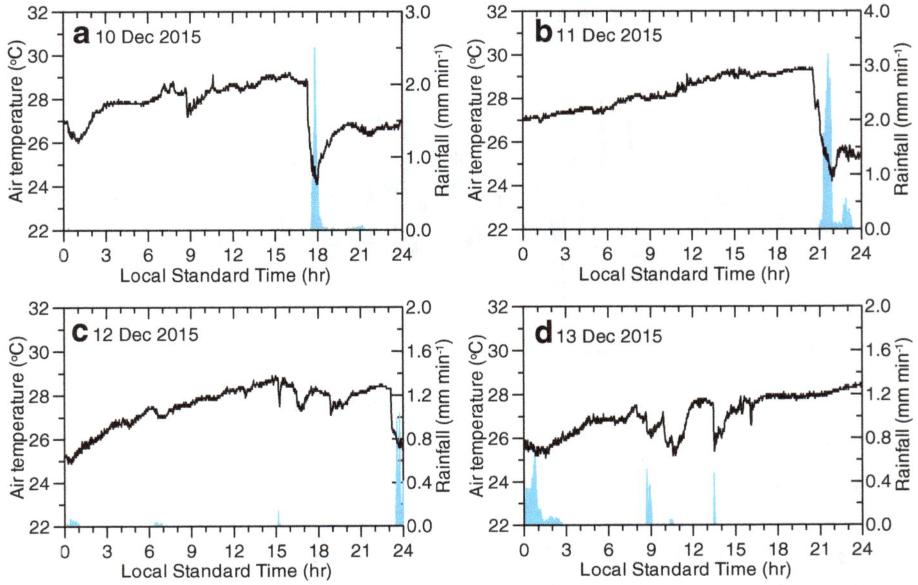

Fig. 5 Time series of surface air temperature and precipitation off the western coast of Sumatra Island observed by the research vessel Mirai for four consecutive days from **a**) 10, **b**) 11, **c**) 12 and **d**) 13 December 2015 at intervals of 1 min

occurred over the sea at nighttime and was caused by precipitation cooling.

Surface air temperature differences between the land and the sea on the western coast of Sumatra Island before and during an active phase of the MJO

A time series of air temperature differences between the land (at Bengkulu) and the adjacent sea (onboard the research vessel Mirai) from 23 November to 17 December 2015 is shown in Fig. 7. Although air temperature decreased during the local active phase of the MJO (Figs. 2, 4, and 6), the land-sea air temperature contrasts during the active MJO (13–17 December) were almost the same as those before the onset of the active MJO. Except for the 2 days of 8 and 16 December, higher temperatures were observed on the island than over the sea during the daytime. On these 2 days, rain occurred on the island in the morning and continued throughout the afternoon, with a daily maximum hourly mean global solar radiation of less than 500 W m^{-2} on the island (Fig. 2). The cloudy conditions prevented the land from heating up rapidly during the daytime, and the sustained

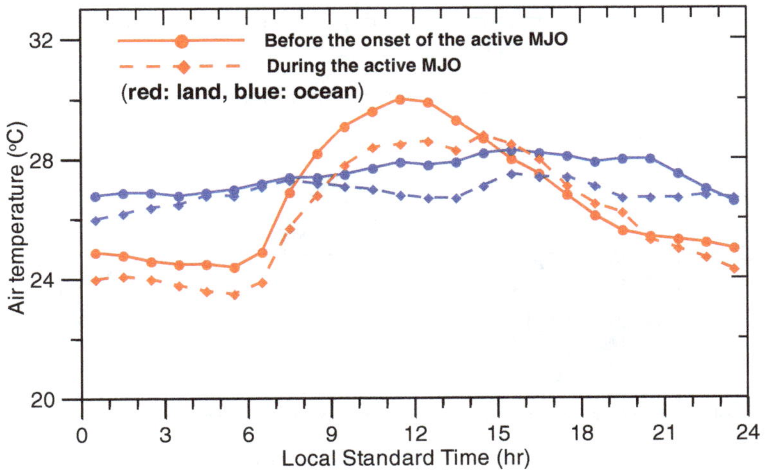

Fig. 6 A comparison between the diurnal changes in the hourly mean air temperature observed at Bengkulu on the western coast (red lines) and over the sea on the research vessel Mirai (blue lines), before the onset of the active MJO (solid lines) and averaged over 23 November to 12 December and during the active MJO (broken lines) averaged over 13 to 17 December 2015

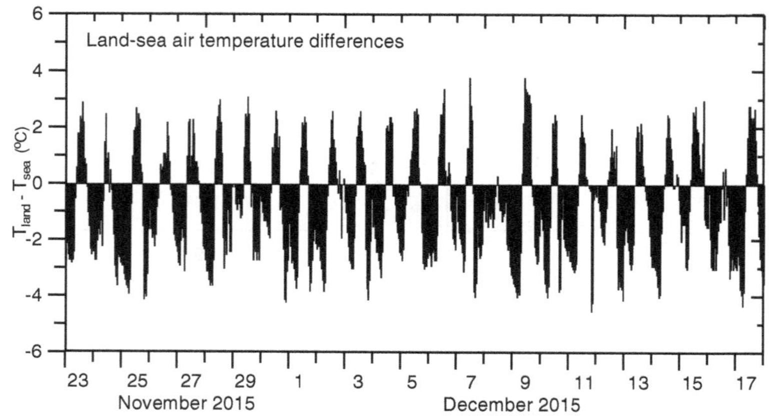

Fig. 7 Time series of air temperature differences between the land (at Bengkulu) and the adjacent sea (on the research vessel Mirai) from 23 November to 17 December 2015. Positive values indicate higher temperatures on the land, and negative values indicate higher temperatures over the sea. The tick marks in the abscissa indicate 00 LT for each of the days within the time period shown

rain cooled the land surface. On the other hand, a higher temperature over the sea occurred from night until the early morning on all days during the observations.

Figure 8 shows a comparison of the diurnal changes in air temperature differences between the land and sea, before and during the local active phase of the MJO. It can be seen that before the onset of the active MJO, a dual maxima of higher surface air temperatures over the sea compared to over the land occurred in the evening and early morning. As described previously, before the local active phase of the MJO, rainfall frequently occurred on the island in the late afternoon and evening.

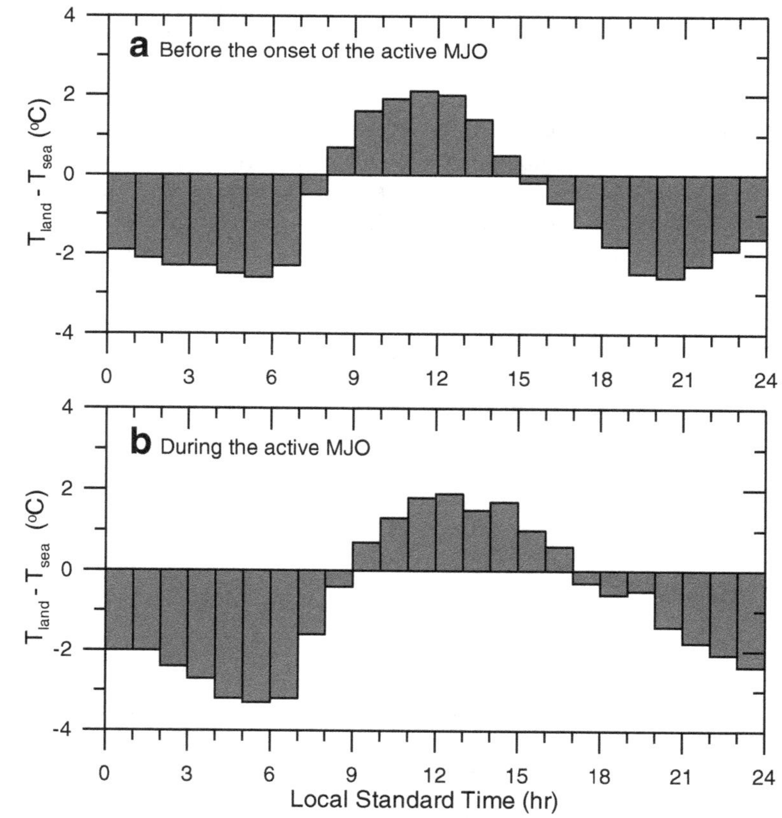

Fig. 8 A comparison of the diurnal changes in surface air temperature differences between the land and sea, **a** before the onset of the active MJO averaged over 23 November to 12 December and **b** during the active MJO averaged over 13 to 17 December 2015

The large decreases in air temperature in the evening on land were caused by rain evaporative cooling of air in the low troposphere and cooling of the land surface by precipitation (Wu et al. 2008; Wu et al. 2009). As a result, there was a higher maximum air temperature over the sea than over the land in the evening. On the other hand, before the onset of the active MJO, the higher maximum temperatures over the sea than over the land in the early morning were considered to result from the nighttime radiative cooling of the land surface.

During the active phase of the MJO, in spite of the cloudy conditions, there were still large temperature contrasts during the daytime and in the morning. The temperature difference in the morning during the active phase of the MJO was even larger than that before the onset of the active MJO, i.e., 3.2 and 2.6 °C, respectively. In an active phase of the MJO, enhanced convection develops. The cloudy sky would then prevent the land from heating up rapidly in the daytime and cooling off rapidly at night. Therefore, it would be expected that the air temperature contrast between the land and the sea would be weaker. This is true in the mid-latitudes, where land-sea temperature differences are the result of daytime solar heating and nighttime radiative cooling of the land surface. However, the situation in the near-equatorial area of western Sumatra Island is a little different. As shown in Fig. 6, air temperatures on the land from late night to early morning during the active MJO were lower than before the onset of the active MJO, over both the land and the sea, due to precipitation cooling and the lower daytime solar insolation. However, there was still a clear diurnal cycle in the air temperatures over the land in the active MJO. On the other hand, because of the higher heat capacity of water, sea surface temperatures greater than 29 °C continued during the active phase of the MJO (Fig. 4). Air temperatures over the sea changed little during the early morning hours in the active MJO while air temperatures decreased over the land. As a consequence, the temperature differences in the morning during the active MJO were larger than those before the onset of the active MJO.

The large temperature differences between the land and sea with lower temperatures occurring on the island during the night to early morning in an active phase of the MJO would induce strong offshore land breezes on the coast, because the strength of the land-sea breezes is directly proportional to the temperature difference between the land and the sea. A stronger offshore land breeze during the active MJO compared to before the onset of the active MJO was observed in the early morning (not shown) during the pre-YMC field campaign, which probably caused wind divergences in the lower troposphere, suppressing convection over the island. Further studies based on numerical experiments with the aid of atmospheric circulation models are needed to reveal the detailed physical mechanism by which the MJO interacts with local processes, such as the sea and land breeze circulations, to influence precipitation over the region.

Conclusions

This study investigated surface air temperature over the land and adjacent sea of the western coast of Sumatra Island and the land-sea surface temperature contrast before and during an active phase of the MJO using observation data obtained from the pre-YMC field campaign from November to December 2015. Over the land, although air temperatures decreased during the local active phase of the MJO, on most days, solar insolation enabled a clear diurnal cycle in air temperature to develop. A sudden drop in surface air temperature frequently occurred on the island in the late afternoon and over the sea at nighttime, due to precipitation cooling. While there was a higher temperature over the island than over the sea during the daytime on most of the days, there was a lower temperature over the island from the night to early morning.

Before an active phase of the MJO, a dual maxima of higher surface air temperatures over the sea compared to over the land occurred in the evening and early morning. The higher air temperature over the sea than over the land in the evening was caused by a decrease in air temperature over the land due to precipitation cooling. Then, during the local active phase of the MJO, under cloudy conditions, there were still large temperature contrasts during the daytime and in the early morning. Unlike in the mid-latitudes, in the near-equatorial area in addition to the nighttime radiative cooling of the land surface, the decrease in air temperature over the land due to precipitation cooling and the lower solar insolation in the local active phase of the MJO caused large land-sea temperature contrasts in the morning. These results suggest that the decrease in air temperature caused by precipitation cooling had a substantial effect on the land-sea surface air temperature contrast on the western coast of Sumatra Island, particularly during an active phase of the MJO.

This study examined the land-sea surface temperature contrast on the western coast of Sumatra Island using surface observation data obtained from the pre-YMC field campaign. Evaporative cooling associated with precipitation is expected to produce deeper layers of cool air in the lower troposphere than radiative cooling of the land surface, which will cause stronger land-sea breezes. Therefore, further studies are needed to elucidate the profile of the land-sea temperature contrast in the lower troposphere caused by precipitation cooling using balloon sounding data from the pre-YMC field campaign.

Abbreviations

BPPT: Indonesian Agency for Assessment and Application of Technology; JAMSTEC: Japan Agency for Marine-Earth Science and Technology; MJO: Madden-Julian Oscillation; SST: Sea surface temperature; YMC: Years of the Maritime Continent

Acknowledgements

We thank Dr. Kunio Yoneyama, Director of the Department of Coupled Ocean-Atmosphere-Land Processes Research of JAMSTEC, for his supervision and encouragement in this study. We also thank Mr. Agus Lacuda, the head of the Meteorological Station of Fatmawati Soekarno Airport in Bengkulu, Indonesia, for his strong support in the observations of this study.

Funding

This work was supported by the JAMSTEC pre-YMC project. PW was partly funded by JSPS KAKENHI Grant Number 15K01167.

Authors' contributions

PW proposed the topic and conceived and designed the study. SM and PW carried out the observations. FS and SM led the observations and collaborated with the corresponding author in the construction of manuscript. All authors read and approved the final manuscript.

Authors' information

PW is a Senior Scientist of the Department of Coupled Ocean-Atmosphere-Land Processes Research (DCOP), Japan Agency for Marine-Earth Science and Technology (JAMSTEC). SM is the Vice Director and a Principal Technical Scientist of DCOP, JAMSTEC. FS is a Principal Investigator at the Center of Regional Resources Development Technology (CRRDT), Agency for the Assessment and Application of Technology (BPPT), Indonesia.

Competing interests

The authors declare that they have no competing interests.

Author details

[1]Japan Agency for Marine-Earth Science and Technology, 2-15 Natsushima-cho, Yokosuka-City, Kanagawa 237-0061, Japan. [2]Indonesian Agency for Assessment and Application of Technology, Jl. Raya Puspiptek, South Tangerang 15314, Indonesia.

References

Araki R, Yamanaka MD, Murata F, Hashiguchi H, Oku Y, Sribimawati T, Kudsy M, Renggono F (2006) Seasonal and interannual variations of diurnal cycles of wind and cloud activity observed at Serpong, West Jawa, Indonesia. J Meteor Soc Japan 84A:171–194. https://doi.org/10.2151/jmsj.84A.171

Birch C, Webster S, Peatman SC, Parker D, Matthews AJ, Li Y, Hassim M (2016) Scale interactions between the MJO and the western Maritime Continent. J Clim 29:2471–2492. https://doi.org/10.1175/JCLI-D-15-0557.1

Deser C, Wallace JM (1990) Large-scale atmospheric circulation features of warm and cold episodes in the tropical Pacific. J Clim 3:1254–1281

Fujita M, Yoneyama K, Mori S, Nasuno T, Satoh M (2011) Diurnal convection peaks over the Eastern Indian Ocean off Sumatra during different MJO phases. J Meteor Soc Japan 89A:317–330

Hamada JI, Yamanaka MD, Matsumoto J, Fukao S, Winarso PA, Sribimawati T (2002) Spatial and temporal variations of the rainy season over Indonesia and their link to ENSO. J Meteor Soc Japan 80:285–310

Hara M, Yoshikane T, Takahashi HG, Kimura F, Noda A, Tokioka T (2009) Assessment of the diurnal cycle of precipitation over the Maritime Continent simulated by a 20 km mesh GCM using TRMM PR data. J Meteor Soc Japan 87A:413–424. https://doi.org/10.2151/jmsj.87A.413

JAMSTEC (2016) R/V Mirai Cruise Report MR15-04: The observational study of the heavy rainfall zone in the eastern Indian Ocean. http://www.godac.jamstec.go.jp/catalog/data/doc_catalog/media/MR15-04_all.pdf. Accessed 5 Feb 2016

Peatman SC, Matthews AJ, Stevens DP (2014) Propagation of the Madden-Julian Oscillation through the Maritime Continent and scale interaction with the diurnal cycle of precipitation. Q J R Meteorol Soc 140:814–825. https://doi.org/10.1002/qj.2161

Rauniyar P, Walsh K (2011) Scale interaction of the diurnal cycle of rainfall over the Maritime Continent and Australia: influence of the MJO. J Clim 24:325–348. https://doi.org/10.1175/2010JCLI3673.1

Saji NH, Goswami BN, Vinayachandran PN, Yamagata T (1999) A dipole mode in the tropical Indian Ocean. Nature 401:360–363. https://doi.org/10.1038/43855

Wheeler M, Hendon H (2004) An all-season real-time multivariate MJO index: development of an index for monitoring and prediction. Mon Wea Rev 132: 1917–1932

Wu P, Hamada J, Mori S, Tauhid YI, Yamanaka MD, Kimura F (2003) Diurnal variation of precipitable water over a mountainous area of Sumatra Island. J Appl Meteorol 42:1107–1115

Wu P, Yamanaka MD, Matsumoto J (2008) The formation of nocturnal rainfall offshore from convection over western Kalimantan (Borneo) Island. J Meteor Soc Japan 86A:187–203

Wu P, Hara M, Hamada J, Yamanaka MD, Kimura F (2009) Why a large amount of rain falls over the sea in the vicinity of western Sumatra Island during nighttime. J Appl Meteor Climatol 48:1345–1361. https://doi.org/10.1175/2009JAMC2052.1

Wu P, Ardiansyah D, Yokoi S, Mori S, Syamsudin F, Yoneyama K (2017) Why torrential rain occurs on the western coast of Sumatra Island at the leading edge of the MJO westerly wind bursts. SOLA 13:36–40. https://doi.org/10.2151/sola.2017-007

Yamanaka MD (2016) Physical climatology of Indonesian maritime continent: an outline to comprehend observational studies. Atmos Res 178-179:231–259. https://doi.org/10.1016/j.atmosres.2016.03.017

6

The dynamics of the mesosphere and lower thermosphere

Robert A Vincent

Abstract

The dynamics of the mesosphere-lower thermosphere (MLT) (60 to 110 km) is dominated by waves and their effects. The basic structure of the MLT is determined by momentum deposition by small-scale gravity waves, which drives a summer-to-winter pole circulation at the mesopause. Atmospheric tides are also an important component of the dynamics of the MLT. Observations from extended ground-based networks, satellites as well as numerical modelling show that non-migrating tidal modes in the MLT are more important than previously thought, with evidence for directly coupling into the thermosphere/ionosphere. Major disturbances lower in the atmosphere, such as wintertime sudden stratospheric warmings, temporarily disrupt the circulation pattern and thermal structure of the MLT. In the equatorial mesosphere, gravity wave driving leads to oscillations in the zonal wind on semiannual time scales, although variability on quasi-biennial time scales is also apparent. Planetary-scale waves such as the quasi-two-day wave temporarily dominate the dynamics of the summertime MLT, especially in the southern hemisphere. Impacts may include short-term changes to the thermal structure and physics of the high-latitude MLT. Here, we briefly review the dynamics of the MLT, with a particular emphasis on developments in the past decade.

Keywords: Atmospheric tides; Gravity waves; Planetary waves; Middle atmosphere; Mesosphere; Lower thermosphere; Wave coupling

Review

Introduction

The mesosphere-lower thermosphere (MLT) is defined as the region of the atmosphere between about 60 and 110 km in altitude. It constitutes the upper part of what is often referred to as the middle atmosphere (10 to 110 km). The MLT is dominated by the effects of atmospheric waves, including planetary waves, tides and gravity waves. The source regions for these waves are lower in the atmosphere. As the waves propagate upward their amplitudes grow exponentially to compensate for the decrease in atmospheric density (e.g. Andrews et al. 1987). Consequently, wave motions often dominate the wind field in the MLT, so considerable averaging may be required to extract the mean flow, especially that of the mean meridional (NS) motions, which are smaller than the zonal (EW) flow. Zonal-mean zonal winds are eastward (westward) in the winter (summer) middle atmosphere, reaching peak values of approximately 60 to 70 ms^{-1} near 70 km and then they reduce in magnitude until they reverse sign at heights between 90 and 100 km.

Large wave amplitudes lead to wave breaking and hence momentum deposition. This, in turn, produces body forces that drive large scale residual circulations. In the winter stratosphere (approximately 20 to 60 km), planetary wave breaking drives a residual circulation from the equator to the winter pole, while in the mesosphere, gravity wave breaking and dissipation drives a circulation from the summer to winter pole (e.g. see Andrews et al. 1987). The combined effects of these large-scale circulation patterns, usually referred to as the Brewer-Dobson circulation, can be profound. For example, the pole-to-pole circulation at the mesopause (approximately 85 to 90 km) leads to rising (descending) motions over the summer (winter) pole with associated adiabatic cooling (heating). These vertical motions produce temperatures that are more than 50K cooler or warmer than would be expected by radiative equilibrium alone. In fact the summer polar mesopause is the coldest place in the Earth's environment with temperatures of 120K or lower. Temperatures below

Correspondence: robert.vincent@adelaide.edu.au
Department of Physics, School of Physical Sciences, University of Adelaide, North Terrace., SA, 5005 Adelaide, Australia

about 130K are sufficient for ice particles to form, leading to a phenomena such as noctilucent clouds.

The mean pole-to-pole flow at the solstices is about 10 ms^{-1} at the mesopause, i.e. at the heights where the shears in the zonal-mean zonal winds are greatest. It is straightforward to calculate the body force required to maintain this meridional flow from the gravity wave flux divergence:

$$\overline{F_u} = -\frac{1}{\rho_o}\frac{\partial\left(\rho_o\overline{u'w'}\right)}{\partial z} = f\overline{v}. \quad (1)$$

Here, ρ_o is the atmospheric density at height z, $\rho_o\overline{u'w'}$ is the mean upward flux of zonal momentum and f is the Coriolis parameter. At the midlatitude mesopause with $\overline{v} \sim 10\ ms^{-1}$ then $|\overline{F}| \sim 100\ ms^{-1}day^{-1}$. The corresponding mean vertical motions in the polar mesosphere are inferred to be about 4 to 5 mm s^{-1} (McIntyre 1989).

Of course, at equatorial latitudes $f \sim 0$ and so, wave driving goes directly into accelerating the mean flow, $\overline{F_u} = \partial\overline{u}/\partial t$. Wave driven phenomena include the quasi-biennial-oscillation (QBO) in the zonal-mean zonal winds in the lower and middle stratosphere (Baldwin et al. 2001) and semiannual oscillations at the equatorial stratopause and mesopause.

Any major disturbances in the stratosphere may significantly modify the gravity wave (GW) fluxes and hence may lead to significant changes in the thermal and wind structure of the MLT. Such disturbances include sudden stratospheric warmings (SSW) where the winds and temperature gradients in the polar stratosphere are reversed on the time scale of days to weeks. Alternatively, other large scale waves can transfer momentum to the MLT to partially offset GW driving. One such phenomenon is the quasi-two-day wave (QTDW), which is a transient wave event that can attain large amplitudes in the MLT.

Here, we review some aspects of middle atmosphere dynamics. Limited space means the review is not exhaustive but focusses on developments in the past decade. It includes discussions on tides, SSW and their effects, the equatorial middle atmosphere, the quasi-two-day wave and on observations of gravity waves, including wave sources, and their impact on the MLT.

Atmospheric tides in the MLT

Tidal oscillations in winds and temperatures are prominent features of the MLT. Periods of 24 (diurnal) or 12 h (semidiurnal) dominate. Careful averaging allows the 8 h (terdiurnal) and even lunar tides with a period of 12.4 h to be observed. Tides are usually classified as either migrating or non-migrating. That is, for a ground-based observer migrating (sun synchronous) tides appear to move westward at the same speed as the sun while non-migrating (sun asynchronous) tides move either eastwards or westwards relative to the sun.

Early work in tidal theory is described in Chapman and Lindzen (1970). Solar heating occurs through the absorption of infrared radiation by water vapour in the lower atmosphere and of UV radiation by ozone in the stratosphere. Classical tidal theory assumes a stationary, isothermal, atmosphere without dissipation. Incorporation of more realistic properties, such as height and latitudinally varying temperatures and background winds, requires numerical techniques to solve the relevant equations. In this regard, the development of mechanistic models such as the Global Scale Wave Model (GSWM) described by Hagan et al. (1999) have proved invaluable for the interpretation of tidal observations. The GSWM continues to evolve to include better heating rates and also the effects of non-radiative heating sources, such as tropospheric latent heat release, and the effect of of clouds (Hagan and Forbes 2002; Oberheide et al. 2002; Zhang et al. 2010a,b). High vertical-resolution climate models such as the Kyushu-GCM (Yoshikawa and Miyahara 2005) and the Canadian Middle Atmosphere Model (Xu et al. 2012) are also powerful tools for studying excitation mechanisms of non-migrating tides and interaction of tides with planetary and gravity waves and the interpretation of data.

Improvements in the understanding of tides came from better and more widespread observations. The establishment of ground-based radars at both equatorial and polar latitudes enhanced the spatial/temporal coverage originally available from the predominately mid-latitude stations that prevailed before 2000 (e.g. Mitchell et al. 2002; Sridharan et al. 2010; Nozawa et al. 2012; Davis et al. 2013). New instruments, notably lidars that provide daytime observations, also came on line (e.g She et al. 2002). In some respects, however, the most notable advances resulted from the deployment of satellite instruments such as the high-resolution Doppler interferometer (HRDI) on the Upper Atmosphere Research Satellite (UARS) in the 1990's (e.g see Burrage et al. 1995) and, in the last decade, the Sounding of the Atmosphere using Broadband Emission Radiometry (SABER) on the Thermosphere, Ionosphere, Mesosphere Energetics and Dynamics (TIMED) satellite (e.g. Mukhtarov et al. 2009).

Perhaps the most interesting finding to emerge from recent modelling and observational studies is that non-migrating tides are often an important component of upper atmosphere dynamics. Figure 1 shows the zonal wind amplitude of the fundamental diurnal westward 1 (DW1) and eastward wave number 3 (DE3) tides derived from the Thermosphere Ionosphere Electrodynamics General Circulation Model (TIME-GCM) forced at is lower boundary by the GSWM in an equinox simulation (Hagan et al. 2009). DW1 propagates from the troposphere and peaks near 100 km altitude and ±25° latitude.

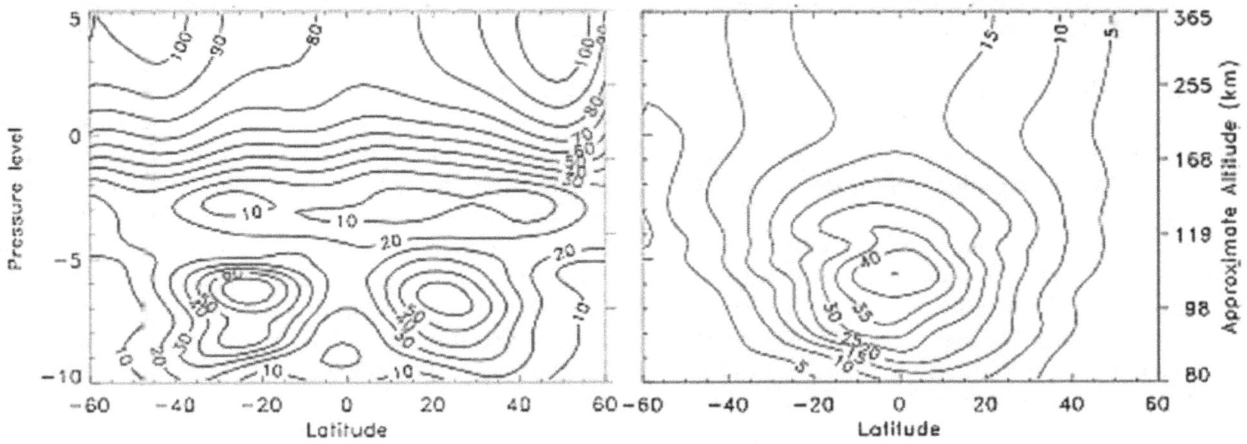

Figure 1 TIME-GCM tidal winds in the mesosphere and thermosphere. Migrating (left) and eastward propagating wave number 3 (right) diurnal zonal wind amplitudes as a function of latitude and pressure level. The scale on the left gives the approximate altitude during a September simulation. After Figure two in Hagan et al. (1999).

The DW1 component above approximately 120 km arises from absorption of solar EUV in the thermosphere and does not propagate energy and momentum vertically. The DE3 is forced in the troposphere and is confined to low latitudes, attaining peak amplitudes near 110 km but penetrating as a propagating tide well into the thermosphere. Interactions between the DW1 and DE3 can lead to secondary tidal and planetary waves (see Hagan et al. 2009 for details and reference to other works). The key point is that the DE3 and other non-migrating modes are substantial features of the upper MLT and can significantly impact the middle and upper atmosphere.

The broader perspective provided by ground-based networks to tidal studies is demonstrated by Murphy et al. (2006). Combining multi-year observations made by radars located around the periphery of the continent and at the South Pole they developed a climatology of the semidiurnal tide over the Antarctic, as shown in Figure 2

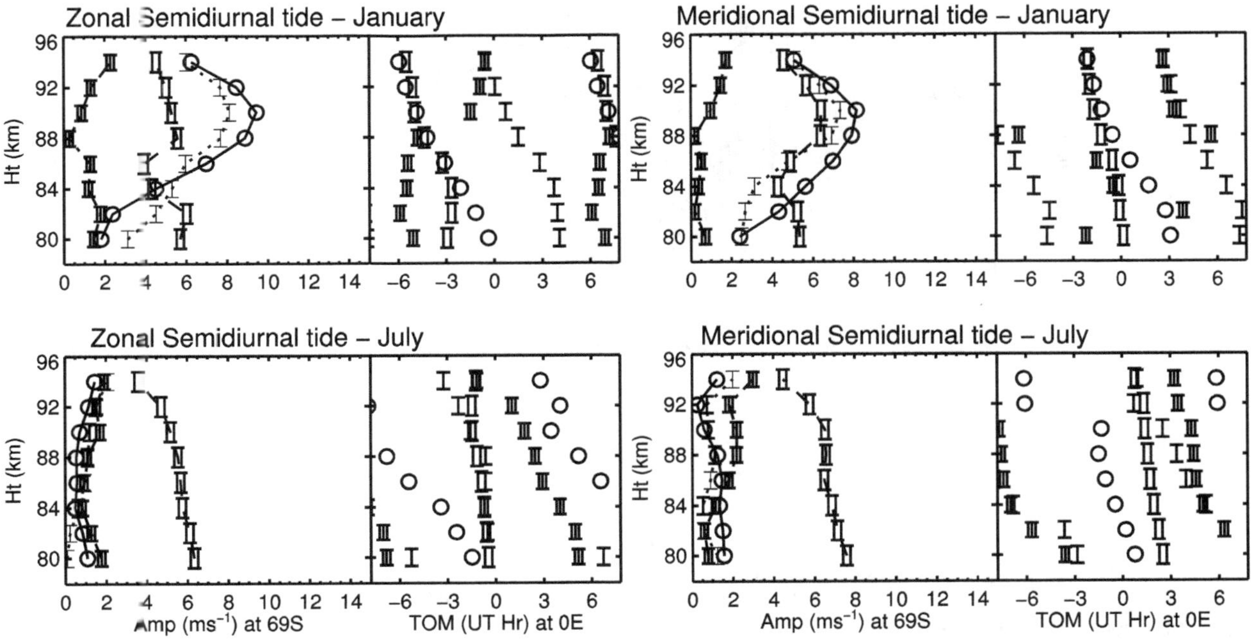

Figure 2 Solar semidiurnal tides in the Antarctic MLT. Climatologies of the amplitude and phase of the zonal (left) and meridional (right) semidiurnal tide in January and July. The symbols O, I, II and III represent the results for the wave numbers 0, 1, 2 and 3 semidiurnal tides, respectively. Derived from Figure six of Murphy et al. (2006).

for the amplitudes and phases in summer (January) and winter (July). Note the strong presence of the SD1 and SD0 tidal modes in summer but their near absence in winter when the semidiurnal migrating SD2 mode is dominant. The excitation mechanisms of the non-migrating modes are not fully understood but probably arise through the interaction between the migrating mode and stationary planetary waves (Murphy et al. 2006).

These are just two examples of many recent studies that show the relative importance of non-migrating tides in the MLT. Their influence extends into the thermosphere/ionosphere. Immel et al. (2006) reported a modulation of *F* region electron densities around the magnetic equator which they ascribed to the effects of non-migrating diurnal tides, such as the DE3 mode shown in Figure 1. However, the mechanisms that underly the coupling into the ionosphere are still a matter of debate. The most commonly proposed mechanism to explain this is an electrodynamic coupling between tides at *E* region altitudes and ion drifts at *F* region altitudes, but multiple-mechanisms involving electrodynamic and electro-chemical effects may be involved (England et al. 2010). Recently, Jones et al. (2013) assessed the relative effects of thermospheric tides generated *in situ* versus the effects of tides generated in the troposphere. Their results suggest that under certain conditions *in situ*-generated non-migrating tidal components dominate some parts of the tidal spectrum and must be taken into account.

Sudden stratospheric warmings and their effects in the MLT

A sudden stratospheric warming is a large-scale disruption of the wind and temperature fields of the wintertime polar stratosphere. SSW are usually classified as major or minor, although a continuum of warmings states may occur (Coughlin and Gray 2009). A major SSW is one in which the zonal-mean eastward winds around the pole cap temporarily reverse to summer-like conditions and the polar stratosphere warms so the NS temperature gradient is reversed from the normal wintertime situation. A minor warming is conventionally defined as one in which the temperature gradient reverses but the winds do not. Major SSW are primarily a northern hemisphere phenomena and occur approximately once or twice each winter. Only one major SSW has been observed in the southern hemisphere, in 2002 (e.g Krüger et al. 2005).

Figure 3 illustrates the impact on 10 hPa temperatures and winds at 60 N of the January 2006 major SSW (Yamazaki et al. 2012). Note the reversal of the zonal wind between 21 January and 10 February. The effects of this and other major warmings can be found throughout the middle and upper atmosphere. Yamazaki et al. (2012) themselves found that the Sq current system driven by

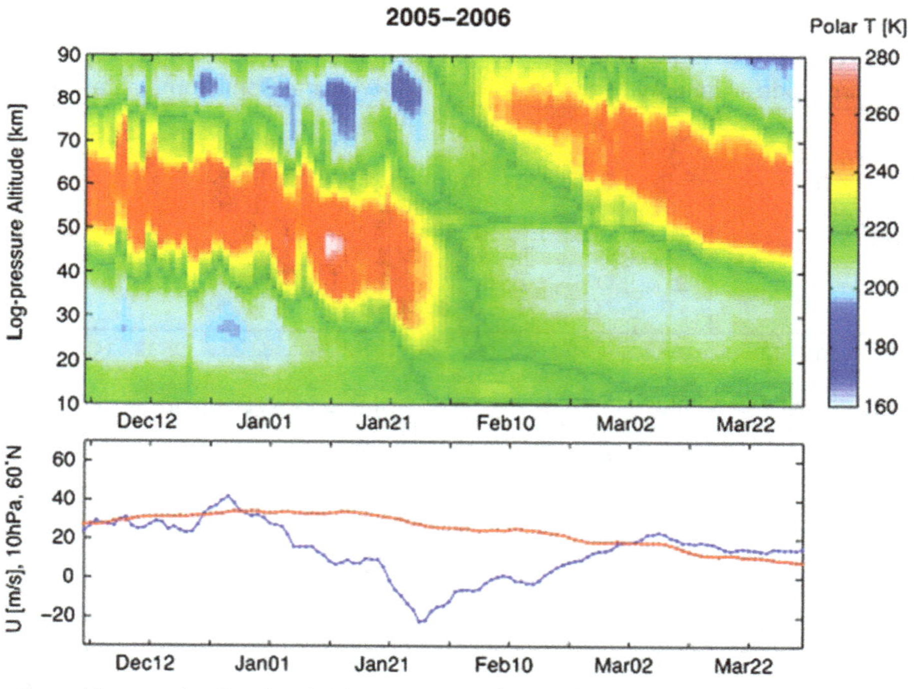

Figure 3 The 2006 major sudden warming. Top: time-height cross-section of temperatures during the major SSW in late January 2006. Note the descent and disappearance of the stratopause at the peak of the warming and its subsequent reappearance and descent from the mesosphere. Bottom: the time variation of the zonal-mean zonal wind at 10 hPa and 60 N. The blue line shows the wind during the stratwarm, while the red line depicts the climatological seasonal cycle. Adapted from Figure one of Yamazaki et al. (2012).

tides in the dynamo region (approximately 100 to 150 km) is profoundly affected by major SSW.

Funke et al. (2010) reported on temperature anomalies in upper atmosphere during the January 2009 SSW. Cooling was observed in the mesosphere (approximately 90 km) and warming at heights near 130 km (see Figure 4). Furthermore, the strength of the anomalies varied as a function of longitude, with a pronounced wave-1 pattern. This effect is ascribed to a large planetary wave in the lower atmosphere imprinting its pattern in the upper atmosphere (Funke et al. 2010), as first suggested by Dunkerton and Butchart (1984). The changing wind and temperature fields in a warming spatially and temporally modulate GW fluxes in the stratosphere. When the waves dissipate, the resulting momentum deposition can leave its imprint in the MLT. This type of mechanism has been verified by Lieberman et al. (2013), who found the MLT circulation in various NH winters is qualitatively consistent with a simple model of wave forcing by drag from gravity waves that have been modulated by stratospheric planetary waves.

Equatorial mesospheric zonal winds

The equatorial middle atmosphere was poorly studied compared with the number of observations made at mid- and high latitudes. What information that was acquired came mainly from long-term radiosonde ascents at a few equatorial stations, such as Singapore (1° N, 104° E), and from semi-regular rocket soundings made from Kwajalein (9° N, 168° W) and Ascension Island (8° S, 14° W). The long sequence of Singapore observations has been particularly important for the study of the QBO, while the rocket measurements led to the discovery and further investigation of semi-annual oscillations in the vicinity of the stratopause (SSAO) and mesopause (MSAO). Garcia et al. (1997) summarized early results.

The installation of radars at equatorial latitudes and space-based observations since 1990 (e.g. Vincent and Lesicar 1991, Burrage et al. 1996) now gives a much better understanding of variability in the at equatorial MLT. For example, Venkateswara Rao et al. (2012b) used approximately 20 years of ground-based radar wind measurements made at various sites in the tropical regions of Asia-Oceania in order to study MLT variability (Figure 5). By grouping observations from stations located at similar latitudes and averaging over all years they studied the equatorial wind field as a function of height and time, as shown in Figure 6. The plots provide more evidence for stronger westward winds in the March equinox than in September. There is also interannual variability on QBO time scales, with a significant enhancement in the March phase of the MSAO at a time when the QBO winds are in their eastward phase. However, there is decadal variability with this so-called MSAO enhancement less strong after 2002.

The causes of the MSAO are not fully understood. Garcia et al. (1997) noted an apparent link to the asymmetry in the strength of the stratospheric

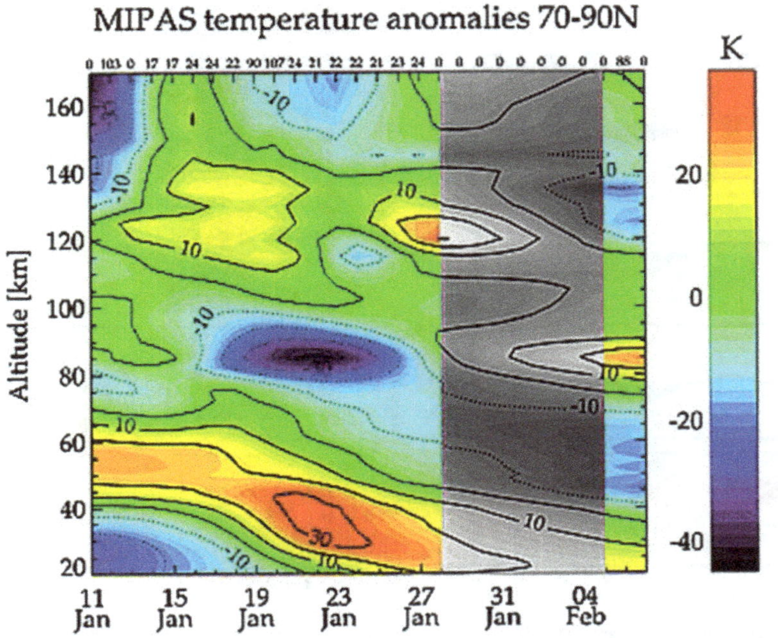

Figure 4 MLT temperature anomalies during the 2009 stratwarm. Temporal evolution of zonally averaged temperature anomalies derived from satellite observations and averaged between 70 and 90 N. The grey area indicates missing data. After Figure one of Funke et al. (2010).

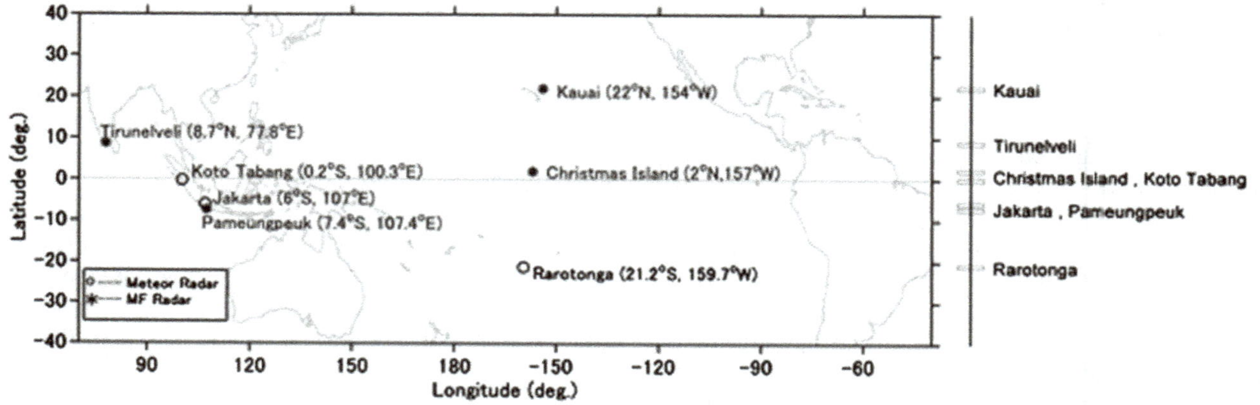

Figure 5 Location of radar stations in the Asia-Pacific. After Figure one of Venkateswara Rao et al. (2012b).

semiannual eastward winds, which also show a QBO-like enhancement at the March equinox. They postulated a role for small-scale gravity wave driving, although other wave modes may also be important. It should be remarked that a QBO-like link to the strength of propagating diurnal tide has been reported (Burrage et al. 1996; Vincent et al. 1998). Recently, Venkateswara Rao et al. (2012b) reported correlations between short-period gravity-wave variances at the mesopause and the strength of the MSAO observed using an MF radar in Indonesia. This result supports the GW momentum deposition drives the MSAO, but the issue is not likely to be fully resolved until momentum flux measurements can be made at the equator.

The quasi-two-day wave and its impact on the MLT

The QTDW is a westward travelling planetary wave that is of considerable interest because it can be one of the largest features of the MLT. It is a transitory phenomenon

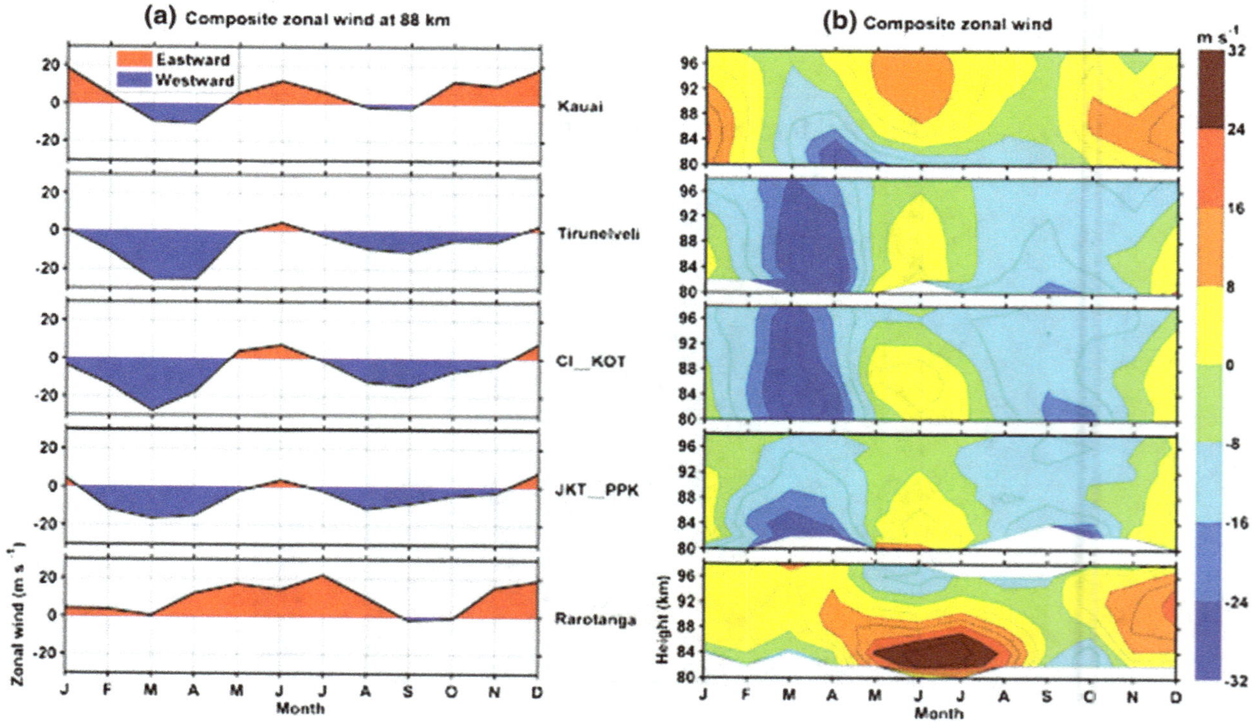

Figure 6 Composite zonal winds in the equatorial MLT. (Left, **(a)**) Plots of seasonal variations of wind at 88 km for various locations between 22° N and 21° S. (Right, **(b)**) Zonal winds between 80 and 98 km. Note that the semi-annual oscillation is particularly strong around the March equinox. After Figure five of Venkateswara Rao et al. (2012b).

appearing after the summer solstice and is primarily a zonal wave number 3 phenomenon in the southern hemisphere but is often a mixture of wave numbers 3 and 4 components in the northern hemisphere. The QTDW is a manifestation of the wave-3 normal mode but amplified by baroclinic instability of the summer westward jet near heights of 60 km (Plumb 1983). Using satellite measurements, Ern et al. (2013) explored how baroclinic instabilities in the mesospheric summer jet are linked to enhanced gravity-wave drag, which decelerates the jet and causes negative potential vorticity gradients.

Figure 7 illustrates many of the features of the QTDW observed in the southern hemisphere MLT. The NS wind measurements (Figure 7a) were made at 92 km with a meteor radar located at Darwin (12° S, 131° E). Figure 7b,c shows the amplitude and phase of the NS winds filtered with a bandpass between 40 and 60 h, with the phase referenced to an oscillation with a period of 48 h. In general, the increase of phase with time indicates a period less than 48 h, but the steady phase observed between about day 21 and 29 shows the period at this time was close to 48 h. The same behaviour was observed

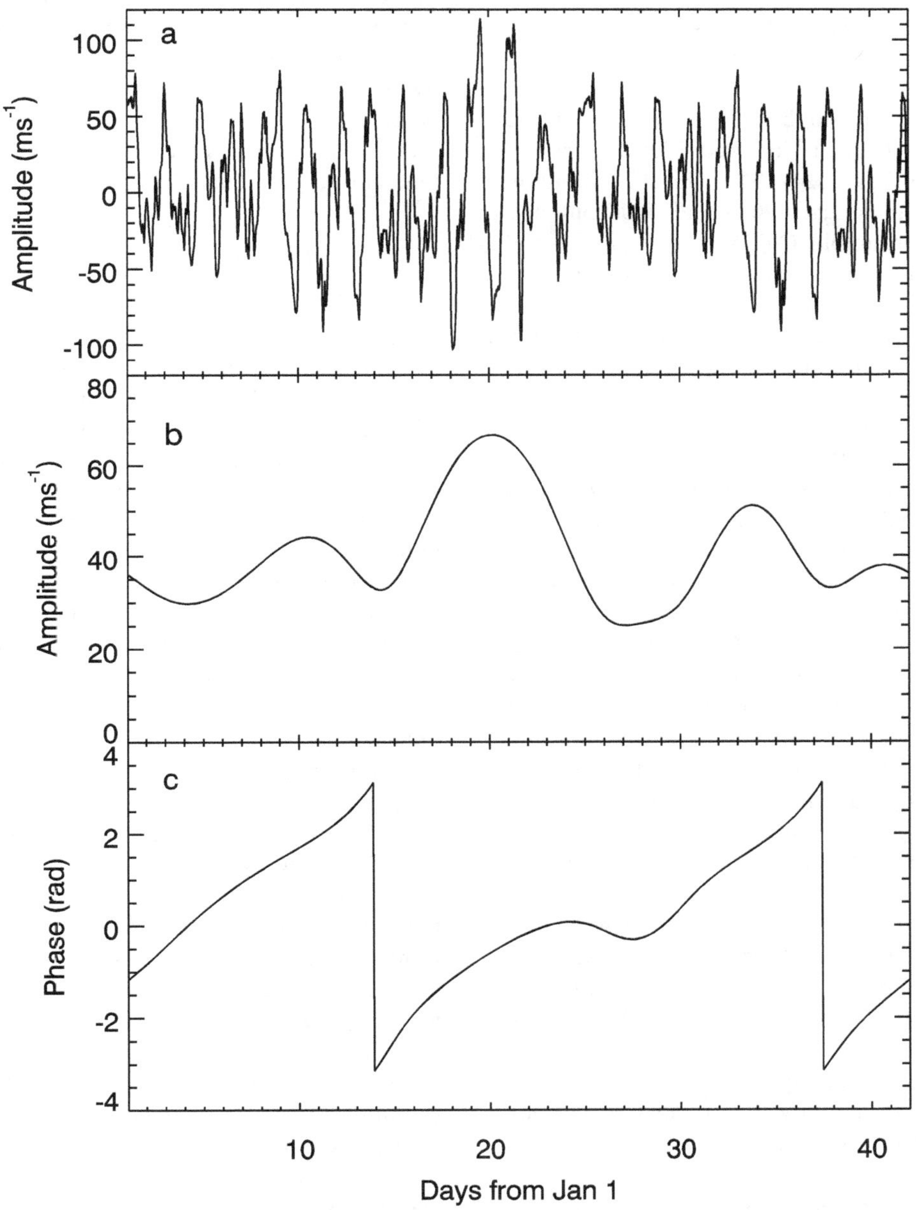

Figure 7 The QTDW in MLT winds at Darwin in 2006. (a) Time series of hourly average NS winds observed at 92 km. **(b)** Amplitude of 48 h component of NS winds. **(c)** Phase of 48 h component.

during the 2006 QTDW event at Adelaide by McCormack et al. (2010) who invoked phase locking through non-linear interactions with the migrating diurnal tide, as proposed by Walterscheid and Vincent (1996). Studies of radar and airglow emissions from the MLT by Hecht et al. (2010) corroborated the phase-locking mechanism. They found the airglow intensity response was much larger than what would be expected from the airglow temperature response, suggesting that the QTDW is causing a significant composition change, possibly due to minor constituent transport. Combining temperature measurements made with the AURA and TIMED satellites Forbes and Moudden (2012) explored interactions between the QTDW and various tidal modes. A key finding was that large secondary waves were produced.

Amplitudes of up to 100 ms^{-1} in the NS wind component are often observed, but there is significant interannual variability (e.g. Siskind and McCormack 2014; Moudden and Forbes 2014). The SH QTDW attained large amplitudes in January 2004 and 2006 but not in 2008 to 2009, for example. QTDW activity in the NH is uncorrelated with that in the SH and smaller in amplitude (Moudden and Forbes 2014).

Due to its large amplitude, the QTDW has significant impact on the MLT and higher. Plumb et al. (1987) first noted that the pulse-like QTDW event in January 1984 caused a temporary but substantial change in the prevailing circulation of the MLT. Short-term effects due to the QTDW may explain the interannual variability of brightness of southern polar mesospheric clouds (PMC), which form as a consequence of the cold mesospheric temperatures at the summer pole (e.g. Pendlebury 2012). In large amplitude QTDW years, the brightness of southern PMC is reduced. Siskind and McCormack (2014) ascribe this effect to the QTDW depositing *westward* momentum that, acting through the Coriolis torque in Equation one, produces a southward flow that partially offsets the GW-driven equatorward, northward wind. Hence, there is a corresponding reduction in the upward motions at the pole with an accompanying reduction in the adiabatic cooling, leading to a temperature increase at the polar mesopause.

Just as the effect of large stratospheric warming events may be felt in the thermosphere/ionosphere, so can the QTDW influence the ionosphere. Yue et al. (2012) used the TIME-GCM to explore how the QTDW modulates the dynamo electric field and map into the ionosphere. Through case studies, Chang et al. (2014) describe how the QTDW drives variability in composition and ionospheric total electron content.

Gravity waves and their effect in the MLT

As noted in the 'Introduction' subsection, momentum deposition by gravity waves plays a crucial role in determining the state of the middle atmosphere. Incorporation of GW effects in climate models is a pressing problem (Alexander et al. 2010). Consequently, a wide range of techniques have been applied to determine wave parameters such as horizontal and vertical wavelengths and phase speeds. Each technique has its own strengths and limitations. Satellites provide a global perspective, but their viewing geometry determines the vertical and horizontal resolutions that can be attained (see Alexander and Barnet 2007). Ground-based radars and radiosonde have good temporal but limited horizontal resolution. Superpressure balloon (SPB) techniques are unique in that they measure gravity wave parameters as a function of intrinsic frequency but only at the float altitude near 18 km. Alexander et al. (2010) review the overall ability of the various techniques to determine important GW parameters.

One way to determine gravity wave activity is to use temperature soundings. The potential energy is then:

$$E_p = \frac{1}{2}\left(\frac{g}{N}\right)^2 \overline{\left(\frac{T'}{T_o}\right)^2} \tag{2}$$

where T' is the GW-induced temperature perturbation from the mean temperature T_o. Using this methodology, Zhang et al. (2012) and Wang et al. (2005) produced maps of wave activity in the stratosphere, while Hoffmann et al. (2013) focussed on locating GW 'hot spots'. They found high stratospheric wave activity over regions of strong convection in summer (e.g. the Amazon Basin) and mountain wave activity in winter (e.g. southern Andes and Antarctic Peninsula). Alexander et al. (2008) used limb-sounding GPS/COSMIC observations to study the temporal evolution of wave activity over the equator. They found evidence for vertically propagating convectively generated gravity waves interacting with the background mean flow. Enhancements in E_p around descending QBO eastward shear lines suggest a wave-mean flow interaction.

An interesting development in the past two decades has been the use of high vertical-resolution radiosondes regularly launched by weather agencies to investigate spatial variability in GW activity (Allen and Vincent 1995). Initially used to study potential energies, radiosonde studies were extended to include wave kinetic energies ($E_k = u'^2 + v'^2$). Using the extensive United States radiosonde network, Wang and Geller (2003) investigated spatial and seasonal variability of wave activity across the USA and its dependencies from the Arctic to South Pacific. This work was important for distinguishing source from propagation effects in determining wave variability. For example, they found that tropospheric GW energy ($E_p + E_k$) maximises over the Rocky Mountains, indicating a mountain wave source. However, in the lower stratosphere, the energy maxima are mostly in the southeastern United States.

QBO and El Nino-Southern Oscillation effects appear to account for much of the observed interannual variability in lower stratospheric wave energy. Their work was extended by Wang et al. (2005) and Gong et al. (2008) to studies of other important GW parameters, such as horizontal wavelengths and source spectra.

GW potential energies can be converted into absolute momentum fluxes if the horizontal (λ_h) and vertical wavelengths (λ_z) are known *viz*:

$$\rho_o|\overline{u'w'}| = \frac{1}{2}\rho_o\frac{\lambda_z}{\lambda_h}\left(\frac{g}{N}\right)^2\overline{\left(\frac{T'}{T_o}\right)^2}. \quad (3)$$

A downward viewing instruments on satellite can be used to determine λ_h by locating similar temperature features on horizontally separated scans. Different instruments have different capabilities, as illustrated in Figure 8, as discussed by Ern et al. (2011). The vertical wavelength is determined from scale of the temperature oscillations on each scan and is obviously limited by the vertical resolution of the instrument. Ern et al. (2011) applied the methodology through the stratosphere and lower mesosphere, with strong fluxes captured at latitudes greater than 50° in local winter and near 25° in the subtropics in summer. Figure 9 illustrates observations made with the SABER instrument on the TIMED satellite. Momentum flux values are enhanced in the polar night jets at mid- and high latitudes and in the subtropics of the summer hemisphere over the monsoon regions, forming a very characteristic zonal triple structure in the northern hemisphere and a triple or quadruple structure in the southern hemisphere. It is noteworthy that the momentum flux distribution at 70 km in the lower mesosphere has a different geographic distribution to that at 30 km.

Absolute fluxes are useful for constraining GW parameterization schemes used in climate models but is better to use signed momentum ($\rho_o\overline{u'w'}$, $\rho_o\overline{v'w'}$) fluxes as a function of intrinsic frequency. Such measurements can be made with SPB, albeit only at one height (Hertzog et al. 2008). Figure 10 shows spectra of the GW momentum flux for SPB flights over the Antarctic during September and October 2010. The float altitude is about 18 km. Note the sharp cut-offs at periods of about 4.5 and 600 min, which correspond to the Väisälä-Brunt and inertial periods, respectively. The peak at a period near 150 min, corresponding to horizontal scales approximately 100 km, is due to orographic waves located over the Antarctic Peninsula, a strong source of such waves (Hertzog et al. 2008). Also note the preponderance of upgoing fluxes, with about 70% of the total flux up-going. This reinforces our understanding that the majority of GW are generated in the lower atmosphere, with the spectrum being modified as the waves propagate upward. Parts of the spectrum will be lost due to wave reflection and dissipation so the spectrum in the MLT is likely to look quite different to that shown in Figure 10.

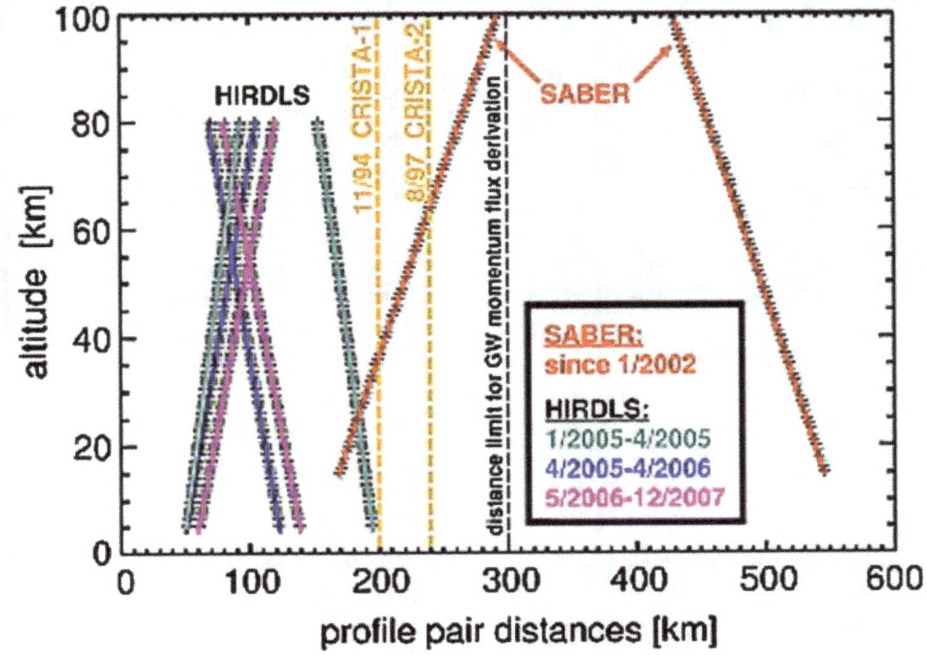

Figure 8 Examples of satellite scan patterns. Horizontal distance between sequential altitude profiles along the measurement tracks for the CRISTA, HIRDLS and SABER instruments. After Figure one of Ern et al. (2011).

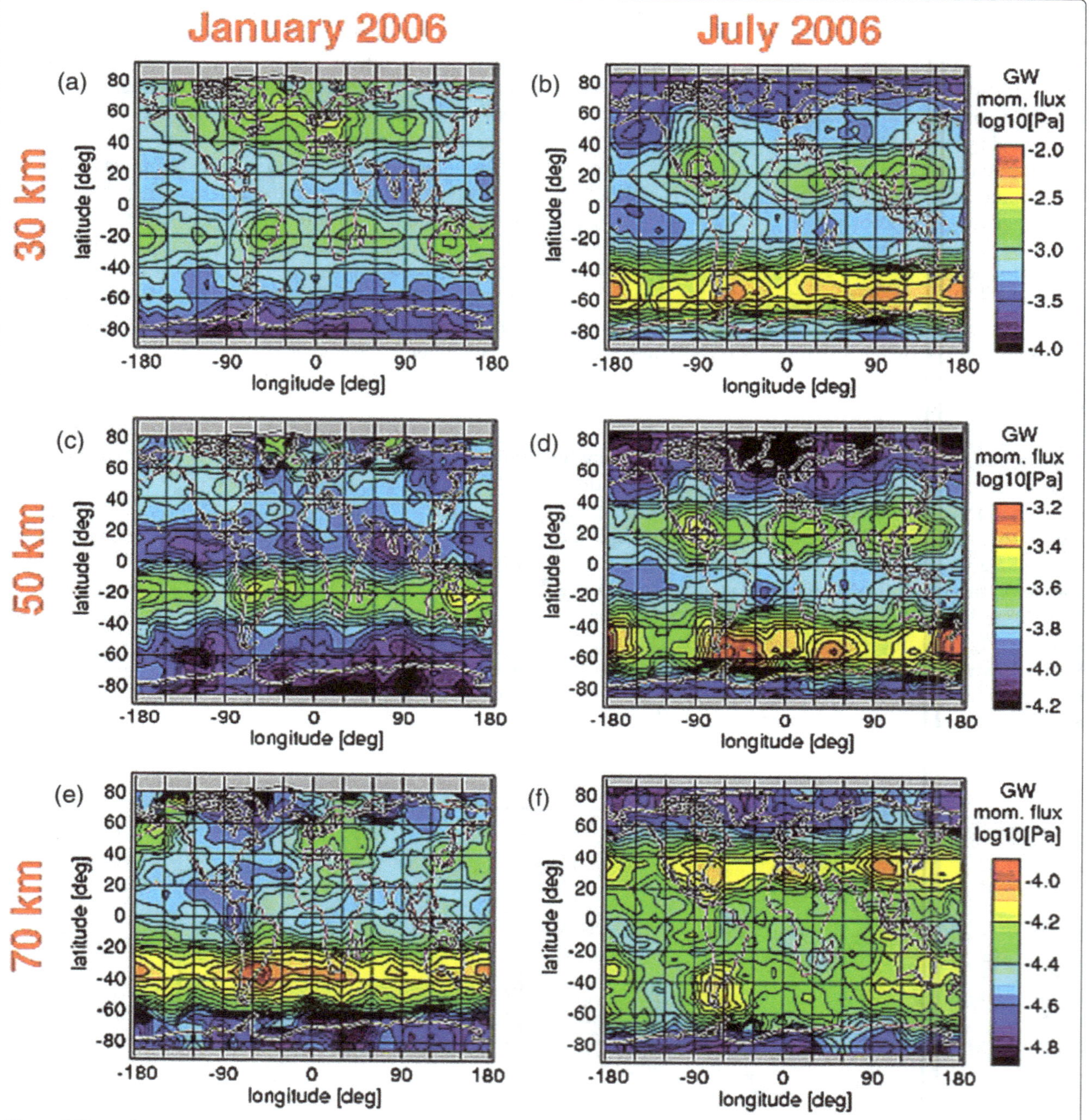

Figure 9 Maps of absolute GW momentum flux as a function of height. Distributions of momentum flux derived from TIMED/SABER observations for January (left, **(a)**, **(c)**, **(e)**) and July (right, **(b)**, **(d)**, **(f)**) 2006 at altitudes of 30, 50 and 70 km. Values are in $\log_{10}$ Pa. After Figure nine of Ern et al. (2011).

Observations of individual wave generation and coupling events through the middle atmosphere give important insights into wave generation and propagation. Hoffmann and Alexander (2009) used satellite data to study waves generated by convection over Darwin and a strong mountain wave over Patagonia. The latter event had a vertical wavelength of approximately 7 to 15 km and could be followed directly through the stratosphere up to 60 km altitude.

Observations of large amplitude GWs raises the question of whether the body force of $\overline{F} \sim 100$ ms^{-1} day^{-1} required in the MLT is maintained by continuous

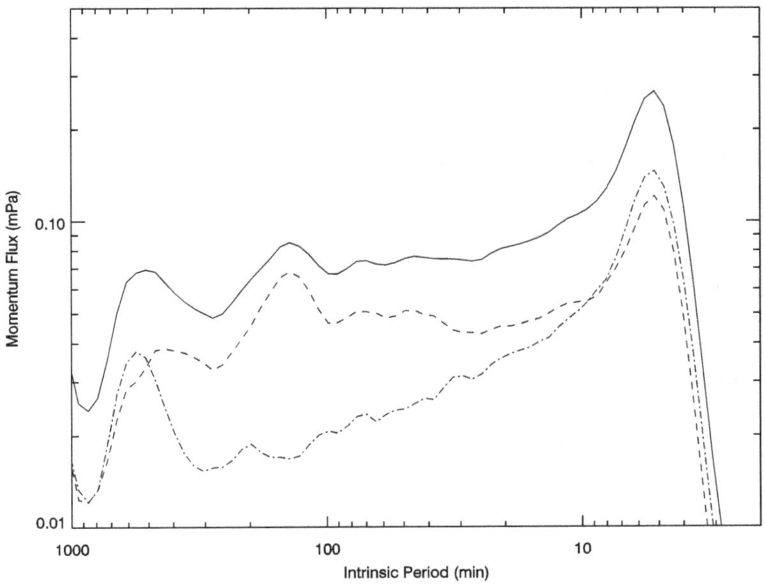

Figure 10 Momentum flux spectrum. Spectra of momentum flux ($\rho_0\overline{u'w'}$) in mPa for total (solid), upward going (dashed) and downward going fluxes (dash-dot) derived from SPB observations over Antarctica, September to October 2010.

gravity wave breaking or by intermittent breaking of large amplitude waves. By considering GW generation by an isolated large convective event over Darwin, Vincent et al. (2013) were able to trace wave propagation through the middle atmosphere to heights where wave breaking created body forces of approximately 200 to 300 ms^{-1} day^{-1} for a duration of a few hours. They noted that there was about 12 such convective events per day over the whole of tropical northern Australia and Indonesia during the monsoon season. Similarly, Hertzog et al. (2008) showed that strong intermittency was a feature of mountain waves observed over the Antarctic Peninsula. Intermittency of strong wave events supports the conjecture of Fritts and Vadas (2002) that the force required to close the mesospheric jets is sporadic in space and time.

Conclusions

In the past decade, there have been considerable improvements in our understanding of MLT dynamics, improvements driven in part by the development and deployment of new instruments and techniques. The importance of waves in the MLT is emphasised here because they dominate the wind and temperature fields of this region, so the MLT is a natural laboratory for the investigation of atmospheric waves and their effects.

Satellites, a wider deployment of ground-based radars and improved modelling capabilities means that the middle atmosphere is now investigated on a global scale and the relative contribution of different waves to the energy and momentum budgets are better understood. While the focus is on observational studies, especially of small-scale gravity waves, a notable advance in understanding has occurred through the development and application of very-high resolution general circulation models. In particular, the KANTO model resolves individual gravity waves in a realistic manner (Watanabe et al. 2008). Using this model, Sato et al. (2009) provide a global view of GW sources and propagation effects and the impact of GW on the momentum balance of the mesosphere.

Oscillations in winds and temperatures due to tides and other planetary-scale waves such as the QTDW are prominent features of the MLT. Wave-wave interactions lead to secondary waves that may also have substantial amplitude and which may be important for phase-locking the QTDW, for example. Tide/gravity wave interactions may produce either growth/acceleration or damping/retardation of tidal amplitudes and phases. Full understanding of the possible effects will come only from better observations and modelling.

Although waves and their effects dominate the MLT, their source regions lie mainly in the lower atmosphere. Consequently, wave impacts on the MLT ultimately depend on source variability and the varying transmission effects in the middle atmosphere. Such effects can lead to 'imprinting', which is especially noticeable at the equator where the mesospheric semiannual oscillation often seems to have a signature of the QBO superimposed upon it. Similarly, wave processing in the MLT will modify the spectrum of waves propagating higher into the thermosphere and ionosphere, leading to the imprinting of MLT effects at higher altitudes.

Investigation of individual wave events leads to a better understanding of wave impacts in the MLT. On planetary scales, large quasi-two-day wave events produce transitory changes to the circulation at the mesopause, leading to changes in the thermal structure at high latitudes with concomitant changes in the micro-physics of this region. At small scales, gravity waves play a preeminent role in the momentum budget of the MLT. It remains to be seen, however, whether the momentum deposition comes from a few sporadic large amplitude wave events or through deposition by many small amplitude waves. Resolution of this issue will only come from future observational and modelling efforts.

Abbreviations

GCM: global climate model; GSWM: Global Scale Wave Model; GW: gravity wave; MLT: mesosphere-lower thermosphere; QTDW: quasi-two-day wave; PMC: polar mesospheric cloud.

Competing interests

The author declares that he has no competing interest.

Acknowledgements

The helpful comments by M. A. Geller on an earlier draft of this paper are appreciated. D. J. Murphy is thanked for his help with Figure 2.

References

Alexander MJ, Barnet C (2007) Using satellite observations to constrain parameterizations of gravity wave effects for global models. J Atmos Sci 64:1652–1665

Alexander SP, Tsuda T, Kawatani Y, Takahashi M (2008) Global distribution of atmospheric waves in the equatorial upper troposphere and lower stratosphere: cosmic observations of wave mean flow interactions. J Geophys Res 113(D24):24115. doi:10.1029/2008JD010039

Alexander MJ, Geller M, McLandress C, Polavarapu S, Preusse P, Sassi F, Sato K, Ern M, Hertzog A, Kawatani Y, Pulido M, Shaw T, Sigmond M, Vincent R, Watanabe S (2010) Recent developments in gravity wave effects in climate models and the global distribution of gravity wave momentum flux from observations and models. QJR Meteorol Soc 136:1103–1124

Allen SJ, Vincent RA (1995) Gravity wave activity in the lower atmosphere: seasonal and latitudinal variations. J Geophys Res 100:1327–1350

Andrews DG, Holton JR, Leovy CB (1987) Middle atmosphere dynamics. Academic Press, London

Baldwin MP, Gray LJ, Dunkerton TJ, Hamilton K, Haynes PH, Randel WJ, Holton JR, Alexander MJ, Hirota I, Horinouchi T, Jones DBA, Kinnersley JS, Marquardt C, Sato K, Takahashi M (2001) The quasi-biennial oscillation. Rev Geophys 39(2):179–229. doi:10.1029/1999RG000073

Burrage MD, Hagan ME, Skinner WR, Wu DL, Hays PB (1995) Long term variability in the solar diurnal tide observed by HRDI and simulated by the GSWM. Geophys Res Lett 22:2641–2644. doi:10.1029/95GL02635

Burrage MD, Vincent RA, Mayr HG, Skinner WR, Arnold PB, Hays N F (1996) Long-term variability in the equatorial mesosphere and lower thermosphere zonal wind. J Geophys Res 101:12847–12854

Chapman S, Lindzen RS (1970) Atmospheric tides. D. Reidel, Dordrecht, Holland

Chang LC, Yue J, Wang W, Wu Q, Meier RR (2014) Quasi two day wave-related variability in the background dynamics and composition of the mesosphere/thermosphere and the ionosphere. Journal of Geophysical Research: Space Physics 119(6):4786–4808. doi:10.1002/2014JA019936

Coughlin K, Gray LJ (2009) A continuum of sudden stratospheric warmings. J Atmos Sci 66(2):531–540

Davis RN, Du J, Smith AK, Ward WE, Mitchell NJ (2013) The diurnal and semidiurnal tides over ascension island (8°S, 14°W) and their interaction with the stratospheric quasi-biennial oscillation: studies with meteor radar, eCMAM and WACCM. Atmos Chem Phys 13(18):9543–9564. doi:10.5194/acp-13-9543-2013

Dunkerton TJ, Butchart N (1984) Propagation and selective tranmission of inertial gravity waves in a sudden warming. J Atmos Sci 41:1443–1460

England SL, Immel TJ, Huba JD, Hagan ME, Maute A, DeMajistre R (2010) Modeling of multiple effects of atmospheric tides on the ionosphere: an examination of possible coupling mechanisms responsible for the longitudinal structure of the equatorial ionosphere. J Geophys Res 115:05308. doi:10.1029/2009JA014894

Ern M, Preusse P, Gille JC, Hepplewhite CL, Mlynczak MG, Russell JM, Riese M (2011) Implications for atmospheric dynamics derived from global observations of gravity wave momentum flux in stratosphere and mesosphere. J Geophys Res 116(D19):19107. doi:10.1029/2011JD015821

Ern M, Preusse P, Kalisch S, Kaufmann M, Riese M (2013) Role of gravity waves in the forcing of quasi two-day waves in the mesosphere: an observational study. J Geophys Res 118:3467–3485. doi:10.1029/2012JD018208

Forbes JM, Moudden Y (2012) Quasi-two-day wave-tide interactions as revealed in satellite observations. Journal of Geophysical Research: Atmospheres 117(D12):12110. doi:10.1029/2011JD017114

Fritts DC, Vadas SL (2002) An estimation of strong local body forcing and gravity wave radiation based on OH airglow and meteor radar observations. Geophys Res Lett 29 26(10):71-1–71-4. doi:10.1029/2001GL013753

Funke B, López-Puertas M, Bermejo-Pantaleón D, García-Comas M, Stiller GP, von Clarmann T, Kiefer M, Linden A (2010) Evidence for dynamical coupling from the lower atmosphere to the thermosphere during a major stratospheric warming. Geophys Res Lett 37:13803. doi:10.1029/2010GL043619

Garcia RR, Dunkerton TJ, Lieberman RS, Vincent RA (1997) Climatology of the semiannual oscillation of the tropical middle atmosphere. J Geophys Res 102:26019–26032

Gong J, Geller MA, Wang L (2008) Source spectra information derived from U.S. high-resolution radiosonde data. J Geophys Res 113:10106. doi:10.1029/2007JD009252

Hagan ME, Burrage MD, Forbes JM, Hackney J, Randel WJ, Zhang X (1999) GSWM-98: results for migrating solar tides. J Geophys Res 104:6813–6828

Hagan ME, Forbes JM (2002) Migrating and nonmigrating diurnal tides in the middle and upper atmosphere excited by tropospheric latent heat release. J Geophys Res 107(D24):4754

Hagan ME, Maude A, Roble RG (2009) Tropospheric tidal effects on the middle and upper atmosphere. J Geophys Res 114:01302. doi:10.1029/2008JA013637

Hecht JH, Walterscheid RL, Gelinas LJ, Vincent RA, Reid M, Woithe JM (2010) Observations of the phase-locked 2 day wave over the australian sector using medium-frequency radar and airglow data. Journal of Geophysical Research: Atmospheres 115(D16):1–22. doi:10.1029/2009JD013772

Hertzog A, Boccara G, Vincent RA, Vial F, Cocquerez P (2008) Estimation of gravity-wave momentum fluxes and phase speeds from quasi-lagrangian stratospheric balloon flights. 2: Results from the Vorcore campaign in Antarctica. J Atmos Sci 65:3056–3070

Hoffmann L, Alexander MJ (2009) Retrieval of stratospheric temperatures from atmospheric infrared sounder radiance measurements for gravity wave studies. J Geophys Res 114:07105. doi:10.1029/2008JD011241

Hoffmann L, Xue X, Alexander MJ (2013) A global view of stratospheric gravity wave hotspots located with Atmospheric Infrared Sounder observations. J Geophys Res 118:416–434. doi:10.1029/2012JD018658

Immel TJ, Sagawa E, England SL, Henderson SB, Hagan ME, Mende SB, Frey HU, Swenson CM, Paxton LJ (2006) Control of equatorial ionospheric morphology by atmospheric tides. Geophys Res Lett 33:15108. doi:10.1029/2006GL026161

Jones M, Forbes JM, Hagan ME, Maute A (2013) Non-migrating tides in the ionosphere-thermosphere: in situ versus tropospheric sources. J Geophys Res 118(5):2438–2451

Krüger K, Naujcot B, Labitzke K (2005) The unusual midwinter warming in the southern hemisphere stratosphere 2002: a comparison to northern hemisphere phenomena. J Atmos Sci 62:603–613

Lieberman RS, Riggin DM, Siskind DE (2013) Stationary waves in the wintertime mesosphere: evidence for gravity wave filtering by stratospheric planetary waves. J Geophys Res 118(8):3139–3149. doi:10.1002/jgrd.50319

McCormack JP, Eckermann SD, Hoppel KW, Vincent RA (2010) Amplification of the quasi-two day wave through nonlinear interaction with the migrating diurnal tide. Geophys Res Lett. doi:10.1029/2010GL043906

McIntyre ME (1989) On dynamics and transport near the polar mesopause in summer. J Geophys Res 94:14617–14628
Mitchell NJ, Pancheva D, Middleton HR, Hagan ME (2002) Mean winds and tides in the arctic mesosphere and lower thermosphere. J Geophys Res 107(A1):1004
Moudden Y, Forbes JM (2014) Quasi-two-day wave structure, interannual variability and tidal interactions furing the 2002-2011 decade. J Geophys Res 119(5):2241–2260. doi:10.1002/2013JD020563
Mukhtarov P, Pancheva D, Andonov B (2009) Global structure and seasonal and interannual variability of the migrating diurnal tide seen in the SABER/TIMED temperatures between 20 and 120 km. J Geophys Res 114:03104. doi:10.1029/2008GL036535
Murphy DJ, Forbes JM, Walterscheid RL, Hagan ME, Avery SK, Aso T, Fraser GJ, Fritts DC, Jarvis MJ, McDonald AJ, Riggin DM, Tsutsumi M, Vincent RA (2006) A climatology of tides in the antarctic mesosphere and lower thermosphere. J Geophys Res 111:23104. doi:10.1029/2005JD006803
Nozawa S, Hall CM, Tsutsumi M, Brekke A, Ogawa Y, Tsuda TT, Oyama S, Fujii R (2012) Mean winds, tides, and quasi-2 day waves above bear island. J Atmos Solar-Terr Phys 0:26–44
Oberheide J, Hagan ME, Roble RG, Offermann D (2002) Sources of nonmigrating tides in the tropical middle atmosphere. J Geophys Res 107(D21):4567
Pendlebury D (2012) A simulation of the quasi-two-day wave and its effect on variability of summertime mesopause temperatures. J Atmos Solar-Terr Phys 80(0):138–151. doi:10.1016/j.jastp.2012.01.006
Plumb RA (1983) Baroclinic instability at the summer mesosphere: a mechanism for the quasi-two-day wave? J Atmos Sci 40:262–270
Plumb RA, Vincent RA, Craig RL (1987) The quasi-two-day wave event of january 1984 and its impact on the mean mesospheric circulation. J Atmos Sci 44:3030–3036
Sato K, Watanabe S, Kawatani Y, Amd K, Miyazaki Y T, Takahashi M (2009) On the origins of mesospheric gravity waves. Geophys Res Lett 36:19801. doi:10.1029/2009GL039908
She CY, Chen S, Williams BP, Hu Z, Krueger DA, Hagan ME (2002) Tides in the mesopause region over Fort Collins, Colorado (41°N, 105°W) based on lidar temperature observations covering full diurnal cycles. J Geophys Res 107(D18):1–12
Siskind DE, McCormack JP (2014) Summer mesospheric warmings and the quasi 2 day wave. Geophys Res Lett 41(2):717–722. doi:10.1002/2013GL058875
Sridharan S, Tsuda T, Gurubaran S (2010) Long-term tendencies in the mesosphere/lower thermosphere mean winds and tides as observed by medium-frequency radar at tirunelveli (8.7°n, 77.8°e). J Geophys Res 115(D8):08109. 0148-0227
Vincent RA, Lesicar D (1991) Dynamics of the equatorial mesosphere: first results with a new generation partial reflection radar. Geophys Res Lett 18:825–828
Vincent RA, S K, Fritts DC, Isler JR (1998) Long-term MF radar observations of solar tides in the low-latitude mesosphere: interannual variability and comparisons with the GSWM. J Geophys Res 103(D8):8667–8683
Vincent RA, Alexander MJ, Dolman BK, MacKinnon AD, May PT, Kovalam S, Reid IM (2013) Gravity wave generation by convection and momentum deposition in the mesosphere-lower thermosphere. J Geophys Res 118:6233–6245. doi:10.1002/jgrd.50372
Venkateswara Rao N, Tsuda T, Kawatani Y (2012b) A remarkable correlation between short period gravity waves and semiannual oscillation of the zonal wind in the equatorial mesopause region. Ann Geophys 30:703–710. doi:10.5194/angeo-30-703-2012
Venkateswara Rao N, Tsuda T, Riggin DM, Gurubaran S, Reid IM, Vincent RA (2012b) Long-term variability of mean winds in the mesosphere and lower thermosphere at low latitudes. J Geophys Res 117(10):1–16. doi:10.1029/2012JA017850
Walterscheid RL, Vincent RA (1996) The tidal generation of phase locked two-day wave in the southern hemisphere by wave-wave interactions. J Geophys Res 101 26567–26576
Wang L, Geller MA (2003) Morphology of gravity-wave energy as observed from 4 years (1998–2001) of high vertical resolution U.S. radiosonde data. J Geophys Res 108 4489. doi:10.1029/2002JD002786
Wang L, Geller MA, Alexander MJ (2005) Spatial and temporal variations of gravity wave parameters. part i: intrinsic frequency, wavelength, and vertical propagation direction. J Atmos Sci 62:125–142
Watanabe S, Kawatani Y, Tomikawa Y, Miyazaki K, Takahashi M, Sato K (2008) General aspects of a t213l256 middle atmosphere general circulation model. J Geophys Res 113:12110. doi:10.1029/2008JD010026
Xu X, Manson AH, Meek CE, Riggin DM, Jacobi C, Drummond JR (2012) Mesospheric wind diurnal tides within the canadian middle atmosphere model data assimilation system. J Atmos Solar-Terr Phys 74(0):24–43
Yamazaki Y, Richmond AD, Liu H, Yumoto K, Tanaka Y (2012) Sq current system during stratospheric sudden warming events in 2006 and 2009. J Geophys Res 117(A12):12313. doi:10.1029/2012JA018116
Yoshikawa M, Miyahara S (2005) Excitations of nonmigrating diurnal tides in the mesosphere and lower thermosphere simulated by the Kyushu-GCM. Adv Space Res 35:1918–1924
Yue J, Wang W, Richmond AD, Liu H-L (2012) Quasi-two-day wave coupling of the mesosphere and lower thermosphere-ionosphere in the TIME-GCM: two-day oscillations in the ionosphere. J Geophys Res 117(A7):07305. doi:10.1029/2012JA017815
Zhang X, Forbes JM, Hagan ME (2010a) Longitudinal variation of tides in the MLT region: 1. Tides driven by tropospheric net radiative heating. J Geophys Res 115:06316. doi:10.1029/2009JA014897
Zhang, X, Forbes JM, Hagan ME (2010b) Longitudinal variation of tides in the MLT region: 2. Relative effects of solar radiative and latent heating. J Geophys Res 115(A6):06317
Zhang Y, Xiong J, Liu L, Wan W (2012) A global morphology of gravity wave activity in the stratosphere revealed by the 8-year SABER/TIMED data. J Geophys Res 117:21101. doi:10.1029/2012jd017676

7

Revisiting ozone measurements as an indicator of tropical width

Sean M. Davis[1,2*], Birgit Hassler[3] and Karen H. Rosenlof[1]

Abstract

The total column ozone (TCO) amount varies with latitude, in part due to the difference in tropopause height between the tropics and midlatitudes. This dependency of TCO on latitude has been used to identify the latitudes of the tropical edges and to compute their variations in time. The previously reported poleward movement of the tropical edge latitudes computed from satellite TCO measurements over the past several decades is greater than 3° latitude per decade. This tropical widening rate is significantly larger than a number of independent estimates and if correct suggests a major deficiency in the representation of tropical widening in models. We revisit the previously used TCO tropical edge latitude diagnostic to extend it forward in time with a new data set and to assess its robustness through comparisons with independent tropical edge diagnostics. We find that the previous TCO-based tropical width timeseries contain a spurious jump, likely due to data inhomogeneities. After removal of this jump using an objective statistical breakpoint identification technique, TCO-based tropical widening is reduced to the point that it is not significantly different than other tropical widening estimates. The strong sensitivity of the TCO method to algorithmic choices, its out-of-phase seasonality in the Northern Hemisphere, and its lack of correlation with well-established tropical width metrics on interannual timescales support our conclusion that the TCO-based tropical width diagnostic as previously implemented is not a robust measure of tropical width.

Keywords: Ozone, Tropical width, Tropical widening, Hadley cell, Stratosphere

Introduction

A number of different methodologies and characteristics of Earth's atmosphere have been used to identify the latitudinal boundary between the tropics and extratropics and compute changes in its position with time (e.g., Davis and Rosenlof 2012; Lucas et al. 2014; Seidel et al. 2008). This so-called "tropical edge" roughly corresponds to the location of the subsiding branch of the Hadley cell in each hemisphere at around 30°. Although the Hadley cell and associated subtropical jet provide the physical underpinning for defining a tropical edge, in practice there have been numerous definitions using features that are assumed to be tied to the Hadley cell or subtropical jet, such as the subtropical break in the height of the tropopause.

These various definitions have been at least part of the reason that estimates of changes in the width of the tropics span such a large range of values, from statistically insignificant and near zero to several degrees latitude per decade of poleward movement and highly significant. Widening based on satellite-observed total column ozone (TCO) by Hudson (2012), hereafter H12) and a series of earlier papers (Hudson et al. 2003; Hudson et al. 2006, hereafter H03 and H06, respectively) are the largest estimates of so-called "tropical widening." For example, H12 provides a value of global tropical expansion (Northern Hemisphere (NH) + Southern Hemisphere (SH)) over the period of 1979–2010 of 3.2° latitude per decade, whereas most other metrics show values < 1° per decade. The few other annual-mean global estimates of tropical widening that are larger than 1° per decade are based on tropopause height (Seidel and Randel 2007) and outgoing longwave radiation (OLR; Hu and Fu 2007). However, subsequent studies have shown that the values from these studies were biased high and that more objective metrics based on tropopause height and OLR give significantly smaller estimates that are below 1° per decade (Birner 2010; Davis

* Correspondence: Sean.M.Davis@noaa.gov
[1]NOAA Earth System Research Laboratory Chemical Sciences Division, 325 S. Broadway, Boulder, CO 80305, USA
[2]Cooperative Institute for Research in Environmental Sciences, University of Colorado at Boulder, Boulder, CO 80303, USA
Full list of author information is available at the end of the article

and Rosenlof 2012). More recent studies have also questioned the robustness of tropopause and OLR metrics and demonstrated that they are not always well correlated with more direct measures of the Hadley cell (Davis and Birner 2017; Solomon et al. 2016; Waugh et al. 2018).

In this paper, we revisit the usage of total column ozone measurements for defining the width of the tropics and quantifying tropical widening rates, with the goal of better understanding the apparent discrepancy between the large TCO-based widening found by H06/H12 and other studies. Towards this end, this paper is organized as follows. In the next section, we give an overview of the previously used TCO datasets and tropical width definitions, and new TCO data considered in this study. In the following section, we attempt to reproduce the H12 results using the same TCO data sets used in the original study, and we test the sensitivity of the H12 tropical widening estimates to the TCO dataset by analyzing a newer merged TCO data set. After this, we compare the seasonality and interannual variability of the H12 tropical width to other tropical width metrics.

Methods/Experimental

This section gives a brief overview of the data sets and previously used methodologies for quantifying a tropical edge latitude using TCO. H03 and subsequent papers divided the NH total column ozone field into four meteorological "regimes:" tropical, midlatitude, polar, and Arctic. In later papers (H06, H12), poleward movement of the tropical-midlatitude boundary, the so-called subtropical front, was quantified. It is this subtropical front boundary that is relevant in this study and to the broader tropical widening literature that has addressed changes in the width of the Hadley cell, tropopause, etc.

In essence, the "ozone meteorological regime" methodologies exploit the fact that TCO has a minimum in the tropics and is higher at midlatitudes in order to identify the subtropical boundary. The physical reasoning for this is that the lower stratosphere makes a large contribution to TCO, so the TCO is in some sense a proxy for the tropopause height. Along with the sharp drop in tropopause height in the subtropics, there is a corresponding increase in TCO.

Total column ozone data

The total column ozone data used in H03, H06, and H12 came from several satellites including the Nimbus-7 total ozone mapping spectrometer (TOMS, 1978–1993, McPeters et al. 1996) and Earthprobe TOMS (1996–2005, McPeters et al. 1998) version 8 data, the Advanced Tiros-N Operational Vertical Sounder (ATOVS) from several of the NOAA polar orbiting satellites, and the ozone monitoring instrument (OMI; Bhartia and Wellemeyer 2002) aboard the NASA Aura-EOS satellite. In all cases, daily gridded data were used. For TOMS and TOVS, the data were gridded at 1.25° lon × 1° lat horizontal resolution, whereas for OMI the data are provided at 1° lon × 1° lat resolution (i.e., the OMTO3d.003 product).

Here, we also use version 3.3 ("patched") of the Bodeker Scientific merged total column ozone data set (hereinafter BODSCI, Bodeker et al. 2005; Muller et al. 2008; Struthers et al. 2009). This data set merges measurements from four TOMS instruments, three different retrievals from the Global Ozone Monitoring Experiment (GOME), four Solar Backscatter Ultra-Violet (SBUV) instruments, and Aura OMI. In this data set, data from the ground-based World Ozone and Ultraviolet Data Center (WOUDC) Dobson and Brewer spectrophotometer network were used to remove offsets and drifts between the different data sets to produce a global homogeneous total ozone column data set. The BODSCI data set is provided at daily temporal resolution and on a 1.25° lon × 1° lat horizontal grid.

Total column ozone "meteorological regime" methods

The basic method for identifying the ozone boundaries in the NH is described in H03, and H06 presents a further modification to the H03 method. The H06 method is also used in H12. Although the methodology described in H03/H06 is for the NH, a similar methodology was also applied to the SH (Ph.D. dissertation, Flores 2004) and included in H12. For simplicity, and since the NH method is documented in the peer-reviewed literature, our primary focus will be on the NH. The reader is referred to H03 and H06 for a more in-depth description of the methods.

The H03 method is an iterative approach that uses daily gridded TCO data and daily isentropic potential vorticity (PV) from the NCEP-NCAR reanalysis (Kalnay et al. 1996) to compute the four ozone regimes (i.e., tropical, midlatitude, polar, and Arctic) and their boundaries. These regimes are intended to characterize regions of quasi-constant TCO that are separated by regions of relatively rapid changes in TCO, and the boundaries between these TCO regimes are supposed to correspond to meteorological phenomena such as the jets. For example, the tropical–midlatitude boundary occurs at the subtropical jet/tropopause break, where the abrupt change in tropopause height causes an abrupt change in TCO values. Similarly, the midlatitude–polar boundary is intended to correspond with the location of the polar front jet, and the polar-Arctic boundary occurs at the edge of the stratospheric polar vortex. By construction, the Arctic regime is the region within the polar vortex, when the vortex is present during winter. The Arctic regime is excluded from subsequent analysis in order to focus on dynamically induced changes to the TCO field and remove air masses that have been potentially affected by heterogeneous ozone chemistry. In H03, the Arctic regime was identified by using the position of the 31.5 PVU (1 PVU = 10^{-6} K m^2 kg^{-1} s^{-1}) boundary on the 450 K isentropic surface. TCO data poleward of this boundary are excluded from further analysis because of their potential to be impacted by ozone loss chemistry. In subsequent studies (H06 and H12), the polar vortex boundary was defined based on the PV value associated with the position of the maximum gradient in PV on the 550 K surface, again using NCEP-NCAR reanalysis data. This procedure was applied using daily data, and a 5-day running mean of the PV threshold was used to reduce noise (Follette 2007).

There are several ambiguities in the vortex definition in H12. First, there are multiple ways one could determine a PV value that is associated with the position of the maximum PV gradient. Also, no discussion is provided of how to define when a vortex is not present, as is the case for most of the year outside of winter months. To avoid these ambiguities, we mask out the polar vortex by using the vortex edge definition given by Nash et al. (1996). As with H12, this definition is applied to daily PV fields at 550 K from the NCEP-NCAR reanalysis, and the 5-day smoothing of the daily PV values at the vortex edge is also applied.

Following H12, after masking out the Arctic regime, a first-guess TCO value for the tropical-midlatitude (i.e., the subtropical front) and midlatitude-polar boundary (i.e., the polar front) is applied to make an initial identification of the three remaining regions. The first-guess TCO thresholds are based on previous work (Karol et al. 1987; Shalamyanskiy and Romashkina 1980) and vary as a function of month (see Fig. 3, H03). A contour program is used to identify the regime boundaries from the first guess, and pixels within each regime are grouped together. Pixels within 2° of latitude and longitude of the boundaries are excluded. The daily mean TCO for each regime is then simply an average of the data within the regime, and the next-guess boundaries are defined as being halfway between the averages for each regime. This iterative process is then repeated until convergence is obtained (i.e., regime area changes by less than 5%) and a final set of TCO boundaries is determined. For the final identification of the regimes, a 30-day running mean of the TCO thresholds is used to reduce noise (Hudson et al. 2003).

In H06/H12, a second iterative process was added to this method to allow for a latitudinal dependence to exist in the regime boundaries. After computing the daily boundaries using the H03 method, these boundaries were then used to separate tropical and midlatitude regimes in each latitude band and come up with a latitudinally varying threshold for the subtropical front. As in H12, the daily timeseries have been averaged to create monthly mean timeseries.

Results and discussion

Replication of previous results

Because separating the meteorological regimes involves a complex algorithm, we first attempt to replicate the results from H12. Figure 1 shows the timeseries of NH subtropical front latitude taken from Fig. 5 of H12 (black line), as well as our implementation of the H12 methodology using the three TOMS data sets and the BODSCI data set.

The most valid comparison between H12 and our replication should be with the TOMS data sets, because they were used by H12 along with the no longer available TOVS_NEURAL to fill in data gaps in the 1990s between the Nimbus-7 and Earthprobe records. Our implementation with TOMS generally reproduces the absolute value and seasonality of H12, particularly during winter, but with significant differences during the summer months. Overall, the H12 values are much farther equatorward during summer than ours. The reason for this is that ozone regime boundaries in the Hudson methodology are extremely sensitive to the polar vortex identification. The size of the polar vortex area masked out impacts not only the ozone boundary values for the polar-midlatitude boundary but also the midlatitude-tropical boundary which is used to identify the latitude of the subtropical front.

The sensitivity to polar vortex masking is shown in Fig. 2, which shows the tropical width time series from the BODSCI data using three methods of identifying the polar vortex, overlaid with the time series from H12. The first method is the Nash method, which is the same method used in Fig. 1 (orange line). The second method identifies the vortex boundary using a PV threshold on the 550 K isentrope. The PV threshold is defined as the median value of PV within the upper second percentile of PV gradients within the NH. This definition reflects our attempt at implementing a robust and objective vortex identification that follows the ambiguous definition offered in H12, namely the "PV value associated with the maximum gradient in PV". The third definition is the same as the second, except that when the PV threshold falls below 40 PVU no vortex is defined. The third definition addresses the fact that for most of the year, the polar vortex does not exist, which is not taken into account by the second definition. As can be seen in Fig. 2, the second method goes much farther south, due to the area masked as "vortex" growing to encompass most of the NH. Since the three ozone regimes are still being defined within an area that is smaller, and the TCO gradients are relatively weak in summer, the TCO boundary values end up being strongly perturbed from their climatological first-guess values. This results in the latitudinal edges being displaced equatorward until they reach the minimum value allowed by the H12 definition (i.e., 25°

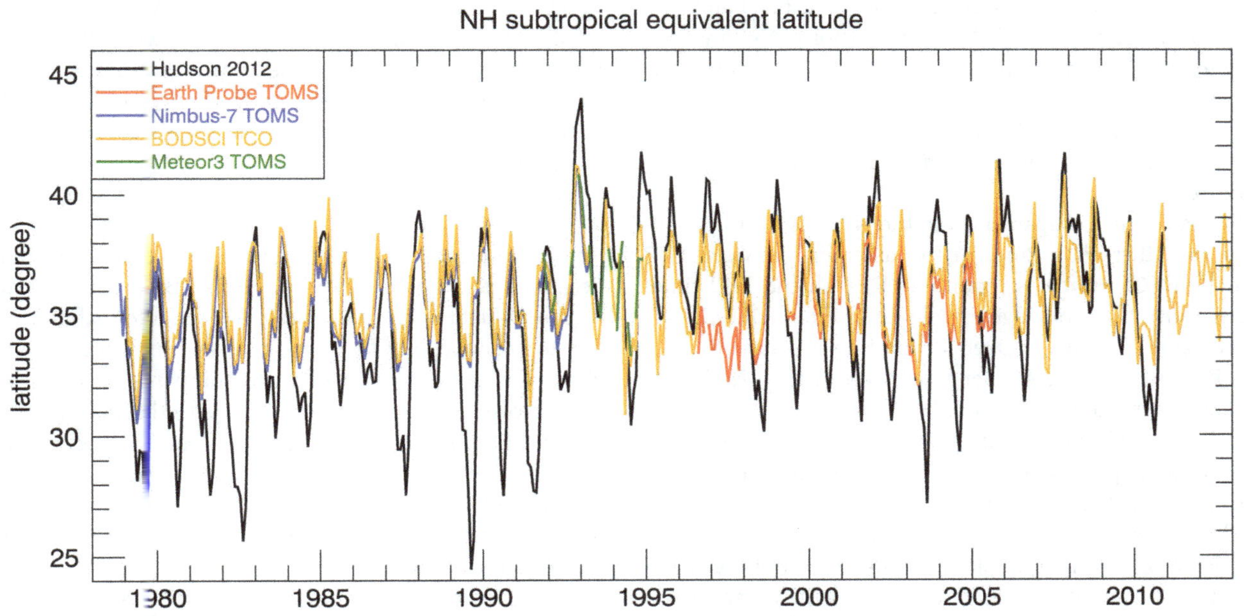

Fig. 1 Timeseries of total column ozone-based subtropical front equivalent latitude in the Northern Hemisphere from five different data sources. Calculations from Hudson 2012 are in black, and the colored lines are our calculations from four different TCO data sets

N). The third definition attempts to mitigate this unphysical growth of the vortex by setting a minimum PV threshold to use for the masking. In all cases, during wintertime when the vortex is present, the three methods are in relatively close agreement.

In addition to the differences between H12 and our implementation during summer months that are caused by vortex identification differences, there is an apparent jump in the H12 latitude in 1992 near the end of the Nimbus-7 TOMS data set. Moreover, at this time, there is also a jump

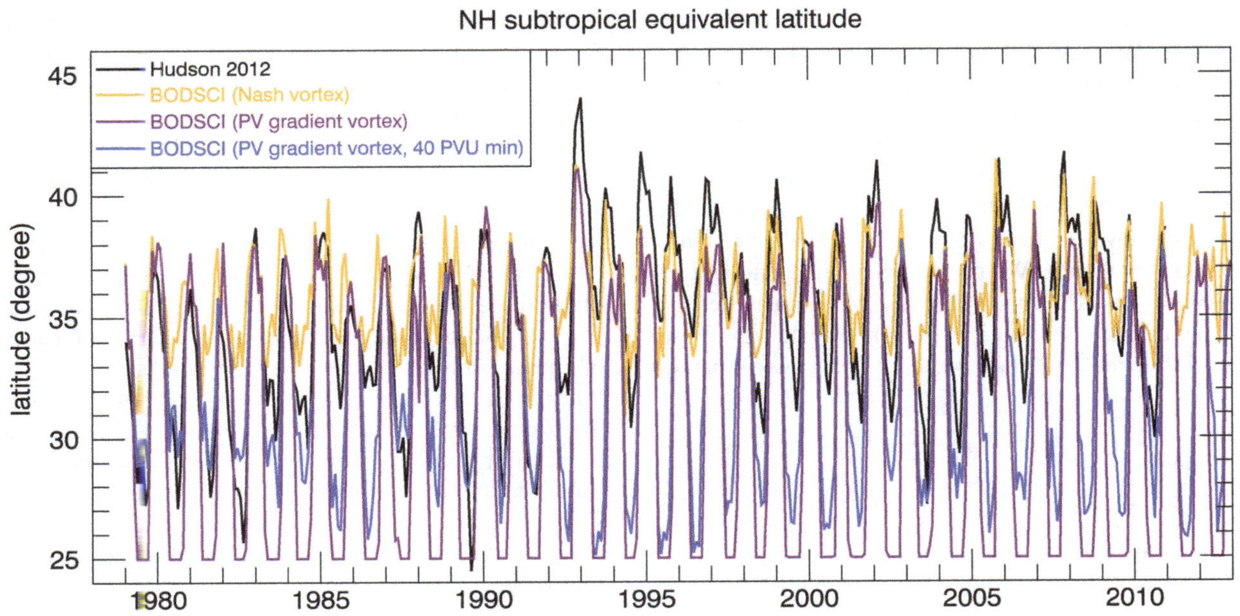

Fig. 2 NH tropical width sensitivity to polar vortex masking. Calculations from Hudson 2012 are in black, and the colored lines show the NH subtropical edge latitude computed from three different vortex identification algorithms

of several degrees in the difference between our Nimbus-7 TOMS timeseries and H12 in 1992. Before this time, our timeseries calculated using Nimbus-7 TOMS TCO generally agrees well with the H12 estimates. After this time, the H12 timeseries is significantly offset poleward of ours by several degrees of latitude, particularly during summer months.

Similar to the comparison with the Nimbus-7 TOMS calculations, there is a jump in the difference between H12 and BODSCI estimates around 1992, with the H12 timeseries moving several degrees poleward of the BODSCI estimates after that time. Overall, our implementation using the BODSCI data agrees very well with that using TOMS, which is expected given that TOMS data are included in the merged BODSCI product. However, the similarity between our implementation of the H12 method using both the raw (unadjusted) TOMS data and the homogenized BODSCI data suggests that the jump is not substantially impacted by data merging in the construction of the BODSCI data set and could be an as-yet unidentified artifact in the data used by H12. As previously noted, H12 used a TOVS data set that is no longer available to fill in some of the data gaps in the 1990s during the transition between TOMS instruments, so it is not possible for us to directly assess the potential impact of data inhomogeneities in their study. Rather, in the next section, we investigate the significance and impact of the jump in the 1990s, and its effect on long-term trends derived in H12.

Trend and breakpoint analysis of tropical width timeseries

In this section, we compare our trends to H12 and assess the significance of the apparent jump in the 1990s in the H12 data. We perform linear trend fits to the anomaly timeseries in each hemisphere and provide 95% confidence intervals on the trend after accounting for lag-1 autocorrelation of the residuals (Santer et al. 2000). Between 1979 and 2010, H12 found poleward movement of the subtropical front of 3.7° ± 0.3° in the NH and 6.5° ± 0.2° in the SH, which is equivalent to a global tropical widening rate of 3.2° ± 0.1° per decade. We performed an independent trend analysis on the H12 data and found very similar trends in each hemisphere (see Table 1), but with much larger uncertainty estimates (~ 1° per decade). The reason that our values are larger is mostly due to us accounting for autocorrelation and quoting a 95% confidence interval (CI) rather than a 1σ estimate. For example, for the global mean trend, we compute a 1σ trend uncertainty of 0.2°, which is similar to that found in H12 but a factor of 5 less than our 95% CI estimate.

To investigate whether or not the potential jump in the 1990s is significant and how it impacts the derived long-term trends, we use the two-phase regression model described by Lund and Reeves (2002). Briefly, this model tests each data point in a timeseries as a potential undocumented changepoint by doing a linear least-squares fit on either side of the changepoint and compares the reduction in sum-of-square errors (SSE) relative to the case of fitting a single line for the entire timeseries. The SSE reduction is quantified using an F statistic, and an undocumented changepoint is said to occur at the maximum value of F if it is greater than a critical value of F (see, e.g., Lund and Reeves 2002, Table 1).

Figure 3 shows the anomaly of the H12 timeseries (i.e., with the seasonal cycle removed) in both hemispheres, as well as our results from the NH using the BODSCI data. As can be seen in this figure, the H12 timeseries contain strong trends in both hemispheres (solid lines), whereas our analysis of the BODSCI calculations in the NH shows very little trend in comparison. The changepoint analysis identifies a statistically significant breakpoint in the H12 timeseries in both hemispheres in the same year and month, July 1992. For the BODSCI NH timeseries, no statistically significant breakpoint is found.

To assess the impact of the breakpoint in the H12 timeseries on trends, we remove its effect by subtracting the latitudinal offset at the breakpoint (i.e., the difference between the dashed lines in Fig. 3a, at July 1992) from the latter portion of the timeseries, and then compute the linear trends on this "corrected" timeseries (Fig. 3b). After removing the breakpoints in the H12 timeseries, the trends in both hemispheres are significantly reduced (see Table 1) and the difference from our NH trend using BODSCI timeseries is not statistically significant (based on the confidence interval method described in Santer et al. 2000). The global tropical widening rate is reduced by more than a factor of three, from over 3° per decade and statistically significant to less than 1° per decade and not statistically significant. It is worth noting that with or without breakpoint removal, the SH trends are greater than those in the NH, in line with many previous studies suggesting the Antarctic stratospheric ozone depletion is one of the main drivers of tropical width changes and that ozone depletion has its greatest impact on the SH tropical width (Polvani et al. 2011; Waugh et al. 2015).

Given that the objective breakpoint analysis reveals a breakpoint at the same date (July 1992) in both

Table 1 Tropical widening trends from the H12 method (° latitude per decade)

O_3 data	H12 published values	H12 our trend calculation	H12 with breakpoint removed	BODSCI
NH	1.2 ± 0.1	1.3 ± 0.9	−0.061 ± 0.7	0.34 ± 0.2
SH	2.0 ± 0.1	2.1 ± 1.3	0.93 ± 1.1	
Global	3.2 ± 0.1	3.3 ± 2.1	0.87 ± 1.6	

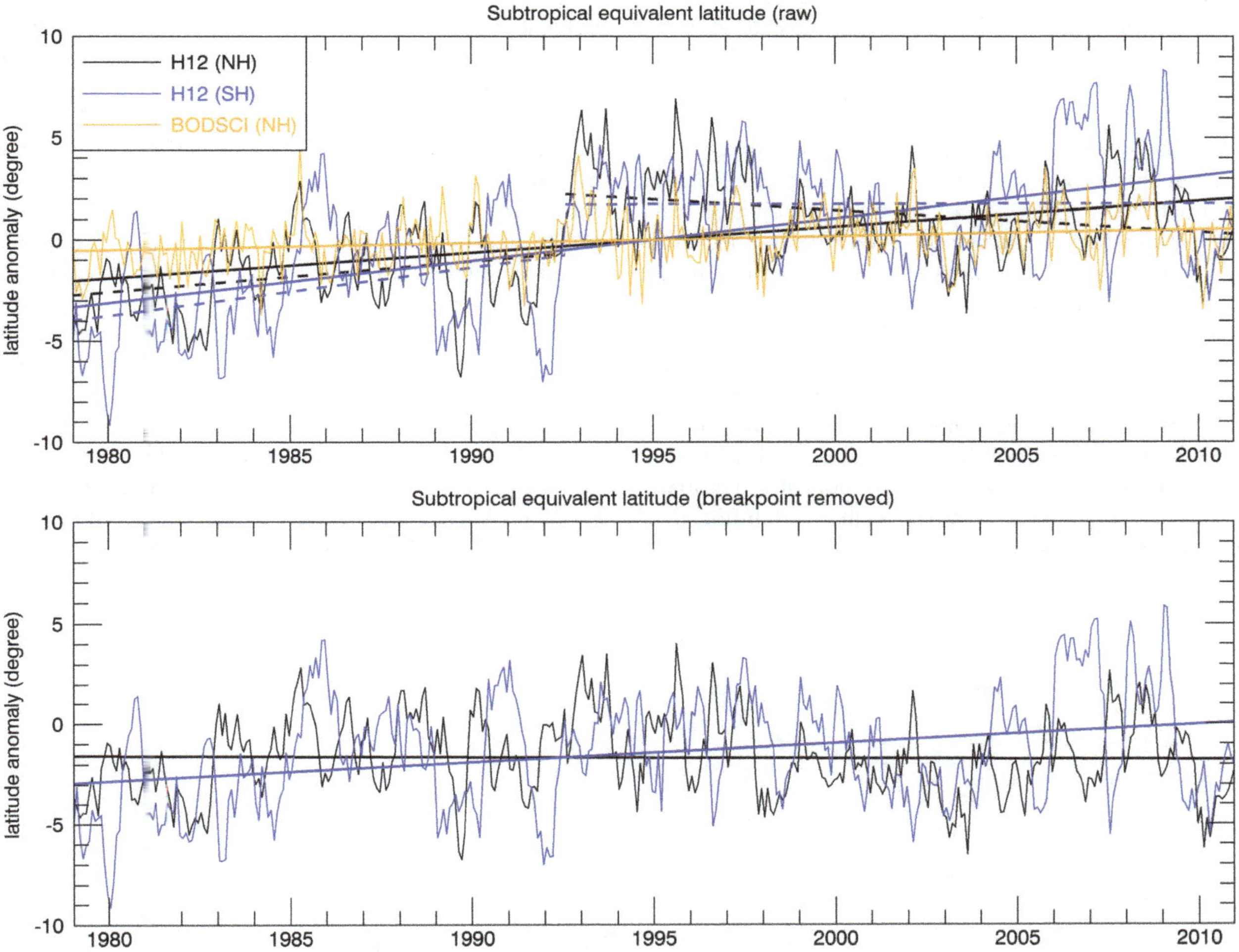

Fig. 3 Top panel Timeseries analysis illustrating an undocumented changepoint in the Hudson 2012 metric (black, blue lines) and our implementation using the BODSCI TCO data (yellow line). Lower panel: the reconstructed timeseries with breakpoint removed

hemispheres in the H12 timeseries, and not in our implementation of the H12 method using the BODSCI TCO data, we hypothesize that the H12 result was impacted by a spurious change related to data merging. Although Nimbus-7 operated until May 1993, it is possible that H12 made the transition to TOVS TCO before the end of the Nimbus-7 lifetime.

Relation to other tropical width metrics

In this section, we investigate how the H12 tropical width metric relates to other tropical width metrics by comparing its seasonality and interannual variability with tropical width calculated over the same time period from the ERA-Interim reanalysis (Dee et al. 2011). The metrics here are a subset of the metrics documented in Adam et al. (2018), and the reader is referred to the discussion there for more thorough definitions. Briefly, we use the poleward edge of the subtropical dry zone, defined by the latitude where zonal mean precipitation equals evaporation (P – E = 0); the poleward edge of the Hadley cell, defined as the zero-crossing of the mean meridional streamfunction at 500 hPa ($\psi_{500} = 0$); the subtropical jet (STJ), defined as the maximum of the mass-weighted wind in the 400–100 hPa layer with the surface wind removed; the subtropical tropopause break, defined by the maximum meridional gradient of the tropopause height; and the eddy-driven jet (EDJ), defined as the latitude of maximum wind on the 850 hPa surface. Although the EDJ is not technically a marker of the "tropics", it has been shown to be well correlated with the position of the Hadley cell. Of these metrics, the so-called "lower" metrics (P-E, ψ_{500}, and EDJ) have been demonstrated to correlate well with one another, whereas the so-called "upper metrics" (STJ and tropopause) are not well correlated with the lower metrics (Davis and Birner 2017; Solomon et al. 2016). Here, we include the upper metrics because there are physical reasons that they may be related to the stratospheric ozone distribution.

Seasonality

If they are measuring similar phenomena, tropical edge latitudes should have similar seasonal cycles. For example, one might expect that TCO- and tropopause-based tropical width metrics would closely follow one another, based on the physical relation between the tropopause height and total column ozone.

Although the reanalysis tropical edge latitudes shown in Fig. 4 have different mean values and different amplitudes in their seasonal cycles, in both hemispheres, they move poleward during summer months and equatorward during winter months. In the SH, the H12 metric roughly follows the seasonal progression of the reanalysis metrics, albeit with a much smaller amplitude than most other metrics. In contrast, in the NH, the H12 metric (and our computation using BODSCI TCO data) is most poleward during winter and most equatorward during summer. Thus, it is out of phase with the rest of the metrics.

As discussed in the "Results and discussion" section, the NH edge latitudes are very sensitive to the vortex masking, which impacts the iteratively determined thresholds for the subtropical front. To see the impact of the iterative process, we also computed the ozone-based tropical width using just the first guess of the subtropical front threshold (i.e., from Fig. 3 of Hudson et al. 2003). Interestingly, the tropical width based on this first guess shows the correct seasonality in the NH, which means that some aspect of the vortex masking or iterative procedure causes the seasonality of the full method to be out of phase. It is possible that a simplified TCO-based tropical edge metric could be defined based on such a seasonally varying fixed TCO threshold, but in such a case chemically induced ozone

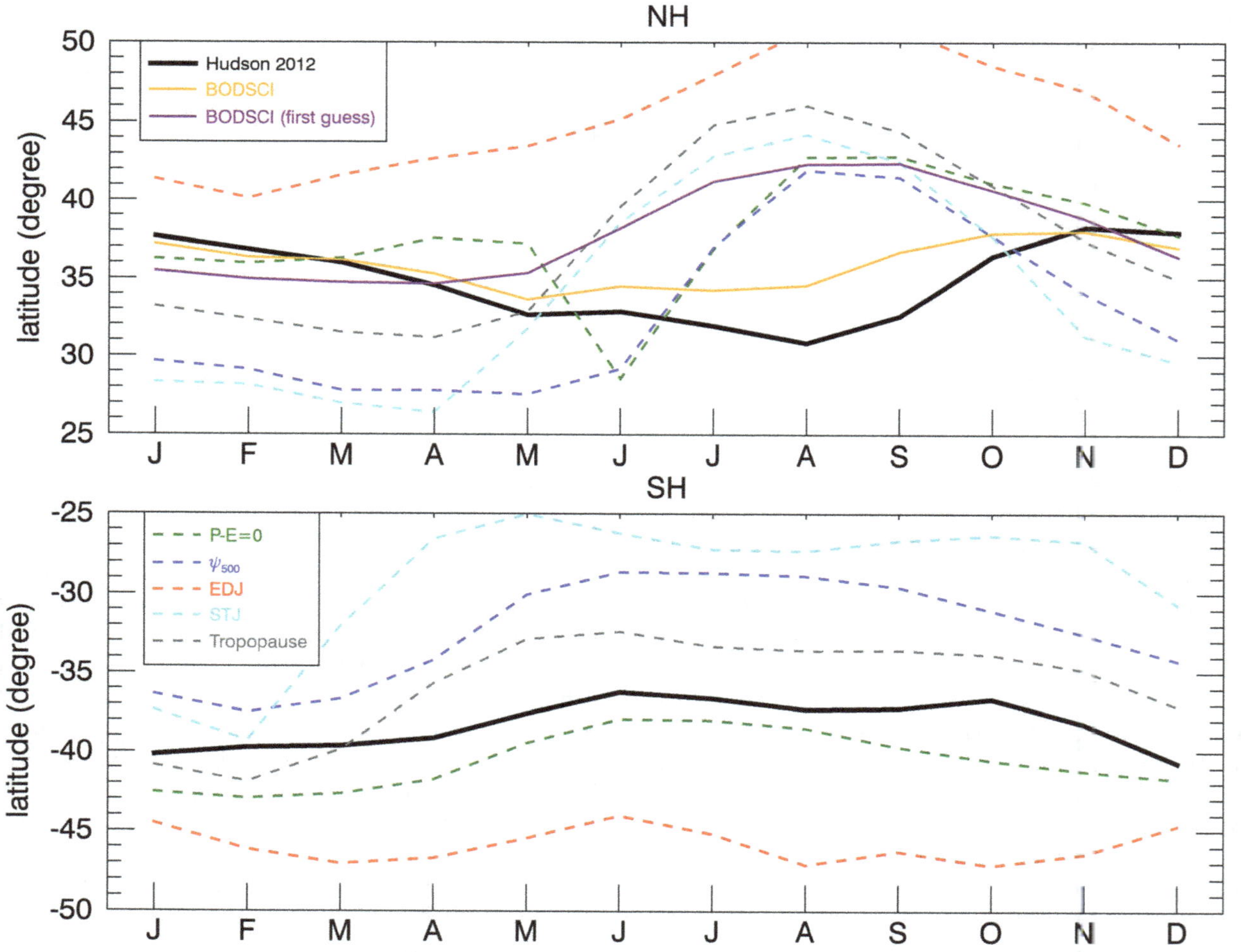

Fig. 4 Seasonal cycle of tropical width metrics from TCO data (solid lines) and ERA-Interim reanalysis (dashed lines) over their common time period (1979–2010). The EDJ metric has been shifted 5° equatorward in the SH

depletion (that is unrelated to dynamical changes) would cause poleward movement of the subtropical front.

Interannual variations

In addition to assessing the seasonality of the H12 metric in comparison to other tropical width metrics, we also consider whether H12 correlates well with the other tropical width metrics on interannual timescales. The idea being tested is that if a pair of metrics are measuring the same physical quantity, they should both be anomalously poleward or equatorward in step with one another. This type of analysis was used to identify the set of "upper" and "lower" tropical width metrics discussed above (Davis and Birner 2017; Solomon et al. 2016; Waugh et al. 2018). Here, we reproduce the results of those previous studies but add in the Hudson 2012 metric in order to identify whether the ozone-based tropical edge correlates with either of these categories of metrics.

Figure 5 shows the correlations between detrended monthly mean tropical width anomalies for all combinations of metrics in each hemisphere, including both reanalysis tropical width metrics from ERA-Interim computed using the Tropical Width Diagnostics package (TropD, Adam et al. 2018) and the uncorrected H12 timeseries, the H12 timeseries with breakpoints removed, and our implementation of the H12 method in the NH data from the BODSCI TCO data. This table shows that the H12 metrics are generally not well

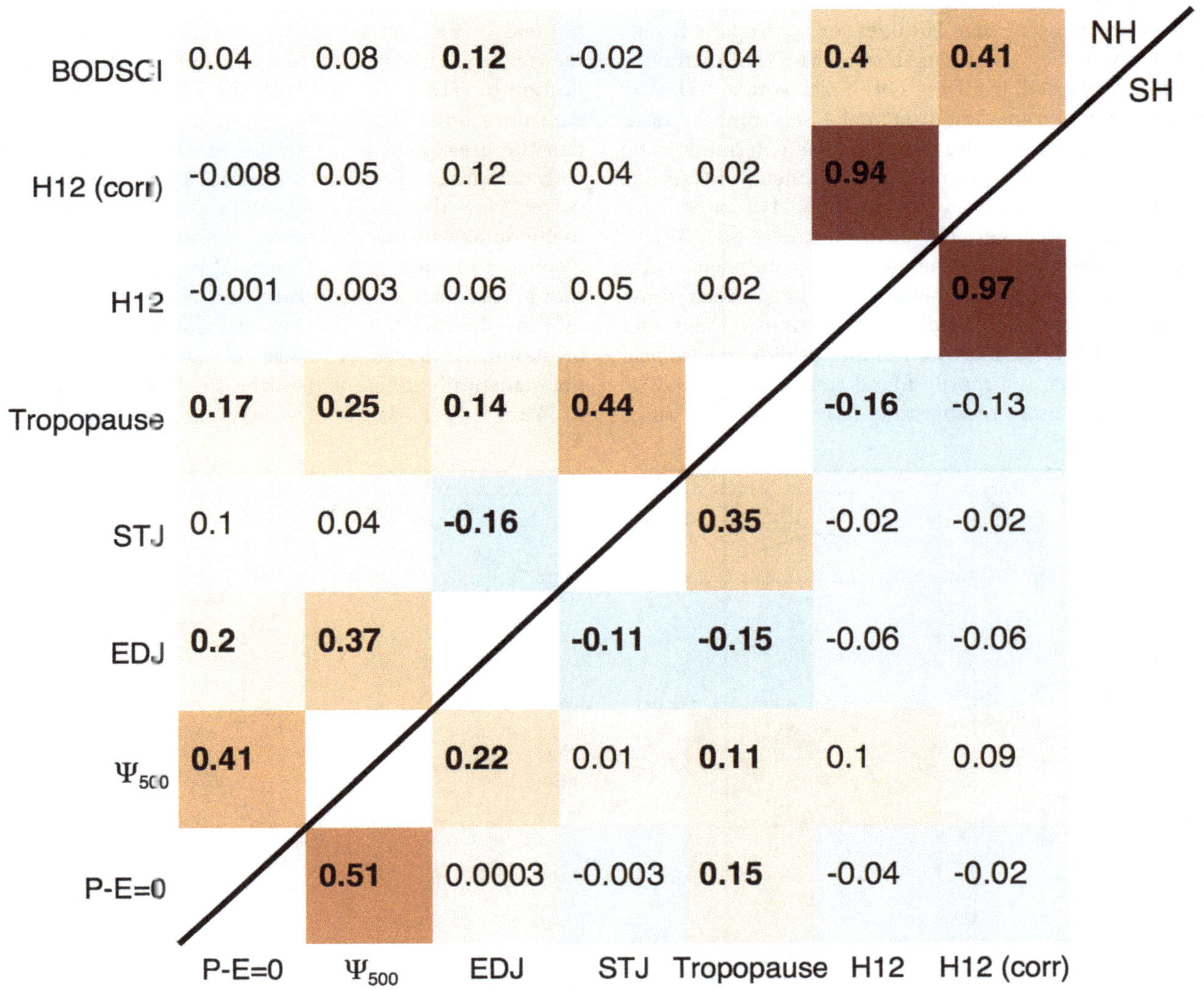

Fig. 5 Correlation coefficients between pairs of detrended monthly mean tropical width metric anomalies (1979–2010) from ERA-Interim reanalysis, calculations from Hudson 2012 (both raw and corrected versions), and our implementation of the Hudson 2012 method using BODSCI TCO data, color-coded by the value. The upper left is NH and the lower right is SH. Statistically significant correlations, taking into account autocorrelation of the timeseries, are shown in bold

correlated with any of the reanalysis tropical width metrics. The only statistically significant correlation is between the ozone metric and the EDJ in the NH, but the magnitude of this correlation is quite small. It is also worth noting that although many of the correlations within the reanalysis metrics are quite low, this is expected given the different categories that have been identified by previous studies.

Overall, the out-of-phase seasonal cycle of the H12 NH metric and the lack of apparent relation to the tropopause height suggests that some methodological detail in the H12 formulation is problematic.

Conclusions

Total column ozone varies with latitude, in part due to tropopause height differences between the tropics and midlatitudes. This physical relation was used by H12 to identify the tropical edge latitudes, or subtropical fronts, and compute their variations in time. The H12 method is rather complicated, involving both polar vortex boundary identification and masking using daily isentropic PV data, and two separate iterative processes for determining the subtropical front TCO thresholds as a function of latitude from daily TCO data. Using this method, H12 found tropical widening over the 1979–2010 of greater than 3° latitude per decade (both hemispheres combined). This tropical widening rate is significantly larger than those published in previous studies. For example, Davis and Rosenlof (2012) considered a number of different tropical edge diagnostics and mostly found trends of less than 1° per decade. A more recent study has reduced this range even further, to ~ 0.1°–0.5° per decade (Staten et al. 2018). To put these numbers in perspective, even tropical widening rates of 1° per decade are outside the range of trends predicted by historical simulations of the twentieth century in models (Grise et al. 2018; Johanson and Fu 2009). Thus, if the extreme tropical widening found in H12 is correct it suggests a major deficiency in the representation of tropical widening in models.

In order to reassess the H12 results, we replicated their methodology in the NH with both the original TOMS data used by H12 and the new homogenized BODSCI TCO data set. Overall, we were able to reproduce the general features of the H12 results, although we noted a large sensitivity to the polar vortex identification process. Furthermore, we found that the H12 tropical width timeseries in each hemisphere contained a statistically significant breakpoint in July 1992, around the end of the lifetime of the Nimbus-7 TOMS instrument when H12 switched to using TOVS data. Our replication of H12 in the NH with the TOMS and BODSCI data does not contain this breakpoint, so we conclude that the breakpoint is a spurious jump, likely associated with data inhomogeneities in the H12 analysis. We then correct for the breakpoint by applying an objective methodology to the H12 timeseries and recompute the tropical widening rates. The resulting global widening rate is less than 1° per decade and is statistically insignificant. In the NH, the H12 widening rate with the breakpoint removed is statistically insignificant and is not statistically different than the small significant trend (0.34 ± 0.2° per decade) we compute based on the

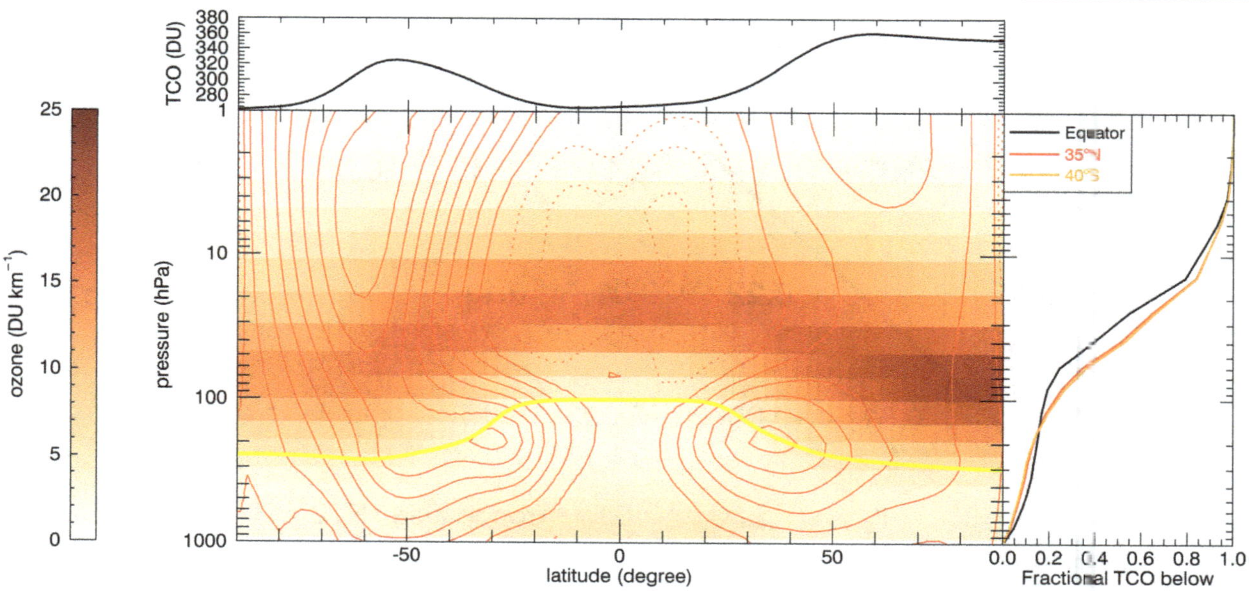

Fig. 6 Climatological ozone distribution (1979–2010) from the ERA-Interim reanalysis, along with the tropopause (yellow line). The right panel shows the fraction of the total column ozone below the given height at three different latitudes

BODSCI data. Thus, we conclude that the large tropical widening trends (i.e., > 1° per decade) reported by H12 were a result of unphysical data jumps. Because the H12 methodology was not documented in H12 and enough detail for us to produce tropical width timeseries in the SH, it is not possible to say definitively what the global tropical widening rate is from this method when applied to the BODSCI data.

In addition to assessing data quality issues in the H12 timeseries, we also assessed the robustness of the H12 method by comparing its seasonality and interannual variability to a number of independent tropical edge latitude diagnostics. While the H12 method generally produces edge latitudes that are within the range of other metrics, both the seasonality and interannual variations indicate that it is not physically well related to other tropical width metrics. In the SH, the H12 seasonal cycle amplitude is smaller than the other metrics, and in the NH, the seasonal cycle is its out-of-phase with the other metrics (i.e., it goes poleward in summer and equatorward in winter). The lack of correlation with well-established tropical width metrics on interannual timescales supports our conclusion that the H12 TCO-based tropical width diagnostic is not a robust measure of tropical width.

One obvious question raised by this conclusion is why the TCO-based width diagnostic is not a good measure of the tropics, given the expected relationship between the tropopause height and total column ozone. One possibility lies in the recent finding that tropopause-based tropical width metrics are themselves not a robust indicator of Hadley cell width, or at least that they do not correlate well with metrics directly related to the Hadley Cell (Davis and Birner 2017; Solomon et al. 2016; Waugh et al. 2018). However, our analysis shows that the ozone-based tropical width metric is not well correlated with subtropical jet metrics, so this explanation is insufficient.

A better explanation for why TCO-based with diagnostics fail to capture the tropopause variability is that variations in TCO are relatively weakly driven by tropopause variability. For example, the fraction of TCO occurring below the overworld stratosphere (≥ 380 K potential temperature, or ~ 90 hPa) is only 15–20% (Fig. 6). Approximately 5–10% of TCO is below the extratropical tropopause (~ 300 hPa), meaning that the amount of ozone in the region that can be impacted by latitudinal variations in the tropopause break (i.e., between the extratropical tropopause and the overworld stratosphere) is only ~ 10%. Given that relatively small lever arm on TCO, it is perhaps unsurprising that defining tropical width based solely on TCO is problematic.

In summary, we have identified several problems with TCO-based tropical width. Broadly, these problems relate to the complex iterative algorithm employed previously, inhomogeneities in the TCO data, and the relatively small component of TCO variability that can be explained by latitudinal variations in the tropopause break. However, in principle, the strong meridional ozone gradient in the lower stratosphere should be a good marker of the location of the tropopause break. Given the potential temporal inhomogeneities in reanalysis fields, there is a justified desire for robust observationally based metrics of tropical width that do not rely on reanalysis output. It is possible that vertically resolved ozone measurements could be used to address this issue, although the duration and reliability of such measurements may be limited to the point that it is not possible to confidently use them for trend detection. Future studies aimed at the suitability of vertically resolved ozone measurements for tropical edge latitude identification should be able to clarify the extent to which existing measurements can be used for tropical widening studies.

Abbreviations

ATOVS: Advanced Tiros-N Operational Vertical Sounder; BODSCI: Bodeker Scientific merged total column ozone data set; EDJ: Eddy-driven jet; GOME: Global Ozone Monitoring Experiment; H03: Hudson et al. 2003; H06: Hudson et al. 2006; H12: Hudson 2012; NASA: National Aeronautics and Space Administration; NCEP-NCAR: National Centers for Environmental Prediction – National Center for Atmospheric Research; NH: Northern Hemisphere; NOAA: National Oceanic and Atmospheric Administration; OLR: Outgoing longwave radiation; OMI: Ozone monitoring instrument; PV: Potential vorticity; SH: Southern Hemisphere; SSE: Sum-of-square errors; STJ: Subtropical jet; TCO: Total column ozone; TOMS: Total ozone mapping spectrometer; TOVS: Tiros-N Operational Vertical Sounder; WOUDC: World Ozone and Ultraviolet Data Center

Acknowledgements

The authors thank Robert Portmann and members of the U.S. CLIVAR Changing Width of the Tropical Belt Working Group and International Space Science Institute Tropical Width Diagnostics Intercomparison Project for helpful feedback and encouragement on this work. The authors also wish to thank Robert Hudson and Melanie Follette Cook for discussions related to the TCO-based tropical width algorithm. The authors thank Eric Nash for providing IDL code for the polar vortex identification. SD gratefully acknowledges the travel grant from Japan Geoscience Union to attend the 2017 JpGU-AGU joint meeting in Tokyo, Japan.

Funding

This work was supported by NASA Grant 12-ACMAP12-0010.

Authors' contributions

KHR and SMD proposed the topic, conceived, and designed the study. BH implemented the TCO algorithm and assisted with the data analysis. All authors collaborated with the corresponding author in the construction of manuscript. All authors read and approved the final manuscript.

Competing interests

The authors declare that they have no competing interests.

Author details

[1]NOAA Earth System Research Laboratory Chemical Sciences Division, 325 S. Broadway, Boulder, CO 80305, USA. [2]Cooperative Institute for Research in Environmental Sciences, University of Colorado at Boulder, Boulder, CO 80303, USA. [3]Deutsches Zentrum für Luft- und Raumfahrt (DLR), Institut für Physik der Atmosphäre, Oberpfaffenhofen, Germany.

References

Adam O, Grise KM, Staten P, Simpson IR, Davis SM, Davis NA, Waugh DW, Birner T (2018) The TropD software package: Standardized methods for calculating Tropical Width Diagnostics. Geosci Model Dev:1–35. https://doi.org/10.5194/gmd-2018-124

Bhartia PK, Wellemeyer CW (2002), OMI TOMS-V 8 total O3 algorithm, algorithm theoretical baseline document: OMI ozone products, II(ATBD-OMI-02, Version 2.0)

Birner T (2010) Recent widening of the tropical belt from global tropopause statistics: sensitivities. J Geophys Res-Atmos 115(D23):13. https://doi.org/10.1029/2010JD014664

Bodeker GE et al (2005) Indicators of Antarctic ozone depletion. Atmos Chem Phys 5:2603–2615

Davis N, Birner T (2017) On the discrepancies in Tropical Belt expansion between reanalyses and climate models and among Tropical Belt width metrics. J Clim 30(4):1211–1231

Davis SM, Rosenlof KH (2012) A multidiagnostic intercomparison of tropical-width time series using reanalyses and satellite observations. J Clim 25(4):1061–1078

Dee DP et al (2011) The ERA-Interim reanalysis: configuration and performance of the data assimilation system. Q J Roy Meteor Soc 137(656):553–597

Flores MFA (2004) Stratospheric ozone trends as determined by regime analysis: the Southern Hemisphere, 154 pp. University of Maryland, College Park

Follette MB (2007) Classification of Northern Hemisphere stratospheric ozone and water vapor profiles by meteorological regime: validation, climatology, and trends, Ph.D. thesis. University of Marlyand, College Park, p 158

Grise KM, Davis SM, Staten PW, Adam O (2018) Regional and seasonal characteristics of the recent expansion of the tropics. J Climate 31(17):6839–6856. https://doi.org/10.1175/JCLI-D-18-0060.1

Hu Y, Fu Q (2007) Observed poleward expansion of the Hadley circulation since 1979. Atmos Chem Phys 7(19):5229–5236

Hudson R et al (2003) The total ozone field separated into meteorological regimes. Part I: defining the regimes. J Atmos Sci 60(14):1669–1677

Hudson RD (2012) Measurements of the movement of the jet streams at mid-latitudes, in the Northern and Southern Hemispheres, 1979 to 2010. Atmos Chem Phys 12(16):7797–7808

Hudson RD et al (2006) The total ozone field separated into meteorological regimes - part II: Northern Hemisphere mid-latitude total ozone trends. Atmos Chem Phys 6:5183–5191

Johanson CM, Fu Q (2009) Hadley cell widening: model simulations versus observations. J Clim 22(10):2713–2725

Kalnay E et al (1996) The NCEP/NCAR 40-year reanalysis project. B Am Meteorol Soc 77(3):437–471

Karol IL et al (1987) Fields of ozone and temperature within the boundaries of air masses. Meteorol Gidrol 10:35–39

Lucas C et al (2014) The expanding tropics: a critical assessment of the observational and modeling studies. Wires Clim Change 5(1):89–112

Lund R, Reeves J (2002) Detection of undocumented changepoints: a revision of the two-phase regression model. J Clim 15(17):2547–2554

McPeters, R., et al. (1996), Nimbus-7 Total Ozone Mapping Spectrometer (TOMS) data products user's guide, NASA Reference Publication 1384, National Aeronautic and Space Administration

McPeters, R., et al. (1998), Earth Probe Total Ozone Mapping Spectrometer (TOMS) data products user's guide, NASA Technical Report TR 1998-206895, National Aeronautic and Space Administration

Muller R et al (2008) Simple measures of ozone depletion in the polar stratosphere. Atmos Chem Phys 8(2):251–264

Nash ER et al (1996) An objective determination of the polar vortex using Ertel's potential vorticity. J Geophys Res-Atmos 101(D5):9471–9478

Polvani LM, Waugh DW, Son S-W (2011) Stratospheric ozone depletion: The main driver of Twentieth-Century atmospheric circulation changes in the Southern Hemisphere. J Climate 24(3):795–812. https://doi.org/10.1175/2010JCLI3772.1

Santer BD et al (2000) Statistical significance of trends and trend differences in layer-average atmospheric temperature time series. J Geophys Res-Atmos 105(D6):7337–7356

Seidel DJ, Randel WJ (2007) Recent widening of the tropical belt: evidence from tropopause observations. J Geophys Res-Atmos 112(D20):D20113

Seidel DJ et al (2008) Widening of the tropical belt in a changing climate. Nat Geosci 1(1):21–24

Shalamyanskiy AM, Romashkina AK (1980) Distribution and variation in the total ozone concentration in various air masses. Izv Atmos Oceanic Phys 16:931–937

Solomon A et al (2016) Contrasting upper and lower atmospheric metrics of tropical expansion in the Southern Hemisphere. Geophys Res Lett 43(19):10496–10503

Staten PW, Lu J, Grise KM, Davis SM, Birner T (2018) Re-examining tropical expansion. Nature Clim Change 8(9):768–775. https://doi.org/10.1038/s41558-018-0246-2

Struthers H et al (2009) The simulation of the Antarctic ozone hole by chemistry-climate models. Atmos Chem Phys 9(17):6363–6376

Waugh DW, Garfinkel CI, Polvani LM (2015) Drivers of the recent tropical expansion in the Southern Hemisphere: Changing SSTs or ozone depletion? J Climate 28(16):6581–6586. https://doi.org/10.1175/JCLI-D-15-0138.1

Waugh DW, Grise KM, Seviour WJM, Davis SM, davis N, Adam O, Son SW, Simpson IR, Staten PW, Maycock AC, Ummenhofer CC, Birner T, Ming A (2018) Revisiting the relationship among metrics of tropical expansion. J Climate 31(18):7565–7581. https://doi.org/10.1175/JCLI-D-18-0108.1

8

Spatial distribution of cold-season lightning frequency in the coastal areas of the Sea of Japan

Daiki Tsurushima[1*], Kiyotaka Sakaida[2] and Noriyasu Honma[3]

Abstract

The coastal areas of the Sea of Japan are a well-known hotspot of winter lightning activity. This study distinguishes between three common types of winter lightning in that region (types A–C), based on their frequency distributions and the meteorological conditions under which they occur. Type A lightning occurs with high frequency in the Tohoku district. It is mainly caused by cold fronts that accompany cyclones passing north of the Japanese islands. Type B, which occurs most frequently in the coastal areas of the Hokuriku district, is mainly caused by topographically induced wind convergence and convective instability, both of which are associated with cyclones having multiple centers. Type C's lightning frequency distribution pattern is similar to that of type B, but its principal cause is a topographically induced wind convergence generated by cold air advection from the Siberian continent. Type A is most frequently observed from October to November, while types B and C tend to appear from November to January, consistent with seasonal changes in lightning frequency distribution in Japan's Tohoku and Hokuriku districts.

Keywords: Lightning, LLS, Cluster analysis, Climatology

Background

The coastal areas of the Sea of Japan are among the most widely known winter lightning activity hotspots in the world (Rakov and Uman 2007). Since winter lightning contains more electrically intensive discharges than summer lightning (Brook et al. 1982; Hojo et al. 1989; Ishii et al. 2012), it often causes serious damage to electrical equipment (transmission lines, wind turbines, etc.).

In Japan, district-wide lightning detection networks have been in place since the 1990s. These networks, generally called lightning location systems (LLS), are able to make highly accurate measurements of lightning geo-locations and the time of lightning occurrences.

LLSs have made it possible to examine detailed patterns of winter lightning activity in the coastal areas of the Sea of Japan. For example, Fujisawa and Kawamura (2005) clarified lightning frequency distributions and seasonal changes (Nov.–Feb.) based on 9 years of data obtained by the LLS in the Hokuriku district. Their results indicated that intensive midwinter lighting activity (Dec.–Feb.) in the Hokuriku district tends to be concentrated within an area tens of kilometers from the coastline. In another study, Sugita and Matsui (2012) examined seasonal and inter-annual variations of lightning frequency around Japan based on 10 years of observations gathered by the Japan Lightning Detection Network (JLDN). Also, Ishii et al. (2014) used data from the LIghtning DEtection Network system (LIDEN) to provide a statistical overview of lighting activity tendencies all around Japan.

However, the meteorological conditions that determine the characteristics of the lightning frequency distribution and its seasonal changes have not been discussed in detail in previous studies. For example, although Fujisawa and Kawamura (2005) gave a statistical overview of the meteorological conditions on days in which intensive lightning activity was observed in the Hokuriku district, their main focus was on how the meteorological conditions differed between lightning and non-lightning days. They did not examine the reasons behind the regional and seasonal differences of lightning frequency in detail. Similarly, Tsurushima et al. (2014) examined the averages of meteorological elements for days on which the daily

* Correspondence: daikitsurushima@gmail.com
[1]IDEA Consultants, Inc., Hayabuchi 2-2-2 Tsuzuki, Yokohama 224-0025, Kanagawa, Japan
Full list of author information is available at the end of the article

lightning frequency exceeded 100 strokes/day during the cold season (Oct.–Mar.) in Tohoku and the eastern part of the Hokuriku district. While the results suggested that cold-season lightning may be primarily caused by synoptic-scale cyclones, variations between individual cases and the relationship between lightning frequency distribution and meteorological disturbances were not thoroughly discussed. Since there are various types of meteorological disturbances (cyclones, fronts, the Japan sea Polar air mass Convergence Zone (JPCZ), upper troughs, etc.) that cause winter lightning in Japan (Kitagawa 1996; Sugita and Matsui 2008), their individual contributions to lightning distribution patterns and their seasonal tendencies are worth investigating.

The aim of this study is to identify major types of winter lightning and to examine the relationships between their lightning frequency distributions and meteorological disturbances based on LLS observational data from the Tohoku district and the eastern part of the Hokuriku district (including Niigata and Toyama Prefectures) (Fig. 1).

In order to identify the major types of winter lightning, hierarchical cluster analysis was applied to lightning frequency distributions of individual cases. We studied some typical cases to clarify the links between lightning frequency distribution and meteorological conditions.

Methods/Experimental

Observational lightning data

This study is based on 17 years worth of lightning observations (the 1994–1995 to 2010–2011 winter seasons) obtained by the LLS of Tohoku Electric Power Company. The LLS consists of nine IMPACT sensors distributed in the Tohoku district as shown in Fig. 2. IMPACT sensors are used to detect electromagnetic pulses emitted from lightning discharges and estimate the geo-location of lightning strokes within the Tohoku area, the eastern part of Hokuriku, and the northern part of the Kanto plane. The LLS detects mainly cloud to ground (CG) and ground to cloud (GC) lightning discharges. The detection efficiency of the LLS for wintertime CG/GC strokes is estimated to be 60%, and the lightning location accuracy is approximately 2 km (Honma et al. 1998; Honma 2012).

In this study, lightning discharges with peak electric currents (I_p) ranging from −10 to +20 kA were omitted from the analysis. Considering the recommendation of Cummins et al. (1998), lighting with small I_p may include cloud discharges that are difficult for the LLS to locate precisely.

Based on the lightning detection range shown in Fig. 2, coastal areas of the Tohoku district and the eastern part of the Hokuriku district including Niigata and Toyama Prefectures (hereinafter referred to as the Hokuriku district) were selected as the study area. Additionally, the LLS in the Tohoku district, originally installed in 1994, was upgraded in 2011, resulting in improved measurement accuracy after 2011 (Honma et al. 2012). Thus, the winter seasons from 1994–1995 to 2010–2011 (17 years) were chosen as the study period in order to avoid discontinuity in data quality.

Meteorological data

Since this study examines detailed meteorological fields associated with lightning activity, high-resolution gridded meteorological data called the Meso-Scale Model Grid Point Value (MSM-GPV) was adopted for the

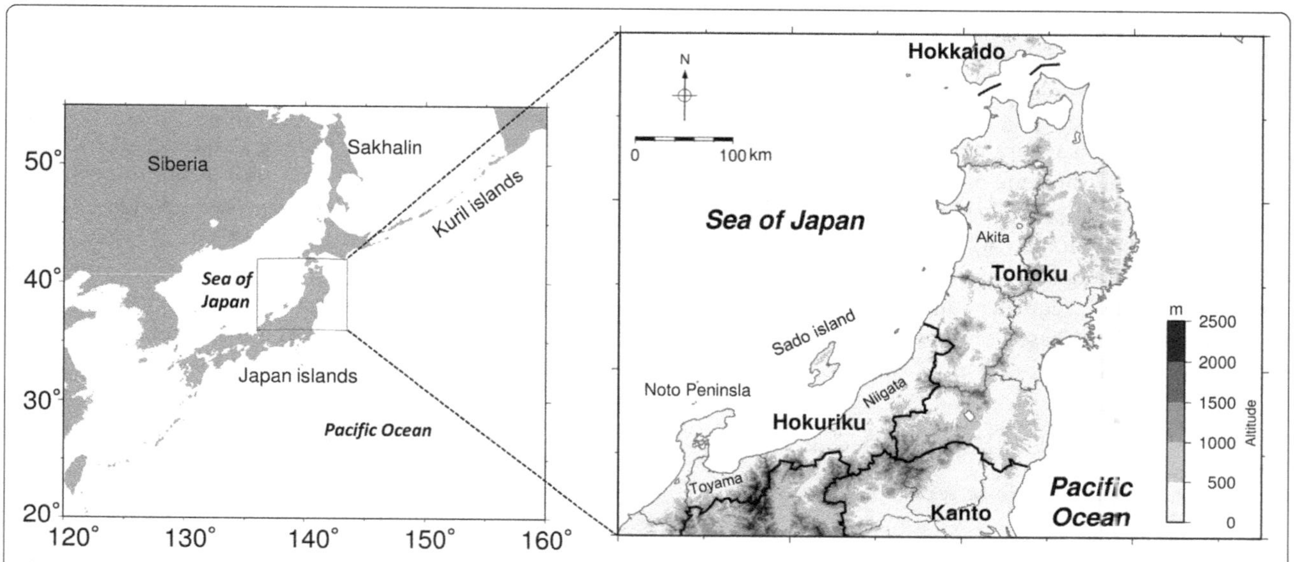

Fig. 1 Maps of the coastal areas of the Sea of Japan

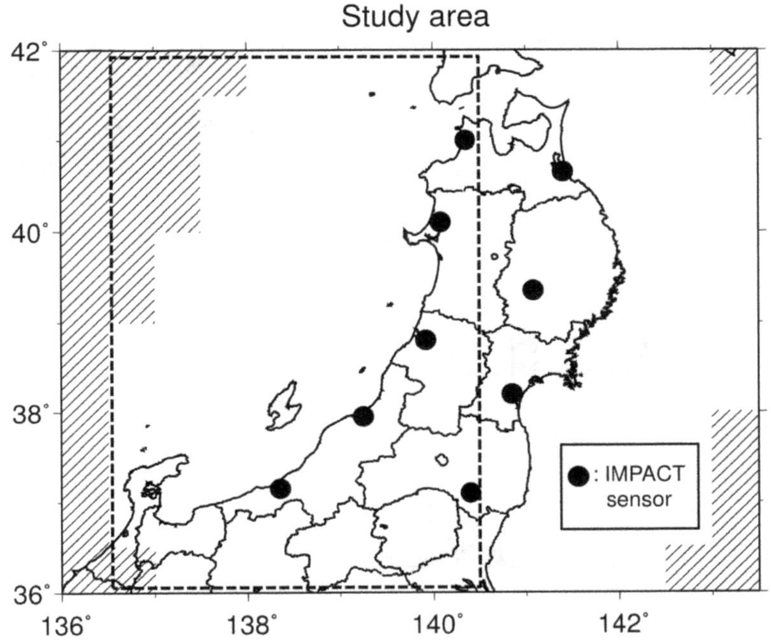

Fig. 2 Locations of IMPACT sensors and the study area. *Meshed areas* indicate zones of undetectable lightning discharges with absolute peak currents below 10 kA

analysis. MSM-GPV data includes objective analysis information calculated by the Meso-Scale Model (MSM) of the Japan Meteorological Agency (JMA). The MSM-GPV data has a spatial resolution of 10 km and a temporal resolution of 3 h.

Selection of cases

The winter lightning incidents (WL cases) to be used in the cluster analysis and case studies were selected based on time series graphs of lightning frequency, hourly lightning frequency maps, and meteorological radar images. We define a WL case as an episode when lightning activity is spatially and temporally isolated as illustrated in Fig. 3. However, cases with fewer than 100 total strokes were omitted since this study focuses on predominant cases that reflect the statistical features of the lightning frequency distribution. WL cases were selected from those occurring from October to March (the cold season), which is the regular winter lightning activity period in the Tohoku and Hokuriku districts (Tsurushima et al. 2014).

During the entire study period (the cold seasons of 17 consecutive years), 430 incidents were selected as WL cases. The number of WL cases in each month is shown in Table 1.

Cluster analysis

The cluster analysis for separating the WL cases into distinct types was conducted following the procedures listed below.

(a) A lightning frequency map of each WL case was created by counting lightning strokes within 50 km × 50 km grid cells in the study area so that small-scale (less than 50 km) fluctuations in the lightning frequency patterns of the WL cases could be neglected.

(b) The lightning frequency maps created in (a) were approximated by linear summations of empirical orthogonal functions (EOFs) as shown in Eq. (1).

$$\mathbf{X} = \sum_{n=1}^{22} \lambda_n \mathbf{f}_n \tag{1}$$

X indicates the lightning frequency in each grid cell expressed in vector format. $\mathbf{f}_n$ and λ_n denote the nth component of the EOF and its corresponding eigenvalue. The EOF summation stops at $n = 22$ since this is sufficient to explain 80% of the variance of the WL cases.

(c) Finally, Ward's clustering was applied to 22 eigenvalues (λ_n) of each WL case obtained in the EOF analysis in (b).

Case studies

In order to clarify the relationship between lightning frequency distribution and meteorological fields, we calculated three meteorological parameters, "surface wind divergence," "convective instability index," and "height at −10 °C,"

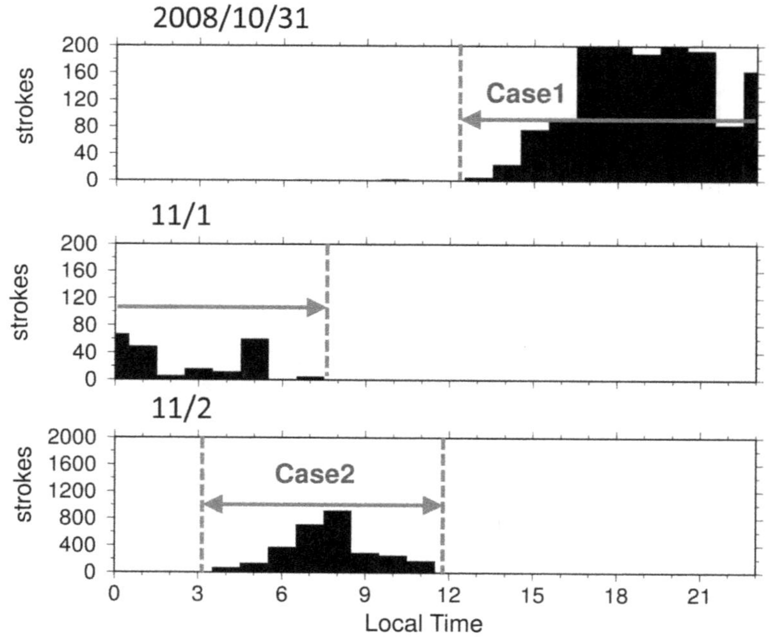

Fig. 3 Example of WL *case* selection based on time series graphs of lightning frequency

based on the MSM-GPV data and then compared them with lightning frequency distributions of the WL cases.

Surface wind divergence (divU) is defined as

$$\text{divU} = \frac{\partial u}{\partial x} + \frac{\partial v}{\partial y} \quad (2)$$

where u and v are the westerly and southerly wind components at the surface. A large negative value of divU indicates a strong updraft near the surface. Previous studies have shown that divU corresponds quite well with lightning activity and is helpful in understanding the relationship between lightning occurrences and surface wind fields (Fujisawa and Kawamura 2005; Tsurushima et al. 2014).

The convective instability index (ΔEPT) is defined as the difference between the equivalent potential temperatures (EPTs) of the lower and upper layers of the atmosphere. ΔEPT is expressed as

$$\Delta\text{EPT} = \text{EPT}_{\text{srf.}-900\ \text{hPa}} - \text{EPT}_{600-400\ \text{hPa}} \quad (3)$$

where $\text{EPT}_{\text{srf.}-900\ \text{hPa}}$ is the vertical mean EPT from the surface to 900 hPa, while $\text{EPT}_{600-400\ \text{hPa}}$ is the vertical mean EPT from 600 to 400 hPa. A large ΔEPT value indicates intensive thermal convection activity from the surface to a height of approximately 5–8 km in the atmosphere, which corresponds to the altitude of winter-time thunderclouds (Kitagawa and Michimoto 1994). Thus, ΔEPT can be regarded as an indicator of the potentiality of convective cloud formation. Although ΔEPT may also be regarded as the strength of resultant convection activity induced by pre-existing thunderclouds, for simplicity's sake, this study adopts the former interpretation.

The values of divU and ΔEPT indicate the strength of updraft and convection activity, respectively, in relation to the formation and development of convective clouds. However, even when these parameters indicate convective cloud formation, a cloud does not produce lighting discharges unless it is electrified. One simple indicator of the possibility that a cloud is electrified is the height of the −10 °C isotherm (H_{10}). According to Michimoto (1993), intensive lightning activity occurs mostly when H_{10} >1.8 km, while weak or no lightning activity is observed when H_{10} <1.8 km. Michimoto (1993) insisted that the dependency of lightning activity on H_{10} could be related to the strength of the updraft around the −10 °C

Table 1 Numbers of WL cases selected by month

Month	Number of WL cases
October	97
November	130
December	99
January	41
February	28
March	35
Total	430

level, where the charge separation is most effective (Takahashi 1984). As H_{10} rises, the maximum updraft velocity around the −10 °C level becomes strong enough to cause charge separation, which eventually produces intensive lightning activity.

Results and Discussion

Average cold-season lightning frequency distributions

Figure 4 displays average lightning frequency distributions in the Tohoku and Hokuriku districts from October to March, during the cold season. The lightning frequency in each grid cell shows the percentage of lightning strokes counted in that cell relative to the total lightning frequency in the whole study area (the total lightning frequency is shown in the bottom-right corner of each map).

During the late autumn season (Oct.–Nov.), the lightning frequency in the study area is the highest (40,023 strokes) with the most intensive lightning activity recorded in the coastal and open sea areas of the Tohoku district. During midwinter (Dec.–Jan.), lightning frequency drops to 6818 strokes and the region of most intense lightning activity moves southward to the Hokuriku district, where most of the lightning tends to be concentrated near the coastline. Late winter (Feb.–Mar.) displays the lowest lightning frequency (2305 strokes). Its lightning frequency distribution is similar to that in midwinter. However, the late winter season can be distinguished by less lightning activity in Niigata Prefecture and more lightning in Akita Prefecture.

Results of the cluster analysis

In order to identify the major types of WL cases that explain the statistical features of lightning activity in Fig. 4, cluster analysis was applied to 430 WL cases. As a result, these WL cases were classified into five clusters (C1, C2, C3, C4, C5) as shown in Fig. 5. Among the five clusters, it is clear that C1 ($N = 156$) and C4 ($N = 144$) are predominant, containing 70% of all the WL cases.

Figure 5 shows that C1 is characterized by high lightning frequency over the coastal and open sea areas of the Tohoku district. This is similar to the average lightning frequency distribution in late autumn (Fig. 4). The lightning frequency distribution of C4, on the other hand, is similar to the midwinter pattern, being characterized by high lightning frequency in the coastal areas of the Hokuriku district.

Figure 6 shows the monthly occurrence frequency of the five clusters. As indicated in Fig. 6, most of the WL cases in late autumn belong to C1, while midwinter is relatively dominated by C4. Thus, C1 and C4 are the predominant types of WL cases in late autumn and midwinter, respectively.

Meteorological conditions for winter lightning occurrence

We examined synoptic-scale atmospheric conditions for C1 and C4 cases (300 in total) on the basis of daily surface weather charts provided by the JMA. The weather charts in Fig. 7 display some typical examples of WL cases. Approximately 79% of C1 and C4 incidents (238 cases) were caused by synoptic-scale or meso-scale cyclones passing in the vicinity of the Japanese islands (Fig. 7a–e). Although these cyclones vary from case to case, they were roughly categorized based on their tracks and geometric shapes.

Firstly, we focused on cyclones with a single center (Fig. 7a–c) and divided them into three categories: SK, SJ, and PO cyclones. The definitions of each category are shown in Fig. 8. Cyclones passing near Sakhalin and the Kuril Islands (northward from the 44° N line) as shown in Fig. 7a are classified as SK cyclones. Cyclones shown in Fig. 7b, c are classified as SJ cyclones or PO cyclones based on whether their centers tended to pass

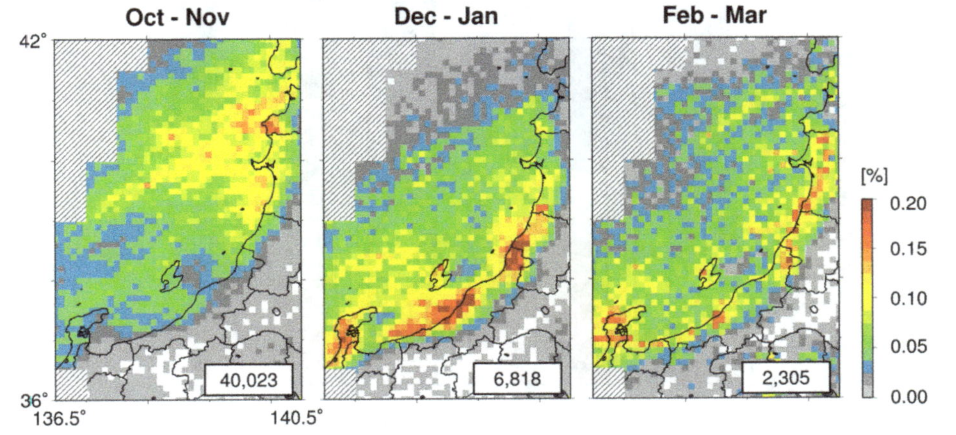

Fig. 4 Seasonal variation of lightning frequency distribution in the study area

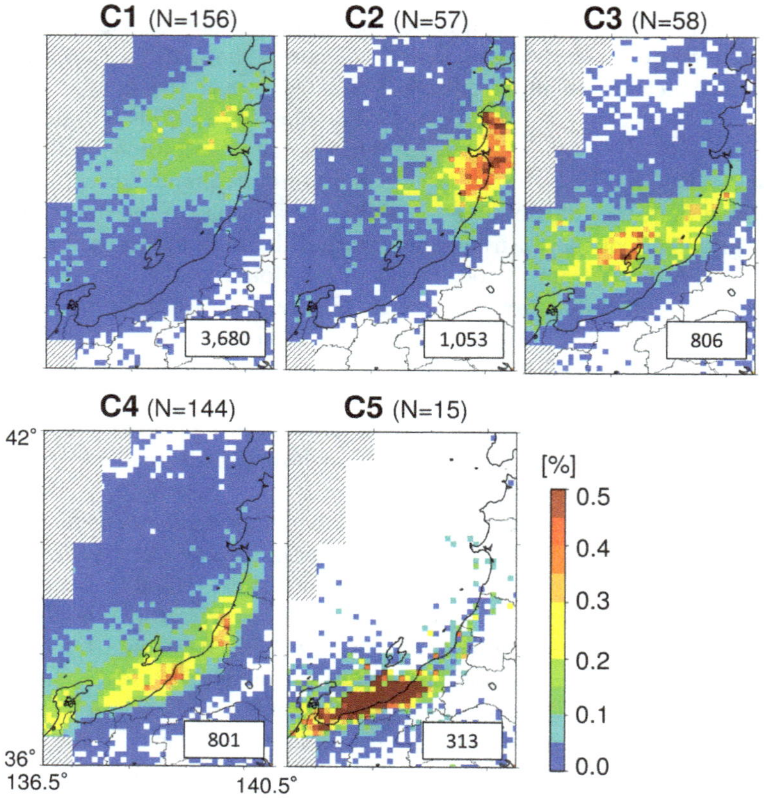

Fig. 5 Lightning frequency distributions for five clusters (*C1, C2, C3, C4, C5*). *N* represents the number of WL *cases* in each cluster

over the Sea of Japan or the Pacific Ocean. It should be noted that there are some cases where SK and SJ cyclones appear simultaneously. We categorized these as SJ cyclones because SJ cyclones pass closer to the study area and may have a greater impact on lightning frequency distribution than SK cyclones.

Secondly, cyclones with multiple centers are labeled as "MC" in Fig. 7d, e. An MC cyclone typically has two centers, one located north of Japan and the other over the Pacific Ocean. MC cyclones were subdivided into two categories: MCI and MCII. When the centers of an MC cyclone pass simultaneously over the SK and PO

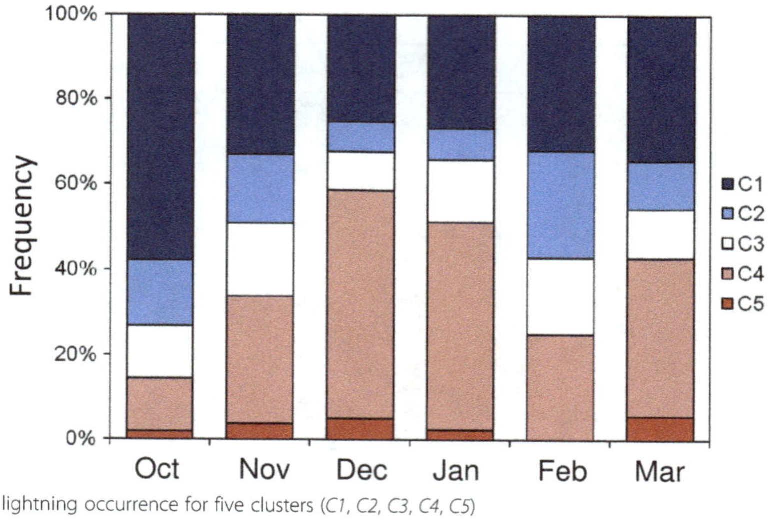

Fig. 6 Seasonal variation of lightning occurrence for five clusters (*C1, C2, C3, C4, C5*)

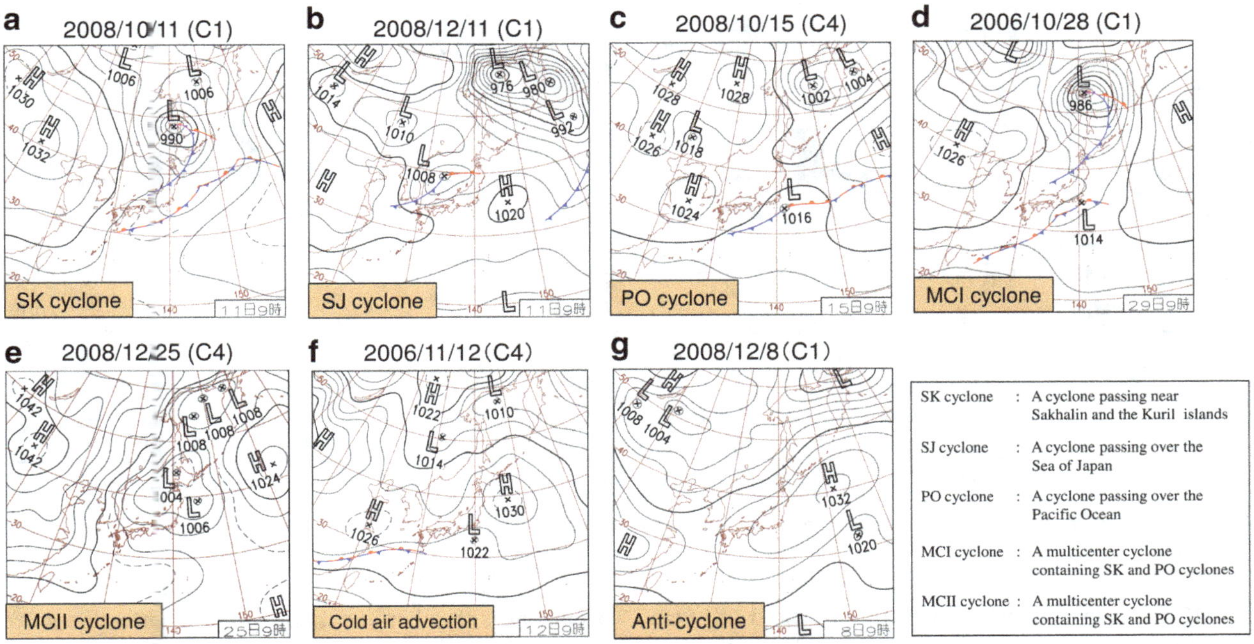

Fig. 7 a–g Examples of surface weather charts corresponding to *C1* and *C4*

cyclone areas indicated in Fig. 8, the cyclone is classified as MCI (Fig. 7d). Similarly, when the cyclone centers pass over the SJ and PO areas simultaneously, the cyclone is classified as MCII (Fig. 7e).

Thirdly, Fig. 7f, g depicts two cases that are not directly associated with cyclone activity. Approximately 13% of the C1 and C4 cases (38 cases) occurred when neither cyclones nor fronts were passing the vicinity of the study area during the entire period of the WL case. For example, Fig. 7f shows a typical case of "cold air advection" when a northerly or northwesterly wind from the Siberian continent is dominant in the study area throughout the entire period of lighting activity. The difference between cold air advection cases and cyclone cases is described in Fig. 9. In the cold air advection cases, the wind direction observed at Sakata (a JMA weather station) is always northwesterly during the lightning activity. In contrast, the wind direction for MCII cyclone cases changes drastically from southeasterly to northwesterly indicating the passing of the cyclone. In addition to daily weather charts, we also examined hourly surface wind directions to distinguish cold air advection cases from cyclone cases. Also, Fig. 7g shows a case in which the study area is at the edge of an anti-cyclone centered over the Pacific Ocean. This is an example of an "anti-cyclone" case, a rarity among the WL cases.

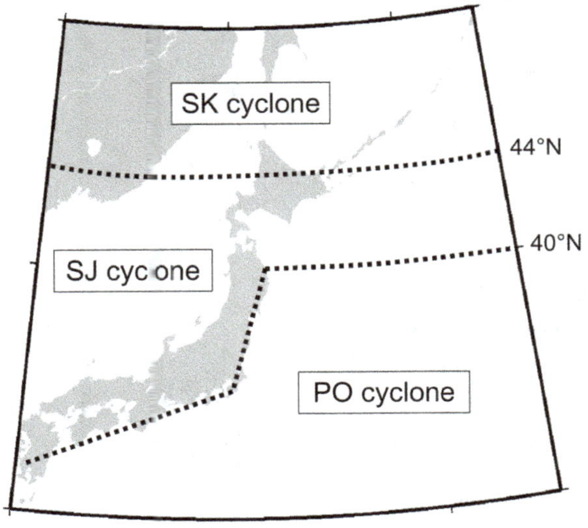

Fig. 8 Definitions of *SK, SJ*, and *PO cyclones*

The remaining 24 (8%) of the C1 and C4 cases cannot be classified into the categories outlined above. These are labeled as "Other" and contain overlaps of several categories. For example, when an SJ cyclone turns into an MCII cyclone while a WL case is occurring, the case would be classified as "Other."

The atmospheric conditions for the C1 and C4 cases are summarized in Table 2. Table 2 indicates that C1 cases are most frequently observed during the late autumn season and are associated with SK or SJ cyclones, which are likely to take relatively northern courses than PO or MC cyclones. Contrarily, C4 cases seem to be most frequent during the midwinter season and are associated

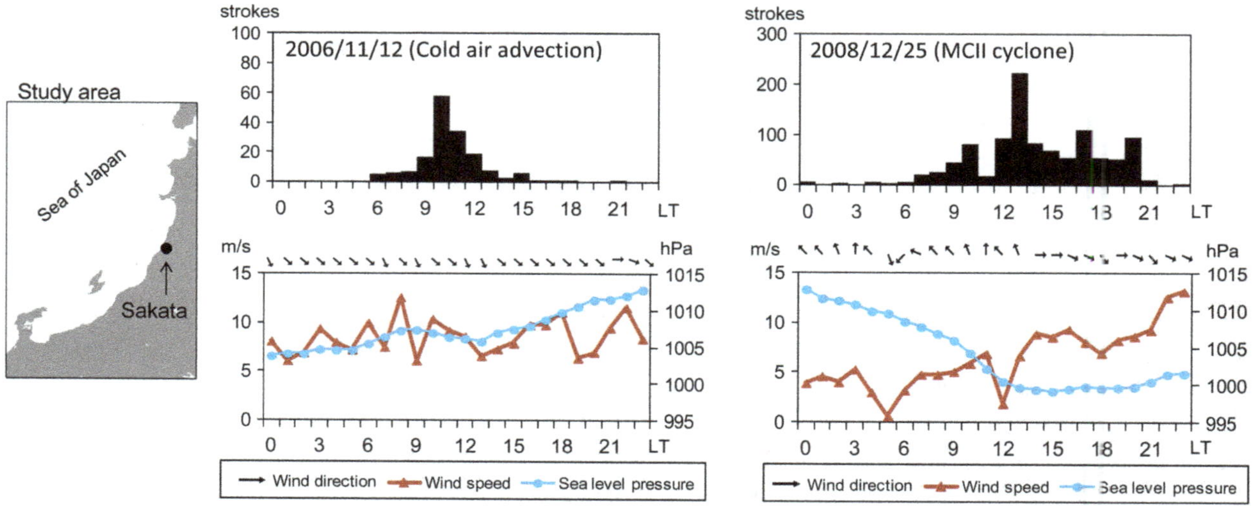

Fig. 9 Comparison of surface winds and sea-level pressure. Observations at Sakata (a JMA weather station) for *cold air advection cases* and *MCII cyclone cases*

with MCII cyclones, which pass further to the south than SK or SJ cyclones. Additionally, Table 2 shows that C4 cases are sometimes caused by cold air advection that is not directly associated with cyclone activity.

WL case types

In summary, three major types of WL occurrences were identified as illustrated in Table 3. Type A incidents are C1 cases frequently observed during late autumn, associated with SK or SJ cyclones. Type B cases are C4 occurrences that happen mainly in midwinter, associated with MCII cyclones. Type C cases, which are less frequent than type A or type B, are C4 cases associated with cold air advection. According to Kitagawa (1996), type C is regarded as "Advection lightning," a typical kind of winter lightning.

Type A case study

As a typical type A occurrence, we chose a WL case observed from 10 October to 11 October 2008 (case A).

Table 2 Numbers of cases in categories C1 and C4 grouped by month and atmospheric conditions. Cell color shows relative frequency (%)

C1

Month \ Type	SK	SJ	MC: MCI	MC: MCⅡ	PO	Cold air advection	Anti-cyclone	Other
Oct.	18	16	11	6	0	1	1	3
Nov.	15	8	5	6	0	1	3	4
Dec.	7	8	1	8	0	0	2	0
Jan.	2	3	1	5	0	0	0	0
Feb.	2	2	1	3	0	0	0	0
Mar.	5	4	1	2	0	0	0	0

(%) 10– / 5–10 / 2–5 / 0–2

C4

Month \ Type	SK	SJ	MC: MCI	MC: MCⅡ	PO	Cold air advection	Anti-cyclone	Other
Oct.	2	1	2	2	1	3	0	1
Nov.	3	6	5	7	0	14	2	2
Dec.	2	14	3	21	1	7	1	4
Jan.	2	5	0	6	0	2	0	5
Feb.	0	5	1	1	0	0	0	0
Mar.	0	4	1	2	1	1	0	4

(%) 10– / 5–10 / 2–5 / 0–2

Table 3 Major WL types

Type	Season	Lighting frequency distribution	Surface weather	Number of cases
A	Oct.–Dec.	C1	SK or SJ cyclone	57
B	Nov.–Jan.	C4	MCII cyclone	34
C	Nov.–Dec.	C4	Cold air advection	21

Time series maps of lightning frequency distribution and synoptic-scale atmospheric conditions for case A are shown in Fig. 10. The locations of cold/warm fronts in Fig. 10b were obtained from JMA's surface weather charts. The lightning frequency distributions for case A indicate that lightning activity commenced over the northwestern part of the Sea of Japan and moved eastward to the coastal areas of the Tohoku district. Figure 10b shows that a cold front associated with an SK cyclone passed over the Sea of Japan on 10 October to 11 October.

In order to clarify the relationship between case A's lightning frequency distributions and meteorological fields, we calculated divU and ΔEPT using Eqs. (2) and (3). Figure 11a shows time series maps of surface winds and divU distributions, while Fig. 11b shows composite maps of ΔEPT distributions and wind convergence lines detected in Fig. 11a. In Fig. 11b, strong wind convergence lines (divU $\leq -20[\times 10^{-5}/\text{s}]$) are expressed by solid lines, while weak convergence lines ($-20[\times 10^{-5}/\text{s}] <$ divU $\leq -10[\times 10^{-5}/\text{s}]$) are expressed by dashed lines. Lastly, Fig. 11c shows composite maps of lightning frequency distributions, wind convergence lines, and regions where ΔEPT exceeds 10 K (highlighted in pink).

On 10 October 21 LT, a strong wind convergence line was formed as a result of westerly and southerly winds blowing over the Sea of Japan in the northwestern part of the study area. Additionally, the gradient of ΔEPT across the convergence line was significantly large, indicating that a boundary existed between the stable air

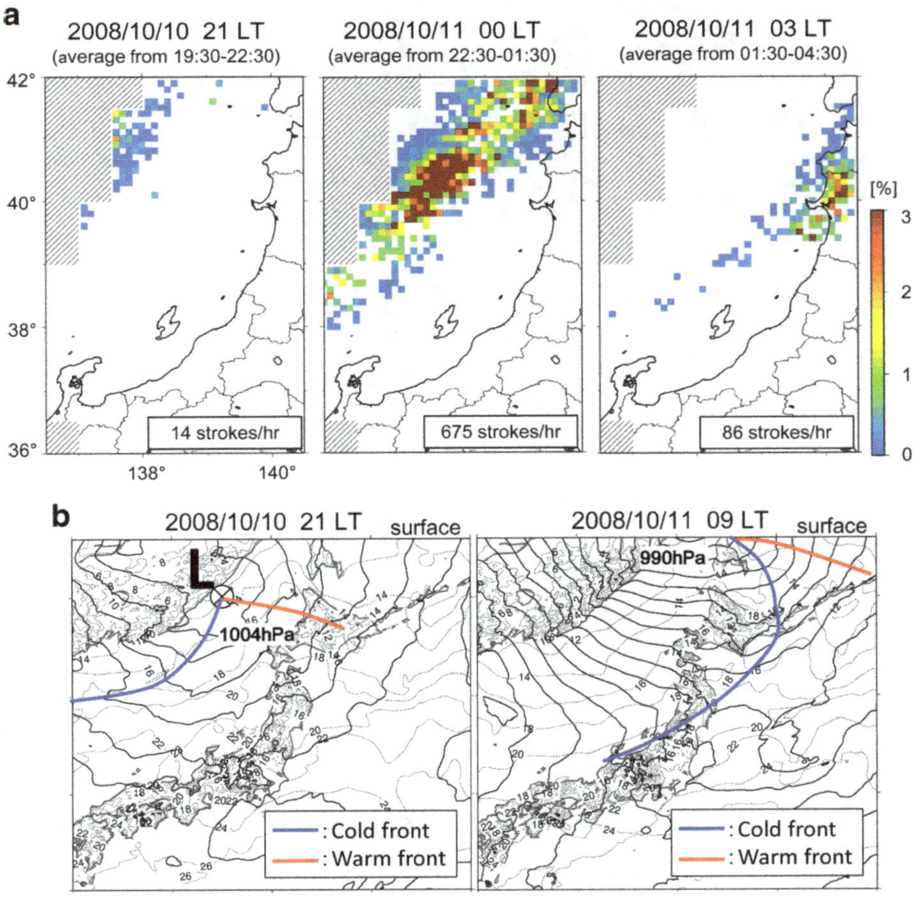

Fig. 10 Lightning frequency distributions and atmospheric conditions at the surface for case A. **a** Time series maps of lightning frequency distribution. The lightning frequency on the map is an hourly average calculated in a 3 h time window around the reference time. **b** Sea-level pressure (*solid lines*) and surface air temperature (*broken lines*) distributions obtained by MSM-GPV. Contour interval of sea-level pressure is 4 hPa. "L" represents the center of the cyclone defined as the point at which sea-level pressure is the lowest. Locations of cold/warm fronts are obtained from JMA's surface weather charts

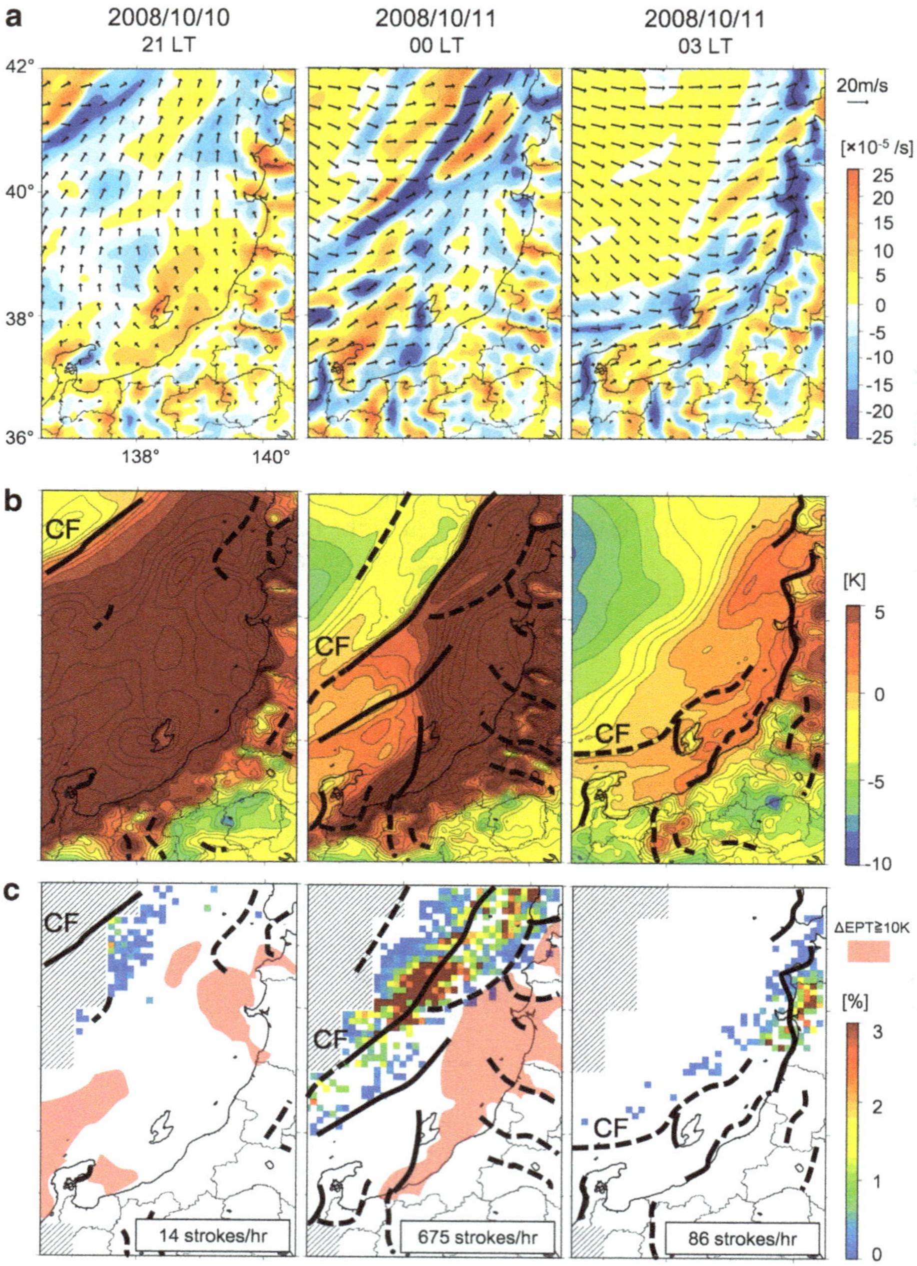

Fig. 11 Comparisons of divU and ΔEPT with lightning frequency distribution for case A. **a** Surface wind and divU distribution. **b** ΔEPT distribution and surface wind convergence lines. Strong wind convergence lines (divU $\leq -20[\times 10^{-5}/s]$) are shown as *solid lines*, while weak convergence lines ($-20[\times 10^{-5}/s] < \text{divU} \leq -10[\times 10^{-5}/s]$) are shown as *dashed lines*. **c** Composite maps of lightning frequency distribution, wind convergence lines, and regions where ΔEPT exceeds 10 K (*highlighted in pink*)

mass and the unstable air mass. These characteristics indicate that the convergence line in the northwestern part of the study area overlapped with the cold front accompanying the SK cyclone (labeled as cold front (CF) in Fig. 11). According to Fig. 11c, lightning strokes were distributed along the convergence line at the cold front.

On 11 October 0 LT, the region of high lightning frequency moved eastward as the cold front moved to the offshore areas of the Tohoku district. Furthermore, 11 October 0 LT exhibited the highest lightning frequency (675 strokes/h) during the entire period.

On 11 October 3 LT, the wind convergence along the cold front appears to have weakened in the north of the Hokuriku district. Additionally, the ΔEPT gradient across the cold front was smaller than during the previous period, also indicating weakening of the cold front. Instead, strong

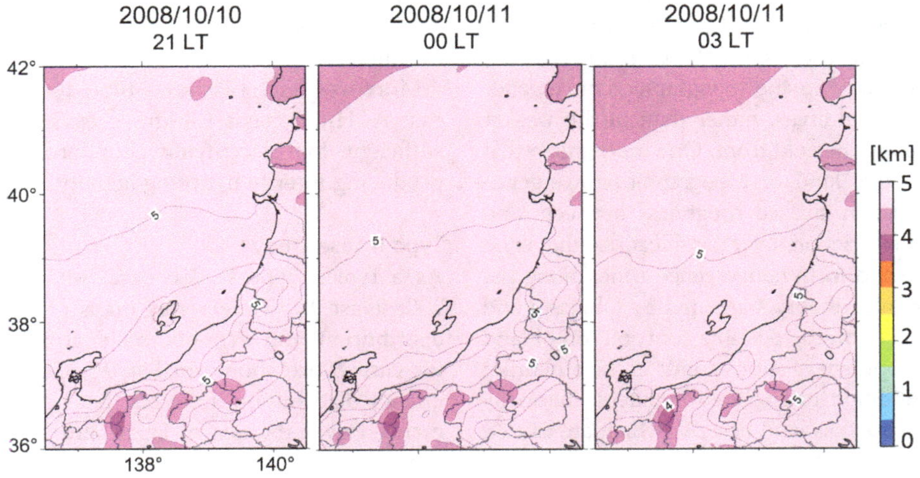

Fig. 12 Distribution of H_{10} for case A

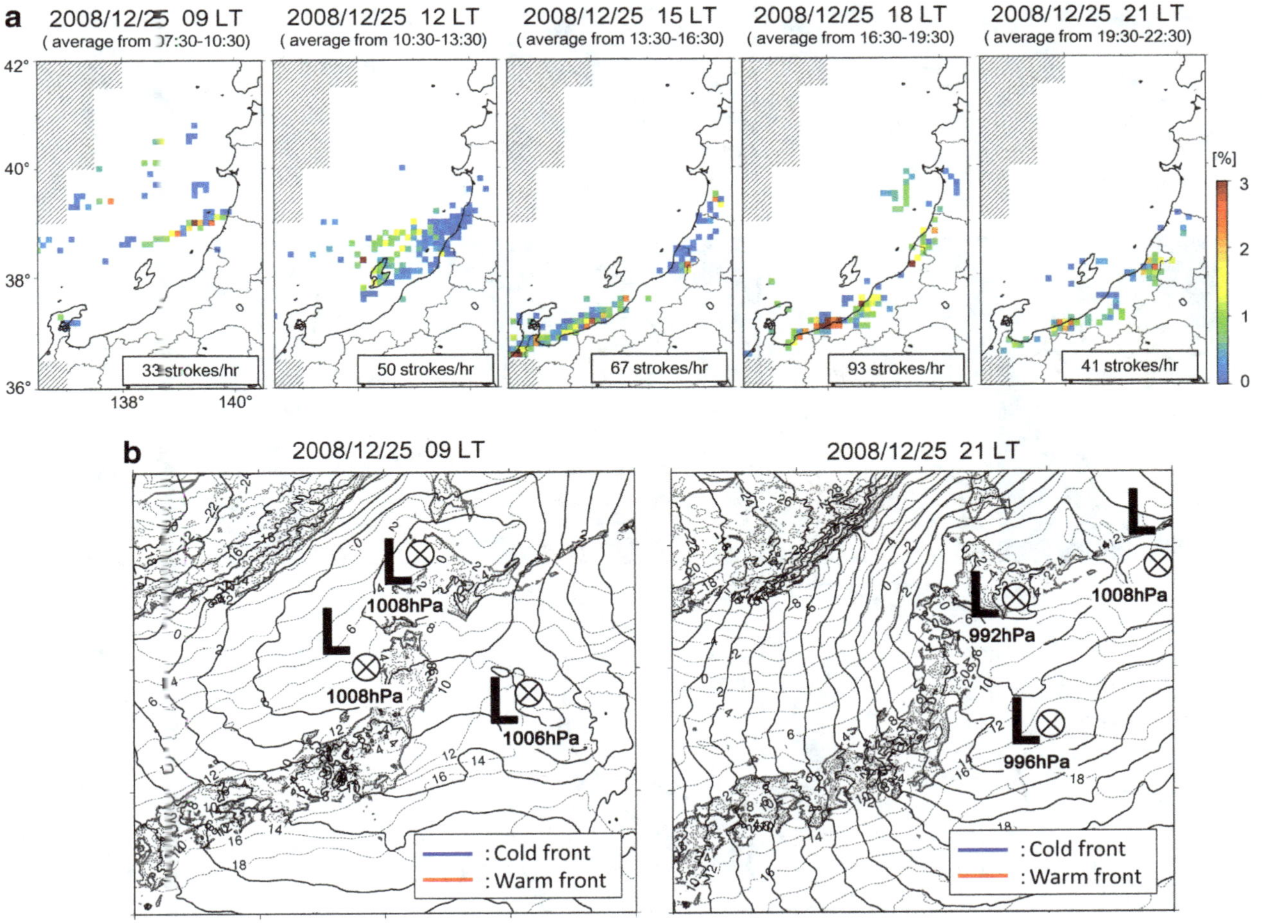

Fig. 13 a, **b** Similar to Fig. 10 but for case B

convergence lines, the result of westerly winds decelerating due to topographic effects, appeared in the coastal areas of the Tohoku district. Figure 11c shows that lightning strokes were mainly concentrated in the topographic convergence lines in Tohoku's coastal areas, rather than in the weaker convergence line along the cold front. One mechanism that could explain why this kind of topographic convergence forms is the contrast in surface roughness between land and sea. The higher friction over land causes the wind speed to decelerate and form convergence zones along the coastline. An alternative theory suggested by Ohigashi and Tsuboki (2005) focuses on the possible contribution of land breezes to the development of snow clouds in the Hokuriku district. Thus, in addition to surface roughness, thermally induced local winds such as land breezes, which are caused by temperature gradients between land and sea, may play a significant role in forming wind convergences near the coastline.

Moreover, Fig. 12 shows time series maps of H_{10}. In case A, H_{10} exceeds 1.8 km across the whole study area, sufficient for electrifying thunderclouds and thus for producing intense lightning activity.

Type B case study

As a typical type B WL case, we chose 25 December 2008 (case B). Time series maps of lightning frequency distribution and synoptic-scale atmospheric conditions for case B are shown in Fig. 13. Case B's lightning frequency distribution indicates that lightning activity commenced over the Sea of Japan and moved southward to

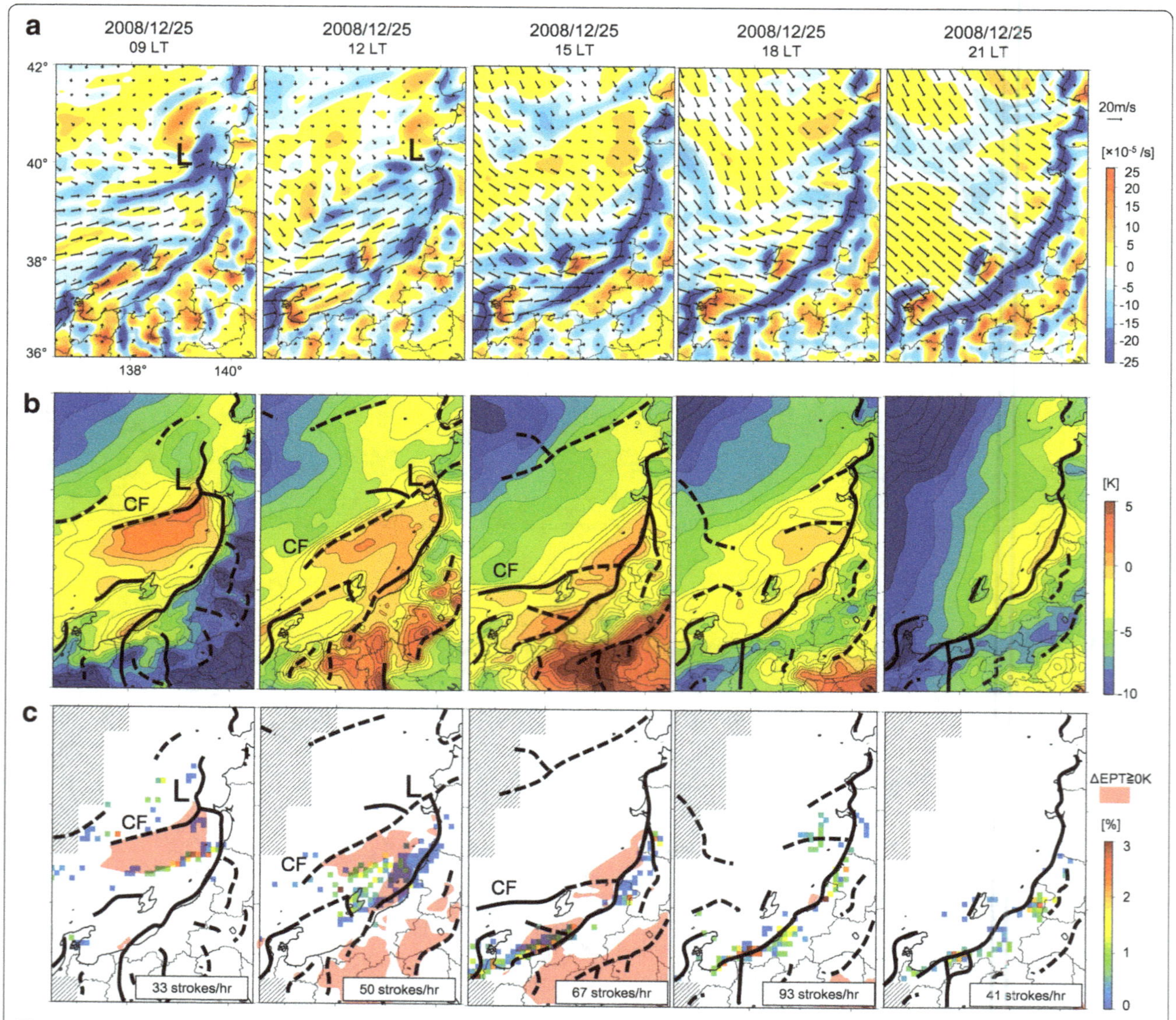

Fig. 14 a–c Similar to Fig. 11 but for case B. "L" represents the center of the cyclone

the coastal areas of the Hokuriku district. Figure 13b indicates that an MCII cyclone passed over the Japanese islands on 25 December.

Figure 14 compares divU and ΔEPT and case B's lightning frequency distribution. From 9 LT to 12 LT on 25 December, a weak wind convergence line (labeled CF) was formed by northwesterly and southwesterly winds blowing over the Sea of Japan in the central part of the study area. Based on surface wind and ΔEPT distributions in Fig. 14a, b, this weak convergence line can be regarded as a CF, although it is not shown on the surface weather charts (Fig. 13b). In addition, a strong convergence line was formed along the Tohoku coastline in connection with the topographic effects described above in the type A case study. Figure 14c shows that lightning strokes were concentrated over the unstable air mass (ΔEPT ≥0 K) and along the topographic convergence line in the coastal areas of the Tohoku district rather than over the weak convergence line along the cold front.

On 25 December 15 LT, as the cold front moved southward to the offshore areas of the Hokuriku district, the unstable air mass seems to have drawn nearer to the Tohoku and Hokuriku coastline. According to Fig. 14c, regions of high lightning frequency formed over the unstable air mass and topographic convergence lines in the coastal areas, whereas the lightning frequency over the cold front was relatively low. It is also noteworthy that a wind divergence zone formed on the leeward side of Sado Island, possibly influenced by a downdraft from the island's mountains. This may also have factored into the lightning frequency being relatively low on the southeastern side of Sado Island.

From 18 LT to 21 LT on 25 December, the cold front disappeared and the study area was dominated by northwesterly winds blowing over the Sea of Japan. Additionally, topographic convergence lines formed along the Tohoku and Hokuriku coastline, while ΔEPT was smaller than in previous periods. Lightning frequency in the area remained high (41–93 strokes/h) with the lightning strokes being concentrated over the topographic convergence lines.

Figure 15 compares H_{10} and corresponding lightning frequency distributions for case B. It shows that H_{10} exceeded 1.8 km in the whole study area when the lightning activity began (09–12 LT on 25 Dec.). However, from 18 LT to 21 LT on December 25, when the study area was dominated by northwesterly winds (see Fig. 14a), the region of high H_{10} was confined to the coastal areas of the Hokuriku district. Since lightning mainly occurs where H_{10} >1.8 km, this indicates that from 18 LT to 21 LT, the coastal areas of the Hokuriku district were more favorable for cloud electrification than were other regions.

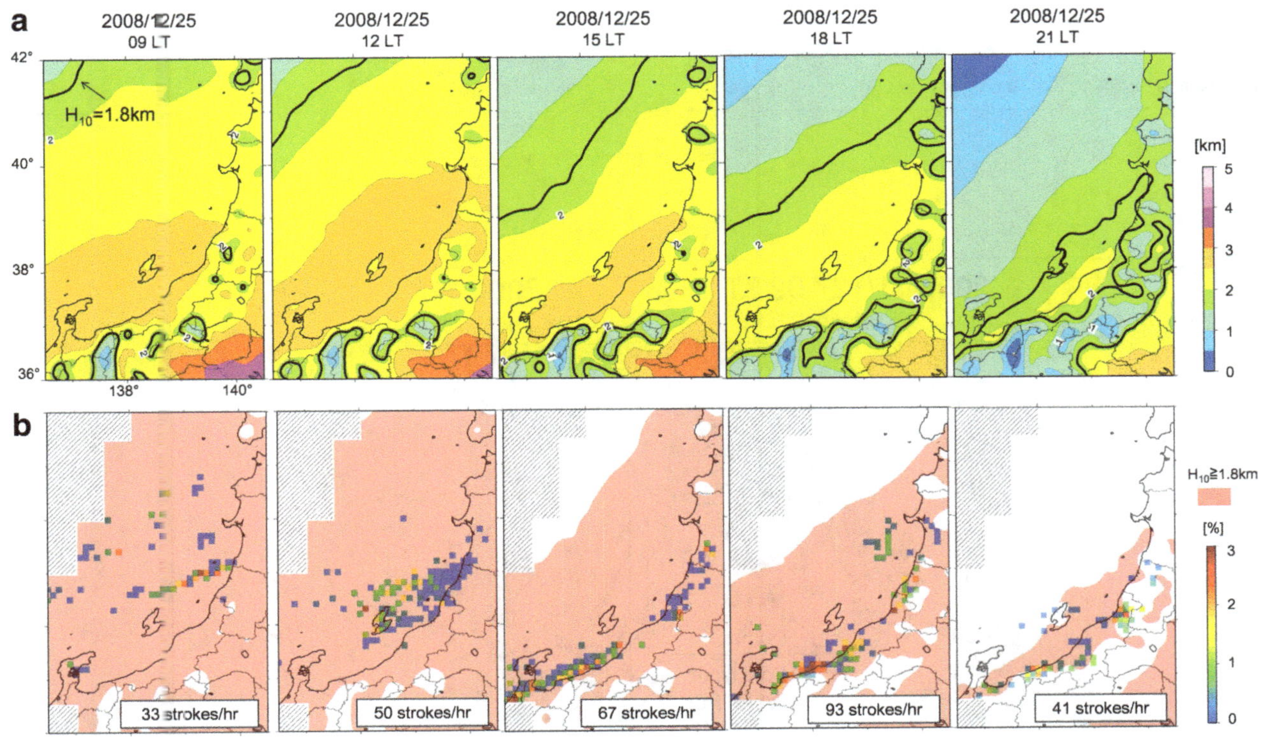

Fig. 15 Comparisons of H_{10} with lightning frequency distribution for case B. **a** H_{10} distribution. **b** Composite maps of lightning frequency distribution and regions where H_{10} exceeds 1.8 km (*highlighted in pink*)

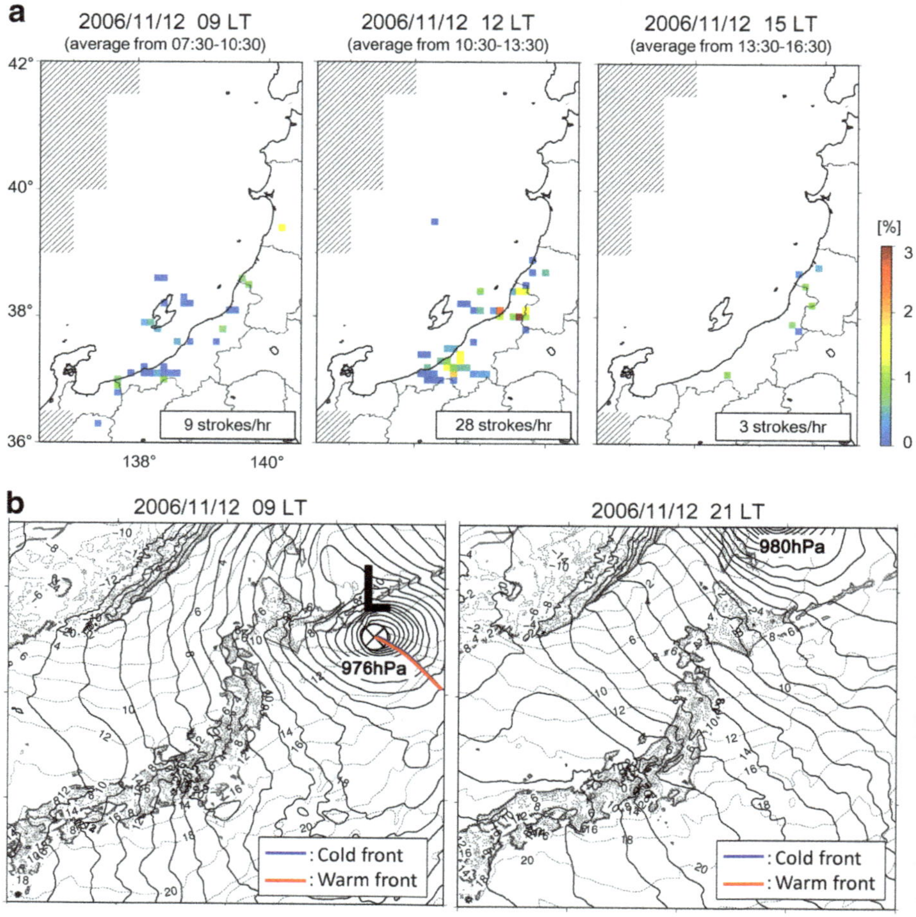

Fig. 16 a, b Similar to Fig. 10 but for case C

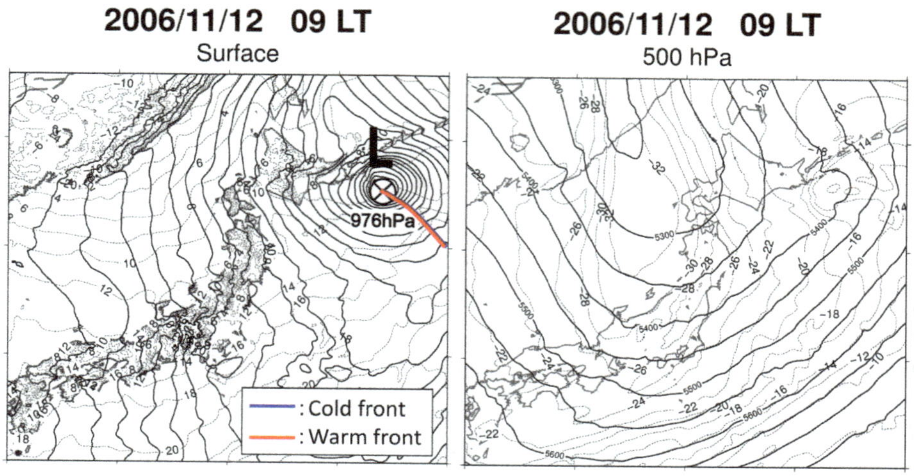

Fig. 17 Atmospheric conditions on 11 November 9 LT, 2006. Sea-level pressure (hPa) (geopotential height (m) at the 500 hPa pressure level) and air temperature distribution (°C)

Type C case study

As a representative type C event, we chose a WL case observed on 12 November 2006 (case C). Time series maps of lightning frequency distribution and synoptic-scale atmospheric conditions for case C are shown in Fig. 16. The lightning frequency distributions of case C suggest that lightning activity was mainly concentrated around the coastal areas of the Hokuriku district during the entire period. Figure 16b indicates that the study area was dominated by cold air advection from the Siberian continent. Although no meteorological disturbance such as a cold front or a cyclone is shown in Fig. 16b, figures for the geopotential height distribution at the 500 hPa pressure level point to the existence of a strong pressure trough as illustrated in Fig. 17.

Therefore, the lightning activity in case C could possibly be caused by the upper trough and resultant convective instability.

Figure 18 compares divU and ΔEPT with the lightning frequency distribution of case C. From 9 LT to 15 LT on

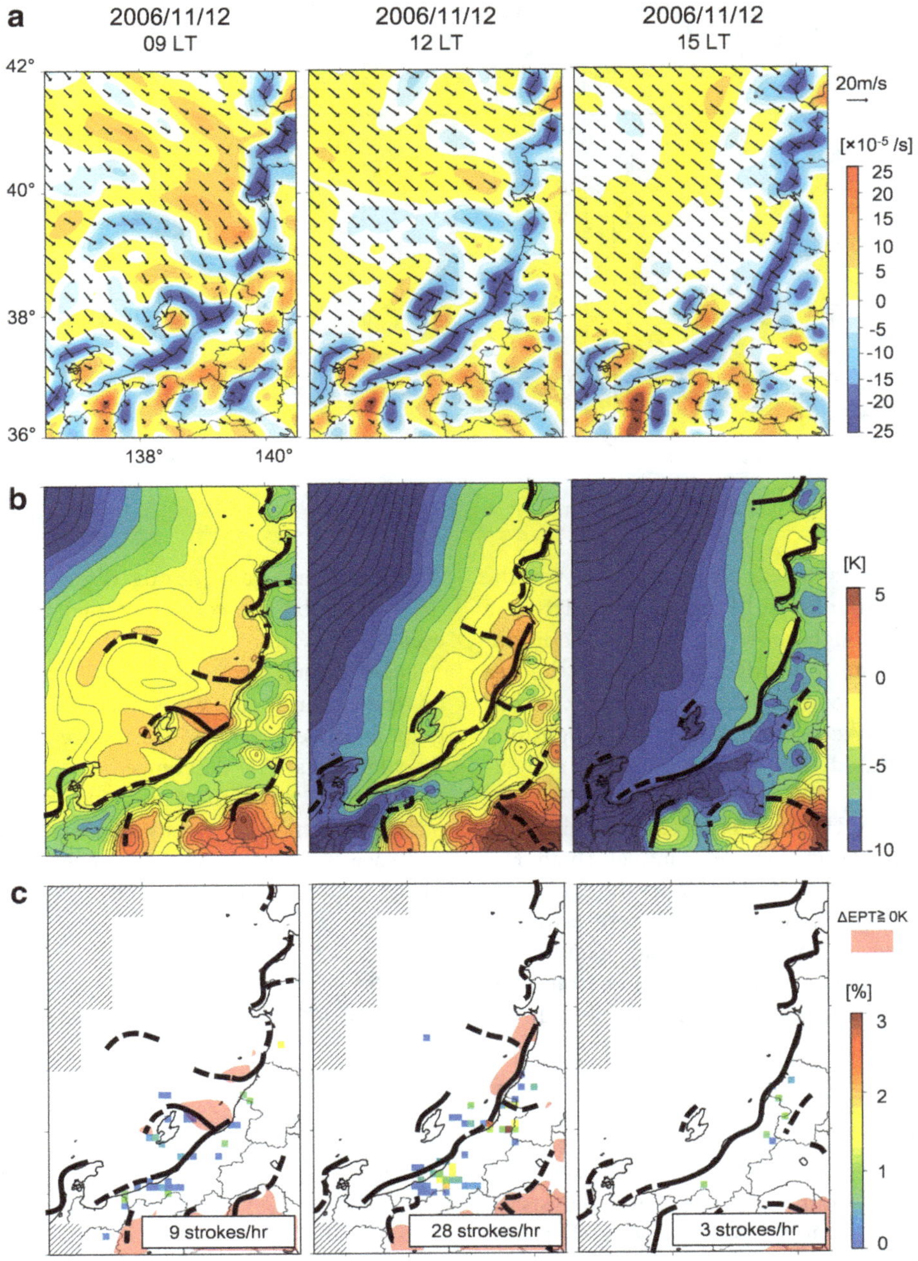

Fig. 18 a–c Similar to Fig. 11 but for case C

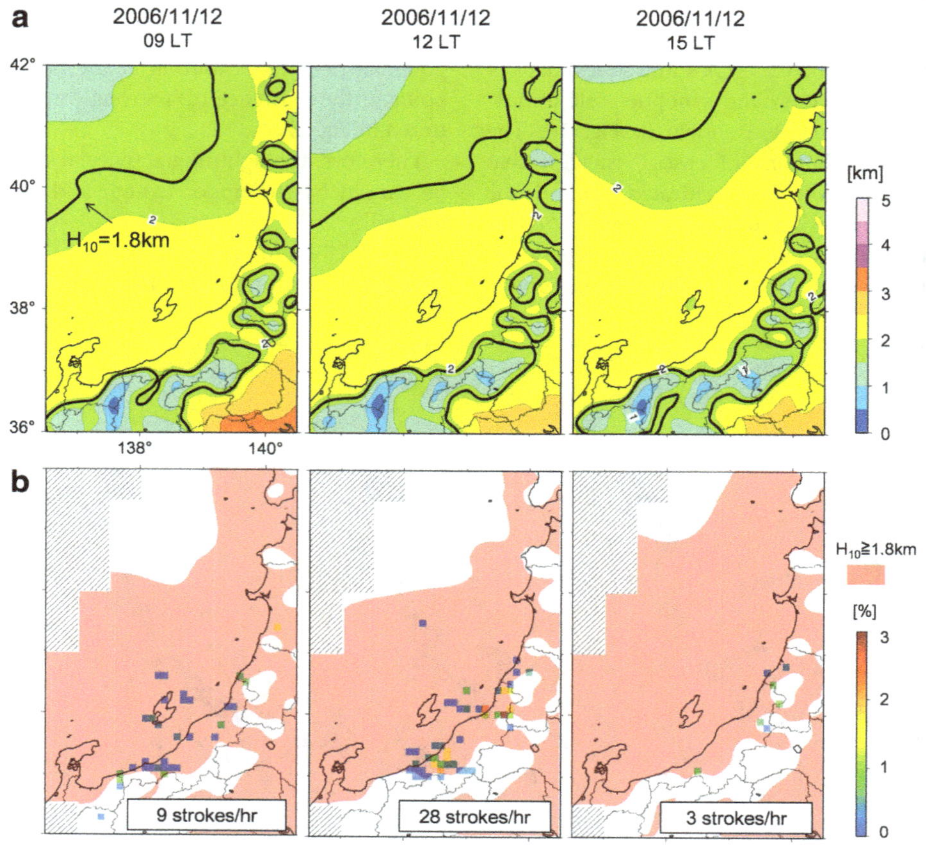

Fig. 19 a, b Similar to Fig. 15 but for case C

12 November, an unstable air mass appeared in the coastal areas of the Tohoku and Hokuriku districts as a result of the upper trough shown in Fig. 17. Additionally, due to northwesterly winds blowing over the Sea of Japan, topographic convergence lines were formed along the coastline. According to Fig. 18c, lightning strokes were distributed over the topographic convergence lines.

Figure 19 compares H_{10} and case C's lightning frequency distribution. As in case B, relatively high H_{10} regions appear along the coastal areas of the Hokuriku district, indicating that these areas are more favorable for cloud electrification than other areas. Figure 19b shows that lightning mainly occurs in the coastal areas of the Hokuriku district, which is consistent with the distribution of H_{10} >1.8 km.

Case study summary

Typical events of types A–C were selected (cases A, B, and C). These were examined based on analysis of surface weather charts and MSM-GPV data. The meteorological features of cases A–C are summarized in Table 4.

The lightning frequency distribution of case A is mainly determined by the cold front that accompanied the SK cyclone.

The lightning frequency distributions of cases B and C, on the other hand, were mainly determined by topographic convergence and convective instability rather than by cold fronts. The difference between case A and case B could be the strength of the cold front. Figure 20 compares the vertical *p*-velocity values at 850 hPa for case A and case B. The updraft near the cold front is weaker in case B (−2 to −4 Pa/s) than in case A (−2 to −8 Pa/s).

Table 4 Case study summary

Case	Type	Date (LT)	Total lightning frequency (stroke)	Meteorological factors determining lightning frequency distribution
A	A	2008/10/10–11	2336	•Cold front •Topographic convergence
B	B	2008/12/25	973	•Topographic convergence •Unstable air mass
C	C	2006/11/12	153	•Topographic convergence

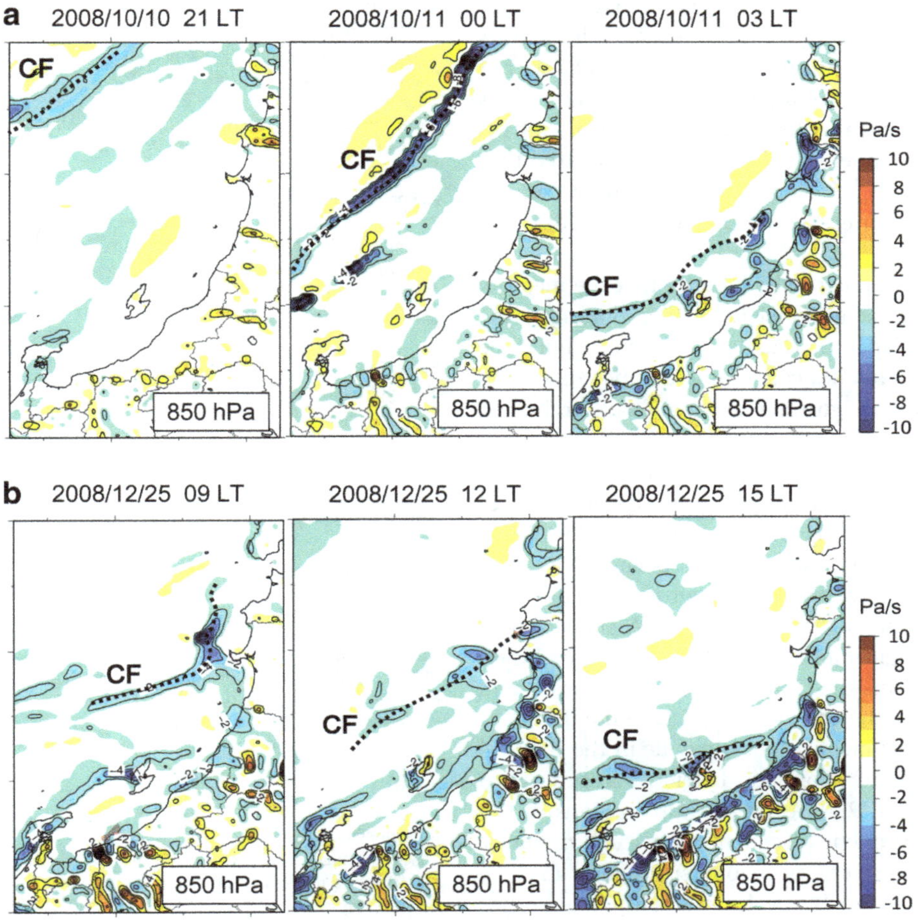

Fig. 20 a, b Comparison of vertical *p*-velocity (Pa/s) at 850 hPa for case A and case B. The vertical *p*-velocity data was obtained from MSM-GPV

Moreover, H_{10} distributions for cases B and C indicate that regions of high H_{10} tend to be confined to the coastal areas of the Hokuriku district, especially when a northwesterly wind dominates the study area. Considering that lightning mainly occurs within areas where H_{10} >1.8 km, the results indicate that lightning frequency distributions in cases B and C may also correlate with the extent of the clouds' electrification.

Possible mechanisms for seasonal changes in lightning frequency distribution

Based on the cluster analysis and case studies, the seasonal changes in lightning frequency distribution (Fig. 4) could be explained as follows.

During the late autumn, lightning frequency distribution seems to be determined by cold fronts accompanying the SK or SJ cyclones. Since SK and SJ cyclones take the northernmost course of all cyclone types, the strong wind convergence line on the cold front may not reach the more southern Hokuriku district. Therefore, in late autumn, high lightning frequency is more often observed in the Tohoku district than in the Hokuriku district.

During the midwinter season, lightning frequency distributions may be more affected by topographic convergence than by cold fronts. Although this study did not examine in detail the mechanisms through which topographic convergences form, they may be induced by differences in surface roughness or elevation between land and sea areas and/or by thermally induced local winds (land breezes). Since midwinter topographic convergences are mostly formed by northwesterly winds blowing over the Sea of Japan, wind convergence becomes the strongest in the coastal areas of the Hokuriku district (since the wind blows perpendicular to the coastline there). Furthermore, northwesterly winds during the midwinter season tend to confine high H_{10} regions to the coastal areas of the Hokuriku district, which means that Hokuriku becomes an area where cloud electrification is most likely to occur. Therefore, midwinter lighting activity tends to concentrate near the coastline of the Hokuriku district.

As mentioned above, this study suggests that midwinter lightning activity is more affected by topographic effects than that of the late autumn season. This fact was not made clear in previous studies, but statistical results presented by Lee et al. (2000) and Ishii et al. (2014) implied a similar tendency. Their results indicated that lightning frequency during the midwinter season shows a significant increase when the land breezes are strong.

Additionally, our study results indicate that the majority of WL cases are probably associated with cyclones passing the vicinity of the Japanese islands. However, it should be noted that previous studies concerning the meteorological conditions of winter lightning in Japan have indicated different results.

For example, Kitagawa (1996) claimed that approximately 60% of winter lightning in the coastal areas of the Sea of Japan (including the Kyushu, Chugoku, Kinki, Hokuriku, and Tohoku districts) is associated with cold air advection from the Siberian continent (advection lightning; type C) rather than by cyclones. A similar result was obtained by Sugita and Matsui (2008), who based their research, which covers the Chugoku, Kinki, Hokuriku, and Tohoku districts, on the observational data of the JLDN.

One cause for this inconsistency may lie in the regional characteristics of the Tohoku district. According to Adachi and Kimura (2007), the offshore regions of the Tohoku district experience frequent wintertime cyclones. Therefore, lightning activity in our study area may be more strongly affected by cyclone activity than other regions in Japan. However, regional differences in meteorological factors affecting lightning activity need to be examined in future research.

Conclusions

This study identified three major types of Japanese winter lightning activity (types A–C) based on lightning frequency distribution and meteorological conditions. Detailed case studies were conducted for typical winter lightning cases (cases A–C) in order to examine the relationship between lightning activity and meteorological disturbances.

The results indicate that lightning frequency distribution during the late autumn is mainly determined by cold fronts accompanying extra-tropical cyclones (case A). On the other hand, midwinter lightning frequency distribution is determined by topographic convergence and convective instability rather than by cold fronts (cases B and C). Finally, we found a few cases of advection lightning, whereas previous studies have regarded it as the dominant type of winter lightning activity in Japan. Although this inconsistency may be explained by regional features of our study area, further discussion and research are needed.

Abbreviations

CF: Cold front; CG: Cloud to ground; EOF: Empirical orthogonal functions; EPT: Equivalent potential temperature; GC: Ground to cloud; JLDN: Japan Lightning Detection Network; JMA: Japan Meteorological Agency; JPCZ: Japan sea Polar air mass Convergence Zone; LIDEN: LIghtning DEtection Network system; LLS: Lightning location system; LT: Local time; MSM-GPV: Meso-Scale Model Grid Point Value; WL: Winter lightning

Acknowledgements

We thank Tohoku Electric Power Company, Inc. for allowing us to use their lightning observation data. We also thank Dr. Masaki Sato, editor of *Progress in Earth and Planetary Science*, and peer reviewers for their thoughtful comments that helped us improve the manuscript. Generic Mapping Tools (GMT) were used for constructing figures.

Funding

Not applicable.

Authors' contributions

DT and SK proposed the topic and conceived and designed the study. NH operated the lightning location system (LLS) and created the lightning stroke datasets. All authors have read and approved the final manuscript.

Competing interests

The authors declare that they have no competing interests.

Author details

[1]IDEA Consultants, Inc., Hayabuchi 2-2-2 Tsuzuki, Yokohama 224-0025, Kanagawa, Japan. [2]Graduate School of Environmental Studies, Tohoku University, Aoba 6-3 Aramaki Aoba, Sendai 980-8678, Miyagi, Japan. [3]R&D Center of Tohoku Electric Power Company, Inc, Nakayama 2-1 Aoba, Sendai 981-0952, Miyagi, Japan.

References

Adachi S, Kimura F (2007) A 36-year climatology of surface cyclogenesis in East Asia using high-resolution reanalysis data. SOLA 3:113–116

Brook M, Nakano M, Krehbiel M, Takeuti T (1982) The electrical structure of Hokuriku winter thunderstorms. J Geophys Res 87:1207–1215

Cummins KL, Murphy MJ, Bardo EA, Hiscox WL, Pyle RB, Pifer AE (1998) A combined TOA/MDF technology upgrade of the U.S. National Lightning Detection Network. J Geophys Res 103:9035–9044

Fujisawa G, Kawamura R (2005) Recent tendencies of winter thunderstorms in the Hokuriku district and associated atmospheric conditions. TENKI 52:13–24 (in Japanese)

Hojo J, Ishii M, Kawamura T, Suzuki F (1989) Seasonal variation of cloud-to-ground lightning flash characteristics in the coastal areas of the Sea of Japan. J Geophys Res 94:13207–13212

Honma N (2012) Performance of the Tohoku IMPACT sensor network in winter lightning detection. IEEJ Trans PE 132(6):579–587

Honma N, Suzuki F, Miyake Y, Ishii M, Hidayat S (1998) Propagation effect on field waveforms in relation to time-of-arrival technique in lightning location. J Geophys Res 103:14141–14145

Honma N, Cummins KL, Murphy MJ, Pifer AE, Rogers T (2012) Improved lightning locations in the Tohoku region of Japan using propagation and waveform onset corrections. IEEJ Trans PE 133(2):195–202

Ishii M, Saito M, Chihara M, Natsuno D (2012) Transferred charge and specific energy associated with lightning hitting wind turbines in Japan. IEEJ Trans PE 132(3):294–295

Ishii K, Hayashi S, Fujibe F (2014) Statistical analysis of temporal and spatial distributions of cloud-to-ground lightning in Japan from 2002 to 2008. J Atmos Electr 34(2):79–86

Kitagawa N (1996) Meteorological aspects of winter thunderclouds along the Japan Sea coast. TENKI 43:7–17 (in Japanese)

Kitagawa N, Michimoto K (1994) Meteorological and electrical aspects of winter thunderclouds. J Geophys Res 99:10713–10721

Lee J, Wada M, Kawasaki Z, Matsuura K, Takeuchi M, Sonoi Y (2000) Lightning activity during winter thunderstorm season observed by SAFIR. Electr Eng Jpn 132(1):811–817

Michimoto K (1993) A study of radar echoes and their relation to lightning discharges of thunderclouds in the Hokuriku district part II: observation and analysis of "single-flash" thunderclouds in midwinter. J Meteor Soc Jpn 71(2): 195–204

Ohigashi T, Tsuboki K (2005) Structure and maintenance process of stationary double snowbands along the coastal region. J Meteor Soc Jpn 83(3):331–349

Rakov VA, Uman MA (2007) Lightning: physics and effects. Cambridge University Press, Cambridge

Sugita A, Matsui M (2008) Examples of winter lightning observed by the JLDN. In: Abstracts of 20th International Lightning Detection Conference, Tucson, Arizona, 21-23 April 2008.

Sugita A, Matsui M (2012) Lightning characteristics in Japan observed by the JLDN from 2001 to 2010. In: Abstracts of 22nd International Lightning Detection Conference, Broomfield, Colorado, 2-3 April 2012.

Takahashi T (1984) Thunderstorm electrification—a numerical study. J Atmos Sci 41:2541–2558

Tsurushima D, Sakaida K, Honma N (2014) Climatological features of winter lightning activity in the coastal areas of Sea of Japan, Tohoku and Hokuriku district. Quart J Geogr 65:189–206 (in Japanese with English abstract)

9

Response of the atmospheric hydrological cycle over the tropical Asian monsoon regions to anthropogenic aerosols and its seasonality

Hiroshi G. Takahashi[1,2*], Shingo Watanabe[2], Makiko Nakata[3] and Toshihiko Takemura[4]

Abstract

This study investigates the impact of anthropogenic aerosols on the atmospheric hydrological cycle over the tropical Asian monsoon region (South Asian, Southeast Asian, and western North Pacific monsoons), using a coupled atmosphere-ocean global climate model (CGCM), Model for Interdisciplinary Research on Climate-Earth System Models. Three-ensemble historical (HIST) and sensitivity (piAERO) experiments for the period 1985–2005 are conducted. The piAERO experiment is the same as HIST, but with anthropogenic aerosol emissions kept at preindustrial values. The results show that, as a whole, the Asian monsoon precipitation is reduced by the increase in aerosol loading during boreal summer and winter. This decrease in precipitation corresponds to a decrease in precipitable water due to the cooling in surface air temperature (SAT), mainly over adjacent oceans. The cooling is explained by the sum of the direct and indirect effects of aerosols. A modulation of the Walker circulation occurs, which can be explained by the east-west horizontal SAT gradient over the tropics due to the spatially heterogeneous increase in aerosols. Concurrent with the modulation of the Walker circulation, the anomalous descending motions over the tropical Asian monsoon region are consistent with the decrease in precipitation. In addition, the changes in local Hadley circulation (or a shift of the inter-tropical convergence zone) are unclear over the Asian monsoon region and thus cannot explain the decrease in precipitation. Moreover, the detailed spatial pattern of the response of the atmospheric hydrological cycle over the Asian monsoon region has distinct seasonality. Interestingly, signals are distinct in regions where tropical disturbance activity is vigorous during both boreal summer and winter However, uncertainties regarding aerosol-cloud-precipitation interactions in current climate models and internal variability in the climate models may have affected the results.

Keywords: Atmospheric hydrological cycle, Asian monsoon, Aerosol, Cloud, Precipitation

Introduction

Although greenhouse gases (GHGs) cause global warming, aerosols cool the Earth's atmosphere. Many previous studies have investigated the global mean value of radiative forcing (e.g., IPCC 2013) and its effects on global climate. It is well known that the time-space distribution of aerosols is much more heterogeneous than that of GHGs, and therefore, the effects of aerosols on climate may vary over time and space. The effects of aerosols are complex because they serve as the source of clouds and thus affect the atmospheric hydrological cycle. In this study, we define the atmospheric hydrological cycle as water vapor, clouds, and precipitation.

The effects of aerosols on global and regional climates can be attributed to two processes: aerosol-radiation interactions, also known as direct aerosol effects (e.g., Ming et al. 2005), and effects through cloud changes, also known as indirect aerosol effects or aerosol-cloud interactions (e.g., Lohmann et al. 2010). Estimates of these indirect effects have high degrees of uncertainty (e.g., Penner et al. 2006) and are still poorly understood (e.g., IPCC 2013).

Compared with the global-mean and zonal-mean impacts of aerosols, the regional impacts of aerosols on

*Correspondence: hiroshi3@tmu.ac.jp
[1]Department of Geography, Tokyo Metropolitan University, 1-1 Minamioosawa, Hachioji, Tokyo 192-0397, Japan
[2]Japan Agency for Marine-Earth Science and Technology (JAMSTEC), Yokohama, Japan
Full list of author information is available at the end of the article

a climatological timescale are not well understood (e.g., Shindell et al. 2012). There are two major sources of anthropogenic aerosols over the Asian monsoon region, i.e., China and India. It is therefore possible that there are clear effects of aerosols. In addition, most studies have investigated the annual mean impact of aerosols. However, seasonal changes in atmospheric circulation can modify the spatial distribution of aerosols and their effects, particularly over the Asian monsoon region, which can have distinct seasonality. Thus, we focus on the tropical Asian monsoon region (South Asian, Southeast Asian, and western North Pacific monsoons).

In India, which is within the Asian monsoon region, a reduction in downward shortwave radiation at the surface (DSRS) due to a large atmospheric concentration of aerosols has been observed. This can result in decreased evaporation and summer rainfall (Ramanathan et al. 2005). The importance of dust and black carbon at the onset of the Indian summer monsoon has been evaluated based on observational data (Lau and Kim 2006). Bollasina et al. (2011) reported that the reduction in the Indian summer monsoon can be attributed primarily to anthropogenic aerosol emissions. These studies, however, focused on the Asian summer monsoon over India, whereas we examined aerosol effects over the entire Asian monsoon region during boreal summer and winter. Bollasina et al (2011) suggested that the mechanism underlying the reduction in precipitation over India was the weakening of the meridional land-sea contrast with weakening of the local Hadley circulation due to the decrease in DSRS over India. However, changes in cloud cover due to an increase in aerosol burden were unclear in the previous studies, which can also change the DSRS. The effects of aerosols on the atmospheric hydrological cycle have not been investigated over the tropical Asian monsoon region other than India. This study focuses on the whole tropical Asian monsoon region.

The present study investigates how aerosols affect cloud cover and precipitation over the tropical Asian monsoon region through aerosol-radiation and aerosol-cloud interactions, including the coupling effects of the ocean-land-atmosphere climate system, using a coupled atmosphere-ocean global climate model (CGCM). We focused on the Asian monsoon region because aerosol emissions over this region are very high and are expected to increase in the future. In addition, it is possible that the distinct seasonal changes in low-level monsoon winds seasonally vary the spatial pattern of aerosols and their effects. Responses of cloud and precipitation to aerosols over the Asian monsoon region are also examined. The numerical experiment is described in the "Methods/Experimental" section, and the results of the impacts of aerosols on atmospheric hydrological cycle over the tropical Asian monsoon region are shown in the "Results" section. Uncertainties of the effects of aerosols are discussed in the "Discussion" section, and the "Conclusions" section is given.

Methods/Experimental

Numerical experiments

To understand the impact of aerosols on regional and global climates, we used historical control experiments (HIST) for the 155-year period from 1851 to 2005, under the protocol for historical experiments of the Intergovernmental Panel on Climate Change-Fifth Assessment Report (IPCC-AR5; Taylor 2012). To clarify the effects of aerosols on the atmospheric hydrological cycle through both the aerosol-radiation and aerosol-cloud interactions, including slow responses, we used the Model for Interdisciplinary Research on Climate-Earth System Models (MIROC-ESM; Watanabe et al. 2011), a CGCM, coupled with the Spectral Radiation-Transport Model for Aerosol Species (SPRINTARS; Takemura et al. 2000; 2002). We performed sensitivity (pre-industrial aerosol; piAERO) experiments to estimate the effects of anthropogenic aerosols; these were the same as HIST, but with the emissions of anthropogenic aerosols fixed at their preindustrial (around 1850) values. The anthropogenic aerosols in our experiments are sulfates, organic carbons, and black carbons. The amount of black carbons is much smaller than that of the other anthropogenic aerosols although the effects of black carbons are quite different from those of the sulfates and organic carbons. We note that there are interannual and long-term variations in natural aerosols, such as dust and volcanic aerosols. We selected the analysis period of from 1976 to 2005 because the emission of anthropogenic aerosols has been increased until the end of the simulation period over the Asian monsoon region (e.g., Ohara et al. 2007). To reduce uncertainty in the simulation, we used three ensembles for each experiment. We used 90 seasonal means to calculate the 30-year seasonal mean. The differences between HIST and piAERO experiments were evaluated using the two-sided Student's t test at the 95% significance level.

Observational datasets

To confirm the low-level atmospheric circulation simulated in the HIST, we used the dataset of the Japanese 55-year Reanalysis Project (JRA-55; Kobayashi et al. 2015; http://jra.kishou.go.jp/JRA-55/index_en.html), including the surface zonal and meridional winds. We also used satellite-derived rainfall data from the Global Precipitation Climatology Project (GPCP; Adler et al. 2003; https://www.esrl.noaa.gov/psd/) for the 30-year period from 1980–2009; this dataset includes observed global monthly precipitation on a 2.5° × 2.5° grid. We also used the 30-year climatology data from 1980 to 2009. Although the averaged period was slightly different from the period of the HIST simulations and the observations, we confirmed that this had a negligible effect on the findings. In addition, we used long-term variations of the observed SAT dataset, the Hadley Centre–Climatic Research Unit Version 4 (HadCRUT4; Morice et al. 2012), for validation.

Results

Impact of aerosols on global climate

This section focuses on the impact of aerosols on the global and regional climates, beginning with the global climate. The performance of the HIST simulation was investigated previously in Watanabe et al. (2011). The following subsection focuses on the reproducibility of the Asian monsoon precipitation, which is important for our purposes. Figure 1 presents the time series of global SAT, precipitation, aerosol optical depth (AOD), precipitable water, and cloud effective radius for the 100-year period from 1906 to 2005. To investigate long-term changes, seasonal components and those with periodicity of less than 10 years were suppressed by using a running mean with a window of 121 months.

It is well known that the global mean SAT has increased due to increased concentrations of GHGs, such as carbon dioxide. All three members of the HIST and piAERO experiments showed similar long-term increases (Fig. 1b), respectively. The simultaneous drops in SAT in the 1960s, around 1980, and around 1991 were considered to be caused by large volcanic eruptions, i.e., the effects of natural aerosols (Fig. 1a), although the drops were unclear due to the effect of the digital filter. These drops in SAT were also simulated in piAERO, as long-term changes in the natural aerosol burden were similar to those in HIST, although the simulated long-term trends in SAT in the piAERO were different from those in the HIST. Compared to the HIST, the increase in global mean SAT was

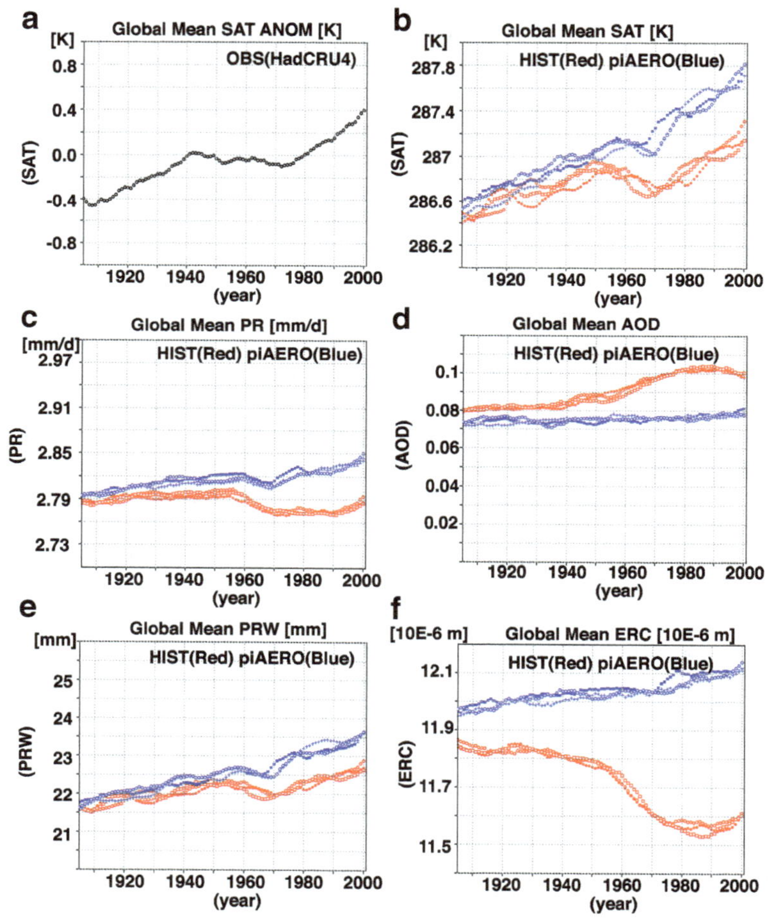

Fig. 1 **a** Observed time series of global mean surface air temperature (SAT) anomaly from the average of the period of 1961–1990. Simulated time series of the global mean **b** SAT, **c** precipitation, **d** aerosol optical depth (AOD), **e** precipitable water, and **f** effective radii of clouds in the three ensemble historical (HIST) and sensitivity (pre-industrial aerosol; piAERO) experiments. To examine the long-term trends, we used a 121-month running mean filter to remove seasonal and shorter interannual variations

larger in piAERO after the 1950s. The timing (from 1955 to 2005) when the differences in global mean SAT between the HIST and piAERO has gradually become larger corresponded to that in AOD (Fig. 1d). The increase in AOD was due to human activity.

In our experiments, during 1905–2005, the global mean SAT increased by approximately 0.6–0.8 K in HIST and by approximately 1.0–1.2 K in piAERO. These simulations indicate that approximately one third to one half of the total increase in global mean SAT was likely suppressed by the increase in anthropogenic aerosol burden. This estimated sensitivity to global warming was greater than that seen in the multi-model results of the IPCC-AR5 (IPCC 2013).

We also examined changes in the atmospheric hydrological cycle. Concurrent with the long-term increase in SAT, over the 100-year period, the global mean precipitation and precipitable water increased in the piAERO experiments (Fig. 1c, e). In general, an increase in global mean SAT results in an increase in precipitation, due to an increase in the global mean precipitable water. However, the global mean precipitation in the HIST decreased after 1950, although the global mean SAT gradually increased. The HIST and piAERO simulations of long-term changes in the global mean cloud effective radius showed clear differences between the HIST and piAERO (Fig. 1f). Smaller cloud effective radii occurred under higher aerosol loading conditions. These results can be explained by the cloud lifetime effects of aerosols (e.g., Twomey et al. 1984; Albrecht 1989).

Reproducibility of the Asian monsoon precipitation and circulation

Before discussing the regional characteristics and seasonality of the aerosol impacts, we examined the reproducibility of precipitation over the Asian monsoon region. Figure 2 presents the climatology based on observations from GPCP and JRA55 and simulations from MIROC-ESM.

The simulation revealed spatial peaks in precipitation during boreal summer (June, July, August; JJA) over the Asian monsoon region (Fig. 2b), similar to the observations (Fig. 2a). However, a major peak in precipitation was simulated over the western coast of the Indian subcontinent while this was observed over the western coast of the Indochina Peninsula. In addition, compared to the observations, weaker monsoon westerlies and a weaker monsoon trough around the Philippines were found in

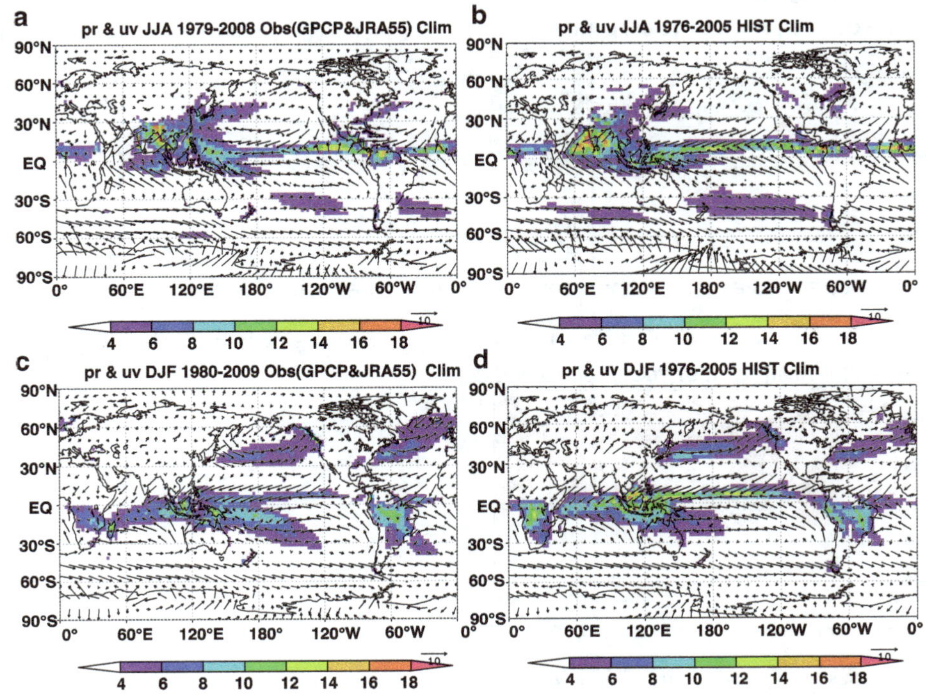

Fig. 2 Climatology in **a** observed and **b** simulated precipitation and surface winds at 10 m above the surface during June, July, and August (JJA). **c** and **d** are the same as **a** and **b**, respectively, except for December, January, and February (DJF). For the observations, the values were averaged over 30 years from 1979 to 2008, and for the HIST simulation, the values were averaged over 30 years from 1976 to 2005. For the simulated climatology, we averaged the three-ensemble members

the HIST experiments. Precipitation over the southern Maritime Continent was also overestimated in HIST.

During boreal winter, although the simulation reproduced the spatial distribution of precipitation and circulation, the precipitation over the equatorial Indian Ocean was greater in HIST than that in the observations. As in simulations of boreal summer, precipitation over the southern Maritime Continent was overestimated, and precipitation over the southern Pacific Ocean was underestimated in HIST.

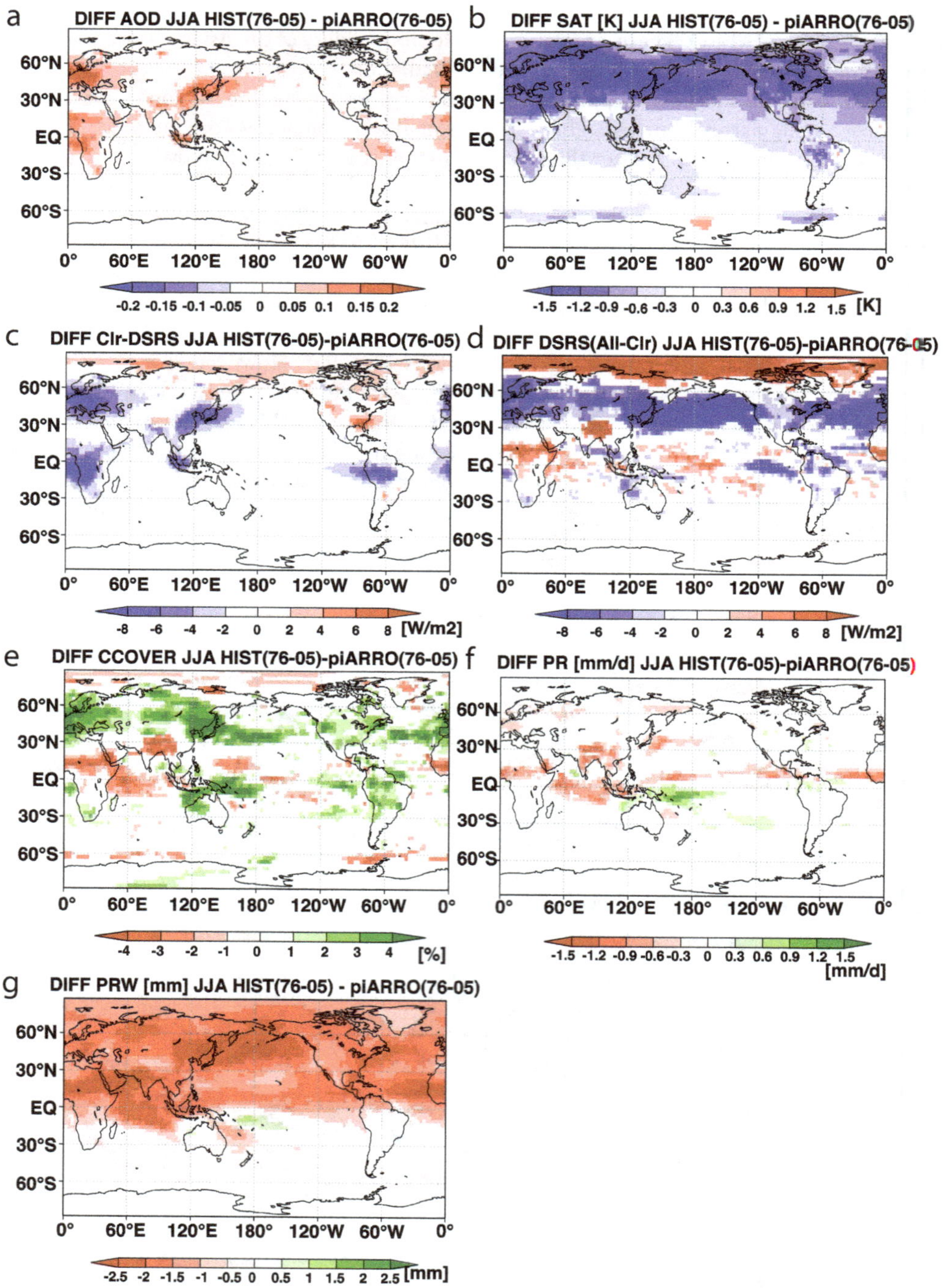

Fig. 3 Differences in 30-year climatological **a** AOD, **b** SAT, **c** clear-sky downward shortwave radiation at the surface (DSRS), **d** the difference between all-sky DSRS and clear-sky DSRS (indirect effect), **e** cloud cover, **f** precipitation, **g** precipitable water during JJA between the ensemble-averaged HIST and piAERO (HIST - piAERO). Both HIST and piAERO values were averaged over the 30-year period from 1976 to 2005. All differences are shown when the local statistical significance level is at least 95%, as determined by Student's *t* test

Although there were some differences between simulated and observed precipitation, the general features and spatial patterns of precipitation and low-level circulation and their seasonal changes were reproduced by HIST.

Impact of aerosols during boreal summer (JJA)

Geographical pattern of AOD changes

To understand how aerosols affect the regional climate of the tropical Asian monsoon, we produced a global map of differences between the HIST and piAERO in AOD, SAT, clear-sky downward DSRS, the difference between all-sky and clear-sky DSRSs, cloud cover, and precipitation (Fig. 3). Because changes outside of the tropical Asian monsoon region may affect the regional climate of the tropical Asian monsoon region, we analyzed the changes globally, focusing on the tropical Asian monsoon region. The difference between all-sky and clear-sky DSRSs will be explained later. To detect statistically significant signals, we used all three ensembles in both experiments.

An increase in AOD during JJA was observed over and around the regions of East Asia, South Asia, Maritime Continent, Western Europe, Western Africa, and South America (Fig. 3a). The signal of AOD was weaker over North America because the timing of improvement air quality problems was different from the other regions. During the period considered, considerable changes have occurred in land use, such as deforestation, conversion from natural vegetation to cropland, and urbanization.

The major source regions of aerosols are located in the Northern Hemisphere. Over this study region of the Asian monsoon, China and India are the major source regions. The aerosols were partly transported downstream of the monsoon flows (see Fig. 2). Notably, over the South Pacific Ocean, there is no major source of aerosols, which induced the meridional contrast of AOD globally, particularly over the Pacific Ocean. The increase in AOD ranged from 0.05 to 0.20.

Surface air temperature response

Over the tropics, statistically significant cooling signals were simulated over the entire Indian Ocean, North Pacific Ocean, and the North Atlantic Ocean (Fig. 3b). Note that the tropical South Pacific Ocean was not cooled by the increase in aerosols. Thus, the east-west inter-oceanic contrast in SAT between the Indian Ocean and Western Pacific Ocean was dominant (Fig. 3b), which may have affected the atmospheric hydrological cycle of the tropical Asian monsoon. In addition, no clear SAT signals were simulated over the Asian monsoon land regions, i.e., the Indian subcontinent and the Indochina Peninsula, indicating that the cooling over land was not systematically simulated. Cooling signals were also simulated over the middle and high latitudes in the Northern Hemisphere (Fig. 3b), whereas less cooling was simulated over the Southern Hemisphere. Notable cooling was simulated over a zonal band from 30° N to 70° N and not only near the source regions of aerosols.

Due to the increases in AOD, negative signals in clear-sky DSRS were observed over the major source regions (Fig. 3c). The spatial distributions of the differences in AOD and clear-sky DSRS were similar, indicating that direct radiation effects resulted in cooling over the major sources of aerosols. However, the changes in clear-sky DSRS due to increased AOD (Fig. 3c) cannot explain all of the changes in SAT (Fig. 3b).

Changes in cloud cover and precipitation

To understand the changes in SAT, we also examined the indirect effects of aerosols by considering changes in cloud cover, which are the parts of the atmospheric hydrological cycle. Of course, we also considered the responses of precipitation and precipitable water. Here, we defined the indirect effects of the aerosols as the difference between the all-sky and clear-sky DSRS (following Mukai et al. 2008; Mukai and Nakajima 2009). For this estimation, the indirect effects were considered to be the changes in cloud cover due to different aerosol conditions. However, this is not simply a sum of the first and second indirect effects; it involves many effects related to changes in the aerosol conditions, so we cannot estimate only the indirect effects of aerosols.

An impact of cloud cover changes on the decrease in SAT due to aerosol-cloud interactions was clear over the tropical regions, such as equatorial Indian Ocean, equatorial Western Pacific, and the Indian subcontinent (Fig. 3d), although significant cooling over the mid-latitudes was also simulated over downstream regions of the aerosol sources. We confirmed that changes due to the indirect effects could be explained by changes in cloud cover because their spatial patterns were similar (Figs. 3d, e). It is noteworthy that the effects of clouds due to the changes in aerosol condition contributed significantly to the creation of inter-oceanic and inter-hemispheric contrasts in SAT over the tropical Asian monsoon region and the Pacific Ocean.

Changes in precipitation were remarkable primarily over the tropical regions (Fig. 3f). An increase in precipitation was simulated over part of the Maritime Continent-southern equatorial Pacific and Amazon regions, whereas a decrease was observed over the Asian monsoon land, the southern equatorial Indian Ocean, and Central Africa. The changes in precipitation can be explained by changes in water vapor (Fig. 3g). The changes in water vapor mostly corresponded to those in SAT. The decreases in water vapor were simulated over the entire Asian monsoon region and the Indian Ocean. Only the equatorial South Pacific Ocean showed a positive signal in water vapor.

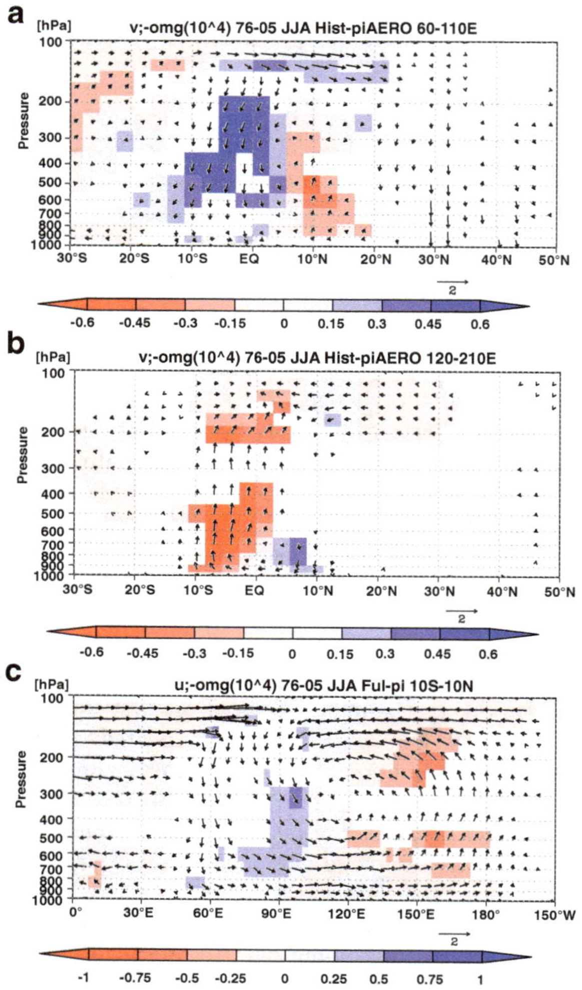

Fig. 4 **a** Latitude-vertical (in pressure) cross-section in meridional and vertical winds over the Asian monsoon region (60° E–110° E) during JJA. **b** Same as Fig. 5a, but over the Pacific Ocean (120° E–150° W) during JJA. **c** Longitude-vertical (in pressure) cross-section in zonal and vertical winds along the equator (10° S–10° N) during JJA. As we used ω, we multiplied ω by -1 for vectors to understand vertical motions. The plotted vectors are statistically significant for horizontal or vertical components, or both. The red (blue) color indicates ascending (descending) motion. The unit of zonal and meridional winds is m s^{-1}. The unit of ω is 10^{-4} Pa s^{-1}. The plotted colors are statistically significant for the vertical component

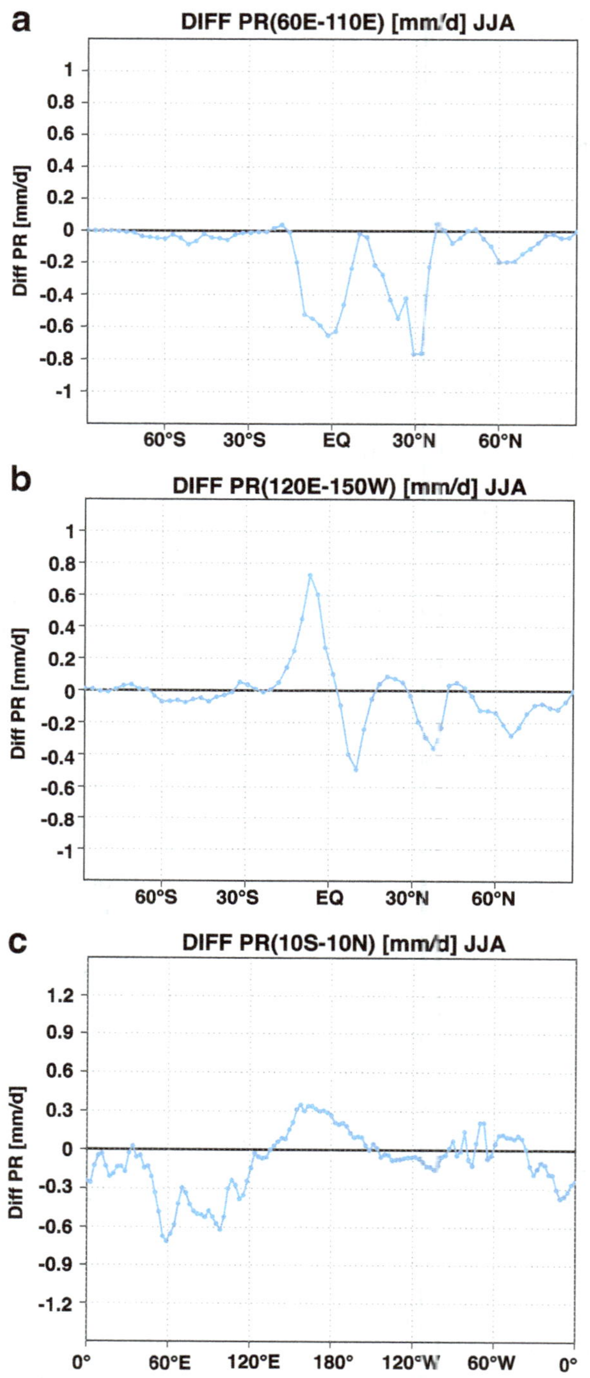

Fig. 5 Meridional distribution of zonally averaged precipitation **a** over the longitudinal band of the Asian monsoon region (60° E–110° E) and **b** over the longitudinal band of the Pacific Ocean (120° E–150° W) during JJA. **c** Zonal distribution of meridional mean precipitation over the latitudinal band of the equator region (10° S–10° N). The blue line indicates the differences in zonal or meridional mean precipitation between 30-year mean HIST from 1976 to 2005 and piAERO from 1976 to 2005. Note that the statistically insignificant values are also plotted in this figure

Changes in atmospheric circulation and coupling with clouds and precipitation

To understand the changes in meridional and zonal circulation on the Asian monsoon regional scale, we produced meridional-vertical cross-sections of winds over the Asian monsoon region and the Pacific Ocean, and a zonal-vertical cross-section of winds along the equatorial region (Fig. 4). To associate atmospheric circulation changes with the precipitation changes, we analyzed zonal and meridional averaged changes in precipitation (Fig. 5).

Over the Asian monsoon region, anomalous descending motions were simulated south of the equator and along the zonal band of 25–35° N (Fig. 4a), consistent with the decreased cloud cover and precipitation (Figs. 3e, f and 5a). These results revealed no changes in local Hadley circulation over the Indian Subcontinent and the Indochina Peninsula (along the zonal band of 10–25° N), which is different from the finding of Bollasina et al. (2011). A weak enhancement of local Hadley circulation over the Asian monsoon region was simulated within 10° S–10° N, which did not cover the Asian landmass (Fig. 4a). Over the Pacific Ocean, only the anomalous ascending motions were simulated over the equatorial regions (Fig. 4b), which is consistent with the changes in precipitation (Fig. 5b). The anomalies in the equatorial zonal circulation showed an eastward shift of the convective center of the Walker circulation or enhancement of the Walker circulation over the Indian-Pacific trans-ocean sector (Fig. 4c), which is consistent with the changes in precipitation (Fig. 5c). This suggested that the changes in equatorial zonal circulation can more coherently explain the large-scale anomalies in cloud cover and precipitation than can changes in local Hadley circulation. The distinct changes in the Walker circulation can be confirmed by the spatial map of surface winds (Fig. 6). These anomalous descending (ascending) motions were consistent with the decreases (increases) in cloud cover and precipitation, making it difficult to determine the cause-effect relationship between the cloud-precipitation and circulation changes.

On the more detailed spatial pattern of the precipitation response, the Asian summer monsoon precipitation decreased along two zonal bands, 10° S to the equator and 20°–30° N (Figs. 3f and 5a). Concurrent with the decreases in precipitation over the Asian monsoon region, an anomalous divergence in surface winds over the equatorial Indian Ocean and an anomalous anti-cyclonic circulation over South Asia and the Tibetan Plateau were simulated (Fig. 6). In addition, an increase in precipitation over the equatorial Pacific was consistent with an anomalous convergence in surface winds. Strong westerly anomalies over the eastern Indian Ocean and Maritime Continent, which modulated the Walker circulation in this region, were simulated. The precipitation signals also corresponded to the anomalous vertical motions (Fig. 4a). This spatial precipitation pattern matched the observed pattern of interannual precipitation variability, i.e., a zonally extended striped pattern (monsoon trough), as indicated in previous observational studies (e.g., Goswami and Mohan 2001; Fujinami et al. 2011; Takahashi et al. 2015). The spatial pattern in precipitation is associated with the tropical cyclone activity (Takahashi et al. 2015; Takahashi and Yasunari 2006), which suggests that the signal is associated with a modulation of tropical cyclone activity. The decrease in precipitable water (Fig. 3g) can weaken low-level water vapor convergence, even if the strength of wind convergence by a tropical disturbance does not change, and may reduce the development of tropical disturbances along the monsoon trough.

Impact of aerosols during boreal winter (DJF)

Geographical pattern of changes in AOD

An increase in AOD during December, January, and February (DJF) was simulated over and around East Asia, Southeast Asia, Western Europe, Western Africa, and North America (Fig. 7a). Over the tropical Asian monsoon region, there were large differences in AOD between JJA and DJF over the eastern region of Eurasia. Although the major sources of aerosols were similar to those during boreal summer, AOD was greater over Southeast Asia and was lower over the mid-latitude North Pacific, compared with those during boreal summer. The differences in spatial distribution of AOD can be explained by the increase in local biomass burning (e.g., van der Werf et al. 2008) over Southeast Asia and by equatorward transportation of the East Asian aerosols due to the climatological winter monsoon winds in the lower troposphere, which contribute to the stronger zonal contrast in AOD between the Indian Ocean-Maritime Continent sector and the equatorial Pacific Ocean in boreal winter.

Surface air temperature response

Over the tropics, cooling was simulated over the tropical Indian Ocean, North Pacific Ocean, and tropical Atlantic Ocean. No clear cooling was simulated over the South Pacific Ocean. This SAT response results in an enhancement of the zonal gradient in SAT due to the increased aerosols (Fig. 7b), consistent with the spatial pattern of AOD (Fig. 7a). These features were similar to those in boreal summer. In addition, cooling in the mid-latitudes was weaker than that in boreal summer, which was consistent with a seasonal change in AOD.

Next, we investigated the aerosol-radiation interaction. Figure 7c shows the differences in clear-sky DSRS during DJF. There was noticeable cooling along the southeastern coast of Eurasia, which corresponded to AOD changes (Fig. 7a). Again, although the geographical pattern of the changes in SAT was not entirely explained by the direct effect of aerosols due to the increased AOD, the sum of direct and indirect effects (Fig. 7c, d) can explain the changes in SAT. It was found that the spatial pattern of cloud cover was similar to that of indirect effects of aerosols. Moreover, the spatial distribution of the indirect effects (Fig. 7d) was different from that seen during JJA (Fig. 7d). The simulated cooling in the mid-latitudes due to the indirect effects of aerosols over the North Pacific Ocean at around 30° N was weaker than that in boreal summer.

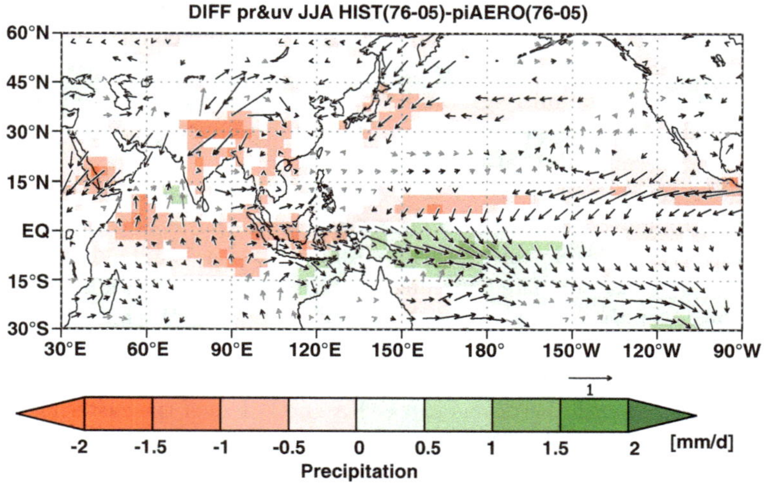

Fig. 6 As in Fig. 3, but for precipitation (color) and 10 m surface horizontal winds (vector) during JJA. The unit of precipitation (wind) is mm day^{-1} (m s^{-1}). Gray vectors represent the spatial pattern of differences in wind, which are not statistically significant. Only stronger differences (> 0.2 m s^{-1}) are plotted

Changes in cloud cover and precipitation

During DJF, east-west contrasts in cloud cover and precipitation were simulated over the tropics (Fig. 7e, f), corresponding to the stronger zonal gradient in SAT and precipitable water over the Indian Ocean and western Pacific Ocean (Fig. 7b, g). Decreases in precipitable water were simulated over the entire tropical region except for the tropical South Pacific Ocean, which can almost entirely explain the decrease in precipitation over the tropical Asian monsoon region. Although the spatial patterns of cloud and precipitation during DJF were somewhat different from those during JJA, the dominant east-west contrasts in cloud cover and precipitation were similar between boreal summer and winter.

Moreover, there were different responses of cloud-precipitation between the tropics and mid-latitudes. In the mid-latitudes, the cyclonic anomalies and enhancement of the winter monsoon outflows over and around Japan were associated with an increase in cloud cover and a decrease in precipitation (Figs. 7 and 8). This response can be explained as the well-known lifetime effect of aerosols. Interestingly, the signs of changes in precipitation corresponded to those of changes in cloud cover over the tropics (Figs. 3 and 7), which were different from the responses in the mid-latitudes. The different responses are discussed in the "Uncertainties of effects of aerosols on climate" section.

Changes in atmospheric circulation and coupling with clouds and precipitation

This subsection discusses changes in cloud cover and precipitation during boreal winter, which are associated with atmospheric circulation. Takahashi and Watanabe (2016) reported that the decreased volcanic sulfate aerosols caused partial enhancement of trade winds over the equatorial western Pacific Ocean from 1991 to 2010, which can be explained by the zonal gradient in SAT. Their study showed that an aerosol forcing can change the large-scale circulation over the tropics, although the forcing of aerosols in our experiment is quite different from that in Takahashi and Watanabe (2016). The differences can be clearly seen from the spatial pattern of SAT. Furthermore, a few observational studies showed co-variability of the aerosol-cloud-precipitation over and around the Maritime Continent on interannual timescales (e.g., Yamaji and Takahashi 2014).

As with the boreal summer, we investigated changes in meridional and zonal circulation (Fig. 9) and changes in precipitation for winter (Fig. 10). The changes in the local Hadley circulation over the Asian monsoon region were not clear (Fig. 9a). Over the Asian monsoon region, anomalous descending motions (Fig. 9a) were consistent with the decreased precipitation (Fig. 10a). Over the Pacific Ocean, anomalous ascending motions were consistent with the increase in precipitation (Figs. 9b and 10b), which can be understood as the southward shift of the inter-tropical convergence zone (ITCZ). In addition, this southward shift of the ITCZ (Figs. 7 and 10) was partly consistent with the annual mean response simulated by Takemura et al. (2005). However, we found that the responses over the Asian monsoon region and the Indian Ocean were quite different.

Changes in the Walker circulation were dominant (Fig. 9c). The anomalous convective center of the Walker circulation was simulated over the western equatorial Pacific Ocean, and this distinct signal was statistically

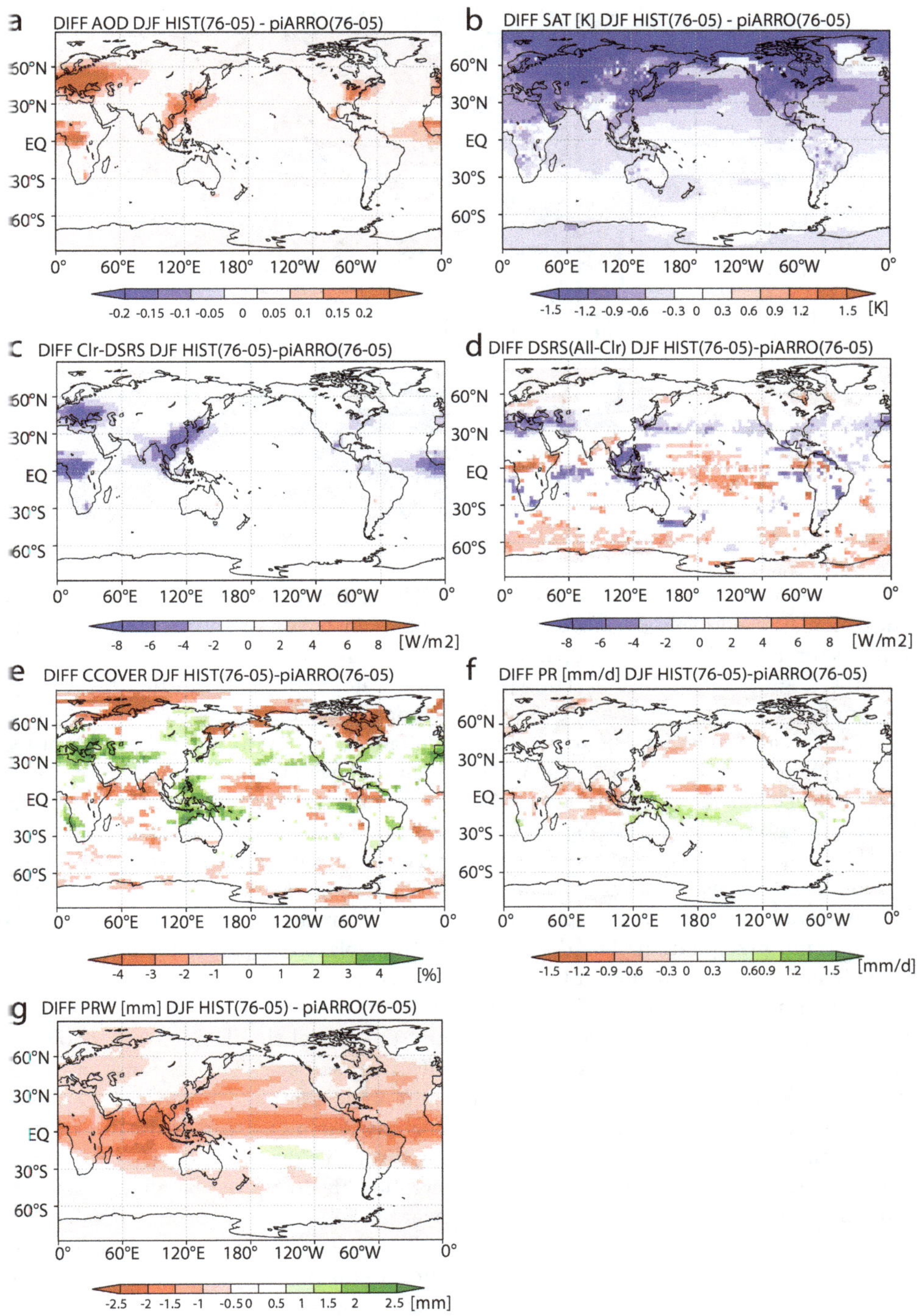

Fig. 7 As in Fig. 3 for both HIST and piAERO, but during DJF averaged over the 30-year period from 1975 to 2004

significant (Fig. 9c and 10c). Thus, changes in part of the Walker circulation, rather than in local Hadley circulation, were strongly coupled with the changes in precipitation over the tropical Asian monsoon region. Note that our results regarding changes of the Walker circulation coupling with cloud and precipitation in association with the changes in zonal gradient of SAT can be explained in a similar manner to Takahashi and Watanabe (2016). Nevertheless, the changed spatial patterns of precipitation, circulation, and SAT are different because of differences in experimental design, such as the treatment of forcing.

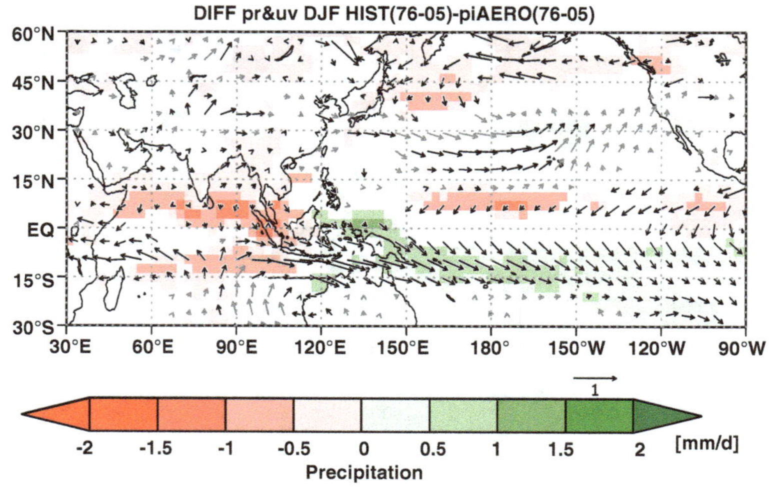

Fig. 8 As in Fig. 6, but during DJF. The unit of precipitation (wind) is mm day^{-1} (m s^{-1}). Gray vectors represent the spatial pattern of differences in wind, which are not statistically significant. Only stronger differences (> 0.2 m s^{-1}) are plotted

The simulated decreases in cloud cover and precipitation coupled with increased aerosols over the Indian Ocean and Asian monsoon regions were symmetric with respect to the equator (Figs. 7e, f and 10a). The spatial pattern of this symmetrical response along 10° S and 10° N over the Indian Ocean corresponded to that of the activity of cyclone pairs that are symmetrical with respect to the equator (e.g., Kiladis and Wheeler 1995; Compo et al. 1999; Ito and Satomura 2009, Takahashi et al. 2011). The activity of cyclone pairs is dominant in this season. This correspondence suggests that modulation of the activity of the tropical disturbances may explain the decrease in cloud cover and precipitation. The decrease in precipitable water (Fig. 7g) due to cooling of the SAT (Fig. 7b) and anomalous descending motions (Fig. 9a, c) can weaken the low-level convergence of water vapor and/or the activity of the tropical disturbances over the equatorial Indian Ocean. Notably, the signals were clear along the major route of the tropical disturbances. These characteristics were common to both boreal summer and winter, although the frequent passage routes of tropical disturbances are different between boreal summer and winter.

Discussion

Uncertainties of effects of aerosols on climate

The previous sections discussed the changes in atmospheric hydrological cycle over the Asian monsoon region. Similar results were also simulated in previous studies. However, as important uncertainties in the effects of aerosols were found in the model experiments, this subsection focuses on two major uncertainties, namely, the aerosol-cloud-precipitation interactions and internal climate variability.

The differences between the mid-latitudes and the tropics in terms of the cloud-precipitation response to increased aerosols can be explained by differences in the total water vapor path. In the tropics, much more water vapor is available for the development of clouds and precipitation. In contrast to the mid-latitudes, the much reduced level of water vapor suppresses the development of clouds and precipitation, which can be understood as the cloud lifetime effect of aerosols.

However, the indirect effects of the aerosol-cloud-precipitation interaction in current climate models are strong compared to the observations (Michibata et al. 2016). Golaz et al. (2011, 2013) reported that the indirect effects in their Geophysical Fluid Dynamics Laboratory (GFDL) model was very sensitive to model turning, and the model simulated strong cooling by the indirect effects of aerosols. The response of clouds to aerosols is quite important, and it can change the sign of the SAT response. An increase in aerosols contributes to SAT cooling through a direct effect. This response may contribute significantly to the weakening of the land-sea contrast over the South Asian monsoon region (e.g., Bollasina et al. 2011). However, the decrease in cloud cover as indirect effects contributes to an increase in DSRS, which can change the sign of the response. This uncertainty may change the response of the Asian monsoon to aerosol forcing, including our current results.

Internal climate variability of the Asian monsoon

The inter-ensemble spread of the changes in precipitation over India was not small (supplemental figures in Bollasina et al. (2011)). The strong association with observed changes can be explained by the aerosol forcing or internal climate variability of the climate system. In

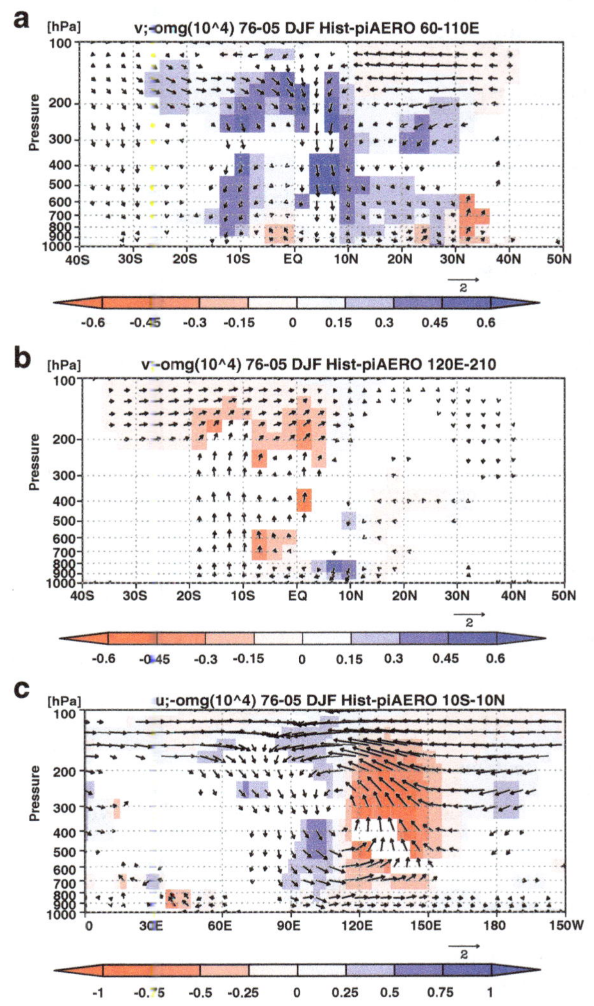

Fig. 9 As in Fig. 4 but for DJF

their three-ensemble experiments designed to understand the impact of aerosols (GHGs), Bollasina et al. (2011) reported that one ensemble member showed a slight increasing trend (decreasing trend), whereas the three-ensemble mean exhibited a decreasing (increasing) trend. These results suggest that the internal climate variability of South Asia was not negligible, although the decreasing trends in precipitation found in multiple observational datasets were reliable. In terms of the inter-ensemble spread in our experiments, the results showed that the changes in precipitation over the Asian monsoon region and changes in the meridional SAT gradient showed the same sign, respectively, although we should have increased the number of ensemble members, to strengthen this conclusion. In addition, Salzmann et al. (2014) supports a possible effect of internal variability on the observed trend in precipitation over South Asia, based on analysis of the Coupled Model Intercomparison Project Phase 5 (CMIP5) dataset. To determine whether aerosol forcing or internal climate variability can explain the decrease in precipitation over the Asian monsoon region, including our current results, many ensemble experiments may be needed in future studies. However,

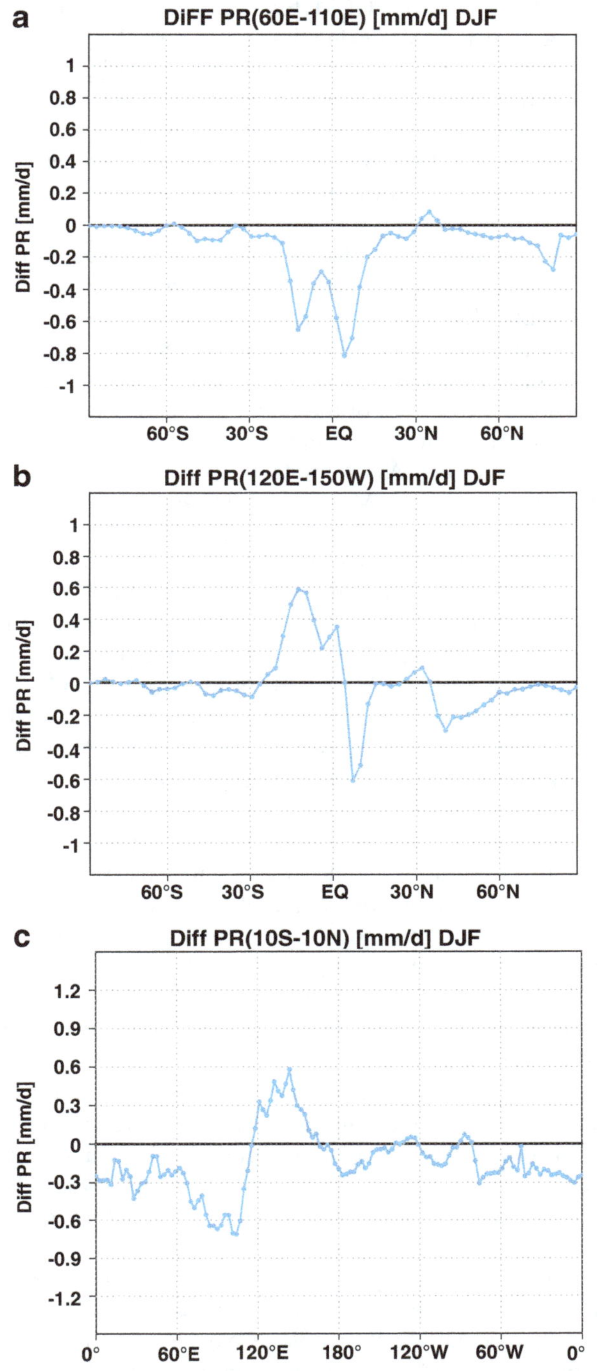

Fig. 10 As in Fig. 5, but for DJF

this approach can only clarify responses of cloud and precipitation to an increase in aerosols in a current CGCM, which cannot essentially improve the understanding of the processes of aerosol-cloud-precipitation interactions. Therefore, both uncertainties should be improved by different approaches.

Conclusions

This study used a CGCM to investigate the effects of increases in aerosols during the period from the preindustrial era to the present. Three-ensemble historical (HIST) and sensitivity experiments (piAERO) were conducted for a 156-year period (1850–2005). The parameters of piAERO were the same as those for HIST, except that the emissions of anthropogenic aerosols were kept at preindustrial values.

The results showed that precipitation decreased over the tropical Asian monsoon region during both summer and winter. This decrease in precipitation corresponded to a decrease in precipitable water due to the cooling in SAT mainly over the adjacent oceans. The cooling was explained by the sum of the direct and indirect effects of aerosols. In addition, the decrease in precipitation over the Indian Ocean/Asian monsoon region, coupled with the increase over the Maritime Continent/western South Pacific, is likely associated with changes in the Walker circulation over the eastern Indian Ocean, Maritime Continent, and western Pacific Ocean. As a part of the Walker circulation, the anomalous vertical motions were consistent with the changes in precipitation. The modulation of the Walker circulation can be explained by the east-west horizontal SAT gradient between the Indian Ocean/Asian monsoon region and the Maritime Continent/western South Pacific due to heterogeneous increases in aerosols.

A southward shift of the ITCZ was simulated only over the equatorial Pacific Ocean particularly in boreal summer and winter, which can also be explained by the meridionally asymmetrical SAT response to aerosols. It was found that the changes in local Hadley circulation were not clear over the Asian monsoon region, which cannot explain the decrease in precipitation over the tropical Asian monsoon region.

The detailed spatial patterns of the decrease in precipitation over the Asian monsoon region were simulated. Interestingly, a decrease in precipitation was found along the major routes of the convectively coupled tropical disturbances in both boreal summer and winter, although the routes vary seasonally.

In this and related previous studies, cloud and precipitation responses to aerosols showed non-negligible uncertainty, particularly in the tropics. Changes in the simulated response of cloud and precipitation due to model uncertainty can be considered as major sources of uncertainty. Therefore, additional studies are required to gain a better understanding of the aerosol effects.

Abbreviations

AOD: Aerosol optical depth; CGCM: Coupled atmosphere-ocean global climate model; CMIP5: Coupled Model Intercomparison Project Phase 5; DJF: December, January, February; DSRS: Downward shortwave radiation at the surface; GFDL: Geophysical Fluid Dynamics Laboratory; GHG: Greenhouse gas; GPCP: Global Precipitation Climatology Project; HadCRUT4: Hadley Centre–Climatic Research Unit Version 4; HIST: Historical experiment; IPCC-AR5: Intergovernmental Panel on Climate Change-Fifth Assessment Report; ITCZ: Inter-tropical convergence zone; JJA: June, July, August; JRA-55: Japanese 55-year Reanalysis Project; MIROC-ESM: Model for Interdisciplinary Research on Climate-Earth System Models; piAERO: Pre-industrial aerosol experiment; SAT: Surface air temperature; SPRINTARS: Spectral Radiation-Transport Model for Aerosol Species

Acknowledgments

This work was partly supported by the Environment Research and Technology Development Fund (S-12) of the Environmental Restoration and Conservation Agency. HGT was partly supported by the Cooperative Program of Atmosphere and Ocean Research Institute, The University of Tokyo. JRA-55 was provided by the Japan Meteorological Agency. GPCP Precipitation data were provided by the NOAA/OAR/ESRL PSD, Boulder, Colorado, USA.

Funding

This work was partly supported by the Environment Research and Technology Development Fund (S-12) of the Environmental Restoration and Conservation Agency.

Authors' contributions

HGT, SW, MN, and TT proposed the topic and conceived and designed the study. WS carried out the experimental study. HGT and MN analyzed the data and SW and TT helped in their interpretation. HGT led the preparation of the manuscript. SW, MN, and TT collaborated with the corresponding author during the construction of the manuscript. All authors read and approved the final manuscript.

Competing interests

The authors declare that they have no competing interests.

Author details

[1]Department of Geography, Tokyo Metropolitan University, 1-1 Minamioosawa, Hachioji, Tokyo 192-0397, Japan. [2]Japan Agency for Marine-Earth Science and Technology (JAMSTEC), Yokohama, Japan. [3]Kindai University, Higashiosaka, Japan. [4]Research Institute for Applied Mechanics, Kyushu University, Fukuoka, Japan.

References

Adler RF, Huffman GJ, Chang A, Ferraro R, Xie P, Janowiak J, Rudolf B, Schneider U, Curtis S, Bolvin D, Gruber A, Susskind J, Arkin P, Nelkin E (2003) The version 2 Global Precipitation Climatology Project (GPCP) monthly precipitation analysis (1979-present). J Hydrometeorol 4:1147–1167

Albrecht BA (1989) Aerosols, cloud microphysics, and fractional cloudiness. Science 245(4923):1227–1230

Bollasina MA, Ming Y, Ramaswamy V (2011) Anthropogenic aerosols and the weakening of the South Asian summer monsoon. Science 334(6055):502–505. https://doi.org/10.1126/science.1204994

Compo GP, Kiladis GN, Webster PJ (1999) The horizontal and vertical structure of east asian winter monsoon pressure surges. QJR Meteorol Soc 125(553):29–54

Fujinami H, Hatsuzuka D, Yasunari T, Hayashi T, Terao T, Murata F, Kiguchi M, Yamane Y, Matsumoto J, Islam MN, Habib A (2011) Characteristic intraseasonal oscillation of rainfall and its effect on interannual variability over Bangladesh during boreal summer. Int J Climatol 31:1192–1204. https://doi.org/10.1002/joc.2146

Golaz J-C, Salzmann M, Donner LJ, Horowitz LW, Ming Y, Zhao M (2011) Sensitivity of the aerosol indirect effect to subgrid variability in the cloud parameterization of the gfdl atmosphere general circulation model am3. J Climate 24(13):3145–3160. https://doi.org/10.1175/2010JCLI3945.1

Golaz J-C, Horowitz LW, Levy H (2013) Cloud tuning in a coupled climate model: impact on 20th century warming. Geophys Res Lett 40(10):2246–2251. https://doi.org/10.1002/grl.50232

Goswami BN, Ajaya Mohan R (2001) Intraseasonal oscillations and interannual variability of the Indian summer monsoon. J Climate 14(6):1180–1198

IPCC (2013) Climate Change 2013: The Physical Science Basis: Working Group I Contribution to the Fifth Assessment Report of the Intergovernmental Panel on Climate Change. Cambridge University Press, Cambridge

Ito M, Satomura T (2009) The interannual variation of intraseasonal oscillation linked with the Indian Ocean dipole mode. SOLA 5:69–72. https://doi.org/10.2151/sola.2009-018

Kiladis GN, Wheeler M (1995) Horizontal and vertical structure of observed tropospheric equatorial rossby waves. J Geophys Res 100(D11):22981–22997

Kobayashi S, Ota Y, Harada Y, Ebita A, Moriya M, Onoda H, Onogi K, Kamahori H, Kobayashi C, Endo H, Miyaoka K, Takahashi K (2015) The JRA-55 reanalysis: general specifications and basic characteristics. J Meteorol Soc Jpn 93(1):5–48

Lau K-M, Kim K-M (2006) Observational relationships between aerosol and Asian monsoon rainfall, and circulation. Geophys Res Lett 33(21). https://doi.org/10.1029/2006GL027546

Lohmann U, Rotstayn L, Storelvmo T, Jones A, Menon S, Quaas J, Ekman AM, Koch D, Ruedy R (2010) Total aerosol effect: radiative forcing or radiative flux perturbation? Atmos Chem Phys 10:3235–3246. https://doi.org/10.5194/acp-10-3235-2010

Michibata T, Suzuki K, Sato Y, Takemura T (2016) The source of discrepancies in aerosol–cloud–precipitation interactions between gcm and a-train retrievals. Atmos Chem Phys 16(23):15413–15424. https://doi.org/10.5194/acp-16-15413-2016

Ming Y, Ramaswamy V, Ginoux PA, Horowitz LH (2005) Direct radiative forcing of anthropogenic organic aerosol. J Geophys Res 110:20208. https://doi.org/10.1029/2004JD005573

Morice CP, Kennedy JJ, Rayner NA, Jones PD (2012) Quantifying uncertainties in global and regional temperature change using an ensemble of observational estimates: The hadcrut4 data set. J Geophys Res 117(D08101). https://doi.org/10.1029/2011JD017187

Mukai M, Nakajima T, Takemura T (2008) A study of anthropogenic impacts of the radiation budget and the cloud field in East Asia based on model simulations with GCM. J Geophys Res 113:12211. https://doi.org/10.1029/2007JD009325

Mukai M, Nakajima T (2009) Potentiality of aerosols in changing the precipitation field in Asia. SOLA 5:97–100. https://doi.org/10.2151/sola.2009-025

Ohara T, Akimoto H, Kurokawa J-I, Horii N, Yamaji K, Yan X, Hayasaka T (2007) An Asian emission inventory of anthropogenic emission sources for the period 1980–2020. Atmos Chem Phys 7(16):4419–4444. https://doi.org/10.5194/acp-7-4419-2007

Penner JE, Quaas J, Storelvmo T, Takemura T, Boucher O, Guo H, Kirkevåg A, Kristjánsson JE, Seland Ø (2006) Model intercomparison of indirect aerosol effects. Atmos Chem Phys 6(11):3391–3405. https://doi.org/10.5194/acp-6-3391-2006

Ramanathan V, Chung C, Kim D, Bettge T, Buja L, Kiehl J, Washington W, Fu Q, Sikka D, Wild M (2005) Atmospheric brown clouds: impacts on South Asian climate and hydrological cycle. Proc Natl Acad Sci 102(15):5326–5333. https://doi.org/10.1073/pnas.0500656102

Salzmann M, Weser H, Cherian R (2014) Robust response of Asian summer monsoon to anthropogenic aerosols in CMIP5 models. J Geophys Res Atmos 119(19): 11321–11337. https://doi.org/10.1002/2014JD021783

Shindell DT, Voulgarakis A, Faluvegi G, Milly G (2012) Precipitation response to regional radiative forcing. Atmos Chem Phys 12:6969–6982. https://doi.org/10.5194/acp-12-6969-2012

Takemura T, Okamoto H, Maruyama Y, Numaguti A, Higurashi A, Nakajima T (2000) Global three-dimensional simulation of aerosol optical thickness distribution of various origins. J Geophys Res 105(D14):17853–17873. https://doi.org/10.1029/2000JD900265

Takemura T, Nakajima T, Dubovik O, Holben BN, Kinne S (2002) Single-scattering albedo and radiative forcing of various aerosol species with a global three-dimensional model. J Climate 15(4):333–352

Takemura T, Nozawa T, Emori S, Nakajima TY, Nakajima T (2005) Simulation of climate response to aerosol direct and indirect effects with aerosol transport-radiation model. J Geophys Res 110:02202. https://doi.org/10.1029/2004JD005029

Takahashi HG, Yasunari T (2006) A climatological monsoon break in rainfall over Indochina-a singularity in the seasonal march of the Asian summer monsoon. J Climate 19(8):1545–1556. https://doi.org/10.1175/JCLI3724.1

Takahashi HG, Fukutomi Y, Matsumoto J (2011) The impact of long-lasting northerly surges of the East Asian winter monsoon on tropical cyclogenesis and its seasonal march. J Meteor Soc Japan 89:181–200. https://doi.org/10.2151/jmsj.2011-A12

Takahashi HG, Fujinami H, Yasunari T, Matsumoto J, Baimoung S (2015) Role of tropical cyclones along the monsoon trough in the 2011 Thai flood and interannual variability. J Climate 28(4):1465–1476. https://doi.org/10.1175/JCLI-D-14-00147.1

Takahashi C, Watanabe M (2016) Pacific trade winds accelerated by aerosol forcing over the past two decades. Nature Clim Change 6(8):768–772. https://doi.org/10.1038/nclimate2996

Taylor KE, Stouffer RJ, Meehl GA (2012) An overview of CMIP5 and the experiment design. Bull Amer Meteor Soc 93(4):485–498

Twomey SA, Piepgrass M, Wolfe T (1984) An assessment of the impact of pollution on global cloud albedo. Tellus B 36(5):356–366

van der Werf GR, Dempewolf J, Trigg SN, Randerson JT, Kasibhatla PS, Giglio L, Murdiyarso D, Peters W, Morton D, Collatz G, Dolman AJ, DeFries RS (2008) Climate regulation of fire emissions and deforestation in equatorial Asia. Proc Natl Acad Sci 105(51):20350–20355. https://doi.org/10.1073/pnas.0803375105

Watanabe S, Hajima T, Sudo K, Nagashima T, Takcmura T, Okajima H, Nozawa T, Kawase H, Abe M, Yokohata T, Ise T, Sato H, Kato E, Takata K, Emori S, Kawamiya M (2011) MIROC-ESM 2010: model description and basic results of CMIP5-20c3m experiments. Geosci Model Dev 4:845–872

Yamaji M, Takahashi HG (2014) Asymmetrical interannual variation in aerosol optical depth over the tropics in terms of aerosol-cloud interaction. SOLA 10:185–189. https://doi.org/10.2151/sola.2014-039

10

A review of atmospheric chemistry observations at mountain sites

Sachiko Okamoto* and Hiroshi Tanimoto*

Abstract

Located far from anthropogenic emission sources, high-altitude mountain stations are considered to be ideal sites for monitoring climatic and environmentally important baseline changes in free tropospheric trace gases and aerosols. In addition, the observations taken at these stations are often used to study the long-range transport of dust as well as anthropogenic and biomass burning pollutants from source regions and to evaluate the performance of global and regional models. In this paper, we summarize the results from past and ongoing field measurements of atmospheric constituents at high-altitude stations across the globe, with particular emphasis on reactive trace species including tropospheric ozone, along with its precursors such as carbon monoxide, nitrogen oxides, total reactive nitrogen, and nonmethane hydrocarbons. Over the past decades, our understanding of the temporal variability and meteorological mechanisms of long-range transport has advanced in tandem with progress in instrumentation and modeling. Finally, the future needs of atmospheric chemistry observations at mountain sites are addressed.

Keywords: Atmospheric observation, High-altitude station, Long-range transport, Seasonal cycle, Long-term trend, Biomass burning, Anthropogenic pollution

Introduction

Trace gases and aerosols present in the atmosphere are responsible for global changes in the environment and climate of the Earth. These changes include global warming, which has caused a temperature increase of 0.85 °C during the period from 1880 to 2012, primarily due to increases in the atmospheric concentrations of greenhouse gases (GHGs) since the pre-industrial era (Intergovermental Panel on Climate Change 2013), poor air quality associated with high levels of photo-oxidants, and fine particulate matter with particle diameters of less than 2.5 μm ($PM_{2.5}$). These changes have been traced to high population densities and hasty industrialization in rapidly developing regions (World Health Organization 2006). Such air pollution problems, which can cause damage to human health and to vegetation growth, are now recognized as important, and robust evidence based on reliable observations is needed to support related policymaking processes.

Located far from anthropogenic emissions, high-altitude mountain stations are ideal sites for monitoring the temporal variations of trace gases and aerosols in the atmosphere at background levels. As a result, atmospheric monitoring programs have been established at a substantial number of mountain stations in order to obtain information on the levels, variability, and trends of trace gases and aerosols in the free troposphere (FT), as well as to assess the influence of anthropogenic and natural emission sources. It is widely accepted that observations at mountain sites are affected by the horizontal and vertical transport of polluted air masses from the ground surface (Henne et al. 2004; Trickl et al. 2003). Based on the analysis of tropospheric ozone (O_3) data collected at 27 stations located in mountainous and rural areas in Western Europe at altitudes ranging from 115 to 3550 m above sea level (a.s.l.), Chevalier et al. (2007) suggested that ground-based stations devoted to the monitoring of background free tropospheric O_3 should be settled, if possible, above 2000 m a.s.l.

Tropospheric O_3, which is an important GHG, has an estimated radiative forcing (RF) of 0.40 ± 0.20 W m^{-2} (Intergovermental Panel on Climate Change 2013). The RF from tropospheric O_3 depends strongly on altitude and latitude due to the coupling of O_3 changes with temperature, water vapor, and clouds (Bowman et al.

* Correspondence: okamoto.sachiko@nies.go.jp; tanimoto@nies.go.jp
Center for Global Environmental Research, National Institute for Environmental Studies, 16-2 Onogawa, Tsukuba, Ibaraki 305-8506, Japan

2013; Worden et al. 2008, 2011). O_3 is the main source of hydroxyl radicals (OH), which drive the oxidizing capacity of the atmosphere. It is widely accepted that tropospheric O_3 has two sources: subsidence transport from the stratosphere and in situ photochemical production, which occurs when carbon monoxide (CO) and hydrocarbons are photo-oxidized in the presence of nitrogen oxides ($NO_x = NO + NO_2$) (Crutzen 1974; Monks et al. 2009). The major loss pathways for O_3 in the troposphere include dry deposition to the Earth's surface, chemical destruction via photolysis to $O(^1D)$ which reacts with water vapor, and chain reaction with hydroperoxyl radicals (HO_2) and OH (Seinfeld and Pandis 2006). The lifetime of O_3 in the troposphere varies strongly in different seasons and locations, ranging from a few days in the tropical boundary layer to 1 year in the upper troposphere. The modeled mean lifetime of tropospheric O_3 has been reported as 22.3 ± 2 days (Stevenson et al. 2006).

Although CO is not a radiatively active gas, it affects the climate indirectly via its interaction with OH radicals, which are the primary CO sink. The oxidation of CO by OH leads to the formation of HO_2, which then can become involved in reactions to produce O_3 in NO_x-rich environments. In contrast, CO oxidation by OH radicals can lead to O_3 destruction through catalytic cycles of HO_x in low-NO_x environments (Kanakidou and Crutzen 1999). The changes in the OH radical concentration due to changes in CO can perturb the concentrations of GHGs such as methane (CH_4) (Thompson and Cicerone 1986) and chlorofluorocarbon (CFC) along with others such as hydrochlorofluorocarbons (HCFCs) and hydrofluorocarbons (HFCs). The major sources of CO emissions into the atmosphere are direct emissions from incomplete combustion of biomass and fossil fuels associated with in situ production via the oxidation of hydrocarbons such as CH_4 and isoprene (C_5H_8). Global estimates for anthropogenic and biomass burning emissions for 2000 were 611 and 459 Tg (CO) year^{-1}, respectively (Lamarque et al. 2010). The global mean tropospheric lifetime of CO is relatively long at approximately 2 months (e.g., Prather 1996) and seasonal and spatial CO lifetime variations have been reported in the study of Duncan et al. (2007). In the Northern Hemisphere winter season, the lifetime tends to be long, so CO from fossil fuels and industry accumulates. Conversely, in the Southern Hemisphere, CO approaches its seasonal minimum in austral summer because its lifetime is short and biomass burning activities are at a minimum. This latitudinal gradient becomes weak as summer approaches because the CO lifetime declines, over a scale of weeks to months, across most of the Northern Hemisphere. Simultaneously, however, the CO lifetime increases at southern mid-latitudes since the biomass-burning season is beginning in the Southern Hemisphere.

The term "background" concentrations have been used with various definitions in the literature over time. Calvert (1990) described "background concentration" as "the concentration of a given species in a pristine air mass, in which anthropogenic impurities of a relatively short lifetime are not present". However, because the troposphere is globally influenced by the long-range transport of pollution, Parrish et al. (2012) adopted the term "baseline," which refers to measurements obtained when local emission influences are determined to be negligible. Since remote coastal or mountainous sites are regarded as having no direct influence from local sources/sinks, such sites are the primary choices when looking to obtain representative data on continental to hemispheric scales. The US Environmental Protection Agency (EPA) defined the "Policy Relevant Background (PRB)" concentration as the concentration that would occur in the USA in the absence of anthropogenic emissions in continental North America (U.S. Environmental Protection Agency, 2006). Hence, since the sources of PRB concentrations include natural emissions and long-range transport of pollutants, they cannot be controlled by domestic regulations in the USA and/or neighboring countries.

Most studies have assumed the representativeness of mountain sites without providing evidence to support the assumption. Nappo et al. (1982) suggested that "representativeness is the extent to which a set of measurements taken in a space-time domain reflects the actual conditions in the same or different space-time domain taken on a scale appropriate for a specific application". However, representativeness not only varies with time, it also strongly depends on the trace gases because the concentration within a certain volume is controlled by vertical/horizontal transport and mixing, chemical transformations, surface deposition, and emissions.

The spatial representativeness of monitoring sites has been assessed by different approaches, primarily in Europe (e.g., Henne et al. 2010; Joly and Peuch 2012; Kovač-Andrić et al. 2010; Spangl et al. 2007; Tarasova et al. 2007). In addition, observations at mountain-based stations provide information of the lower FT, which is located between the planetary boundary layer (PBL) and the tropopause. Many species in the FT, such as O_3, have longer lifetimes than those in the PBL owing to the lower temperatures and lack of deposition in the FT, which is also where most of the transport of chemical species within the atmosphere occurs. The combination of long-range transport and longer chemical lifetimes in the FT indicates that the chemistry of this zone is an important factor when determining the chemical composition of regions remote from pollutant sources (Hemispheric Transport of Air Pollution 2007).

However, observations at ground-based stations can often be influenced by local sources. The objective of data selection is to differentiate the concentrations representative of well-mixed air masses near the site from those substantially affected by local sources or meteorology. Data selection can be performed on the basis of independently measured parameters such as wind direction and speed, statistical approaches for rejecting particular data, or combined meteorological and statistical filters. For example, statistical approaches have been applied to trend analyses of trace gases (e.g., Novelli et al. 1998, 2003; Ruckstuhl et al. 2012; Schuepbach et al. 2001; Thoning et al. 1989; Zellweger et al. 2009) in order to evaluate source regions and their emission estimates (e.g., Greally et al. 2007; Prinn et al. 2001) and to model the long-range transport of trace gases (Balzani Lööv et al. 2008).

When using measurements at mountainous sites, it is necessary to consider upslope flow, which often imposes pollution from local anthropogenic sources located in the foothills. This source is the most likely factor in the contamination of clean air masses arriving at the site from the FT. Conversely, in the absence of emission sources near the site, the upslope flow can transport clean air masses from the boundary layer that have atmospheric compositions that are different from that from the FT (e.g., Hahn et al. 1992; Oltmans et al. 1996b; Peterson et al. 1998).

Upslope flow on isolated mountains results from two primary mechanisms: mechanically forced lifting and buoyant upslope flow. Mechanically forced lifting is caused by the deflection of strong winds by a mountain slope. The importance of this mechanism differs among mountains and depends mainly on the height of the mountain and the mean speed of the wind. Buoyant upslope flow is caused by the daytime solar heating of air near the surface of the mountain. During the day, warmed air masses rise toward the summit. At night, flows down the mountain occur due to radiative cooling of the ground surface. Both the upslope and downslope flows are important for mountain meteorology and the transport of local pollution (e.g., Furger et al. 2000). In practice, however, determination of baseline concentrations is difficult because the nature and magnitude of such local influence depend on the mountain topography, local wind patterns, and the characteristics of local emission sources.

The World Meteorological Organization (WMO) launched the Global Atmosphere Watch (GAW) Programme in 1989 to promote systematic and reliable observations of the chemical composition of the atmosphere. Currently, the GAW has the most extensive measurement program. Its focal areas include aerosols, GHGs, selected reactive gases, O_3, ultraviolet (UV) radiation, and precipitation chemistry. In October 1990, WMO designated the Japan Meteorological Agency (JMA) in Tokyo as the World Data Centre for Greenhouse Gases (WDCGG) and tasked it with collecting, archiving, and providing data related to GHGs and reactive gases in the atmosphere and oceans from observation sites throughout the world that participate in the GAW and other scientific monitoring programs.

Since its establishment, WDCGG has provided users with data and other information through its regular publications and the WDCGG website (http://ds.data.jma.go.jp/gmd/wdcgg/). In addition, observational data is currently provided by many regional monitoring programs worldwide including the European Monitoring and Evaluation Programme (EMEP; http://www.emep.int) in the European region, the Clean Air Status and Trends Network (CASTNET; http://www.epa.gov/castnet), the Interagency Monitoring of Protected Visual Environments (IMPROVE; http://vista.cira.colostate.edu/improve/) in North America, which monitors visibility conditions, tracks changes in visibility, and works to identify sources and causes of regional haze in class I areas (including national parks, wilderness areas, and national memorial parks), and the Acid Deposition Monitoring Network in East Asia (EANET; http://www.eanet.asia/).

In this paper, we describe past and current atmospheric observations conducted at high-altitude stations on mountains, the data of which are available in public databases or published papers. First, overviews of the individual mountain stations above 1500 m a.s.l. are given. Next, we review temporal variations, particularly the long-term trends and seasonal variations of reactive species including tropospheric O_3, and its precursors such as CO, NO_x, total reactive nitrogen (NO_y), and nonmethane hydrocarbons (NMHCs), but excluding long-lived GHGs. Since observations of aerosol optical properties at high-altitude stations are well described by Andrews et al. (2011), short-lived species—with the exceptions of carbon dioxide (CO_2) and mercury (Hg) described in association with the analysis of reactive species—are mainly discussed here. We will review previous observations of the long-range transport of pollutants from anthropogenic and biomass burning sources, desert dust, and volcanic plumes from eruption events, along with the downward transport of stratospheric air masses into the troposphere as important sources of O_3. Moreover, we also review the modeling studies using photochemical box models and chemical transport models.

Review

High-elevation mountain sites

The ground-based stations for atmospheric chemistry observations presented in this study are located at high altitudes above 1500 m a.s.l., as shown in Fig. 1; detailed site information is listed in Table 1. In total, the list contains 31 stations. These stations were selected because

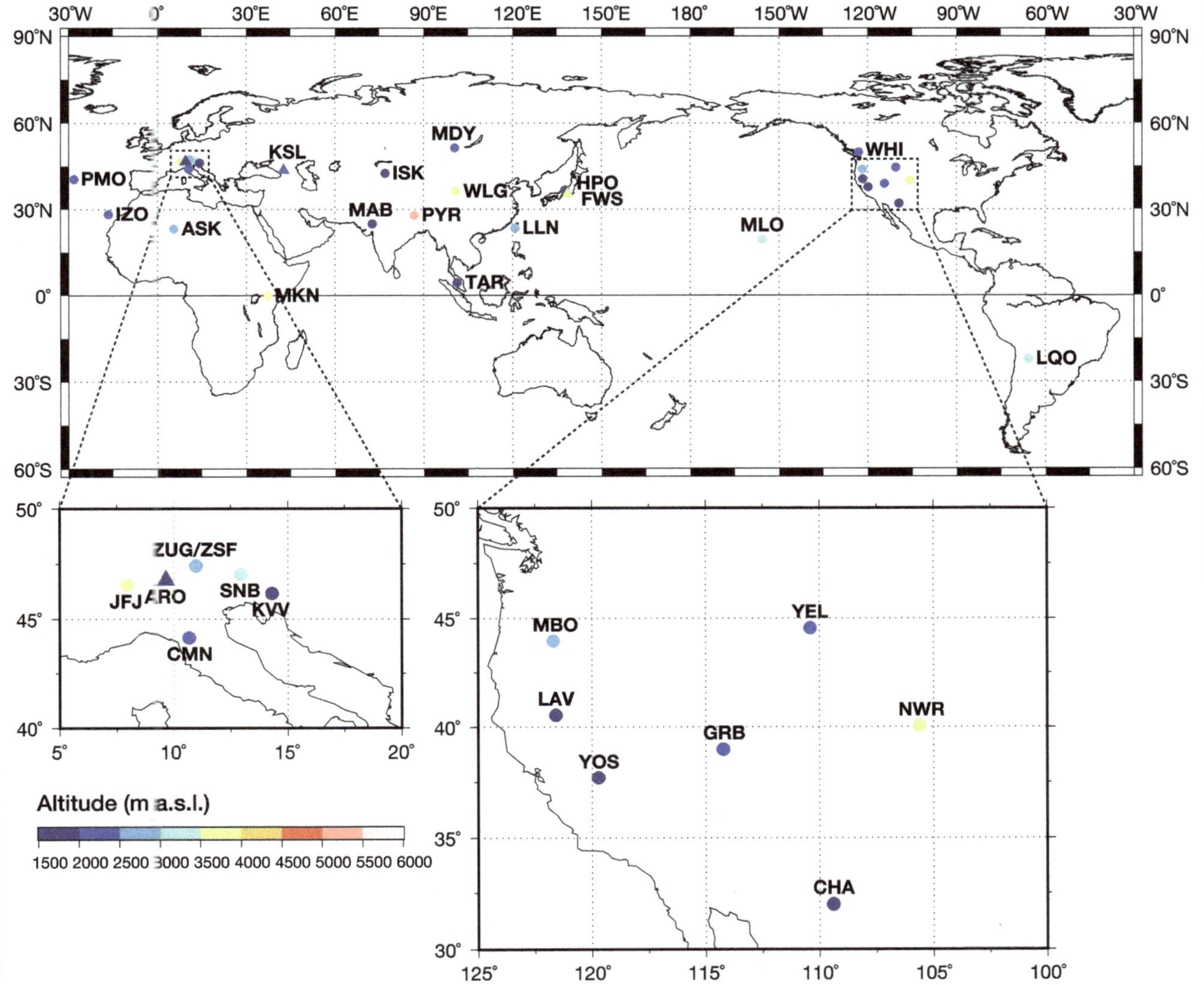

Fig. 1 Map showing the distribution of high-altitude mountain stations discussed in this paper. *Circles with abbreviations* denote stations providing the data presented in this paper. *ASK* Assekrem, *CHA* Chiricahua NM, *CMN* Monte Cimone, *GRB* Great Basin NP, *HPO* Mt. Happo, *ISK* Issyk-Kul, *IZO* Izāna, *JFJ* Jungfraujoch, *KVV* Krvavec, *LAV* Lassen Volcanic NP, *LLN* Lulin, *LQO* La Quiaca Observatorio, *MBO* Mt. Bachelor Observatory, *MKN* Mt. Kenya, *MDY* Mondy, *MLO* Mauna Loa, *NWR* Niwot Ridge, *PMO* Pico Mountain Observatory, *PYR* Nepal Climate Observatory–Pyramid, *SNB* Sonnblick, *TAR* Tanah Rata, *WHI* Whistler Mountain, *WLG* Mt. Waliguan, *YEL* Yellowstone NP, *YOS* Yosemite NP, *ZSF* Zugspitze–Schneefernerhaus, *ZUG* Zugspitze–Gipfel. *Triangles with abbreviations* denote stations for which data are not presented but are reviewed in this paper. *ARO* Arosa, *FWS* Mt. Fuji, *KSL* Kislovodsk

they provide reliable observational data that have been utilized in numerous previous studies. Furthermore, we tried to cover the entire world except for Arctic and Antarctic Polar Regions. Many stations are located in northern mid-latitudes, particularly in central Europe and western North America. Compared to the Northern Hemisphere, significantly fewer stations are in the Southern Hemisphere, and no high-altitude stations are present in the Oceania region.

Region I: Africa

Assekrem (ASK), Algeria ASK is a GAW Global station operated by the Déportent Météorologique Régional Sud of the Office National de la Météorologie (ONM), Algeria, and is located on the summit of the second highest point of the Hogger Mountain Range in the Sahara Desert (Zellweger et al. 2007). It plays an important role in capturing air pollutants from continental Africa.

Mt. Kenya (MKN), Kenya MKN is a GAW Global station that has been designated for long-term measurements of various chemical compounds and physical and meteorological parameters in the lower troposphere. The main office is the Kenya Meteorological Department (KMD). MKN is located on the northwestern slope of Mt. Kenya

Table 1 List of stations mentioned in this paper

Station	ID	Country	Latitude	Longitude	Altitude (m)	Data source
Region I: Africa						
Assekrem	ASK	Algeria	23.27° N	5.63° E	2710	WDCGG[a]
Mt. Kenya	MKN	Kenya	0.06° S	37.30° E	3678	WDCGG[a]
Izaña (Tenerife)	IZO	Spain	28.31° N	16.50° W	2373	NMHCs: Schmitt and Schultz et al. (1998) Other: WDCGG[a]
Region II: Asia						
Mt. Waliguan	WLG	China	36.29° N	100.90° E	3810	WDCGG[a]
Issyk-Kul	ISK	Kyrgyzstan	42.62° N	76.98° E	1640	WDCGG[a]
Nepal Climate Observatory–Pyramid	PYR	Nepal	27.96° N	86.81° E	5079	WDCGG[a]
Lulin	LLN	Taiwan	23.47° N	120.87° E	2862	LLN website[b]
Mondy	MDY	Russian Federation	51.65° N	100.92° E	2006	EANET[c]
Mt. Abu	MAB	India	24.65° N	72.78° E	1680	Naja et al. (2003)[d]
Mt. Happo	HPO	Japan	36.70° N	137.80° E	1850	CO: Kato et al. (2000); Narita et al. (1999); Pochanart et al. (2004)[e] NMHCs: Sharma et al. (2000) Other: EANET[c]
Mt. Fuji	FWS	Japan	35.37° N	138.73° E	3776	–
Region III: South America						
La Quiaca Observatorio	LQO	Argentina	22.10° S	65.60° W	3459	WDCGG[a]
Region IV: North and Central America						
Whistler Mountain	WHI	Canada	50.06° N	122.96° W	2182	Macdonald et al. (2011)[f]
Great Basin NP	GRB	USA	39.01° N	114.22° W	2060	CASTNET[g]
Mt. Bachelor Observatory	MBO	USA	43.98° N	121.69° W	2743	Gratz et al. (2015); Weiss-Penzias et al. (2007)[h]
Niwot Ridge	NWR	USA	C1: 40.04° N T-van: 40.05° N	C-1: 105.54° W T-van: 105.59° W	C-1: 3021 T-van: 3523	WDCGG[a]
Chiricahua NM	CHA	USA	32.01° N	109.39° W	1570	CASTNET[g]
Lassen Volcanic NP	LAV	USA	40.54° N	121.58° W	1756	O_3: CASTNET[g] Aerosol: IMPROVE[i]
Yellowstone NP	YEL	USA	44.57° N	110.40° W	2430	CASTNET[g]
Yosemite NP	YOS	USA	37.71° N	119.71° W	1605	CASTNET[g]
Region V: South–West Pacific						
Tanah Rata	TAR	Malaysia	4.48° N	101.37° E	1545	WDCGG[a]
Mauna Loa	MLO	USA	19.54° N	155.58° W	3397	NO_x and NO_y: Ridley et al. (1998) Aerosol: IMPROVE[i] Other: WDCGG[a]

Table 1 List of stations mentioned in this paper *(Continued)*

Region VI: Europe						
Sonnblick	SNB	Austria	47.05° N	12.96° E	3106	WDCGG[a]
Zugspitze–Gipfel	ZUG	Germany	47.42° E	10.99° E	2962	WDCGG[a]
Zugspitze–Schneefernerhaus	ZSF	Germany	47.42° N	10.98 °E	2671	WDCGG[a]
Monte Cimone	CMN	Italy	44.17° N	10.68° E	2165	NMHCs: EBAS[j] Other: WDCGG[a]
Kislovodsk	KSL	Russian Federation	43.73° N	42.66° W	2070	–
Krvavec	KVV	Slovenia	46.18° N	14.32° E	1720	WDCGG[a]
Arosa	ARO	Switzerland	46.78° N	9.67° E	1840	–
Jungfraujoch	JFJ	Switzerland	46.55° N	7.99° E	3580	Aerosol: EBAS[j] Other: WDCGG[a]
Pico Mountain Observatory	PMO	Portugal	38.47° N	28.40° W	2225	NMHCs: WDCGG[a] NO_x and NO_y: Val Martin et al. (2008b)

[a]World Data Centre for Greenhouse Gases (http://ds.data.jma.go.jp/gmd/wdcgg/)
[b]Data available at http://lulin.tw
[c]Acid Deposition Monitoring Network in East Asia (http://www.eanet.asia/)
[d]Data obtained from M. Naja
[e]Data obtained from S. Kato
[f]Data obtained from A. M. Macdonald
[g]Clean Air Status and Trends Network (http://www.epa.gov/castnet)
[h]Data obtained from D. Jaffe
[i]Interagency Monitoring of Protected Visual Environments (http://vista.cira.colostate.edu/improve/)
[j]EBAS (http://ebas.nilu.no)

and in a protected portion of the Mount Kenya National Park. The climate of equatorial East Africa is dominated by seasonal displacement of the intertropical convergence zone (ITCZ). During boreal summer, the ITCZ is situated far to the north, resulting in southerly to southeasterly winds over Kenya. From September onward, the ITCZ begins to retreat southward, followed by a rainy period from mid-October to December over Kenya. Throughout boreal winter, the ITCZ is situated south of the equator. As a result, a northeasterly monsoon circulation dominates East Africa. When the ITCZ begins to progress northward again, it brings a second rainy period from mid-March to the beginning of June to equatorial East Africa. One noteworthy feature of the boreal monsoon circulation is the East African low-level jet (EALLJ), which develops from May onward in the southeast trade winds over Madagascar. With the northward transition of the ITCZ, the EALLJ strengthens and extends northward, reaching its maximum extent in June or July. Its counterpart, blowing from the north along the East African coast during boreal winter, is less pronounced (Henne et al. 2008).

Izaña (IZO), Spain IZO is a GAW Global station operated by the Agencia Estatal de Meteorología (AEMET) for long-term measurements of various chemical compounds and physical and meteorological parameters. The station is located on the Island of Tenerife roughly 300 km west of the African coast. As typically occurs in subtropical oceanic atmospheres to the west of continents, the low troposphere in this region is strongly stratified owing to the synoptic-scale subsidence conditions. Two air masses are well differentiated by the presence of a temperature inversion layer: the marine boundary layer (MBL) and the FT (Rodríguez et al. 2009).

In the moist and cool MBL, the north to northeast trade wind blows. Because the condensation level is usually lower than the inversion level under the trade wind regime, the top of the MBL, which is frequently located just below the temperature inversion layer, is characterized by a layer of stratocumulus clouds formed by the condensation of water vapor onto available pre-existing particles. The presence of this stratocumulus layer creates a quasi-continuous foggy and rainy regimen throughout the year at altitudes between 800 and 2000 m a.s.l. on the island.

In the FT, northwest subsiding dry airflows dominate throughout the year, except in summer, when they frequently alternate with southeast airflows laden with Saharan mineral dust particles (Chiapello et al. 1999). In winter (January to March), air masses originate from the North Atlantic, North America, and the Sahara (Cuevas et al. 2013). In spring (April to June), a pattern similar to that in winter is observed. The contributions from the Sahara and the southern part of the North Atlantic are weaker, and the transport from North America is more well-defined than that in winter. In summer (July to September), essentially only two geographical sectors of the air mass pathways are identified: that from North America and that from the Sahara and Northern Sahel. Air masses from North America and the North Atlantic travel at relatively high altitudes, whereas those from the Sahara originate from low levels.

The observational records of a Fourier transform infrared (FTIR) spectrometer that began routine operation in March 1999 have been utilized in many studies (García et al. 2012; Schneider et al. 2005a, b, 2008; Sepúlveda et al. 2012; Viatte et al. 2011). The instrument is part of the Total Column Carbon Observation Network (TCCON), which is a network of ground-based Fourier transform spectrometers designed to retrieve precise and accurate column abundances of atmospheric constituents including CO_2, CH_4, nitrous oxide (N_2O), hydrogen fluoride (HF), CO, and water vapor isotopologues (Wunch et al. 2011). The TCCON was established in 2004 and is currently affiliated with 26 sites. In addition, other remote sensing observations have been conducted by lidar (Welton et al. 2000), multipass optical absorption spectroscopy (MOAS) (Armerding et al. 1997; Comes et al. 1995, 1997), differential optical absorption spectrometry (DOAS) (Carslaw et al. 1997; Gil et al. 2000, 2008), and multi axis-DOAS (MAX-DOAS) (Puentedura et al. 2012).

Region II: Asia

Mt. Waliguan (WLG), China WLG is a GAW Global station operated by the China Meteorological Administration (CMA). This station is located at the edge of the northeastern boundary of the Qinghai–Tibetan Plateau. Mt. Waliguan is 90 km from Xining and 260 km from Lanzhou, which are the main industrial and most populated regions of central–eastern China. The wind patterns near WLG are controlled by the Qinghai–Tibetan Plateau monsoon and are thus associated with seasonal variations. The predominant wind directions are from the southwest in cold seasons and from the east in warm seasons. Locally, WLG is affected by the mountain–valley breezes (Fu et al. 2012). Xue et al. (2011) identified three major types of air masses at WLG in summer: those originating from the east and southeast having passed over populated regions of China, those originating from the west having passed over remote regions of central Asia and Xinjiang, and those originating mainly from the north having passed over regions of Siberia and Mongolia. During a recent 10-year period (2000 to 2009); the transport regime at WLG in summer was dominated by air originating from the east that may have passed over populated regions in China.

Issyk-Kul (ISK), Kyrgyzstan ISK is located in the northern Tien-Shan Mountains along the shoreline of Lake Issyk-Kul. The Issyk-Kul lake region is surrounded by two giant mountain chains including Kungey Ala-Too to the north and Terskey Ala-Too to the south, which together form a closed area. The mountain chains impede the transport of the polluted air from Chu Valley in Bishkek, the capital of Kyrgyzstan, and from Almaty, a huge industrial center in Kazakhstan (Semenov et al. 2005). Observations of a long-term total O_3 column (Aref'ev et al. 1995; Visheratin et al. 2006), total NO_2 column (Aref'ev et al. 1995, 2009; Ionov et al. 2008), total CO column (Aref'ev et al. 2013), and height mean concentrations of CO_2 (Kashin et al. 2007, 2008) have been reported along with atmospheric spectral transparency observations (Aref'ev and Semenov 1994, Aref'ev et al. 2008; Semenov et al. 2005).

Nepal Climate Observatory–Pyramid (PYR), Nepal PYR is a GAW Global station installed within the framework of the Stations at High Altitude for Research on the Environment (SHARE) project of Everest-K2 National Research Council (Ev-K2-CNR) and the atmospheric brown cloud (ABC) project of the United Nations Environmental Programme (UNEP) in order to obtain information on the atmospheric background conditions in this region (Bonasoni et al. 2008). This station is located at the confluence of the secondary Lobuche valley and the main Khumbu valley in Sagarmatha National Park in the eastern Nepal Himalaya.

In the Himalayan region, a 3 km thick brownish layer of pollutants known as Asian brown clouds has been observed to extend from the Indian Ocean to the Himalayan range (Ramanathan and Crutzen 2003). This phenomenon strongly affects the air quality, visibility, and the energy budget of the atmosphere over the entire Indian subcontinent. The high-altitude meteorology in the Himalayas is strongly influenced by the Asian monsoon circulation and by local mountain wind systems (Bollasina et al. 2002). The mountain–valley wind system is predominant, with a strong valley wind from the south to southwest during daytime and a weak mountain wind at night. In summer, the valley wind prevails throughout the day.

Nepal, including its borders with India and Bangladesh, is the most frequent origin of air masses transported in the PBL that reach PYR throughout the year (Gobbi et al. 2010). The atmospheric conditions in the Himalayas are influenced by the transport of polluted air masses from South Asia and the Indo–Gangetic Plains, the northwest-to-northeast region extending from eastern Pakistan across India to Bangladesh and Myanmar. Occurrences of brown clouds over the Himalayan foothills and Northern Indo–Gangetic Plains have been identified by high aerosol optical depth (AOD) values >0.4 and are associated with high concentrations of pollutants measured at PYR, as characterized by an up-valley breeze circulation (Bonasoni et al. 2010).

Lulin (LLN), Taiwan LLN is located inside Jade Mountain National Park, which is part of the Central Mountain Range. This station is operated by National Central University (NCU), Taiwan. Owing to the mountain–valley circulation, LLN is frequently located in the FT, particularly during the winter, and is generally free from surface pollution in the boundary layer (Ou-Yang et al. 2014; Sheu et al. 2010; Wai et al. 2008). Taiwan is located at the edge of the western Pacific Ocean and is southeast of the East Asian continent. The southwest monsoon prevails in summer, whereas the northeast monsoon prevails between late fall and spring. Additionally, since the Westerlies prevail at a higher elevation in spring (Ou Yang et al. 2012), LLN is affected by pollution outflows from Southeast Asia and the Asian continent.

Mondy (MDY), Russian Federation MDY is located in central Eastern Siberia near the Russia–Mongolia border, which is known to be a very lightly populated continental area of the world. MDY is operated by the Limnological Institute, Russian Academy of Sciences, and contributes to the EANET program. Air masses reaching MDY are typically classified into four groups: long-range transport from Europe, Siberia east of the Ural Mountains, high-latitude polar regions, and much shorter-range transport usually less than several hundred kilometers from the southwest (Pochanart et al. 2003).

Mt. Abu (MAB), India MAB is located at a mountaintop known as Guru Shikhar, which is the highest peak in the southern end of Aravalli Mountain Range in western India. In this region, northwest winds dominate in winter (November to February) and southwest winds dominate in summer (May to August) (Kumar and Sarin 2010; Naja et al. 2003; Ram et al. 2008; Ram and Sarin 2009; Rastogi and Sarin 2005, 2008). Diurnal O_3 concentration variations in spring and summer at MAB show a unique pattern, whereas those in autumn and winter show a pattern typical of high-latitude sites (Naja et al. 2003). Naja et al. (2003) suggested that this seasonal change in diurnal variation might be caused by changes in the wind pattern in different seasons and changes in the boundary layer mixing height during the day and at night.

Mt. Happo (HPO), Japan HPO is operated by the Ministry of the Environment (MOE) of Japan, which contributes to the EANET program. This station is located in a mountainous area near the coast of the Japan Sea. During fall to spring, westerly winds are dominant (Pochanart et al. 1999), and most of the air masses pass over the Asian continent, particularly over Mongolia and northwestern China, driven by the Siberian High (Liu et al. 2013). The North Pacific high-pressure system is strong in summer, and the Siberian High reverses to a low-pressure system. This behavior is typical of the summer monsoon in East Asia. Associated with the summer monsoon, pristine air masses from the Pacific are transported to HPO more frequently, particularly in the boundary layer. Domestic pollution plumes have only been observed during limited periods in summer, particularly in August, when the land–sea air circulation over Tokyo Bay occasionally transports pollution from Tokyo to cause a late-afternoon daily maximum (Chang et al. 1989).

Mt. Fuji, Japan (FWS), Japan FWS was operated by the JMA until 2004, after which the nonprofit organization (NPO) Valid Utilization of Mount Fuji Weather Station (http://npofuji3776-english.jimdo.com) assumed responsibility for its operation. Mt. Fuji, which is the highest mountain in Japan, is located in the central part of the main Japanese island of Honshu. Its summit is positioned in the FT for most of the year (Igarashi et al. 2004). Westerly winds prevail at FWS, particularly strong winds in winter, owing to the southward shift of the polar jet stream (Nakazawa et al. 1984).

Region III: South America

La Quiaca Observatorio (LQO), Argentina LQO is operated by the Servicio Meteorológico Nacional (SMNA). This station is located in the Puna–Altiplano Plateau (15–26° S, 65–69° W) in the central portion of the Andes, which is an important dust activity region (Gaiero et al. 2013; Prospero et al. 2002). The Altiplano is situated between the hyper-arid Pacific coastal desert to the west and the moist continental lowlands to the east. The prevailing flows in the middle and upper troposphere over the Altiplano are westerly from May to October in the dry season and easterly from December to March in the rainy season (Garreaud et al. 2003).

Region IV: North and Central America

Whistler Mountain (WHI), Canada WHI is located on the summit of Whistler Mountain in a part of the Coast Mountain Range approximately 100 km north of Vancouver. The station was established in 2002 to measure aerosols and trace gases in the lower troposphere and is operated by Environment and Climate Canada (Macdonald et al. 2011). In this region, synoptic scale climatology is controlled by the Aleutian Low and the Pacific High. In winter, the zonal flow is generally stronger, and the Aleutian Low causes a more southwesterly flow over British Columbia. In summer, the intensified Pacific High causes more a northwesterly flow over the south coast of British Columbia (Klock and Mullock 2001). PBL influence was found to occur mostly during spring and summer and less frequently in late autumn and winter (Gallagher et al. 2011, 2012). Biogenic processes characterized during the Whistler Aerosol and Cloud Study 2010 (WACS 2010), which was a large measurement campaign with a focus on aerosols and clouds, have also been shown to be important at the station during late spring and summer (e.g., Ahlm et al. 2013; Lee et al. 2012; Pierce et al. 2012; Wainwright et al. 2012; Wong et al. 2011).

Mt. Bachelor Observatory (MBO), USA MBO is located on the summit of an extinct volcano in the Cascade Ridge of central Oregon. This station is operated by the University of Washington. At this altitude, the flow is predominantly from the southwest to the northwest. Owing to its elevation, the prevailing westerly winds, regional topography, and the lack of large urban/industrial emission sources upwind, MBO is well positioned for sampling FT air masses near the West Coast of the USA with minimal influence from local anthropogenic emissions (Weiss-Penzias et al. 2006). Three distinct pollution influences have been identified at MBO: Asian long-range transport, North American biomass burning, and North American industry (Jaffe et al. 2005; Weiss-Penzias et al. 2007; Reidmiller et al. 2010).

Niwot Ridge (NWR), USA NWR is located approximately 35 km west of Boulder, Colorado, on the Continental Divide for North America with runoff on the two sides destined for the Colorado and Mississippi Rivers, respectively. The station is situated on land owned by the Mountain Research Station of the University of Colorado and is used as a Long-Term Ecological Research (LTER) site with support from the National Science Foundation. This site is also used by the National Oceanic and Atmospheric Administration (NOAA) Global Monitoring Division for long-term monitoring, flask sampling, and for the collection of air that is used for preparing calibration standards of trace gases. Continuous O_3 measurement has been conducted at two sites, data from both of which are available at the WDCGG website. We used the O_3 data from the C-1 research site located at 3021 m a.s.l. in a subalpine forest 10 km east of the Continental Divide of the Americas. For CO data, we used observations from the treeline van (T-van) research site at 3523 m a.s.l. located in an alpine fellfield. Weekly and bi-weekly flask samplings have been also operated at the T-van. Wind patterns at NWR are dominated by Westerlies (Haagenson 1979), although, at times, upslope flows from the south or southeast transport air from Denver to the site (Hahn 1981).

Chiricahua National Monument (CHA), USA

CHA is located in Arizona and is operated as a Clean Air Markets Program site by the EPA. CHA contributes to the CASTNET program (https://www3.epa.gov/castnet/site_pages/CHA467.html), which monitors air quality and deposition including O_3 concentration, and to the IMPROVE program.

Great Basin National Park (GRB), USA GRB is located on the northeastern flank of the Snake Range in Nevada. This station is operated by the Great Basin National Park and is sponsored by the National Park Service (NPS) Air Resources Division. This agency administers an extensive air monitoring program that measures air pollution levels in national parks for visibility, gaseous pollutants, and atmospheric deposition. GRB contributes to the CASTNET (https://www3.epa.gov/castnet/site_pages/GRB411.html) and IMPROVE programs.

Lassen Volcanic National Park (LAV), USA LAV is located on the northwest flank of Mt. Lassen, the southernmost active volcano in the Cascade Mountain Range in northern California. This station is operated by the Lassen Volcanic National Park and is sponsored by the NPS Air Resources Division. LAV contributes to the CASTNET (https://www3.epa.gov/castnet/site_pages/LAV410.html) and IMPROVE programs.

Yellowstone National Park (YEL), USA YEL is located in Wyoming. It is operated by the Yellowstone National Park and is sponsored by the NPS Air Resources Division. YEL contributes to the CASTNET (https://www3.epa.gov/castnet/site_pages/YEL408.html) and IMPROVE programs.

Yosemite National Park (YOS), USA YOS is operated by Yosemite National Park and is sponsored by the NPS Air Resources Division. YOS is located on the western slope of the Sierra Nevada Mountains in northern California. YOS contributes to the CASTNET program (https://www3.epa.gov/castnet/site_pages/YOS404.html) and IMPROVE programs.

Region V: South-West Pacific

Tanah Rata (TAR), Malaysia TAR and the co-located Cameron Highlands Meteorological Station are operated by the Malaysian Meteorological Department (MMD) (Toh et al. 2013) and participate in the EANET program. Malaysia's climate can be categorized into four seasons: the northeast monsoon (winter monsoon) from November to early March, the southwest monsoon (summer monsoon) from late May to September, and two transition periods between the monsoons (Yonemura et al. 2002). The mean wind direction at TAR depends on the monsoon season. The prevailing winds during the northeast and southwest monsoons are from the northeast and southwest, respectively. The precipitation pattern has two maximum and two minimum periods, and temperatures are approximately constant throughout the year.

Mauna Loa (MLO), USA MLO is a GAW Global station operated by the Global Monitoring Division (GMD), NOAA Earth System Research Laboratory (NOAA/ESRL). This station is located on the slope of the active Mauna Loa volcano and is free from continental pollution sources. Atmospheric constituents have been continuously monitored since the 1950s. Currently, MLO is well known for its measurements of rising anthropogenic CO_2 (e.g., Keeling et al. 1995). Winds at MLO are controlled by several factors including the strength of synoptic scale or trade winds, the height and strength of the trade wind inversion, the flow of the synoptic winds around the island mountain topography, and the daily heating and cooling cycle on the island. At night, the station is above the inversion layer in the free tropospheric atmosphere with minimal influence from local emissions. A long-term trend analysis of aerosol optical properties (Collaud Coen et al. 2013) has shown significant negative trends for the backscatter fraction and Ångström exponent and positive trends for the scattering and absorption coefficients.

Region VI: Europe

Sonnblick (SNB), Austria SNB is located on the highest peak among the main ridges of the Austrian Alps. This station is organized by the Environment Agency, Austria, and is one of the high-altitude stations of the Monitoring Network in the Alpine Region for Persistent and other Organic Pollutants (MONARPOP). Since its foundation in 1886, most of the main meteorological parameters have been measured. These data serve as a unique long-term climate record (Schöner et al. 2012). The origins of pollution reaching SNB are typically industrial areas in southern and central Germany, most parts of northeastern Europe, and the northern parts of Italy near the Milan agglomeration (Holzinger et al. 2010). Local pollution sources reach SNB by convective mixing, particularly in the daytime, when the mixing height surpasses the altitude of SNB. Henne et al. (2010) assessed the parameters reflecting site representativeness in consideration of population (emission proxy), deposition, and transport at 34 sites in western and central Europe. They classified SNB as a "mostly remote" site where the total population influence, population variability, and total deposition influence are small.

Zugspitze–Gipfel (ZUG) and Zugspitze–Schneefernerhaus (ZSF), Germany Zugspitze is the highest mountain of the Wetterstein Mountains in the Bavarian Alps. Monitoring activities began at ZSF and were later moved to ZUG in 2001 and 2002. ZUG and ZSF are both GAW

Global stations and are located at the summit and on the southern slope of Zugspitze, respectively.

ZSF has been jointly operated by the Federal Environmental Agency (Umweltbundesamt, UBA) and the German Weather Service (Deutscher Wetterdienst, DWD). Because ZUG/ZSF, which is located at the northern flank of the Alps, is more distant from the central Alps and at a lower elevation compared with SNB, Henne et al. (2010) classified ZUG/ZSF as a "weakly influenced, constant deposition" site where the total population influence, population variability, and deposition variability are systematically larger than those at "mostly remote" sites. ZUG is a TCCON site, at which total columns of GHGs have been measured by means of FTIR spectroscopy (e.g., Angelbratt et al. 2011a, b; Rinsland et al. 2003; Vigouroux et al. 2008).

Monte Cimone (CMN), Italy CMN is a GAW Global station operated by the Institute of Atmospheric Sciences and Climate of the National Research Council of Italy (ISAC-CNR). Mt. Cimone is the highest peak of the northern Italian Apennines and is the only high mountain station for atmospheric research located between the Southern Alps and the northern Mediterranean Sea. The Apennines are the first mountain chain affected by air masses from the Sahara on their way to Europe (Bonasoni et al. 2004). The station is considered to be representative of the baseline conditions of the Mediterranean FT (Bonasoni et al. 2000a; Fischer et al. 2003). Even during the warm seasons, the influence of the boundary layer air is substantial owing to convection and the mountain breeze (Fischer et al. 2003; Van Dingenen et al. 2005). The Mediterranean basin is located at the boundary between the tropical zone and northern midlatitudes. It serves as a crossroad for air masses originating from Europe, Asia, and Africa, where anthropogenic emissions encounter natural emissions (Kanakidou et al. 2011; Lelieveld et al. 2002). Henne et al. (2010) classified CMN as a "weakly influenced, constant deposition" site.

Kislovodsk (KSL), Russian Federation KSL is organized by the Oboukhov Institute of Atmospheric Physics (IFA), Russian Academy of Sciences. This station is located on the mountain plateau at the northern slope of the side ridge in the North Caucasus. The main features of air transport to KSL in the summer are the prevailing northern components, which include the town of Kislovodsk and the Caucasus foothills, and the northeastern components, which include the northern Caspian. In the winter, the main air transport features are the prevailing southern components, which include the regions behind the Caucasus and Asia Minor deserts, and the southwestern components, which include the Mediterranean and the deserts of North Africa (Tarasova et al. 2003).

During the warm season from April to October, the mountain–valley air circulation affects the O_3 concentration (Elansky et al. 1995; Senik and Elansky 2001). Simultaneous measurements of O_3 concentration were conducted at KSL and at the nearest town located 18 km to the north (Senik et al. 2005). That study indicated that surface O_3 at KSL is only slightly influenced by the pollutants accumulated within the atmospheric boundary layer and revealed information relating primarily to the regional and global state of the troposphere. Observations of the total O_3 and NO_2 columns have also been reported (Elansky et al. 1995).

Krvavec (KVV), Slovenia KVV, managed by the Hydrometeorological Institute of Slovenia (HIS), is located on the slope of Mt. Krvavec in the Kamniško–Savinjske Alps, which are part of the southern margin of the Alps. At this station, continuous measurements of black carbon (BC) and O_3 have been conducted within the framework of the EMEP, which is a science-based and policy-driven program of the Convention on Long-Range Transboundary Air Pollution (CLRTAP) program set up to facilitate international cooperation in solving transboundary air pollution problems.

Arosa (ARO), Switzerland ARO is located within the eastern part of the Swiss Alps at the end of Plessur Valley. This station is surrounded by mountains with heights ranging from 2500 to 3000 m a.s.l. The transport of emissions from large cities such as Chur and Davos to ARO is rare due to a channeling effect (Li et al. 2005). The longest available total O_3 record in the world is from ARO, where total O_3 measurements have been conducted continuously since late 1926. The total O_3 data are based on measurements of ultraviolet (UV) radiation by Dubson spectrophotometers and have been homogenized (Staehelin et al. 1998a). These data have been utilized in numerous studies (e.g., Brönnimann et al. 2003; Hoegger et al. 1992; Krzyścin 2000; Rieder et al. 2010a, b; Scarnato et al. 2010; Staehelin et al. 1998b).

Jungfraujoch (JFJ), Switzerland JFJ is a GAW Global station located on a mountain crest on the northern edge of the Swiss Alps. As part of the station is incorporated into the Swiss National Air Pollution Monitoring Network (NABEL), a wide range of in situ trace gas observations is performed by the Swiss Federal Laboratories for Materials Science and Technology (Empa) in association with the Swiss Federal Office for the Environment (FOEN). First sporadic aerosol observations at JFJ started in the 1970s, but the majority of the operational long-term measurements of aerosols and aerosol optical properties were initiated by the Paul Scherrer Institute (PSI) in the mid-1990s (Bukowiecki et al. 2016; Collaud Coen et al. 2007; Fierz-Schmidhauser et al. 2010;

Nyeki et al. 2012). The site has been shown to be excellent for studies on new particle formation (Bianchi et al. 2016) or aerosol–cloud interaction, including mixed-phase and glaciated clouds (Hoyle et al. 2016; Verheggen et al. 2007). In addition, remote sensing investigations have been performed by FTIR spectroscopy, which allows for quasi-simultaneous investigation of total column concentrations of a large number of gaseous constituents (e.g., Barret et al. 2002, 2003a, b; De Mazière et al. 1999; Dils et al. 2011; Duchatelet et al. 2009, 2010; Krieg et al. 2005; Mahieu et al. 1997; Melen et al. 1998; Rinsland et al. 1991, 1992, 1996, 2000, 2008; Vigouroux et al. 2007; Zander et al. 2008, 2010). JFJ is exposed mostly to FT air masses in autumn and winter. In late spring and summer, it is intermittently influenced by vertically exported polluted air masses transported in the PBL over Europe (e.g., Zellweger et al. 2000, 2003). Henne et al. (2010) classified JFJ as a "mostly remote" site.

Pico Mountain Observatory (PMO), Portugal PMO is located on the summit caldera of Mt. Pico on the Azores Islands in the central North Atlantic. PMO was jointly established in 2001 by scientists from the University of the Azores and Michigan Technological University and has been operated by these two groups, along with scientists from the University of Colorado, Boulder, since then. The Azores Islands are often affected by continental outflow. Emissions from North America are transported in the lower FT to the Azores. Episodically, emissions exported from the eastern US are transported by warm conveyor belts associated with convection followed by subsidence and reach the Azores within 5 to 7 days (Owen et al. 2006). In addition, the Azores region is affected by transport from high latitudes. Trace gases and aerosols emitted from boreal wildfires in Canada, Alaska, and Siberia are transported to the Azores over long periods ranging from 6 to 15 days (Honrath et al. 2004; Lapina et al. 2008; Val Martin et al. 2008a). PMO also has the advantage of being located in the lower FT most of the time due to the mechanically forced upslope flow resulting from the deflection of strong winds by the mountain slope (Kleissl et al. 2007). In addition, the influence of local sources on the island via upslope flows is relatively small (Helmig et al. 2008; Kleissl et al. 2007; Val Martin et al. 2008b).

Instrumentation and measurement techniques for ozone and its precursors

O_3 concentrations can be measured by UV absorption, gas-phase titration, or conventionally by iodometric (KI) titration. The GAW Programme established a guideline for O_3 measurement that recommends the use of UV absorption in O_3 analyzers as a routine measurement technique for continuous in situ observation (World Meteorological Organization 2013). This measurement method is based on the absorption of radiation at 253.7 nm by O_3 in the gas cells of the instrument. An O_3 photometer is used as the primary standard, and traceability is ensured through instrument comparisons (Tanimoto et al. 2007).

The CO concentrations reported in this paper were determined by collection of discrete air samples in flasks with subsequent analysis in the laboratory, in addition to continuous measurements. GC/HgO, which refers to gas chromatography combined with mercury oxide (HgO) reduction detection, was used for measurements of CO concentrations in flasks. Non-dispersive infrared radiometry (NDIR), wavelength-scanned cavity ring down spectroscopy (WS-CRDS), or gas chromatography with flame ionization detection (GC-FID) were used for in situ measurements. All measurement methods require reference gases. In the GAW Programme, the gas standards for CO measurements have been established and maintained by the NOAA/ESRL (World Meteorological Organization 2010).

The NMHC concentrations reported in this paper were determined by collection of discrete air samples in flasks with subsequent analysis in the laboratory, in addition to continuous measurements. Gas chromatography with mass spectrometry (GC-MS) was used for in situ measurements, and GC-FID was used for in situ and laboratory measurements.

NO concentrations can be measured by ozone chemiluminescence detection (CLD), which utilizes the gas-phase reaction of ambient NO with O_3 added in excess to the air sample. NO_2 concentrations can be also measured by CLD via reduction to NO by photolytic conversion. CLD combined with a thermal Au catalytic convertor has been used for NO_y measurements. Intercomparisons of NO_y measurements were performed at IZO (Zenker et al. 1998) and MLO (Atlas et al. 1996). The NO_2 instrument at HPO uses a molybdenum (Mo) converter to thermally decompose NO_2 into NO. It has been shown that oxidized nitrogen compounds such as nitrous acid (HONO), peroxyacetyl nitrate (PAN), and nitric acid (HNO_3) are also converted to NO with high efficiency by the Mo convertor, and therefore contribute to the overestimation of NO_2 concentration under certain conditions (Fehsenfeld et al. 1987). In rural and remote sites, where NO_x is small compared to NO_y, the interference is negligible. Therefore, we regard NO_x concentrations at HPO as NO_y concentrations.

Seasonal and long-term changes

Tropospheric ozone

Stratospheric vs. tropospheric processes Numerous observations in the Northern Hemisphere have enabled the analysis of the seasonal and long-term variations in tropospheric O_3. For a long period of time, stratosphere–troposphere exchange (STE) followed by subsiding transport to the middle and lower troposphere was considered to be the primary source of tropospheric O_3 (e.g., Danielsen 1968; Junge 1962). Junge (1962) found a uniform seasonal cycle of tropospheric O_3 with a maximum in spring and a minimum in winter within the Northern Hemisphere. Because the maximum was later than that of stratospheric O_3 by 2 months, this delay was thought to be caused by the response time that depends on the rate of O_3 destruction within the troposphere.

Seasonal cycles of radionuclides were also investigated as evidence for STE. Beryllium-7 (^{7}Be), beryllium-10 (^{10}Be), and lead-210 (^{210}Pb) radionuclides, with half-lives of 53.3 days, 1.5×10^6 years, and 22 years, respectively, are useful tracers for tracking the movement of airflow and convective activity in the atmosphere. ^{7}Be and ^{10}Be are produced by cosmic ray spallation reactions with N and O atoms in the stratosphere and upper troposphere. More than half (~67%) of the ^{7}Be and ^{10}Be sources are located in the stratosphere. The remainder are located in the troposphere, particularly in the upper troposphere (Lal and Peters 1967).

In addition to the independent use of ^{7}Be and ^{10}Be, the ^{10}Be/^{7}Be ratio has been considered as a stratospheric tracer because its value in the stratosphere is significantly higher than that in the troposphere because of the markedly longer radioactive half-life of ^{10}Be than that of ^{7}Be (Koch and Rind 1998). ^{210}Pb is a daughter nuclide of radon-222 (^{222}Rn), which is emitted from soils and has a half-life of 3.8 days (Turekian et al. 1977). Therefore, high ^{7}Be concentrations with respect to low ^{210}Pb concentrations could indicate strong air subsidence from upper altitudes, which can increase surface ^{7}Be and O_3 concentrations, because these components are produced mainly in the stratosphere/upper troposphere (Cristofanelli et al. 2006, 2015; Cuevas et al. 2013; Tsutsumi et al. 1998; Zanis et al. 1999).

Contrastingly, O_3 produced photochemically within polluted air masses in the troposphere can be traced by a high concentration of ^{210}Pb. The combined use of ^{7}Be and ^{210}Pb can clarify information about the origin of the air masses (Graustein and Turekian 1996; Heikkilä et al. 2008; Lee et al. 2007; Zanis et al. 2003b). Gros et al. (2001) utilized CO and its isotopic composition (^{13}C, ^{18}O, and ^{14}C) together with ^{7}Be as an STE tracer because ^{14}CO is produced by the neutron–proton exchange reaction in nitrogen-14 (^{14}N) prior to its oxidation. Approximately three fourths of all ^{14}CO in the atmosphere is of direct cosmogenic origin. Gros et al. (2001) determined the proportion of stratospheric air in an air mass by examining the background and stratosphere levels of ^{14}CO at SNB.

The origins of air masses associated with high-O_3 episodes and the contributions from the stratosphere and the troposphere have previously been discussed in detail, along with the correlations of O_3 with water vapor and anthropogenically emitted compounds such as CO, CO_2, NMHCs, and total atmospheric Hg (Ambrose et al. 2011; Bonasoni et al. 2000b; Cristofanelli et al. 2006, 2015; Kajii et al. 1998; Tsutsumi et al. 1998; Wang et al. 2006). Cristofanelli et al. (2006) investigated the influence of stratospheric intrusion by using ^{7}Be, relative humidity, potential vorticity (PV), and total column O_3. Based on the aforementioned methods, Cristofanelli et al. (2015) reconstructed the frequency of days influenced by stratospheric intrusion at CMN for the period from 1996 to 2011. Liang et al. (2008) utilized the oxygen isotopic ratio of atmospheric CO_2 ($\delta^{18}O(CO_2)$) as a tracer at WLG. The $\delta^{18}O(CO_2)$ and $\Delta^{17}O(CO_2)$ values were found to be enhanced by CO_2 originating in the stratosphere and to have a seasonal cycle that differed from those of surface sources (Assonov et al. 2010).

PV ($1\ \text{PVU} = 10^{-6}\ m^2\ s^{-1}\ K\ kg^{-1}$) analysis has also been widely used to detect the STE (Cristofanelli et al. 2006, 2009a, 2015; Cuevas et al. 2013; Ding and Wang 2006; Kentarchos et al. 2000; Leclair De Bellevue et al. 2006; Schuepbach et al. 1999; Tsutsumi et al. 1998). PV is the product of absolute vorticity and thermodynamic stability. Because PV is conserved in adiabatic, frictionless flow, the dynamical tropopause is generally considered to be a well-defined continuous surface with a quasi-material character. Air parcels can only cross the dynamical tropopause when diabatic or other non-conservative processes (such as friction) change the PV. In the atmosphere above 350 hPa, PV rapidly increases with altitude, reaching typical values from 1.0 PVU (Danielsen 1968) to 3.5 PVU (Hoerling et al. 1991).

On the other hand, during the Los Angeles smog episodes in the 1950s, it was revealed that O_3 was produced via photochemical reactions in the polluted air containing NO_x and hydrocarbons (Haagen-Smit 1952). Later in the 1970s and 1980s, photochemistry involving the oxidation of CO and hydrocarbons in the remote atmosphere had been regarded as the major source of tropospheric O_3 (e.g., Chameides and Walker 1973). The seasonal cycles of tropospheric O_3 have been argued as a clue to understanding the contributions from the stratospheric influence and the tropospheric photochemistry. The spring O_3 maximum has widely been observed at many remote sites across the mid-latitudes in the Northern Hemisphere (Monks 2000; Oltmans et al. 2004). Arguments have been presented on the possible mechanisms, sources, and processes of this variation, particularly the roles of stratospheric intrusion and tropospheric photochemistry (Monks 2000; Scheel et al. 1997).

As we discussed above, the former was considered to be more important in 1960, while the latter was considered to play a key role in later years, triggered by the finding of a springtime PAN maximum, which is an unambiguous marker for tropospheric photochemistry. It has been suggested that the onset of photochemical O_3 production from the accumulation of O_3 precursors such as CO and NMHCs over the winter may be related to the springtime maximum of O_3 (Penkett and Brice 1986). Additionally, the lifetime of O_3 in winter is long enough to allow its accumulation once produced from the precursors emitted from anthropogenic sources (Fenneteaux et al. 1999; Liu et al. 1987). Intensive campaigns such as the Mauna Loa Observatory Photochemistry Experiment (MLOPEX) were made to gain an understanding of the detailed photochemical mechanisms of tropospheric O_3 in the pristine marine atmosphere, thereby suggesting an important role of in situ photochemical production of O_3 (e.g., Ridley and Robinson 1992).

In more recent years, chemistry transport models, evaluated against the seasonal cycles of tropospheric O_3, predicted that photochemical production and stratospheric influence were 4974 ± 223 and 556 ± 154 Tg(O_3) year^{-1}, respectively, thereby highlighting the importance of photochemical production of O_3 (e.g., Stevenson et al. 2006; Sudo and Akimoto 2007). Based on these state-of-science models, the contribution of stratospheric O_3 to the spring maximum is thought to be minor. Nagashima et al. (2010) showed the seasonal variations of the contributions from the stratosphere, PBL, and FT at 25 selected sites including ZSF, ISK, MDY, HPO, NWR, and MLO and concluded that the stratospheric contributions from surface sites, including the interior of the Eurasian continent (ISK and MDY), had a minimum in summer and autumn.

However, arguments on the role of the stratospheric O_3 in the tropospheric O_3 distributions and variations continue, with emphasis on the differences in the season, region, and altitude. Škerlak et al. (2014) compiled a global 33 year climatology of STE for the period 1979 to 2011 by using the ERA-Interim reanalysis dataset from the European Centre for Medium-Range Weather Forecasts (ECMWF) and showed the geographical and seasonal distributions of stratospheric O_3 fluxes across the tropopause and down to the top of PBL. They reported that western North America (from March to May) and the Tibetan Plateau (in except for the period from September to November) are the regions where surface O_3 levels were most likely affected by the STE, with a maximum value of more than 120 kg km^{-2} month^{-1}. This could have significant implications for surface O_3 air quality. However, these estimates could still have large uncertainties resulting from uncertainties and/or biases in the reanalysis datasets used.

Seasonal variations The seasonal cycle of O_3 is controlled by the interaction of photochemical and dynamical processes and shows clear dependences on the latitude, altitude, and availability of precursors. Figure 2 shows the seasonal cycles of O_3 at the 26 high-altitude stations around the globe. Here, we show only data that were available on the public websites or by the referred publications; the data providers are summarized in Table 1. The aforementioned typical small springtime O_3 maxima have been reported at numerous stations on remote islands such as IZO, with peak concentration of ~55 parts per billion (ppb) (Cuevas et al. 2013; Rodrıguez et al. 2004; Schmitt and Volz-Thomas 1997; Fig. 2f); MLO, at ~50 ppb (Oltmans et al. 1996a, 2006; Fig. 2q); and PMO (Kumar et al. 2013). In a later study, Hauglustaine et al. (1999) utilized a photochemical box model in conjunction with the MLOPEX 2 measurements to evaluate the processes and species controlling photochemistry and to provide insight into the O_3, odd nitrogen, and odd hydrogen budget in the FT over the Pacific. Their analysis indicated a well-balanced production and destruction of O_3 during all seasons. The net production rate of O_3 was slightly negative, indicating that this region or the tropical Pacific troposphere acted as a net photochemical sink for O_3.

In continental Europe, including stations ZSF, SNB, JFJ, KVV CMN, and KSL, the seasonal cycle of O_3 at high altitudes exhibits a minimum in winter and a broad maximum beginning in spring and extending to summer. This behavior represents a combination of the spring maximum generally observed in the Northern Hemisphere and a secondary summer maximum, which is attributed to enhanced regional pollution associated with persistent anticyclonic weather (Chevalier et al. 2007). Such broad spring/summer maxima in O_3 have been reported at SNB (~55 ppb) (Gilge et al. 2010; Fig. 2b), ZSF (~55 ppb) (Gilge et al. 2010; Schuepbach et al. 2001; Fig. 2a), CMN (~60 ppb) (Bonasoni et al. 2000a; Campana et al. 2005; Fig. 2e), KSL (Elansky et al. 1995; Senik and Elansky 2001; Senik et al. 2005; Tarasova et al. 2003, 2009), KVV (~55 ppb; Fig. 2d), ARO (Campana et al. 2005; Pochanart et al. 2001; Staehelin et al. 1994), and JFJ (~60 ppb; Balzani Lööv et al. 2008; Gilge et al. 2010; Schuepbach et al. 2001; Zanis et al. 2007; Fig. 2c). During three intensive observations known as the FREE Tropospheric Experiment (FREETEX), which covered the period of O_3 buildup from late winter to late spring, the concentrations of peroxy radicals (HO_2+RO_2) showed a gradual increase from February to May (Zanis et al. 2003a). The calculated net O_3 production rate showed an increasing seasonal trend from late winter to late spring. Zanis et al. (2003a) suggested that in situ photochemistry is important and might have played a dominant role in controlling the O_3 accumulation from winter to spring in the lower FT over the European Alps.

In East Asia, a distinct spring maximum associated with a summer minimum with a small secondary peak in autumn was reported at HPO (~70 ppb; Kajii et al.

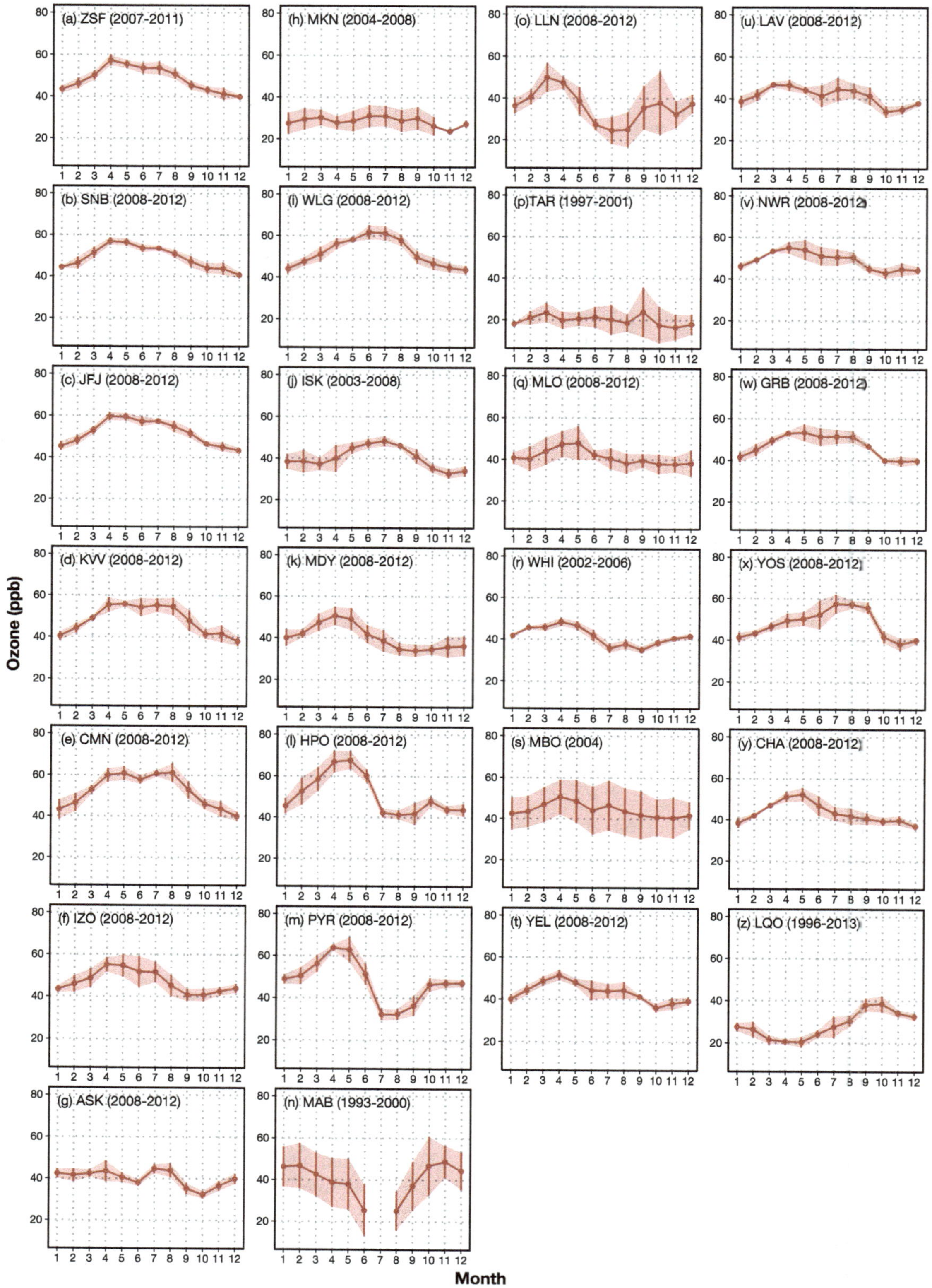

Fig. 2 Seasonal variations in tropospheric O_3 concentration at 26 mountain stations (**a**–**z**). The *lines with circles* denote the average observed O_3 concentrations during a particular period. The *years in parentheses* denote the averaging periods. The *red shaded bands* and *error bars* denote the standard deviations

1998; Pochanart et al. 2004; Fig. 2l) and FWS (Tsutsumi et al. 1994) in Japan and at LLN (~50 ppb; Ou Yang et al. 2012; Fig. 2o) in Taiwan. This seasonal pattern and magnitude are controlled by a combination of two factors. Although O_3 produced by photochemical activity accumulates from winter to summer, the East Asian monsoon brings pristine air masses from the Pacific Ocean during summer, yielding the spring maximum and the summer minimum (Tanimoto et al. 2005). It should be noted that the mean concentration at the spring peak during the period from 2008 to 2012 at HPO is 70 ppb, which is higher than those at European and North American sites owing to the substantial effects of anthropogenic activity in East Asia. It is considered likely that air mass intrusion from the stratosphere might not contribute directly to the high springtime O_3 concentration at HPO, according to analysis of the correlation between O_3 and absolute humidity (Kajii et al. 1998). In East Siberia, the seasonal cycle of O_3 at MDY showed a spring maximum and summer minimum (~50 ppb; Pochanart et al. 2003; Fig. 2k). This summer minimum is not caused by a continental–marine air mass exchange because MDY is located in the middle of Siberia. Rather, it is caused by an O_3 sink that includes surface deposition and more active vegetation (Pochanart et al. 2003). Wild et al. (2004) reported that the mean European influence on the O_3 concentration at MDY varied between 3.5 ppb in spring and 0.5 ppb in summer. In South Asia, the seasonal cycle at PYR showed a spring maximum and a summer minimum (~65 ppb; Bonasoni et al. 2010; Fig. 2m), which is similar to the case in East Asia. This likely occurred because PYR is under a greater influence from the South Asian monsoon, which brings clean maritime air masses during summer (Bonasoni et al. 2010). In addition, STE events at PYR showed frequency maxima during the dry winter and pre-monsoon spring seasons, which affected the O_3 concentration (Bracci et al. 2012; Cristofanelli et al. 2009a). The estimated average O_3 increase with respect to the average mean values by STE events for the period from 2006 to 2008 was 24 % (Bracci et al. 2012). The seasonal cycle of O_3 at MAB also depends on the South Asian monsoon. However, springtime O_3 at MAB did not show a value as high as that at PYR, and the seasonal cycle was characterized by the summer minimum and late autumn–winter maximum in the period from October to February of ~50 ppb (Naja et al. 2003; Fig. 2n). It is considered likely that one of the reasons for the disagreement is the altitude difference of approximately 3400 m. Because the stratospheric influence generally increases with an increase of altitude, PYR was frequently influenced by STE in spring (Bracci et al. 2012; Cristofanelli et al. 2009a).

In central Asia, a broad summer maximum was found in the surface O_3 concentration observed at ISK (~50 ppb; Fig. 2j) in Kyrgyzstan and at WLG on the Qinghai–Tibetan Plateau (~60 ppb; Ding and Wang 2006; Lee et al. 2007; Liang et al. 2008; Ma et al. 2005; Wang et al. 2006; Xu et al. 2015; Xue et al. 2011; Zhu et al. 2004; Fig. 2i). Analyses of the summer maximum at WLG have yielded contradicting results that have led to two competing hypotheses on its cause: long-range transport of polluted air masses from central–eastern China and intrusion of O_3-rich air masses from the stratosphere. In a three-dimensional (3D) regional chemical transport model study, Zhu et al. (2004) concluded that the seasonal O_3 transitions associated with the Asian monsoon system and transport from eastern–central China, central–South Asia, and even Europe are responsible for the distinct seasonal O_3 cycle at WLG. An alternative theory is that downward transport of air masses from the upper troposphere–lower stratosphere plays a key role in controlling the seasonal cycle of O_3 at WLG (Ma et al. 2005). An analysis of O_3 measurements and tropospheric and stratospheric tracers such as water vapor, CO, NMHCs, PV, and $\delta^{18}O$ (CO_2) attributed the summertime O_3 enhancements at WLG to stratospheric intrusion rather than the transport of anthropogenic pollution (Ding and Wang 2006; Liang et al. 2008; Wang et al. 2006). From measurements of ^7Be and ^{210}Pb, Lee et al. (2007) suggested that the combined processes of the downward transport of the stratospheric–upper tropospheric air masses and the long-range transport of polluted air masses from eastern–central China caused the summertime O_3 peak. A model calculation using the Nested Air Quality Prediction Modeling System (NAQPMS) estimated the contributions of the regional transport of photochemically produced O_3 and stratospheric O_3 at 10–25 and ~5 ppb, respectively (Li et al. 2009). An analysis of backward trajectories for the period from 2000 to 2009 revealed that air masses from central–eastern China dominated the airflow at WLG in summer, suggesting strong impacts of anthropogenic forcing on the surface O_3 and other trace constituents such as CO, NO, NO_2, NO_y, PAN, HNO_3, and particulate nitrate (NO_3^-) on the Tibetan Plateau (Xue et al. 2011). Using a photochemical box model constrained by observations, Xue et al. (2013) calculated the photochemical budgets of O_3 and radicals and showed that O_3 production was significantly greater than O_3 destruction during daytime in both spring and summer, which is indicative of net O_3 production in both seasons. In contrast, Ma et al. (2002) reported that the net O_3 production is positive in winter but negative in summer. Xue et al. (2013) determined that the key factor leading to this disagreement was the difference in NO levels and suggested that the transport of anthropogenic pollution from central–eastern China might perturb the chemistry in the background atmosphere over the Tibetan Plateau.

The circumstances described above indicate that the summer maximum at WLG can be attributed to the transport of anthropogenic pollution. Based on the source-receptor analysis, Nagashima et al. (2010) suggested that the summer maximum of O_3 at ISK was mainly due to the enhanced contribution of O_3 produced in the surrounding regions.

In North America, three different types of seasonal cycle were found in surface O_3. A spring maximum was observed at WHI (~50 ppb; Macdonald et al. 2011; Fig. 2r) and CHA (~55 ppb; Fig. 2y). A broad spring–summer maximum was observed at GRB (~50 ppb; Fig. 2w), LAV (~45 ppb; Fig. 2u), MBO (~50 ppb; Gratz et al. 2015; Weiss-Penzias et al. 2007; Fig. 2s), NWR (~55 ppb; Oltmans and Levy 1994; Fig. 2v), and YEL (~50 ppb; Fig. 2t). A distinct summer maximum was observed at YOS (~60 ppb; Fig. 2x). It is possible that these types of seasonal cycle in western North America reflect a difference in the influence of the boundary layer at each site. All high-O_3 events at MBO were observed between early March and late September and were caused by the Asian long-range transport, the subsidence of O_3-rich air from the upper troposphere–lower stratosphere, and the mixed influence of these two mechanisms (Ambrose et al. 2011). Therefore, the seasonal cycle at MBO might have been caused by the spring peak of background O_3 in the Northern Hemisphere and high O_3 events occurring between early spring and late summer.

Compared with the Northern Hemisphere, O_3 seasonal cycle observations have not been conducted extensively in the Southern Hemisphere and the equatorial regions, and the O_3 seasonal cycles observed at MKN and TAR, which are located in the tropics, appear to be rather flat relative to those at other mid-latitude sites. The O_3 concentrations at MKN exhibit little variation, with a minimum in November and a broad maximum in boreal summer (Henne et al. 2008). This cycle can be explained by the seasonal variation of monsoon flow over equatorial East Africa, which is driven by seasonal displacement of the ITCZ. Although the seasonal variations in O_3 at TAR were not pronounced (Fig. 2p), based on observations from 2006 to 2008, Toh et al. (2013) reported a substantial seasonal cycle associated with a minimum that occurred twice a year in the November–December and March–April timeframes, along with a maximum that occurred in the summer monsoon season in late May–September, which that coincided with the regional biomass burning period. This discrepancy is likely to be due to the interannual variability of O_3 in tropical regions. The seasonal cycle at ASK, which is located in the Sahara Desert of Algeria between IZO and MKN, has an autumn minimum and a summer maximum in the July–August timeframe (~45 ppb; Fig. 2g). In South America, the seasonal cycle of O_3 observed at LQO is characterized by an autumn minimum in the March to May period and a spring maximum in September–October timeframe (~40 ppb; Fig. 2z). It should be noted that while these O_3 maximum periods coincide with the August to October biomass burning season in the Southern Hemisphere (Duncan et al. 2003), the actual mechanism involved in the seasonal cycles has not been investigated.

The latitudinal gradient in tropospheric O_3 has also been discussed in previous research. For example, Scheel et al. (1997) found that the O_3 seasonal cycle amplitude was smaller in the northwest area than in the southeast area within Europe, and Winkler (1988) showed that the surface O_3 concentration over the Atlantic Ocean, which was twice as high in the Northern Hemisphere as that in the Southern Hemisphere, generally increased with latitude, and then dropped rapidly at 70° N. Figure 3 shows the annual mean concentrations and the amplitude (annual maximum and minimum) of the seasonal cycles of tropospheric O_3 observed at the mountain sites as a function of latitude or altitude for the period of 2008 to 2012. As can be seen in this figure, the mean O_3 concentrations and the seasonal amplitude generally increased with latitude for the range of 0°–40° N and decreased or stabilized at higher latitudes (Fig. 3c). The means were ~50 ppb in Europe and 30–45 ppb in Africa. Those in Asia were 35–50 ppb depending on the location, although three maxima were noted above 45 ppb at WLG, PYR, and HPO. In particular, the maxima at HPO and PYR were higher than those at other sites. Although one of the reasons for such a high value at PYR is believed to be its altitude at 5079 m a.s.l., no altitude dependency was observed for the annual mean O_3 concentration (Fig. 3a). Therefore, the high maximum values at PYR and HPO are associated with their seasonal cycle spring maxima. The minima were in the range of 30–50 ppb for all of the mid-latitude stations except for LLN. The tropical sites differed from other mid-latitude sites, showing small variability over the course of the year. Insufficient data were available to discuss the latitudinal gradient south of ~20° N. The seasonal amplitude for the 22 stations was 10–20 ppb, and again, the magnitude was generally dependent on the latitude. Moreover, the seasonal amplitude at MAB, LLN, HPO, and PYR, at ~25–30 ppb (Fig. 3d) was substantially larger than that at the other stations. This result was caused by either a larger spring or autumn maximum due to intense photochemical production, and a smaller summer minimum due to the Asian monsoon, than those reported at other sites. No altitude dependency was observed for the amplitude of the O_3 concentration (Fig. 3b).

Long-term trends The long-term trends for tropospheric O_3 have been actively examined and summarized in many studies (e.g., Cooper et al. 2014; Oltmans et al.

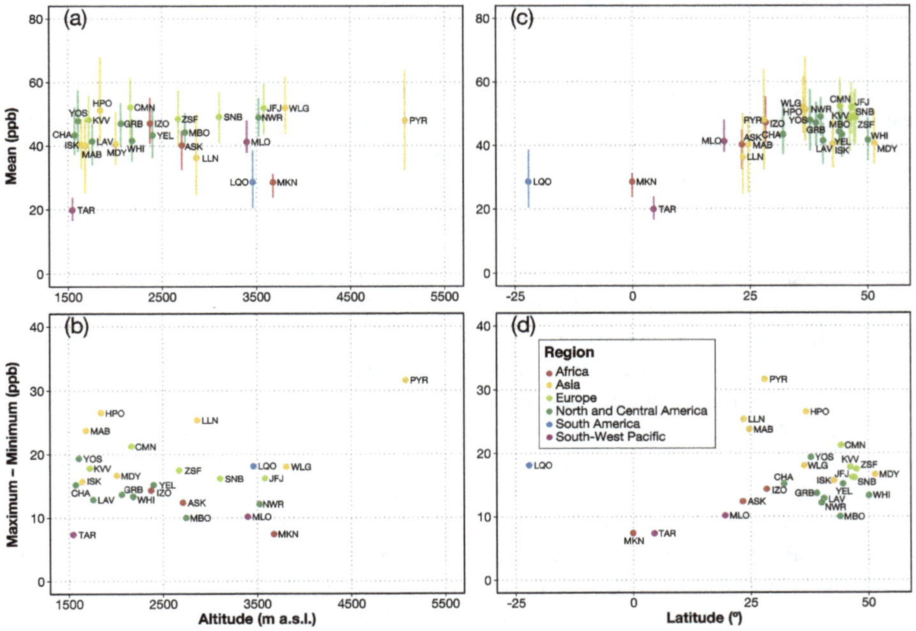

Fig. 3 Annual mean concentration and range (maximum–minimum) of seasonal cycles of tropospheric O_3 observed at mountain sites with respect to latitude and altitude of station (**a**–**d**). The *error bars* denote the maxima and minima

2006, 2013; Vingarzan 2004). The abundance of O_3 was first observed in the 19th century. The relevancies of old measurement data were examined and an apparent increase in the background O_3 levels was observed throughout the 20th century in Europe (Marenco et al. 1994; Volz and Kley 1988). More recently, Gilge et al. (2010) calculated the long-term trends for surface O_3 at SNB, ZUG/ZSF, and JFJ to be 0.22 ± 0.17 ppb year^{-1} (1989–2007), 0.42 ± 0.09 ppb year^{-1} (1975–2007), and 0.26 ± 0.16 ppb year^{-1} (1986–2007), respectively. During the past 40 years, the O_3 concentrations at ZUG/ZSF showed significant increases in the 1980s that slowed in the 1990s (Gilge et al. 2010; Logan et al. 2012; Oltmans et al. 2013; Parrish et al. 2012). In the 1990s, O_3 increases at European stations including SNB, CMN, ARO, ZUG/ZSF, and JFJ were broadly observed, indicating that the increasing trend of O_3 is a regional-scale phenomenon. The O_3 concentrations at European stations peaked in the late 1990s or early 2000s, followed by the current decreasing trend (Brönnimann et al. 2002; Cristofanelli et al. 2015; Cui et al. 2011; Gilge et al. 2010; Logan et al. 2012; Parrish et al. 2012; Oltmans et al. 2013; Staehelin et al. 1994). For the period from 2000 to 2009 Logan et al. (2012) calculated the mean rate of O_3 decrease at seven European alpine sites including SNB, ZUG/ZSF, and JFJ to be -0.16 ± 0.14 ppb year^{-1}. They discussed possible contributing factors to the O_3 changes in central Europe, including changes in stratospheric input and in O_3 precursors. In addition, they showed that emissions of NO_x from both the USA and Europe were almost constant in the 1980s but decreased in the 1990s. Model studies have suggested that variations in domestic emissions should have the largest influence on O_3 in summer, whereas variations in emissions in North America and Asia should have the largest influence in spring and late autumn over Europe (Fiore et al. 2009; Jonson et al. 2010). Such studies concluded that CO and CH_4 emissions should have only a minor influence on O_3 trends. However, the relationship between the observed O_3 trends and anthropogenic emission changes is complex. Although total anthropogenic NO_x emissions at northern mid-latitudes increased by only ~14 % for the period from 1980 to 2010, long-term O_3 trends showed remarkable increases (Parrish et al. 2014). The long-term trend for surface O_3 at KSL in Caucasus differed from that in central and southern Europe (Senik and Elansky 2001; Tarasova et al. 2003, 2009) with the former showing a decreasing trend in the 1990s (Senik and Elansky 2001; Tarasova et al. 2003). For the period from 1991 to 2001, a strong concentration change of -0.91 ± 0.17 ppb year^{-1} was observed, after which the O_3 changes decreased to -0.37 ± 0.14 ppb year^{-1} for the period from 1997 to 2006 (Tarasova et al. 2009). Tarasova et al. (2009) reported that the O_3 surface trend difference was caused by differences in the station locations relative to the source areas affecting the O_3 variations. The decreasing O_3 trend at KSL was caused by both substantial emission decreases in the 1990s due to the breakdown of the former USSR and emission control measures implemented in Europe.

In the North Atlantic, the surface O_3 record at IZO showed a slightly positive trend at 0.09 ± 0.02 ppb year^{-1} for the period from 1988 to 2009 with a rapid increase between 1996 and 1998 (Cuevas et al. 2013). Cuevas et al. (2013) suggested that the abrupt changes in the O_3 levels reflect a phase shift of the North Atlantic Oscillation (NAO), which altered the transport pattern to IZO. The predominant high positive phase of the NAO changed from 1995 to 1996. IZO is influenced by flow from both the mid-latitudes and the subtropics and NAO changes have resulted in an increased flow of the Westerlies in the mid-latitude subtropical North Atlantic, thus favoring the transport of O_3 and its precursors from North America. This process increased the tropospheric O_3 in the subtropical North Atlantic region after 1996.

In East Asia, a positive O_3 surface trend at 0.25 ± 0.17 ppb year^{-1} was observed at WLG for the period from 1994 to 2013 (Xu et al. 2015), and an especially strong increase was found in autumn, followed by spring. HPO in Japan experienced a generally large increase in springtime O_3 compared with the increases over Europe and North America during the 2000s (Tanimoto 2009; Parrish et al. 2012). The mean growth rate in springtime O_3 was ~1 ppb year^{-1} for the period from 1998 to 2006 (Tanimoto et al. 2009). Tanimoto et al. (2009) showed that variations in NO_x emissions and satellite observations of tropospheric NO_2 for East Asia were consistent with a remarkable increase of O_3 at HPO. Although this increasing trend is consistent with the increasing anthropogenic emissions in East Asia, it can only explain half of the observed increase by the model calculation. In the western USA, positive trends in springtime O_3 concentrations have been reported (e.g., Cooper et al. 2014; Jaffe et al. 2003b; Jaffe and Ray 2007; Parrish et al. 2012; Vingarzan 2004). Jaffe and Ray (2007) reported positive trends in O_3 concentration at LAV and at YEL in all seasons with growth rates of 0.33 and 0.50 ppb year^{-1}, respectively, based on the analysis of deseasonalized monthly mean O_3 concentrations (daytime data only) for the period from 1987 to 2004. Moreover, they proposed possible explanations for these O_3 trends that include (1) increases in regional emissions, (2) changes in the distribution of emissions, (3) increases in biomass burning in the western USA, and (4) increases in global background O_3. Jaffe et al. (2008) showed that the increase in fires is largely responsible for the increase in summertime O_3, but fires could not explain the trend in O_3 for other seasons. Jaffe (2011) found statistically significant correlations between springtime O_3 at 11 sites in the western USA including CHA, GRB, and YEL and free tropospheric O_3, as measured by an ozonesonde in Boulder. This correlation indicates a downward transport of enhanced O_3 from the FT to the surface. In summer, free tropospheric and surface O_3 concentrations and the number of days exceeding the US National Ambient Air Quality Standards (NAAQS) threshold were all significantly correlated with emissions from biomass burning in the western USA. This also indicates that wildfires had a large impact on O_3 concentrations in the western USA. The mean springtime O_3 at MBO showed significant increases of 0.73 ± 0.54 ppb year^{-1} for the period from 2004 to 2013, whereas the annual mean O_3 did not show a significant trend (Gratz et al. 2015). An analysis of transport patterns by Gratz et al. (2015) suggested that long-range transport from Asia has a dominant influence on the increasing springtime O_3. At HPO and MBO, which are both strongly influenced by springtime outflow from East Asia, the O_3 levels showed larger increases at higher percentile levels than those at lower percentile levels (Gratz et al. 2015; Tanimoto 2009). A statistically significant positive trend in the mean maximum daily 8-h average (MDA8) O_3 concentration at 0.12 ppb year^{-1} was observed at CHA for the period from 1989 to 2010 (Chalbot et al. 2013). The surface O_3 recorded at MLO in the Northern Pacific has shown a significant increase of 0.16 ± 0.04 ppb decade^{-1} (Oltmans et al. 2013). Oltmans et al. (2013) showed an increasing trend into the early 2000s followed by a decline in later years. In the earliest period, the increase was observed primarily in spring, but in the most recent decade, increases during autumn and early winter were observed to be greater. This change is attributed to a shift in the transport pattern associated with greater outflow from higher latitudes north of 30° N (Oltmans et al. 2006). Such a seasonal shift in the O_3 maximum has not been observed at most stations in the mid-latitudes, but shifts in the seasonal cycle to an earlier time of the year were observed at ZUG/ZSF, at 5.1 ± 3.5 days decade^{-1} (Parrish et al. 2013); JFJ, at 5.6 ± 4.1 days decade^{-1} (Parrish et al. 2013); and LAV, at 14 ± 19 days decade^{-1} (Cooper et al. 2014; Parrish et al. 2013). Additional studies are needed to determine why the seasonal O_3 cycle shifts across northern mid-latitudes are only seen at some sites and not others. Parrish et al. (2014) showed comparisons of long-term O_3 records observed at northern mid-latitude sites with three chemistry–climate model results. Some disagreement was noted between the observed and modeled results, suggesting that only limited confidence can be placed in the estimates of the present RF of tropospheric O_3 derived from modeled historic concentration changes and in the prediction of future O_3 concentrations.

Some differences between the changes in O_3 concentrations at high-altitude sites and those at low-latitude sites have been observed. The temporal behavior at Mace Head, which is a remote site in western Ireland, was similar to that at the alpine site, even though the O_3 increase in the late 1990s occurred earlier at ZUG/ZSF than at Mace Head (Logan et al. 2012). Parrish et al. (2012) showed the long-term O_3 trends of 11 datasets regarded as baseline conditions. These datasets covered the range from low level to FT level, and included ZUG/ZSF, ARO, JFJ, HPO, and LAV data. They showed that both absolute O_3 concentrations and positive trends in O_3 generally increased with the altitude of the measurement site. In addition, they suggested that the changes in O_3 concentration showed a high degree of longitudinal uniformity over the last half of the 20th century at northern mid-latitudes.

Carbon monoxide

In the Northern Hemisphere, the seasonal maximum of CO occurs in late winter and early spring, and the seasonal minimum occurs in summer (Novelli et al. 1998); the maximum occurs 3 to 4 months after the seasonal minimum in OH radicals. The summer minimum in CO is explained well by the maximum occurrence of OH radicals, which are a CO main sink in the troposphere. The time lag for the CO maximum in late winter–early spring reflects the mixed effects of the seasonal variations in OH radical concentrations and CO emissions. The slower chemical destruction of CO by OH leads to a longer CO lifetime in winter, which allows more accumulation. Because the lifetime of CO is long enough for it to be transported across the globe, the long-range transport of industrial and biomass burning emissions originating on the Asian and North American continents can have substantial impacts on seasonal variations, depending on the location of the observation site (Novelli et al. 2003; Pfister et al. 2006; Yurganov et al. 2004, 2005).

Figure 4 shows seasonal cycles of CO at 16 high-altitude stations around the globe. The aforementioned, well-known seasonal cycles of CO have been reported at numerous stations on remote islands such as at IZO, with a peak concentration of ~115 ppb (Gomez-Pelaez et al. 2013; Fig. 4f); PMO (Kumar et al. 2013), and MLO at ~105 ppb (Greenberg et al. 1996; Jaffe et al. 1997; Fig. 4m). The seasonal cycle at ASK, at ~115 ppb (Fig. 4g), is similar to that at IZO.

In continental Europe including ZUG, SNB, JFJ, KVV, and CMN, the seasonal cycle of CO at high altitudes exhibits a maximum in late winter–early spring and a minimum in summer. The levels, amplitude, and timing differ slightly among sites, with SNB at ~160 ppb (Gilge et al. 2010; Gros et al. 2001; Fig. 4b), ZUG at ~180 ppb (Gilge et al. 2010; Fig. 4a), CMN at ~150 ppb (Cristofanelli et al. 2013; Fig. 4e), KVV at ~180 ppb (Fig. 4d), and JFJ at ~140 ppb Balzani Lööv et al. 2008; Dils et al. 2011; Forrer et al. 2000; Gilge et al. 2010; Zellweger et al. 2009; Fig. 4c).

Maxima in late winter–early spring and minima in summer also occur in North America, with WHI at ~180 ppb (Macdonald et al. 2011; Fig. 4n), NWR at ~135 ppb (Turnbull et al. 2007; Fig. 4p), and MBO at ~150 ppb (Weiss-Penzias et al. 2007; Fig. 4o).

In East Asia, the timing of the maximum shifts late toward the spring in association with a summer minimum and shows a small secondary peak in autumn, as observed at HPO with ~225 ppb (Kajii et al. 1998; Kato et al. 2000, 2002; Narita et al. 1999; Pochanart et al. 2004; Fig. 4j), and LLN with ~215 ppb (Cheng et al. 2013; Lin et al. 2010; Ou-Yang et al. 2014; Fig. 4l). Similar to that of O_3, the seasonal pattern and magnitude are controlled by a combination of two factors: the impact of continental outflow from the Asian continent from winter to spring and the East Asian monsoon that brings pristine air masses from the Pacific Ocean during summer. It should also be noted that the mean concentration of the spring peak during the period from 2008 to 2012 at HPO and LLN was between 200 and 250 ppb, which is higher than those at European and North American sites, due to substantial impacts from anthropogenic activities in East Asia in the case at HPO (Pochanart et al. 2004), and Southeast Asia biomass burning activities in the case at LLN (Cheng et al. 2013; Ou-Yang et al. 2014). The seasonal cycle at MAB, at ~180 ppb (Naja et al. 2003; Fig. 4k), shows a different pattern, with the maximum occurring after the monsoon season, as observed with O_3, due to transported pollution from the northeast.

In inland Asia, two types of seasonal cycle were found. The seasonal cycle at MDY is a typical pattern with a spring maximum and a summer minimum (Pochanart et al. 2003). Wild et al. (2004) reported that the mean European influence on the CO concentration at MDY varied between 30 ppb in spring and 10 ppb in summer. However, the seasonal pattern at WLG (Zhang et al. 2011) has a broad spring and summer maximum, which is similar to that of O_3 at WLG, thereby suggesting that both O_3 and CO are controlled by the same mechanism. At WLG, the CO data plotted in the present study showed a different seasonal cycle from that found by Zhang et al. (2011), which showed a late winter and early spring maximum. This discrepancy is likely caused by large interannual variability and variance in the flask sampling data (Fig. 4i).

In the tropics, the seasonal cycle of MKN, located in the equatorial region, differs from that at other stations. The maximum CO levels occur in January to March, followed by a sharp decrease toward April and May and an increase by the end of the year (Henne et al. 2008). Henne et al. (2008) determined that this seasonal cycle is associated with two maxima in February and August and is caused by the oscillation of monsoon circulation. The winter and spring maximum is caused by advection of CO-enriched air masses in the Northern Hemisphere during boreal

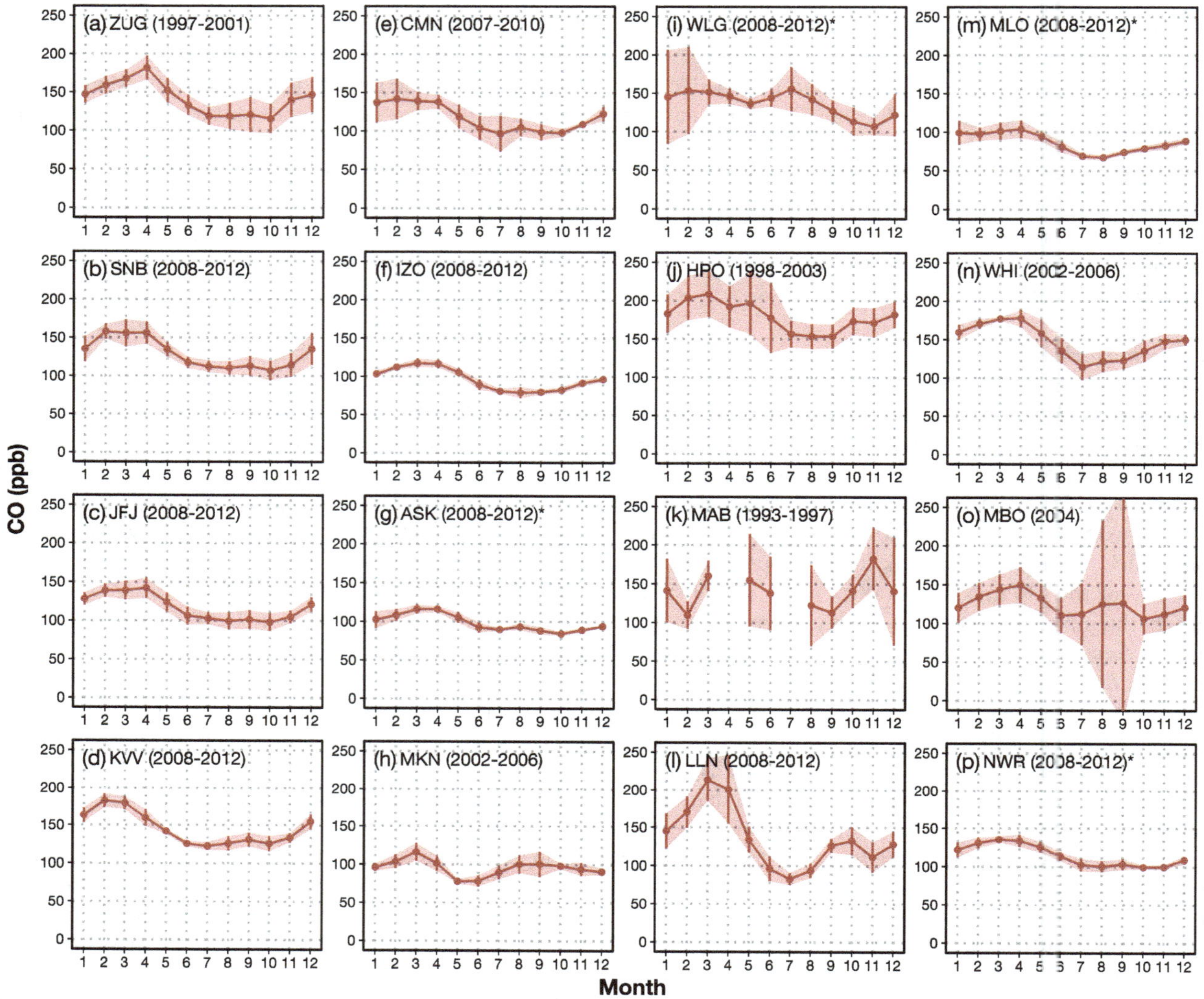

Fig. 4 Seasonal variations in CO concentration at 16 mountain stations (**a**–**p**). The *lines with circles* denote the average observed CO concentrations during a particular period. The *years in parentheses* denote the averaging periods. The *asterisks* denote CO data obtained by flask sampling. The *red shaded bands* and *error bars* denote the standard deviations

winter. In contrast, the summer maximum was observed during advection of air masses from the Southern Hemisphere that were affected by biomass-burning emissions in southern Africa.

Similar to O_3, CO also shows a gradient from south to north and greater seasonal amplitude in the Northern Hemisphere than that in the Southern Hemisphere. Based on the analysis of the global distribution of surface CO determined by flask sampling, Novelli et al. (2003) revealed a gradient between the Northern and Southern Hemispheres, with a greater seasonal amplitude in the Northern Hemisphere than that in the Southern Hemisphere and greater interannual variations in CO concentrations in the Southern Hemisphere than that in the Northern Hemisphere. Figure 5 shows the annual mean concentrations and amplitude of the seasonal cycles of tropospheric CO observed at the mountain sites as a function of altitude or latitude for the period from 2008 to 2012. The annual mean concentrations and amplitude (annual maximum/minimum) of the seasonal cycles of CO generally increased with latitude (Fig. 5c, d). This is a reverse latitudinal distribution of tropospheric OH (e.g., Spivakovsky et al. 1990). The annual mean concentrations and amplitude of the seasonal cycles of CO generally decreased with altitude (Fig. 5a), reflecting the fact that the emission sources of CO are located at the ground surface. The means were from 120 to 180 ppb in Europe and Asia and 100 ppb in Africa, and the maxima at HPO

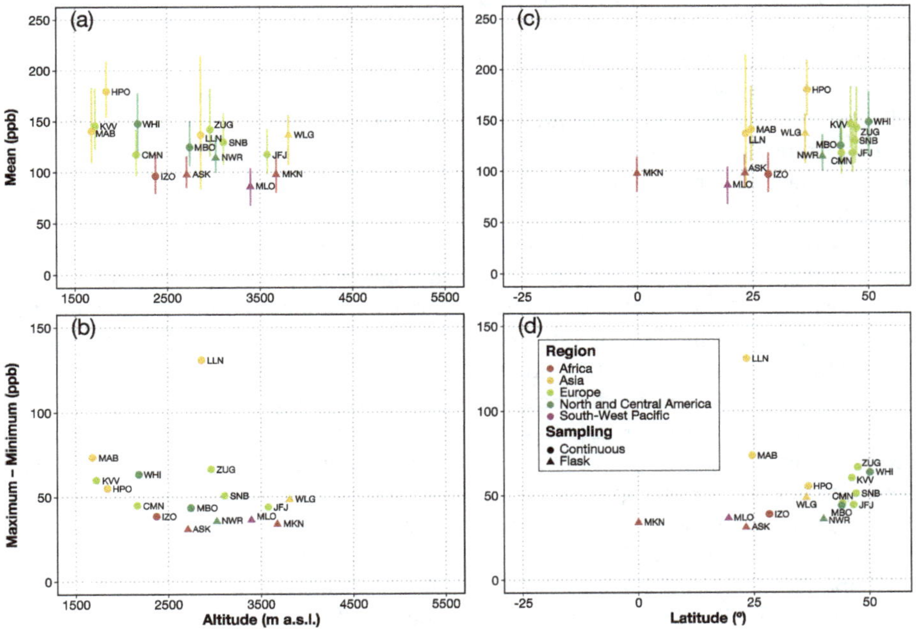

Fig. 5 Annual mean concentration and range (maximum–minimum) of seasonal cycles of tropospheric CO observed at mountain sites with respect to latitude and altitude of station (**a**–**d**). The *error bars* denote the maxima and minima

and LLN were higher than those at other sites. One of the reasons for such a high value at HPO is believed to be its averaging of the period from 1998 to 2003. The seasonal amplitude for the 15 stations except for LLN fell into the range of 85 to 180 ppb (Fig. 5b, d). The large seasonal amplitude at LLN was caused by a high spring maximum of ~215 ppb due to biomass burning activity in Southeast Asia (Ou-Yang et al. 2014), rather than the low CO of ~85 ppb in summer due to the Asian monsoon, because the summertime CO concentration at LLN was not significantly lower than that at other stations.

The CO levels increased in the Northern Hemisphere from the 1950s to the 1980s and then decreased in the late 1980s (Khalil and Rasmussen 1994; Novelli et al. 1998, 2003; Zander et al. 1989). Based on observations from NOAA Climate Monitoring and Diagnostics Laboratory sites, Novelli et al. (2003) estimated that the global mean CO concentrations decreased at a rate of -0.52 ± 0.10 ppb year^{-1} for the period from 1991 to 2001. Worden et al. (2013) examined satellite observation data for total column CO recorded by the Measurements of Pollution in the Troposphere (MOPITT) instrument from 2000 through 2011 and reported a continued decreasing trend in the Northern Hemisphere. They stated that further studies are needed in order to understand the trends and the contributions from anthropogenic emissions and possible changes in atmospheric chemistry through reactions with OH radicals. Few records of continuous observations of CO are available beginning in the 1990s at high-altitude stations other than MLO (Novelli et al. 1998, 2003) and JFJ (Dils et al. 2011; Gilge et al. 2010; Zellweger et al. 2009). The trend of CO at JFJ for the period 1996–2007 has been determined to be -3.36 ± 1.08 ppb year^{-1} (Gilge et al. 2010) and -2.65 ± 0.04 ppb year^{-1} (Zellweger et al. 2009) calculated using background data, respectively. At MBO in the western USA, the mean springtime CO showed a significant decrease for the period from 2004 to 2013, whereas the annual average CO did not show a significant trend (Gratz et al. 2015).

Other trace gases: NMHCs, NO_x, and NO_y

NMHCs Atmospheric NMHCs show considerable variation on spatial and temporal scales, but it is difficult to determine these variations on a regional basis because of the limited observational data on NMHCs available from high-altitude stations. In the troposphere, NMHCs are predominantly transformed not only by the chemical processes of photolysis and reaction with OH radical typically during daytime but also by reaction with the NO_3 radical during nighttime, and reaction with O_3 for unsaturated species, in addition to coastal and marine area reactions with chlorine (Cl) atoms, during daytime (Atkinson 2000). The reaction rate constants increase with molecular weight within a given class of NMHCs, causing lighter, saturated NMHCs to exhibit slower atmospheric decay and longer lifetimes (Atkinson and Arey 2003). NMHC concentrations are determined by the strength of the emission sources and their atmospheric removal processes. Generally, the seasonal cycles of NMHCs show a maximum in winter and a minimum in summer owing to reaction with OH radicals. NMHC concentration ratios are useful indicators of the degree of photochemical processing or aging because NMHCs are removed at different rates, and their concentration patterns change during transport due to reaction rate differences, which span several orders of magnitude (e.g., Helmig et al. 2008, 2015).

Figure 6 shows the seasonal cycles of NMHCs at eight high-altitude stations. In this paper, only relatively long-lived species that are subject to intercontinental transport are shown, based on the lifetime or rate constant data reported by Atkinson and Arey (2003). Ethane (C_2H_6) with a lifetime of ~47 days showed a distinct seasonal cycle with a winter–spring maximum and a summer minimum with peak concentrations of ~1870 ppt at JFJ (Fig. 6a), ~1390 ppt at PMO (Helmig et al. 2008, 2015; Fig. 6b), ~1910 ppt at HPO (Sharma et al. 2000; Fig. 6c), ~1240 ppt at IZO (Schmitt and Volz-Thomas 1997; Fig. 6d), ~1210 ppt at ASK (Fig. 6e), and ~950 ppt at MLO (Greenberg et al. 1996; Fig. 6f) in the Northern Hemisphere. The seasonal cycle at MKN in the equatorial region showed two minima in April and October and two maxima in January and August, with a peak concentration of ~690 ppt (Fig. 6g). This pattern corresponds to the climate pattern in equatorial East Africa induced by the displacement of the ITCZ. Acetylene (C_2H_2), with a lifetime of ~16 days, showed a distinct seasonal cycle with a winter maximum and summer minimum at CMN with ~340 ppt (Fig. 6h), IZO with ~250–280 ppt (Schmitt and Volz-Thomas 1997; Fig. 6j), and MLO with ~260–320 ppt (Greenberg et al. 1996; Fig. 6k). The seasonal cycle at HPO showed a summer maximum with ~800 ppt, which might be the result of a local source (Sharma et al. 2000; Fig. 6i). However, based on data that eliminated the polluted air masses, Sharma et al. (2000) showed a distinct seasonal cycle of C_2H_2 with a maximum in winter and a minimum in summer. Propane (C_3H_8), with a lifetime of ~11 days, also shows a distinct seasonal cycle with a maximum in late winter–spring and a minimum in summer with ~720 ppt at JFJ (Fig. 6l), CMN with ~770 ppt (Fig. 6m), PMO with ~380 ppt (Helmig et al. 2008, 2015; Fig. 6n), HPO with ~790 ppt (Sharma et al. 2000; Fig. 6o), ~440 ppt at IZO (Schmitt and Volz-Thomas 1997; Fig. 6p), ~265 ppt at ASK (Fig. 6q), and ~140 ppt at MLO (Fig. 6r). The seasonal cycle at MKN showed two minima in April and October and two maxima in January and June at ~95 ppt (Fig. 6s). Benzene (C_6H_6), with a lifetime of ~10 days, showed a distinct seasonal cycle with a winter maximum and summer minimum at JFJ with ~105 ppt (Fig. 6t), CMN with ~150 ppt (Fig. 6u), IZO with ~90–110 ppt (Fig. 6v), and MLO with ~40–60 ppt (Fig. 6w). Generally speaking, the amplitudes of the seasonal cycles of C_2H_6 and C_3H_8 depend on the latitude, with smaller amplitudes at the lower-latitude sites and higher amplitudes at the higher-latitude sites (Fig. 7c, d). This feature is caused by a larger seasonal contrast of photochemical removals at higher latitudes than at lower latitudes. Li et al. (2005) showed that the concentrations of NMHCs, including C_6H_6 at ARO, were higher than those at JFJ because the contributions from the boundary layer emissions were generally higher at ARO due to the lower altitude. The annual maximum concentrations of C_2H_6 and C_3H_8 at higher-altitude stations are higher than those at lower-altitude stations within each latitudinal band (Fig. 7a, b).

NO_x and NO_y NO_x in the troposphere are important precursors for tropospheric O_3 and also affect OH and peroxy radicals (RO_x) concentrations, which govern the lifetimes of various gases in the troposphere (e.g., Levy 1971). In the troposphere, nitrogen species are primarily emitted in the form of NO_x from anthropogenic sources and are subsequently oxidized to other reactive nitrogen species along numerous different pathways (Fahey et al. 1986). The atmospheric lifetime of NO_x is generally in the order of hours to a day (Seinfeld and Pandis 2006). NO_y, which includes the sum of NO_x and its oxidation products such as PAN, dinitrogen pentoxide (N_2O_5), nitrous acid (HNO_2), HNO_3, and particulate nitrate (NO_3^-), plays an important role in the overall oxidizing capacity of the troposphere (Logan 1983).

In European countries, NO_x emissions declined significantly in the early 1990s due to the mandated installation of catalytic converters in vehicles and other emission reduction measures. Long-term measurements of NO_y, NO_x, and NO have been conducted at JFJ (Gilge et al. 2010; Pandey Deolal et al. 2012). For the period from 1998 to 2009, the overall annual averages of O_3 and NO_y were the highest in 2003, which was the hottest year in recorded European history caused by an exceptionally prolonged summer heat wave (e.g., Tressol et al. 2008). For the period from 1998 to 2002, the NO_y trend was positive but not significant, whereas in the period from 2004 to 2009, a significant decreasing tendency was observed at a rate of -0.048 ± 0.012 ppb year^{-1}. For the period from 1998 to 2009, no significant trend was found (Pandey Deolal et al. 2012). The comparison of PAN concentrations between the period from 1997 to 1998 and from 2009 to 2010 has indicated slightly lower values in recent measurements (Pandey Deolal et al. 2013).

NO_y measurements showed a broad spring and summer maximum at SNB, with a peak concentration of ~1.4 ppb (Kaiser 2009; Fig. 8f); ZUG, at ~1.5 ppb (Fig. 8g); and JFJ, at ~1.1 ppb (Gilge et al. 2010; Pandey Deolal et al. 2012; Zanis et al. 2007; Fig. 8h). On the other hand, NO_x showed a minimum during these seasons at SNB, with a peak concentration of ~0.6 ppb

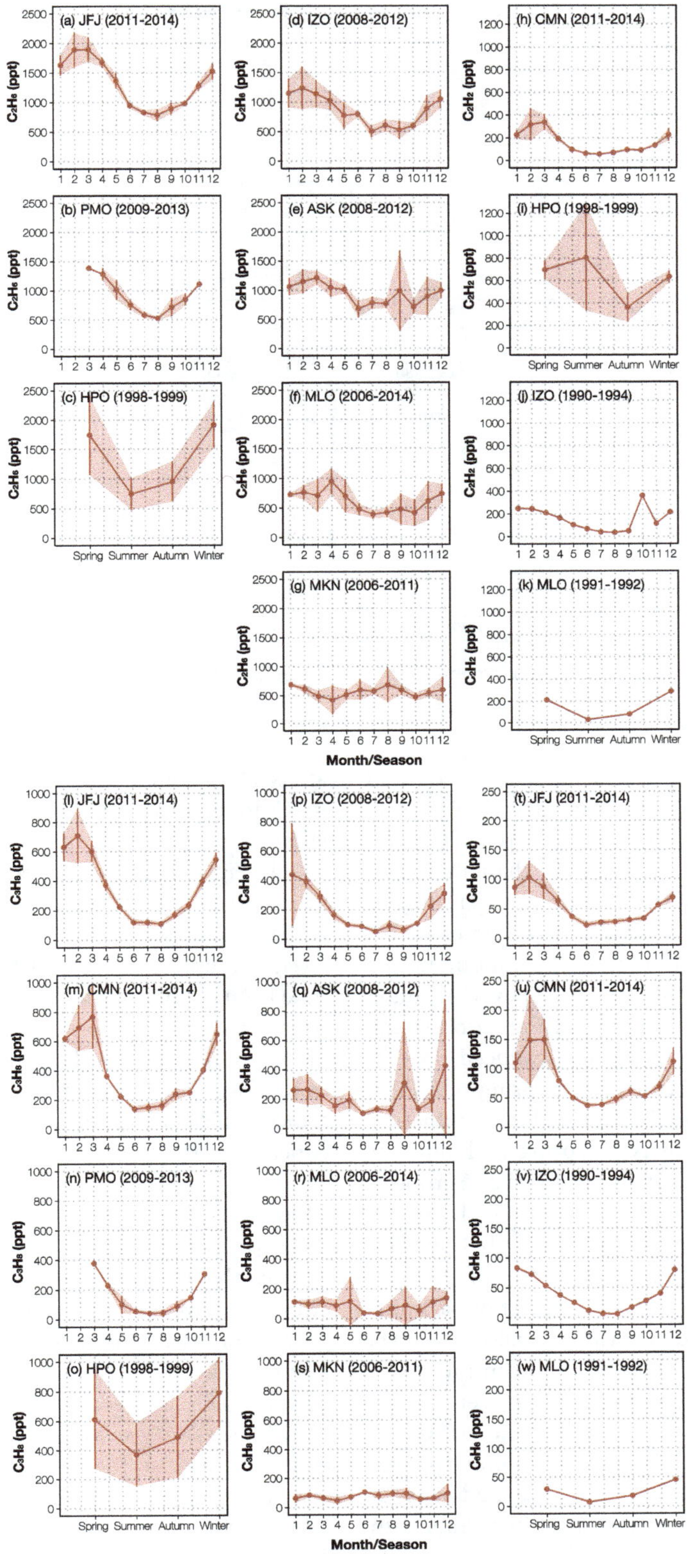

Fig. 6 Seasonal variations in C_2H_6, C_2H_2, C_3H_8, and C_6H_6 concentrations at eight mountain stations (**a**–**w**). The *lines with circles* denote the average NMHCs for observations during a particular period. The *years in parenthesis* denote the averaging periods. The *red shaded bands* and *error bars* denote the standard deviations

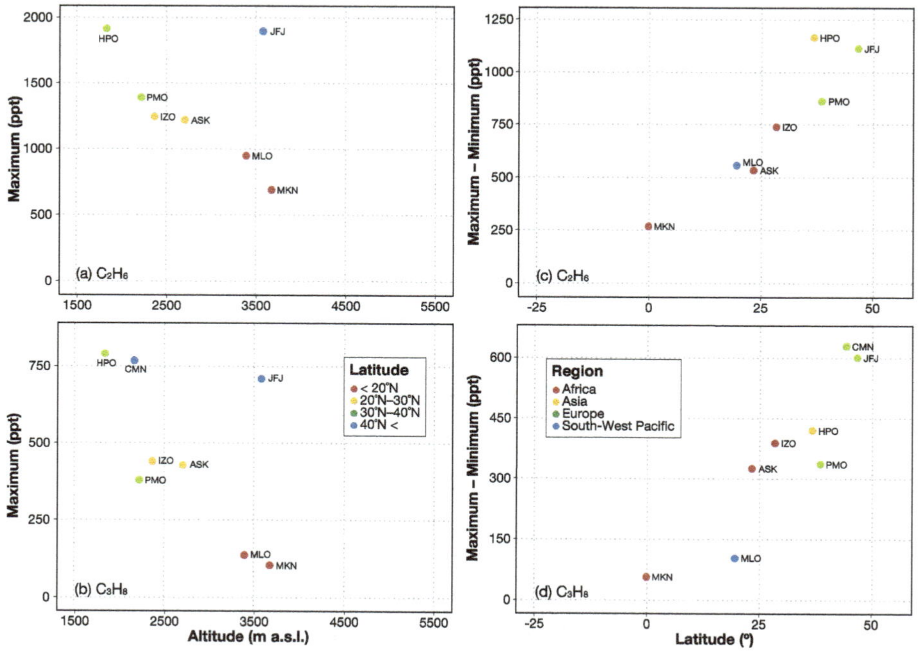

Fig. 7 Annual maximum concentration and range (maximum–minimum) of seasonal cycles of C_2H_6 and C_3H_8 observed at mountain sites with respect to latitude and altitude of station (**a**–**d**)

(Kaiser 2009; Fig. 8a); ZUG, at ~0.7 ppb (Fig. 8b); and JFJ, at ~0.4 ppb (Gilge et al. 2010; Pandey Deolal et al. 2012; Zanis et al. 2007; Fig. 8c). Because the NO_x lifetime is significantly shorter in summer than in winter (e.g., Schaub et al. 2007), NO_x concentrations are higher during the cold season. At JFJ, PAN measurements also showed a broad spring and summer maximum in the study of Pandey Deolal et al. (2013). They concluded that the mechanism of hydrocarbon accumulation in the FT in winter and the photochemically produced PAN in spring is not a dominant factor of the springtime PAN maximum. Rather, their analysis indicated a combination of strong photochemical activity in spring and summer with enhanced vertical transport from the European PBL. The seasonal cycle of NO_y at HPO showed a spring–early summer maximum at ~2.0 ppb (Fig. 8k). This cycle and level are similar to those of NO_y in Europe. At WLG, the NO_2 level did not show a minimum in summer in the study of Meng et al. (2010). Ma et al. (2002) also reported higher NO_x concentrations in July than in January. The NO_x and NO_y concentrations are low on remote islands such as IZO and MLO. At MLO, the seasonal cycle of NO_x, averaged only for free tropospheric observations, showed a spring maximum at ~0.04 ppt, and NO_y showed a maximum in spring, with a peak concentration of ~0.4 ppb (Ridley et al. 1998; Fig. 8d, h). NO_x/NO_y ratios were fairly invariant over the year, which implied that the partitioning of NO_y was near steady state (Ridley et al. 1998). NO_x observations at PMO exhibited a distinctive seasonal cycle with a July to October peak concentration of ~0.04 ppb that was larger than those in other months (Val Martin et al. 2008b; Fig. 8e). Val Martin et al. (2008b) concluded that in situ sources of NO_x such as PAN decomposition and potential photolysis of HNO_3 are required to provide NO_x levels because the lifetime of NO_x is shorter than its transit time from continental source regions. NO_y observations exhibited a distinct seasonal cycle with a peak concentration of ~0.4 ppb, which was larger from June to September than from November to March due to boreal forest wildfires and a more efficient export and transport from eastern North America (Val Martin et al. 2008a; Fig. 8j).

Long-range transport from continental sources

Anthropogenic emissions

During the past several decades, global anthropogenic emissions have increased markedly, following the dramatic growth of socioeconomic activity and energy use (e.g., Bond et al. 2007; Lamarque et al. 2010) particularly in China, which has become the largest contributor of Asian emissions. The growth rates for China's anthropogenic emissions are also the highest because of the unremitting increase in energy consumption, economic activity, and infrastructural development (e.g., Kurokawa et al. 2013). Trace gases and aerosols emitted from anthropogenic sources in East Asia have been observed at a number of mountain sites in Asia and North America including WLG (Cheng et al. 2007; Fu et al. 2012; Mu et al. 2007;

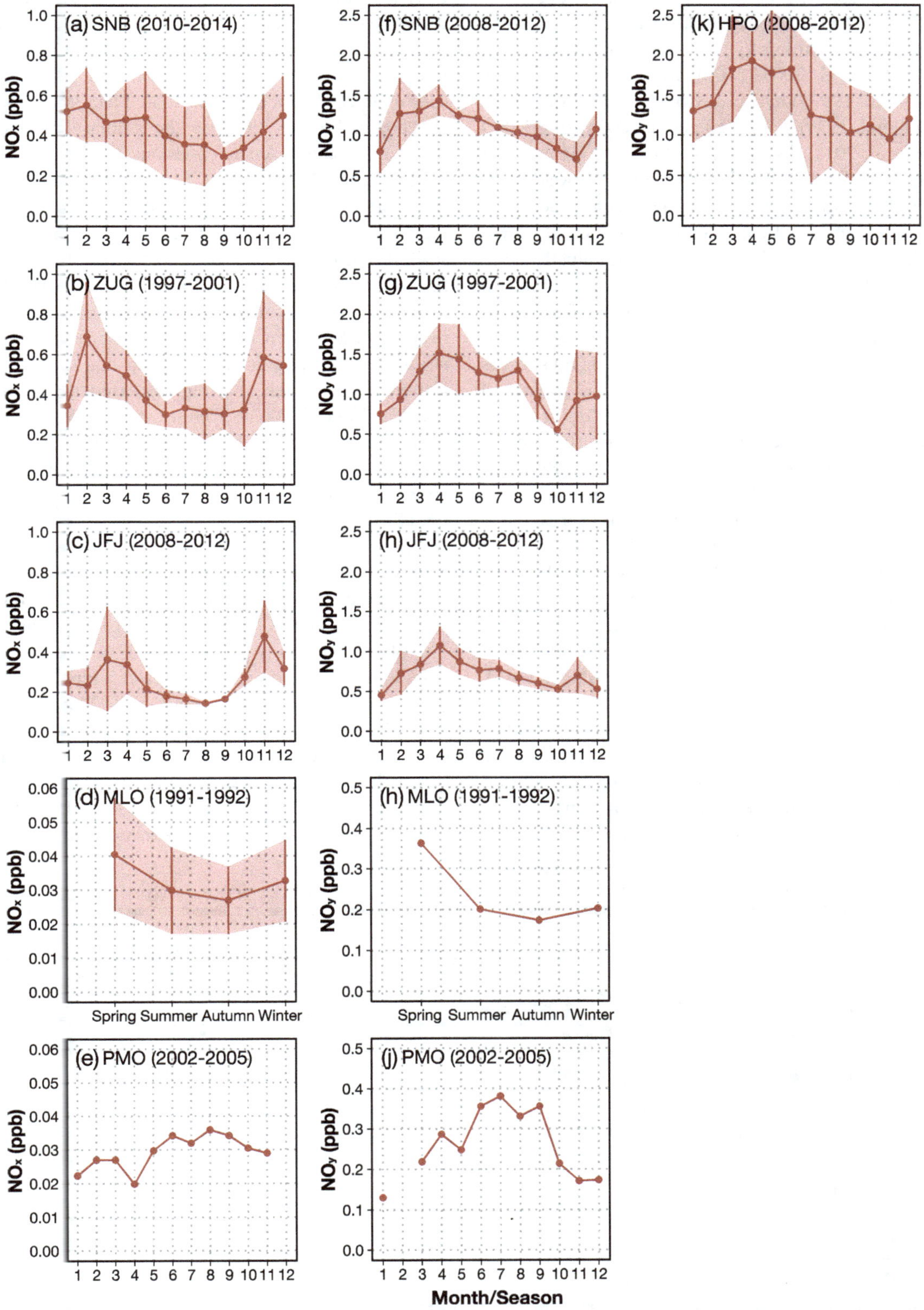

Fig. 8 Seasonal variations in NO_x and NO_y concentrations at six mountain stations. The *lines with circles* denote the average observed NO_x and NO_y concentrations during a particular period. The *years in parenthesis* denote the averaging periods. The *red shaded bands* and *error bars* denote the standard deviations. The concentrations of NO_x and NO_y were measured using photolytic (**a–e**), thermal Au (**f–j**), and Mo (**k**) convertors, followed by chemiluminescence detection

Wang et al. 2006; Xue et al. 2011, 2013; Zhang et al. 2009a, 2011), HPO (Liu et al. 2013; Pochanart et al. 2004; Sharma et al. 2000; Tanimoto 2009; Tanimoto et al. 2009), FWS (Igarashi et al. 2004, 2006), MLO (Brasseur et al. 1996; Jaffe et al. 1997; Levy and Moxim 1989; Luria et al. 1992; Perry et al. 1999), LAV (Jaffe et al. 2003a), and MBO (Ambrose et al. 2011; Fischer et al. 2010a, b; Genualdi et al. 2009; Jaffe et al. 2005; Primbs et al. 2008a, b; Reidmiller et al. 2009, 2010; Strode et al. 2008; Swartzendruber et al. 2006; Timonen et al. 2013; Weiss-Penzias et al. 2006, 2007; Wolfe et al. 2007). These emissions are illustrated in Fig. 9. HPO, which is located near the Sea of Japan, is an appropriate site for assessing the influence of air mass transport from the continent. Wild et al. (2004) estimated that local sources in Japan and regional East Asian sources contributed about 15 and 8 %, respectively, to O_3 observed at HPO in April 1996. The mass concentration of BC at HPO in a study by Liu et al. (2013) showed the highest value in spring and the lowest value in winter. By using the 3D Community Multiscale Air Quality (CMAQ) chemical transport model combined with the Process, Age, and Source Region Chasing Algorithm (PASCAL) source and process apportionment method, Liu et al. (2013) estimated that approximately 53 % of the observed BC was transported from northern China. Tanimoto et al. (2009) examined the contribution of Asian anthropogenic emissions to the observed O_3 trends at HPO by using CMAQ. However, a discrepancy was found between the observation and model results, thereby suggesting a significant underestimation of the actual growth of the Asian anthropogenic emissions or incompleteness of the modeling of pollution exported from continental Asia. MBO has proven to be effectively positioned for observing the transpacific transport of Asian plumes. Ambrose et al. (2011) described the detailed mechanism of Asian long-range transport to MBO. Surface emissions are lifted from the Asian boundary layer into the FT by convective, frontal, and orographic lifting (Bey et al. 2001; Cooper et al. 2004; Liang et al. 2004; Liu et al. 2003a). Frontal lifting is the primary mechanism for delivering eastern Asian emissions to the FT (Liu et al. 2003a). Once lifted to the FT, pollutants can be efficiently transported eastward toward North America by the prevailing westerly winds. Stohl et al. (2002) estimated the time scale of the trans-Pacific transport in the FT to be from 4 to 10 days. Subsidence wind over the northeast Pacific delivers air masses bearing Asian emissions to the lower FT over the US West Coast. This eastward export of Asian emissions is most efficient in spring due to intense frontal activity coupled with a strong westerly jet over the central Pacific (Liu et al. 2003a; Yienger et al. 2000). A calculation using the Goddard Earth Observing System chemical transport model (GEOS-Chem) revealed that, during the Intercontinental Chemical Transport Experiment Phase B (INTEX-B) period (Zhang et al. 2009b), most of the O_3 production in Asian O_3 pollution events occurred over East Asia, with maxima over northeast China and southern Japan in the spring of 2006. This model predicted significant O3 production over the North Pacific during plume transport. The plume on May 1 passed a more northerly and higher-altitude route than that on May 10, resulting in less O_3 production over the Pacific (Zhang et al. 2008). Both plumes showed a secondary O_3 production maximum at the West Coast of the USA, where the subsidence of air masses caused PAN decomposition and more O_3 production (e.g., Zhang et al. 2008). In addition to the trans-Pacific transport, regional pollution from North America has been observed

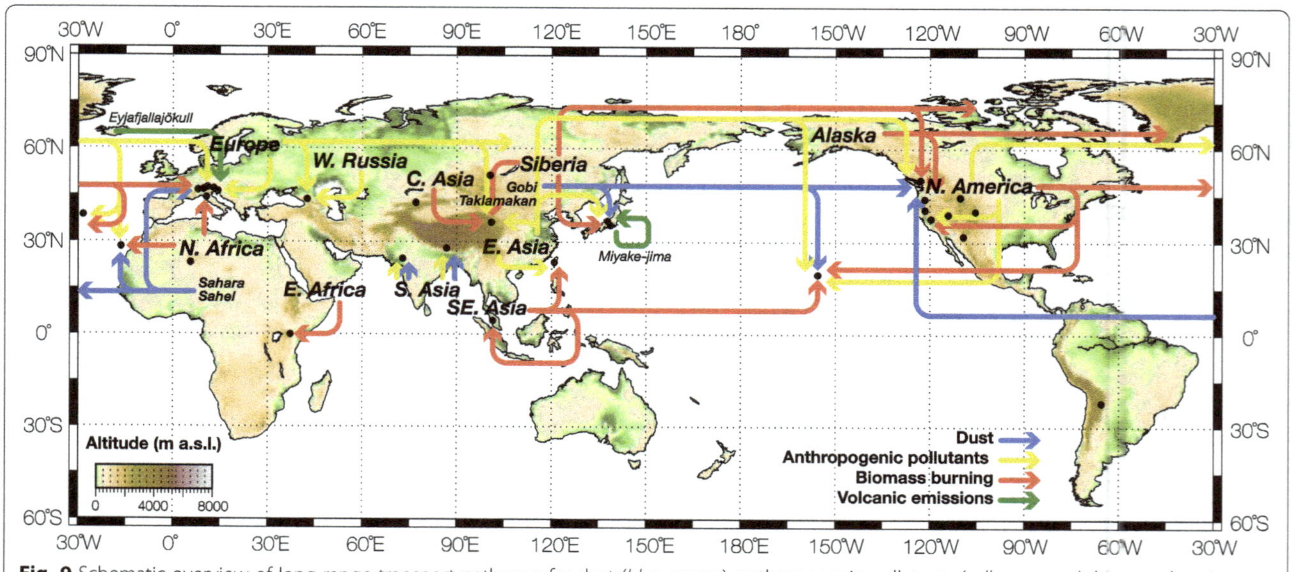

Fig. 9 Schematic overview of long-range transport pathways for dust (*blue arrows*), anthropogenic pollutants (*yellow arrows*), biomass burning (*red arrows*), and volcanic emissions (*green arrows*), which have been reported in the literature over time

at MBO (Fischer et al. 2010b; Primbs et al. 2008a, b; Reidmiller et al. 2010; Strode et al. 2008; Weiss-Penzias et al. 2006) and WHI (Macdonald et al. 2011). The influence of Asian anthropogenic emissions on MLO has been identified through observations of CO (Jaffe et al. 1997) and aerosols (Perry et al. 1999). Jaffe et al. (1997) concluded that the majority of the elevated CO events occurred during spring and were usually associated with transport from Asia. Perry et al. (1999) estimated that at least one third of the sulfate (SO_4^{2-}) species including sulfuric acid (H_2SO_4), ammonium bisulfate ($(NH_4)HSO_4$), and ammonium sulfate ($(NH_4)_2SO_4$) detected during spring from 1993 to 1996 were anthropogenic in origin. This result is consistent with the seasonal cycles of bromine (Br), lead (Pb), and zinc (Zn), high concentrations of which are found in the fine fractions of fly ash from coal-fired power plants. Levy and Moxim (1989) concluded that the NO_y maximum in the summer was caused by fossil fuel combustion in the USA, and that the Asian emissions contributed to only a small increase in spring.

Air masses affected by anthropogenic emissions have been observed at WLG by the enhancement of O_3, CO, NO_y, total gaseous mercury (TGM), and volatile organic compounds (VOCs). Backward trajectory analysis at WLG for the summers of 2000 to 2009 suggests the dominance of anthropogenic pollution transported from central and eastern China in most years (Xue et al. 2011). Fu et al. (2012) identified potential source regions of anthropogenic pollutants at WLG as eastern Gansu, eastern Qinghai, western Shanxi, Ningxia, and northern India, while Zhang et al. (2009a) described the transport mechanisms for polluted air from Lanzhou, the capital of Gansu province. They found that the plumes from Lanzhou were advected by strong easterly winds, which were vented to the FT by a belt-like mountain and were transported to WLG, either directly or by mixing, by nighttime mountain breezes. In addition, the high concentrations of organochlorine pesticides (OCPs) such as hexachlorocyclohexane (γ-HCH), dichlorodiphenyltrichloroethanes (DDTs), and polybrominated biphenyls (PBDEs) were found to be related mainly to air masses passing over Russia and Kazakhstan (Cheng et al. 2007). The particle number concentration at WLG was higher during the summer and lower during the rest of the year in the study of Kivekäs et al. (2009). Their back trajectory analysis revealed that air masses transported from the east are associated with higher particle number concentrations than those transported from other directions. As a result, they concluded that the eastern air masses might be affected by regional pollution sources such as Xining and Lanzhou.

The atmospheric conditions at PYR can be influenced by the transport of polluted air masses from South Asia and the Indo–Gangetic Plains and the ABCs extend from the Indian Ocean to the Himalayan ridge (Bonasoni et al. 2010). Bonasoni et al. (2010) showed that a mountain valley can provide an efficient channel favoring the transport of pollutants to high altitudes when the brown clouds extend as far as the Himalayan foothills under active valley breezes. They found that the most frequent direct transport of brown clouds occurred during the pre-monsoon season, occupying 20 % of total pre-monsoon days. Aerosol total number concentration, aerosol optical properties, PM_1, PM_{1-10}, and BC showed a maximum during the pre-monsoon season and a minimum during the monsoon season (Marcq et al. 2010; Marinoni et al. 2010; Sellegri et al. 2010). The monsoon minimum was attributed to wet scavenging caused by frequent clouds and precipitation. The pre-monsoon maximum was linked to an increase in anthropogenic emissions in the Indian subcontinent coupled with dry meteorological conditions, and to the large vertical extent of the regional PBL on the Indo–Gangetic Plains. At MAB, ammonium (NH_4^+), NO_3^-, and non-sea-salt SO_4^{2-} (nss-SO_4^{2-}) aerosol concentrations showed a distinctive seasonal cycle, with a decrease (increase) during the wet (dry) season (Rastogi and Sarin 2005). They attributed the low levels of NH_4^+, NO_3^-, and nss-SO_4^{2-} concentrations during the wet season to the predominance of the oceanic air mass and the high levels during the dry season to the enhanced downwind transport of pollutants and the lack of wet removal. The relative enrichment of NH_4^+ and nss-SO_4^{2-} and their linear relationship in $PM_{2.5}$ suggests that their predominant source is anthropogenic in northern India in winter (Kumar and Sarin 2010). Francis (2012) identified a high sulpher dioxide (SO_2) region surrounded by (30° N, 75° E) north of MAB by using the GEOS-Chem model and concluded that the observed spikes in SO_2 concentration at MAB were caused by the transport from that high-SO_2 region.

Emissions transported from North America have been observed at IZO (Cuevas et al. 2013; Schultz et al. 1998), PMO (Honrath et al. 2004), ARO (Huntrieser et al. 2005), ZUG/ZSF (Huntrieser et al. 2005), and JFJ (Guerova et al. 2006; Huntrieser et al. 2005; Pandey Deolal et al. 2013). Most of the elevated CO events observed at PMO during the summers of 2001 to 2003 were attributed to the outflow of North American pollution and the long-range transport of biomass-burning emissions, whereas O_3 and CO concentrations were strongly correlated with the flow from the eastern USA when the biomass-burning effects were absent (Honrath et al. 2004). The concentration ratios of NMHCs have been used as indicators of the degree of photochemical processing or aging, mainly via reaction with OH radicals, because the ratios evolve as atmospheric processing occurs depending on the individual OH rate constants (e.g., Parrish et al. (2007). Honrath et al. (2008) demonstrated that continuous measurements of NMHCs are a useful tool for identifying and characterizing pollution transport events. Owen et al. (2006) described the transport mechanisms of North American CO emissions to PMO in the lower FT over the central North Atlantic. They found that two mechanisms were responsible for the export of pollution from the North American boundary layer that led to the enhanced CO events at PMO: direct advection from the continental boundary layer eastward over the Atlantic Ocean and uplift ahead of cold fronts. In another study, NO_y levels were significantly increased in the presence of North American anthropogenic emissions (Val Martin et al. 2008a). Zhang et al. (2014) analyzed the chemical evolution of two pollution plumes transported from North America to PMO in the summers of 2009 and 2010. They concluded that the enhanced change in O_3 relative to CO ($\Delta O_3/\Delta CO$), which has been frequently used as an indicator of O_3 production in transported plumes, might not reflect O_3 chemistry alone, and that the CO destruction in the plume could also explain the higher values. Kumar et al. (2013) showed significant decreasing trends of CO and O_3 for the period from 2001 to 2011 by in situ measurements at PMO and model simulations using GEOS-Chem. They concluded that the most important factor contributing to these decreasing trends was the decline in anthropogenic emissions from North America. Under the framework of the Convective Transport of Trace Gases into the Middle and Upper Troposphere over Europe (CONTRACE) project, aged plumes of North American pollution that passed over Europe were observed by Huntrieser et al. (2005) at ARO, ZUG/ZSF, and JFJ. They used the FLEXible PARTicle (FLEXPART) Lagrangian particle dispersion model to simulate the transport pathway of the pollution plume over the Atlantic and its dispersion over Europe. The passage of the North American plumes over Europe required almost 3 days. The time elapsed since the pollutant emission, derived by the NO_y/CO slope in the polluted plumes, was in good agreement with that simulated by the FLEXPART age spectra. Furthermore, a positive correlation between O_3 and CO in the plumes suggested that photochemical O_3 production occurred. A comparison of the model results with observations at JFJ revealed that the GEOS-Chem model tended to underestimate the O_3 concentrations (Guerova et al. 2006). Because the stratospheric contribution to the estimates based on the $^7Be/^{10}B$ ratio was shown to be reasonable (Zanis et al. 2003b), Guerova et al. (2006) suggested that this discrepancy resulted from an underestimation of the European contribution. They found that a substantial North American contribution over Europe did not always result in O_3 enhancement in summer. Cui et al. (2009) estimated the contribution of stratospheric intrusion and intercontinental transport from the North American PBL to JFJ in 2005 by using FLEXPART and the Lagrangian analysis tool (LAGRANTO) trajectory model. They determined that JFJ was influenced by stratospheric intrusion during 19 and 18 % of the total annual time and by intercontinental transport from North American PBL during 13 and 12 % of the total annual time, as indicated by FLEXPART and LAGRANTO, respectively. At IZO, a good correlation between O_3 and CO in winter and spring was observed, suggesting long-range transport of photochemically generated O_3 from North America (Cuevas et al. 2013). Sholkovitz et al. (2009) estimated that from 12 to 30 % of the annual apportionment of the soluble aerosol Fe fluxes was contributed by anthropogenic aerosols. IZO has also been influenced by North African pollution. During the summer, IZO remains in the altitude of the Saharan Air Layer (SAL), which is exported westward at low tropical latitudes (<15° N) in winter and at higher subtropical latitudes in summer (15–30° N; Rodríguez et al. 2011). Rodríguez et al. (2011) suggested that North African industrial pollutants, including emissions from crude oil refineries, the phosphate-based fertilizer industry, and power plants located at the Atlantic coast of Morocco, northern Algeria, eastern Algeria, and Tunisia might be mixed with desert dust and before being exported to the North Atlantic in the SAL. Verver et al. (2000) reported pollution outbreaks both from the Iberian Peninsula and northern or central Europe to IZO during the Second Aerosol Characterization Experiment (ACE-2) campaign.

The Alps, located in the center of Europe, are surrounded by regions of intensive air pollutant emissions. Transport of anthropogenic emissions in Europe to the high-altitude Alpine stations have been observed at SNB (Gros et al. 2001; Jabbar et al. 2012; Kaiser 2009; Kaiser et al. 2007; Karl et al. 2001a, b; Seibert et al. 1998), ZUG/ZSF (Jabbar et al. 2012; Kaiser et al. 2007), CMN (Bonasoni et al. 1997; Cristofanelli and Bonason 2009, 2013), KVV (Kaiser et al. 2007), ARO (Campana et al. 2005; Pochanart et al. 2001), and JFJ (Collaud Coen et al. 2011; Forrer et al. 2000; Kaiser et al. 2007; Kuebler et al. 2001; Lanz et al. 2009; Legreid et al. 2008; Lugauer et al. 1998; Pandey Deolal et al. 2013; Seibert et al. 1998; Tuzson et al. 2011). The Po Basin is a densely populated and highly industrialized region located immediately adjacent to the Alps that is known to have a relatively high level of air pollution. Emissions transported from the Po Basin have been observed at ARO (Campana et al. 2005), CMN (Bonasoni et al. 1997; Cristofanelli and Bonason 2009), SNB (Seibert et al. 1998), and JFJ (Lanz et al. 2009; Seibert et al. 1998). By using trajectory-based methods, Kaiser et al. (2007) showed potential source regions of air pollutants and air flow regimes connected with pollution events at ZUG, SNB, JFJ, and KVV. They determined that the influence of emissions from the Po Basin was strongest in the west Alps (JFJ), whereas the influence of emissions from Eastern Europe was strongest at SNB. In another study, the influences of meteorological processes on the regional transport to JFJ were subdivided into two categories: thermally driven transport, which occurs on a local scale, and transport at regional (foehn) or synoptic (front) scales (Forrer et al. 2000). Thermally induced processes were observed mainly in spring and summer, and processes at the regional and synoptic scales were observed during the entire year. Continuous measurements of the total and backward scattering coefficients, absorption coefficient, condensation nuclei concentrations, epiphaniometer signals, total suspended particulate matter (TSP) mass concentrations, and major PM_1 and TSP compounds at JFJ have shown distinct seasonal cycles with a summer maximum and a winter minimum (Baltensperger et al. 1997; Cozic et al. 2008; Henning et al. 2003; Lugauer et al. 1998; Nyeki et al. 1998a, b). The seasonal cycles of SO_4^{2-}, NO_3^-, and NH_4^+ of aerosols at SNB also showed a summer maximum and a winter minimum (Kasper and Puxbaum 1998). During winter, JFJ and SNB are influenced mainly by the clean FT, resulting in low concentrations of most aerosol components. Higher mass concentrations have been found during summer when convection occasionally transports polluted air masses to JFJ and SNB. At different non-urban sites located along a west–east transect in Europe including SNB, Oliveira et al. (2007) showed year-round variations of lipophilic particulate organic compounds emitted from vehicle exhaust constituents, meat smoke tracers, phytosterols of higher photosynthetic plants, and wood smoke components. They determined that the concentrations at SNB maximized during summer, whereas an absence of seasonal variation was observed at the oceanic site. At ARO, the O_3 concentrations were found to depend significantly on the residence times of air masses over the polluted region of Europe during spring and summer (Pochanart et al. 2001). An increase in O_3 concentrations has been observed in the aged air masses, which is the result of direct photochemical accumulation of O_3 when the air masses pass over large-scale anthropogenically influenced regions in Europe. South foehn is a typical Alpine meteorological phenomenon characterized by south to north advection and is caused by a strong pressure gradient across the Alps that leads to a descending air stream northward (Seibert 1990). Campana et al. (2005) observed enhanced O_3 concentrations during south foehn events during spring and summer at ARO. In the case of the south foehn, the continental-scale influence of pollutant emissions on O_3 appeared to be far less important than the direct influence of the Po Basin emissions. The main sources of NO_x and CO are air masses with long residence times over the European continent (Kaiser et al. 2007). Kaiser et al. (2007) showed the differences between the main NO_x source regions, including northwest Europe and the region covering East Germany, the Czech Republic, and southeast Poland, and the main CO source regions including the central, northeastern, and eastern parts of Europe. The statistics for O_3 concentrations show strong seasonal effects. Near-ground air masses are poor (rich) in O_3 in winter (summer). The main source of high O_3 concentrations in winter is air masses that subside from higher elevations. During summer, the Mediterranean constitutes an important additional source of high O_3 concentrations, particularly for air masses that have crossed the Po Basin. The transport of polluted air masses from Europe and other continents can exert an influence on southern Europe and the Mediterranean basin. In particular, anthropogenic pollutants emitted from continental Europe are transported toward southern Europe and the Mediterranean basin during summer (e.g., Duncan et al. 2008; Henne et al. 2005). Based on modeled CO concentrations, Folini et al. (2009) quantified the regions of influence (ROIs) of 13 European rural background and high-altitude stations including SNB, CMN, ZUG/ZSF, and JFJ. The ROI definition of a measurement site is the area near the station from which surface emissions contribute substantially to the short-term variability of the measured trace gas concentrations. Folini et al. (2009) defined an individual site as one that roughly covers a circular area of 400 km in radius and said that large cities or other strong emission sources within 100 km from the site should be avoided. The seasonal cycle of BC at CMN was characterized by the presence of a winter minimum (December–January) and maxima in spring (April–May) and late summer (August–September), reflecting a higher efficiency of uplift during warmer months and the subsequent transport of polluted air masses from the boundary layer. In addition, a secondary minimum was observed in June, when wet deposition is expected to peak owing to maximum rainfall (Cristofanelli et al. 2013). They indicated that CMN was affected by air masses rich in anthropogenic CO during spring and summer. In addition, Cristofanelli et al. (2013) identified three main regimes of the seasonal correlation among CO, O_3, and BC at CMN: negative correlation between CO and O_3 during the period from October to December, significant positive correlation between CO and O_3 during the period from May to September, and almost no correlation between CO and O_3 and BC enhancements mainly associated with relatively old anthropogenic emissions during the period from January to April. They considered it likely that the negative correlation between O_3 and CO was attributed to the combined processes concurrent in enhanced O_3 with low CO (i.e., STE) and O_3 titration with NO in polluted air masses along with low photochemical activity. The seasonal cycle of O_3 at KSL in the Caucasus, as well as at other stations in the Alps, showed a broad spring–summer maximum. Although the aforementioned stations are under the effects of European emissions, particularly from Po Basin, KSL is influenced by semi-polluted air masses from the regions of the northern Caspian, southern Urals, and Volga, which causes photochemical O_3 production in these air masses, thereby resulting in seasonal O_3 maxima (Tarasova et al. 2003).

Biomass burning

Biomass burning is a significant global source of gaseous and particulate species emissions to the troposphere. Emissions from biomass burning are known to be an important source of GHGs and chemically active gases on regional and global scales (e.g., Andreae and Merlet 2001). Interannual variations in biomass burning can be dramatic, depending on natural and socioeconomic factors such as rainfall and political incentives to clear land. In some regions of the world, biomass burning occurs throughout the year, even though on the global scale it usually peaks in March and again in September (Duncan et al. 2003). The peak in December to April during the dry season in the Northern Hemisphere is predominately associated with burning in the tropical regions of North Africa and Southeast Asia, with much smaller contributions from Central America and Mexico and northern South America. The peak in August to October during the Southern Hemisphere dry season is associated with burning in South Africa and Brazil, with smaller contributions from Indonesia and Malaysia. Although the Indonesia and Malaysia region straddles the equator, the burning season generally occurs in the latter half of the year. Figure 9 shows a schematic overview of the long-range transport of trace gases and aerosols emitted from biomass-burning sources.

MKN is rarely influenced directly by African biomass burning emissions, which are most prominent in the western part of the continent. Henne et al. (2008) attributed the boreal summer maximum of CO at MKN to advection of Southern Hemispheric air loaded with emissions from biomass burning in South Africa. They analyzed fire counts and fire radiative power (FRP) recorded by Moderate Resolution Imaging Spectroradiometer (MODIS) sensors and concluded that the interannual variability of this summer maximum could be mostly explained by a combination of changes in transport patterns and biomass burning intensity.

Biomass burning is a seasonally important source of emissions in Central Europe (Lanz et al. 2010). Cristofanelli et al. (2009b) identified the transport of a biomass-burning plume with mineral dust from North Africa by means of air mass circulation analysis and aerosol chemical characterization. Cristofanelli et al. (2013) determined that biomass-burning effects on CMN were maximized during the warm months from July to September. Moreover, they analyzed the source regions of biomass burning events by using FLEXPART and identified transport at the local scale in the central Alps, North Africa, southern Italy, and Greece; that at the regional scale in Russia; and that at the global scale in North America and western equatorial Africa.

CO and BC influences from Siberian boreal forest fires have been observed at HPO (Kato et al. 2002; Liu et al. 2013; Pochanart et al. 2004), and aerosol influences have been observed at FWS (Kaneyasu et al. 2007). It is considered likely that biomass burning contributed to the mass concentration of BC values at HPO, particularly in spring (Liu et al. 2013). An increase in the observed mass concentration of BC values in April and May 2008 suggested substantial effects of very active biomass burning in Siberia. Wang et al. (2006) investigated the influence of biomass burning activities on the levels of trace gases at WLG. They concluded that forest fires in central Asia affected WLG observations during spring. In contrast, in eastern Siberia, forest fires occurred mainly during summer, and those air masses were rarely transported to WLG.

LLN has been influenced by Southeast Asian biomass burning activities in spring (Chang et al 2013; Chen et al. 2013; Chi et al. 2010; Lee et al. 2011; Lin et al. 2010; Ou Yang et al. 2012, 2014; Sheu et al. 2010). Spring is the major biomass-burning season in the Indochina Peninsula (Streets et al. 2003). Lee et al. (2011) determined that $PM_{2.5}$ at LLN was highest during the biomass-burning season, particularly in March. They also concluded, based on back trajectory analysis, that the observed high $PM_{2.5}$ levels were attributed mainly to the biomass-burning events. Backward trajectory analysis by Lin et al. (2010) showed that most air masses observed by LLN originated from India, the Indochina Peninsula, and southern coastal China, which together accounted for 85 % of the total trajectory. A different transport mechanism that favors CO transport from Indochina to the downwind Taiwan area was identified via a Weather Research and Forecasting (WRF) meteorological model simulation (Cheng et al. 2013). The results indicated the formation of a thermal forcing-induced low-pressure system in Indochina and that the high terrain located in the northern part of Southeast Asia further forced the uplift of biomass-burning emissions. Contrastingly, a sufficiently strong northeasterly monsoonal flow intruding into Indochina would hinder the development of the thermal low and would weaken the upward movement, which would, in turn, prevent the transport of biomass-burning emissions from Indochina to the Taiwan area. During the dry period of the year, the biomass burning in Sumatra normally begins around May and lasts until September or October. The mass concentration of PM_{10} at TAR in the study of Streets et al. (2003) showed a cycle similar to that of O_3, with two minima in March or April and November or December, and a maximum during the summer monsoon, thereby indicating that the O_3 and PM_{10} precursors had a common source. Toh et al. (2013) showed that the transport of air pollutants from biomass burning in Sumatra released large amounts of precursors and led to increases in O_3 and PM_{10}.

Occasionally, MLO has been influenced by biomass burning plumes transported from Southeast Asia and the Indian subcontinent across the Pacific (Karl et al. 2003), and from North America (Perry et al. 1999). Karl et al. (2003) measured the levels of methanol (CH_3OH), acetonitrile (CH_3CN), acetone (C_3H_6O), C_5H_8, methyl vinyl ketone (MVK), and methacrolein (MACR) in spring 2001 and found high correlations among CO, C_3H_6O, and CH_3CN during the event.

WHI has been influenced by regional biomass burning, particularly in northern California, Oregon, Alaska, and Yukon Territory (Macdonald et al. 2011; McKendry et al. 2010, 2011). MBO has also experienced numerous events associated with biomass-burning emission originating from Alaska (Weiss-Penzias et al. 2007) and the northwestern USA including California and Oregon (Finley et al. 2009; Primbs et al. 2008a, b; Weiss-Penzias et al. 2007). In July and August 2008, two regional biomass-burning events induced increases of CO and O_3 concentrations at both WHI and MBO (McKendry et al. 2011). Takahama et al. (2011) analyzed the organic functional groups of aerosols collected during those biomass-burning periods and showed a large contribution of ketones, at 26 % on average, to the organic aerosol mass in addition to sharp peaks of ketones attributed to plant wax. Macdonald et al. (2011) examined the influence of different transport and source regions on the O_3 and CO concentrations at WHI. They noted little difference between trans-Pacific air masses and the clean background in summer. The air masses from North America showed the greatest increases in O_3 and CO concentrations above the background values during years when the largest areas of western North America area burned. Finley et al. (2009) reported that regional wildfire events produced larger enhancement of CO (ΔCO), aerosol scatter coefficient ($\Delta\sigma_{sp}$), and NOy (ΔNO_y) values than the Asian long-range transport events due to the significantly shorter transport times from source to receptor, which meant less plume dilution, photochemical removal, and wet particle deposition. Jaffe et al. (2008) showed that the area of summer burning in the western USA was significantly correlated with daytime summer O_3 concentrations at LAV and YEL, which indicates that the presence of fire played an important role in increasing the summer O_3 in the western USA. Chalbot et al. (2013) indicated that regional fire incidents might trigger high O_3 episodes. They estimated a 3 ppb increase in the MDA8 O_3 concentration at CHA in Arizona by such incidents within 400 km of the sites. McKendry et al. (2010) emphasized the importance of permanent lidar installations in mountainous locations for understanding the vertical transport process and the analysis of the chemical time series.

In past decades, significant intercontinental transport of air masses containing abundant biomass burning emissions from North America to Europe and the North Atlantic have been reported in many studies. Emission transport from North American and Siberian boreal wild fires has substantially affected the concentrations of trace gases at PMO (Helmig et al. 2008; Honrath et al. 2004; Lapina et al. 2008; Pfister et al. 2006; Val Martín et al. 2006; Val Martin et al. 2008a, b). The NO_x/CO emission ratio is related to the relative amounts of smoldering and flaming combustion because CO concentrations typically show maxima during smoldering combustion while NO_x concentrations show maxima during flaming combustion (Lobert et al. 1991). The NO_x/CO emission ratios from boreal fires had higher values in early summer and lower values in late summer, which is consistent with the smoldering combustion variation (Lapina et al. 2008). Lapina et al. (2008) suggested that this change in fuel properties affects the results for the O_3 production rate in model calculations because it impacts the relative properties of the species released from fires and an increase in the overall fuel consumption. During summer 2004, extensive wildfires in Alaska and western Canada released large amounts of trace gases and aerosols into the atmosphere. For example, CO emitted from mid-June to August was in the order of anthropogenic CO emissions for the entire continental USA during the same period (Pfister et al. 2005; Turquety et al. 2007). Intense boreal wildfires plumes were observed at PMO during the International Consortium for Atmospheric Research on Transport and Transformation (ICARTT) study (Pfister et al. 2006; Val Martín et al. 2006). This long-range transport of boreal wildfire emissions resulted in large enhancements of CO, BC, NO_y, and NO_x (Val Martín et al. 2006). Emissions from these wildfires frequently affected PMO downwind 6 to 15 days later. The enhancement ratios relative to CO were variable in the plumes sampled due to variations in wildfire emissions and removal processes during transport. Analyses of ΔBC/ΔCO, $\Delta NO_y/\Delta$CO, and $\Delta NO_x/\Delta$CO ratios indicated that NO_y and BC were, on average, efficiently exported in these plumes, and suggested that the decomposition of PAN to NO_x was a significant source of NO_x. High levels of NO_x suggested the continuing formation of O_3 in these well-aged plumes. Pfister et al. (2006) simulated the O_3 production from boreal forest fires based on a case study of wildfires in Alaska and Canada in summer 2004 by using a chemistry transport model, the Model for O_3 and Related Chemical Tracers (MOZART-4). Their results indicate that fires in the boreal region could have a significant impact on the O_3 production over large parts of the Northern Hemisphere.

Dust and volcanic emissions

Every year, massive amounts of mineral dust are uplifted into the atmosphere (1950 to 2400 Tg year^{-1}; Ginoux et al. 2004). The desert region of North Africa (65 % of the global dust emissions) is the largest source of soil dust suspended in the atmosphere while Asia's deserts are the second largest source (25 %). Saharan dust has significant climatic and environmental impacts in the North Atlantic and in Europe (Fig. 9). Mineral dust is transported from North Africa along three pathways: westward across the Atlantic to the Americas, northward across the eastern Mediterranean to Europe, and eastward across the eastern Mediterranean to the Middle East and Asia (Middleton and Goudie 2001). Saharan dust events have often been observed at IZO (Alastuey et al. 2005; Arimoto et al. 1995; Díaz et al. 2006; Kandler et al. 2007; Maring et al. 2000; Rodríguez et al. 2011; Salisbury et al. 2006; Schmitt and Volz-Thomas 1997; Smirnov et al. 1998), CMN (Bonasoni et al. 1997, 2004; Cristofanelli and Bonason 2009; Cristofanelli et al. 2009b; de Reus et al. 2005; Gobbi et al. 2003; Hanke et al. 2003; Marenco et al. 2006; Putaud et al. 2004; Van Dingenen et al. 2005), and JFJ (Chou et al. 2011; Collaud Coen et al. 2004; Cozic et al. 2008; Fierz-Schmidhauser et al. 2010; Nyeki et al. 1998b; Schwikowski et al. 1995; Zieger et al. 2012). Furthermore, long-range transport of Saharan dust over a pathway spanning from Asia and the Pacific to western North America has been identified at WHI by McKendry et al. (2007).

Mineral dust affects the gas phase chemistry in the troposphere by providing a reactive surface that is able to support heterogeneous trace gas reactions (e.g., Usher et al. 2003). Hanisch and Crowley (2003) investigated the heterogeneous reaction between O_3 and Saharan dust surfaces in a Knudsen reactor and observed the destruction of O_3 on the dust surface and the formation of O_2. The phenomena of low O_3 and high dust concentrations have been observed in the field (e.g., Bonasoni et al. 1997; Prospero et al. 1995). As part of the Mineral Dust and Tropospheric Chemistry (MINATROC) project, which was an intensive field measurement campaign conducted at CMN from June 1 to July 5, 2000 (Balkanski et al. 2003; Bauer et al. 2004; Bonasoni et al. 2004; de Reus et al. 2005; Fischer et al. 2003; Gobbi et al. 2003; Hanke et al. 2003; Putaud et al. 2004; Van Dingenen et al. 2005) and at IZO from July 15 to August 15, 2002 (de Reus et al. 2005; Salisbury et al. 2006; Umann et al. 2005). Their aim was to fully characterize gas and aerosol species with a complete set of measurements in order to identify the heterogeneous reactions of dust. Bonasoni et al. (2004) showed that the O_3 concentrations during the dust events at CMN were 4 to 21 % lower than the de-trended monthly mean dust transport values, thus suggesting that Saharan dust has a substantial influence on the O_3 levels, even in a polluted region of Southern Europe. During the Saharan dust advection period, mixing of fine particles from anthropogenic particles and coarse particles from the Saharan region was observed at CMN (Putaud et al. 2004). Putaud et al. (2004) suggested that NO_3^- was shifted toward the aerosol super-μm fraction in the presence of dust. Hanke et al. (2003) and Umann et al. (2005) observed an efficient uptake of gaseous HNO_3 by mineral dust aerosol particles, thereby suggesting that mineral dust might be a significant sink for HNO_3 in the troposphere. Hence, it could affect the photochemical cycles and O_3 production. Bauer et al (2004) estimated the impact of heterogeneous chemical reactions at the surface of mineral dust on global O_3 chemistry using a global general circulation model, coupled online with atmospheric chemistry and mineral aerosol modules that were evaluated based on observational data during the MINATROC campaign. The model simulated a decrease in global tropospheric O_3 mass by ~5 % due to heterogeneous reactions, and implied that the most important heterogeneous reaction was the uptake of HNO_3 on the dust surface. Salisbury et al. (2006) observed the concentrations of the organic compounds CO, CH_3OH, formaldehyde (HCHO), acetaldehyde (CH_3CHO), C_3H_6O, C_3H_8, C_5H_8, and toluene ($C_6H_5CH_3$) under both normal conditions and during large dust storms, in order to investigate the effects of mineral dust on oxygenated organic species. However, no clear production for CH_3CHO or HCHO on dust particles was ascertained. By using the Module Efficiency Calculating the Chemistry of the Atmosphere (MECCA) chemistry box model, de Reus et al. (2005) concluded that the reduced RO_x and NO_x concentrations in the Saharan dust plume contributed to a reduced net O_3 production rate, which likely explains the observed relatively low O_3 concentrations.

After reaching CMN, Saharan dust plumes often move farther northward and eastward to eventually reach northern Italy, the Alps, and central Europe. At JFJ, Saharan dust events have been detected as a pronounced change in aerosol properties compared with the background conditions (Chou et al. 2011; Collaud Coen et al. 2004; Cozic et al. 2008; Fierz-Schmidhauser et al. 2010; Nyeki et al. 1998b; Schwikowski et al. 1995; Zieger et al. 2012). Collaud Coen et al. (2004) summarized the annual distribution of Saharan dust events in Europe, the Mediterranean basin, West Africa, and over the Atlantic Ocean and explained that Saharan dust events frequently occurred during spring and less frequently in July and August. Figure 10a shows the seasonal variation of Ca^{2+} (a component of mineral dust) concentration in total suspended particles (TSP) at JFJ. Associated with relatively large monthly variability caused by occasional intense events, seasonal variations are not very distinct. However, the annual maximum is seen in February (~0.10 μg m^{-3}) associated with large variability.

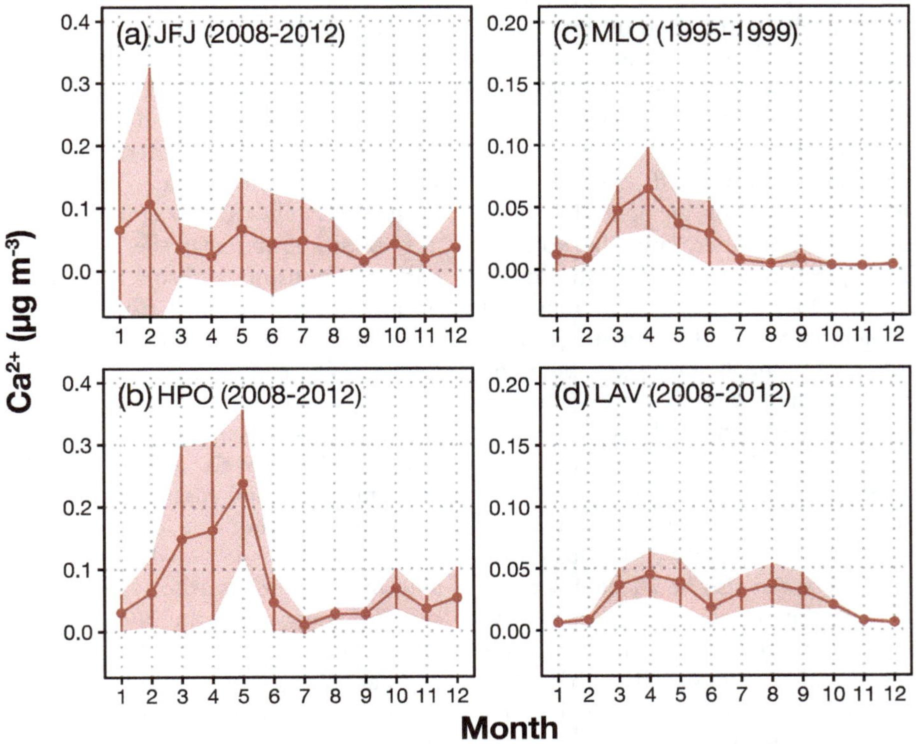

Fig. 10 Seasonal variations in Ca^{2+} concentrations at four mountain stations. The *lines with circles* denote the average observed Ca^{2+} concentrations during a particular period. The *years in parenthesis* denote the averaging period. The *red shaded bands* and *error bars* denote the standard deviations. The concentrations of Ca^{2+} in TSP were measured at JFJ (**a**) and HPO (**b**) and those in $PM_{2.5}$ were measured at MLO (**c**) and LAV (**d**)

Asian dust particles known collectively as "Kosa" are generated when the surface soil particles in the arid and semi-arid regions of inland Asia (such as Taklamakan and Gobi deserts) are lifted by the winds and transported over long distances by strong westerly winds. Intense Asian dust events are observed each spring in China, Korea, Japan, and even occasionally in Hawaii and North America (Fig. 9). This phenomenon normally occurs in spring. Weak dust layers occur in the FT over Japan during periods with no evident dust outbreaks or even in seasons other than spring (e.g., Iwasaka et al. 1983). Asian dust events have been observed at FWS (Suzuki et al. 2008, 2010), MLO (Bodhaine 1995; Clarke and Charlson 1985; Darzi and Winchester 1982; Holmes and Zoller 1996; Hyslop et al. 2013; Parrington et al. 1983; Parrington and Zoller 1984; Perry et al. 1999; VanCuren and Cahill 2002; Zieman et al. 1995), LAV (Jaffe et al. 2003b; Liu et al. 2003a; VanCuren 2003; VanCuren and Cahill 2002; VanCuren et al. 2005), and WHI (Leaitch et al. 2009; McKendry et al. 2008). A distinct seasonal cycle with a spring maximum of Ca^{2+} concentration can be seen at HPO (~0.25 μg m^{-3}; Fig. 10b) and MLO (~0.05 μg m^{-3}; Fig. 10c). At LAV, two peaks can be seen in spring (~0.05 μg m^{-3}) and summer (Fig. 10d). The concentrations at two IMPROVE stations (MLO and LAV) and at HPO were determined by $PM_{2.5}$ and TSP analyses, respectively. VanCuren and Cahill (2002) found that the tropical wind bands were displaced to the south of Hawaii and that strong westerly circulation can reach the islands to deliver Asian aerosols, particularly at MLO from December to May. Other studies have reported that in January or February, many elements associated with continental crust experienced an increase in the mass of samples. The aerosol concentration of crustal elements remained elevated until June or July, suggesting that the dust season is from late winter to mid-summer (Parrington et al. 1983; Zieman et al. 1995). VanCuren et al. (2005) concluded that Asian aerosols influence LAV throughout spring, summer, and fall.

Mineral dust is a major component of PM_{10} observed at PYR by Decesari et al. (2010). They suggested that the increase in mineral components in the pre-monsoon season is due mainly to an increase in the background concentration. This increase is likely related to changes in their long-range transport from the arid regions of western India, possibly to upwind regions such as the Arabian deserts. Aerosols at MAB are dominated by mineral dust (Kumar and Sarin 2009, 2010; Ram et al. 2008; Rastogi and Sarin 2005). Kumar and Sarin (2009) showed that the mass concentration of $PM_{10-2.5}$ at MAB is influenced by the uniform contribution of dust (60 to 80 %) throughout the year. The predominant of sources of MAB are the Thar and Arabian deserts during summer months and the Indo–Gangetic Plain during winter months (Kumar and Sarin 2009).

Volcanic emissions such as ash clouds, aerosols, and volcanic gases often affect local or regional air quality (e.g., Huebert et al. 2001). Furthermore, the volcanic aerosols released by explosive eruption affect the stratospheric aerosol layer (Deshler 2008), which is important for radiation (e.g., Solomon et al. 2011) and atmospheric chemistry, particularly O_3 depletion (e.g., Solomon 1999). In the later part of the twentieth century, the signatures of major volcanic eruptions such as El Chichón in 1982 and Mt. Pinatubo in 1991 were identified in the total O_3 records at ISK (Visheratin et al. 2006) and ARO (Rieder et al. 2010a, b) and in the total column NO_2 records at JFJ (De Mazière et al. 1998). The oxidizing capacity of the troposphere can also be affected because the amount of UV radiation reaching the troposphere decreases due to the enhanced stratospheric aerosols and SO_2. Bândă et al. (2015) showed that the sulfur emissions from Pinatubo significantly affected the CH_4 budget using a 3D global chemistry model.

In spring 2010, eruptions of Eyjafjallajökull volcano (63.63° N, 19.62° W) in Iceland had a major impact on aviation because ash from the April–May summit eruption was transported with northwesterly winds toward Europe and the North Atlantic. The volcanic plume resulting from that eruption was observed at CMN (Sandrini et al. 2014), KVV (Beeston et al. 2012), ZSF (Flentje et al. 2010; Schäfer et al. 2011), and JFJ (Bukowiecki et al. 2011). Moreover, enhanced PM_{10} and SO_2 concentrations were observed at ZSF and JFJ. Monthly mean SO_2 concentrations in the spring of 2010 at JFJ were higher than those in other years (Fig. 11a). These observations suggest that the large volcanic eruption had a significant impact on the sulfur cycle on a regional scale.

On July 8, 2000, Miyake-jima volcano (34.09° N, 139.53° E), located in the northwest Pacific Ocean, began to erupt. In winter, volcanic emissions were transported southeastward to the Pacific Ocean due to the prevailing northwesterly wind, while in summer volcanic emissions were transported southwestward to Japan, Korea, and Taiwan due to the subtropical high-pressure system (Kajino et al. 2004). At FWS, Naoe et al. (2003) showed that the large fraction of the SO_4^{2-} concentrations in aerosols was most likely produced in the volcanic plume from Miyake-jima. At HPO, aerosols, precipitation, and gaseous pollutants including O_3, SO_2, NO, and NO_2 were observed prior to the eruption, and short-term aerosol sampling was conducted by Satsumabayashi et al. (2004). They reported unchanged O_3 levels, a fourfold enhancement of SO_2 concentration, and increased SO_4^{2-} and NH_4^+ levels in aerosols following the eruption. When the volcanic plume was transported directly, the atmospheric SO_2 and SO_4^{2-} concentrations in the aerosols increased simultaneously, and the NO_3^- and chlorine (Cl^-) concentrations decreased. Monthly mean SO_2 concentrations at HPO for the period from the summer of 2000 to the spring of 2001 were higher than those in other years (Fig. 11b). Kajino et al. (2005) quantitatively estimated the indirect acidification of nonvolcanic NO_3^- and Cl^- as a result of the increase in SO_4^{2-} aerosol concentration from the eruption. They concluded that the contribution of indirect acidification could not be negligible and that this effect should be considered in general air pollution because SO_2 emissions could be expected to continue increasing—especially in Asia.

Recommendations for future atmospheric monitoring

This paper has provided a comprehensive review on the seasonal and long-term variations and long-range transports of tropospheric O_3 and its precursors based on past observations made at mountain sites. The main

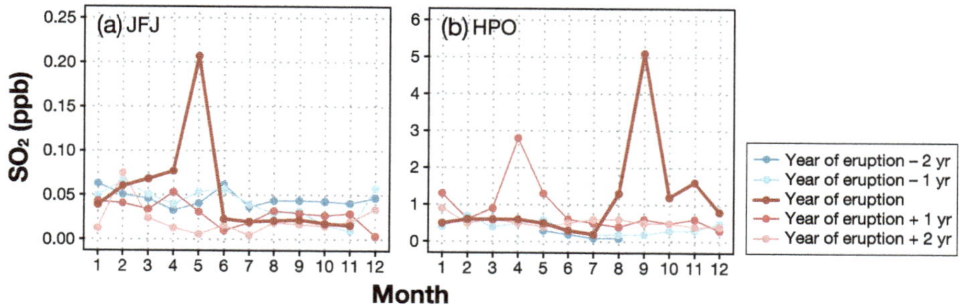

Fig. 11 Seasonal variations in SO_2 concentrations at two mountain stations. The *thick* and *thin lines with circles* denote the monthly average SO_2 in the year of the volcanic eruption and in the pre- and post-eruption years, respectively. The eruptions of Eyjafjallajökull volcano and Miyake-jima volcano occurred in 2010 (**a**) and in 2000 (**b**), respectively

recommendations for future atmospheric monitoring are summarized as follows:

1) Expanding the continuous measurements of trace species other than O_3 at high-altitude stations.

The observed species and temporal coverage differ among stations and research programs for which individual stations are operated. For example, O_3 and CO data are available, even though the data of other components are limited. Continuous measurements of NO_x and NMHCs are carried out mainly in Europe. For example, the number and mass concentrations, AOD, and visibility of aerosols have been reported extensively, but observations of their chemical properties are currently limited to high-altitude stations. Because of this bias in observations of aerosol parameters at high-altitude stations, it is still unfeasible to provide a regionally consistent summary on their long-term trends and seasonal variations. Accordingly, observations of trace species and parameters ancillary to tropospheric O_3 such as NO_x, SO_2, NMHCs, and aerosols should be increased in the future.

2) Expanding the global atmospheric observational capability through the establishment of high-altitude stations.

The geographical locations of the current high-altitude stations are not well distributed across the globe. Many are located in the northern mid-latitudes including North America, Europe, and Asia, and several are located in Africa. In contrast, few stations are located in Siberia, in spite of the importance of its role. In the Southern Hemisphere, there are severe data limitations because only a few sites are located in South America; and none are located in Oceania regions. Recently, two new stations were established in the South Hemisphere. One is located on Mt. Mäido, Réunion Island, which is a volcanic island located in the southwestern part of the Indian Ocean (Baray et al. 2013). The second is located in a high-altitude basin known as the Altiplano, Bolivia (Rose et al. 2015). Sofen et al. (2016) indicated that the current monitoring network of surface O_3 (including low-altitude stations) is insufficient for measurements in key regions (China, India, Amazon, Africa, tropical oceans, and Southern Ocean) and suggested the establishment of 20 possible sites (10 island and 10 continental). A strategy should be implemented to facilitate observations of key species in areas in which few observatories are located. It should also be noted that continuous observational data at mountain sites are useful for checking the consistency between sondes and regular aircraft, which cannot perform continuous ambient sampling (Tanimoto et al. 2015), when making globally consistent datasets for tropospheric species such as tropospheric O_3.

3) Building an objective and homogenous classification for assessment of the spatial representativeness of specific pollutants.

Remote mountainous sites are important because they are designed to provide representative data on continental to hemispheric scales and data for the lower FT. However, except for some European stations, most sites have never been assessed. In general, chemical species with strong surface sources or sinks, and with short atmospheric lifetimes due to photochemistry and deposition, show stronger spatial variability. Therefore, these species have smaller areas of representativeness than species with weak surface fluxes and long lifetimes. It is beneficial for any analysis, such as validation of satellite data, comparison of model outputs, and evaluation of long-term trends to consider the representativeness of atmospheric observational data.

4) Building a filtering method that can be applied to the observational data at baseline sites in the same way.

Filtered data are the most ideal for the analysis of long-term changes. Depending on the purpose, different filtering methods should be applied. For the estimation of "background" concentrations, the influences of local pollution (e.g., buoyant upslope flow) and long-range transport should be excluded, while for the estimation of "baseline" concentrations, only local pollution should be excluded. In most cases, unfiltered data have been used for trend analysis on regional or global scales because it is difficult to make determinations on baseline concentrations at individual stations (e.g., Logan et al. 2012; Parrish et al. 2012). The specifications of each observational site should be determined before handling the data, which primarily represent the regional and intercontinental scale concentrations. For example, Logan et al. (2012) confirmed that local influence was not of great concern for their analysis based on the amplitudes of diurnal variation and comparison to ozone sonde data. Meanwhile, Parrish et al. (2012) showed that the long-term trends derived from the unfiltered data were similar to those from filtered data analyzed by different methods at seven elevated sites including JFJ, ZUG/ZSF, and LAV, even though the trend for unfiltered data was different from that for filtered data at Mace Head, which receives air masses recirculated from the local continent. As the influences of local pollution levels were negligible, these two studies examined the trends for baseline concentrations. However, this does not mean that filtering is unnecessary. Calculated trends for CO at JFJ for the period from 1996 to 2007 were different between background data (-2.65 ± 0.04 ppb $year^{-1}$; Zellweger et al. 2009) and unfiltered data (-3.36 ± 1.08 ppb $year^{-1}$; Gilge et al. 2010).

It is important to consider all of the above recommendations because observational data at remote sites can contribute to improvements in satellite observations, model simulations, and emission inventories. Observations at mountain sites have often been used for comparison with the outputs of global and regional chemistry-transport models when evaluating their performance levels. This tests our understanding of key processes and mechanisms in the troposphere such as the long-range transport of dust and biomass burning from the source regions, photochemistry in the FT, and the budget of reactive species. Although chemistry-transport models are useful for obtaining a better understanding of the mechanisms causing spatiotemporal variations in trace gases and aerosols, and for deriving their global/regional budget and assessing the impacts to human health and ecosystems, the current models still have limitations (e.g., Fiore et al. 2014). Disagreements between observation and model results might lead to improvements in existing models, although such guidance remains unclear. In addition, the role of mountain-based observations in filling the gap between the views from the ground and those from space should be enhanced in the future. Satellite observations have emerged as a useful technique for the long-term monitoring of tropospheric compositions of species such as O_3, CO, and NO_2 (e.g., Chehade et al. 2014; van der A et al. 2008; Worden et al. 2013). Its capability to make reliable observations in the FT has been demonstrated over the last decades. The spatial and temporal resolution of satellite observations has been improved in recent years, and the number of observable species has increased from basic chemical species such as O_3 and CO to other minor but important species such as NO_2, HCHO, and CH_3CN (Streets et al. 2013). It is important to expand the mountain-based platforms in a systematic manner to complement satellite observations, and thus to improve our understanding of the chemistry and dynamics in the FT.

The greatest advantage of models and satellite observations is their global coverage. However, this fact does not necessarily indicate a lack of interest in mountain-based observations. Indeed, observations in mountainous regions have several advantages. The uncertainty is significantly lower for in situ observations at mountainous sites than that for satellites. In addition, mountain-based observations have the capability to observe multiple components simultaneously with a higher time-resolution than is possible by satellites, and in most cases, those measurements are continuous. This advantage enables the capturing of occasional events as well as climatological features, which helps us to examine the chemical and physical processes in greater detail. Moreover, direct sampling for the analysis of ultra-trace components such as isotopes is possible at mountainous sites. Therefore, even though it is difficult to construct, operate, and maintain stations on remote mountains, the assistance they provide in tackling the above issues makes the effort and expense worthwhile.

Conclusions

Herein, we summarized the findings from past and ongoing field measurements and analyses of atmospheric constituents conducted at mountainous sites across the globe. Observations at mountain-based stations could provide representative data at continental to hemispheric scales and the lower FT. Many species in the FT have longer lifetimes than those in the PBL due to the lower temperatures and the lack of deposition in the FT. In addition, most of the transport of chemical species within the atmosphere occurs in the FT. The combination of long-range transport and longer chemical lifetimes indicate that the chemistry of the FT is important in determining the chemical compositions of regions remote from pollutant source regions. We identified observations at 31 mountain stations on the continents of Europe, North and South America, Eurasia, and Africa and islands in the Pacific, Indian, and Atlantic Oceans. These stations were selected because of their reliable observational data and because many previous studies have utilized these data. Furthermore, we adopted the stations located in regions in which few atmospheric observations have been conducted thus far, thereby covering the entire world, except for the Antarctic and the Arctic regions. Particular emphasis was placed on reactive species including tropospheric O_3 and its precursors such as CO, NO_x, NO_y, and NMHCs.

Numerous studies have shown that the tropospheric O_3 maximum occurs during spring in wide regions across the mid-latitudes in the Northern Hemisphere, including the remote islands in the North Pacific and North Atlantic Oceans. The seasonal cycle of O_3 at European stations has a minimum in winter and a broad maximum in spring–summer that represents a combination of a spring maximum with a secondary maximum in summer. This broad summer maximum is caused by enhanced regional pollution in Europe. Largely influenced by the Asian monsoon, the seasonal cycle of O_3 in the margin of East Asia and Himalayan region has a deep minimum in summer. In contrast, surface O_3 over the Qinghai–Tibetan Plateau and central Asia exhibits a summertime maximum. In North America, the seasonal cycle of O_3 shows a broad spring–summer peak or two peaks in spring and summer. The seasonal cycle of O_3 in Africa shows a minimum in boreal autumn and a broad maximum in boreal summer. In the Northern Hemisphere, the mean concentration of O_3 and the amplitude of the seasonal cycle depend on latitude, with larger values at higher latitudes. The mean concentrations are ~50 ppb in Europe and 30 to 45 ppb in Africa. The tropical sites differ from other mid-latitude sites, with small variability shown over the course of the year. The seasonal amplitudes in East Asia and South Asia are substantially larger than those found at other stations, likely due to the intense photochemical production in spring and the Asian monsoon influence in summer.

In the Northern Hemisphere, the seasonal maximum and minimum for CO occur in late winter–early spring and in summer, respectively. The summer CO minimum is caused by the maximum occurrence of OH radicals, which are the main sink for CO in the troposphere. The CO maximum in late winter–early spring reflects the combined effects of the minimum in OH radical concentration and the enhancement of winter CO emissions. The maximum and minimum concentrations of CO and the amplitude of the CO seasonal cycle depend on latitude, with lower values found at higher altitudes. This indicates that the emission sources for CO are located at the ground surface. The means are between 120 and 180 ppb in Europe and Asia and 100 ppb in Africa. The maxima in East Asia are higher than those at other sites owing to intense springtime air pollution. The seasonal amplitude at the 15 stations falls into the range of 85 to 180 ppb, except for LLN in Taiwan. This large seasonal amplitude for CO can be attributed to the combination of a high spring maximum and summer low, at ~215 ppb, caused by the effect of the Asian monsoon and biomass burning activity in Southeast Asia.

Unlike O_3 and CO, NMHCs, NO_x, and NO_y have only been observed continuously at very few stations. The seasonal cycles of C_2H_6, C_2H_2, C_3H_8, and C_6H_6 in the Northern Hemisphere generally show a maximum in winter and a minimum in summer owing to reaction with the OH radical. The seasonal cycle at MKN in the equatorial region shows a different cycle with two minima and two maxima corresponding with the climate pattern in equatorial East Africa induced by the displacement of the ITCZ. Although NMHC concentration ratios are useful indicators of the degree of photochemical processing or aging (e.g., Helmig et al. 2008), such studies have been rarely conducted at mountainous sites. Generally, the amplitudes of the seasonal cycles of C_2H_6 and C_3H_8 depend on the latitude, with smaller amplitudes at the lower-latitude sites and higher amplitudes at the higher-latitude sites. The annual maximum concentrations of C_2H_6 and C_3H_8 at higher-altitude stations are higher than those at lower-altitude stations within each latitudinal band.

NO_y measurements show a broad spring and summer maximum, whereas NO_x shows a minimum during these seasons at SNB, ZUG, and JFJ. The seasonal cycle of NO_y at HPO show a spring and early summer maximum. The NO_x and NO_y concentrations are low on remote islands such as IZO and MLO. At MLO, the seasonal cycle of both NO_x and NO_y shows a spring maximum, but their concentrations are fairly invariant over the year, which implies that the partitioning of NO_y is almost steady state. NO_x observations at PMO exhibit a distinct seasonal cycle from July to October that is larger than those in other months, while NO_y observations exhibit a distinct seasonal cycle which is larger from June to September than from November to March, primarily due to boreal forest wildfires and more efficient export and transport from eastern North America. Despite the importance of the photochemistry of tropospheric O_3, continuous measurements of these species, which are precursors and by-products of O_3 photochemistry at high-altitude stations, are very limited.

The observations at European stations in the 1990s show increases in tropospheric O_3. The O_3 concentrations at these stations peak in the late 1990s or early 2000s before exhibiting a decreasing trend. The long-term trends for surface O_3 observed at SNB, ZUG/ZSF, and JFJ are 0.22 ± 0.17 ppb year^{-1} (1989–2007), 0.42 ± 0.09 ppb year^{-1} (1975–2007), and 0.26 ± 0.16 ppb year^{-1} (1986–2007), respectively. The mean rate of O_3 decrease at the seven European alpine sites including SNB, ZUG/ZSF, and JFJ for the period from 2000 to 2009 is -0.16 ± 0.14 ppb year^{-1}. In the North Atlantic, the surface O_3 record at IZO shows a slightly positive trend with a rate of 0.09 ± 0.02 ppb year^{-1} for the period from 1988 to 2009. In particular, a rapid increase is found between 1996 and 1998 that coincides with the phase shift of the NAO. In East Asia, a positive trend of surface O_3 at 0.25 ± 0.17 ppb year^{-1} is observed at WLG for the period from 1994 to 2013. Of particular note, the strongest increase is found in autumn. HPO in Japan experiences a generally large increase in spring O_3 of ~1 ppb year^{-1} for the period from 1998 to 2006. In the western USA, the mean spring O_3 at MBO shows a significant increase of 0.73 ± 0.54 ppb year^{-1} for the period from 2004 to 2013. At HPO and MBO, which are significantly influenced by East Asian outflow in spring, larger increasing trends in the levels of O_3 occur in higher percentiles than in lower percentile levels.

In the Northern Hemisphere, the CO levels observed at mountainous sites increase from the 1950s until the 1980s, and then decrease beginning in the late 1980s. The decreasing rate of the global mean CO concentrations was estimated to be -0.52 ± 0.10 ppb year^{-1} for the period from 1991 to 2001. The trend for CO at JFJ for the period from 1996 to

2007 is about –3 ppb $year^{-1}$. At MBO in the western USA, the mean spring CO shows a significant decrease for the period from 2004 to 2013, whereas the annual average CO did not show a significant trend. This result is consistent with satellite observations of total column CO, which show a continued decreasing trend from 2000 through 2011 in the Northern Hemisphere.

There are few long-term measurements of NO_y and NO_x at high-altitude stations. In European countries, NO_x emissions declined significantly in the early 1990s due to the mandated installation of catalytic converters in vehicles and other emission reduction measures. For the period from 1998 to 2002, the trend in NO_y was positive but not significant, whereas in the period 2004 to 2009, a significant decreasing tendency was observed at a rate of -0.048 ± 0.012 ppb $year^{-1}$. The overall annual averages of O_3 and NO_y were the highest in 2003, which was the hottest year in recorded European history due to an exceptional prolonged summer heat wave.

List of stations

ARO, Arosa
ASK, Assekrem
CHA, Chiricahua National Monument
CMN, Monte Cimone
FWS, Mt. Fuji
GRB, Great Basin National Park
HPO, Mt. Happo
ISK, Issyk-Kul
IZO, Izãna
JFJ, Jungfraujoch
KSL, Kislovodsk
KVV, Krvavec
LAV, Lassen Volcanic National Park
LLN, Lulin
LQO, La Quiaca Observatorio
MAB, Mt. Abu
MBO, Mt. Bachelor Observatory
MDY, Mondy
MKN, Mt. Kenya
MLO, Mauna Loa
NWR, Niwot Ridge
PMO, Pico Mountain Observatory
PYR, Nepal Climate Observatory–Pyramid
SNB, Sonnblick
TAR, Tanah Rata
WHI, Whistler Mountain
WLG, Mt. Walliguan
YEL, Yellowstone National Park
YOS, Yosemite National Park
ZSF, Zugspitze–Schneefernerhaus
ZUG, Zugspitze–Gipfel

Abbreviations

ABC: Atmospheric brown cloud; ACE-2: Second Aerosol Characterization Experiment; AEMET: Agencia Estatal de Meteorología; CASTNET: Clean Air Status and Trends Network; CLRTAP: Convention on Long-Range Transboundary Air Pollution; CMA: China Meteorological Administration; CONTRACE: Convective Transport of Trace Gases into the Middle and Upper Troposphere over Europe; DWD: German Weather Service (Deutscher Wetterdienst); EANET: Acid Deposition Monitoring Network in East Asia; ECMWF: European Centre for Medium-Range Weather Forecasts; EMEP: European Monitoring and Evaluation Programme; Empa: Swiss Federal Laboratories for Materials Science and Technology; EPA: US Environmental Protection Agency; ESRL: Earth System Research Laboratory; Ev-K2-CNR: Everest-K2 National Research Council; FOEN: Swiss Federal Office for the Environment; FREETEX: Free Tropospheric Experiments; GAW: Global Atmosphere Watch; GMD: Global Monitoring Division; HIS: Hydrometeorological Institute of Slovenia; ICARTT: International Consortium for Atmospheric Research on Transport and Transformation; IFA: Oboukhov Institute of Atmospheric Physics; IMPROVE: Interagency Monitoring of Protected Visual Environments; INTEX-B: Intercontinental Chemical Transport Experiment Phase B; ISAC-CNR: Institute of Atmospheric Sciences and Climate of the National Research Council of Italy; JMA: Japan Meteorological Agency; KMD: Kenya Meteorological Department; LTER: Long-Term Ecological Research; MINATROC: Mineral Dust and Tropospheric Chemistry; MLOPEX: Mauna Loa Observatory Photochemistry Experiment; MMD: Malaysian Meteorological Department; MOE: Ministry of the Environment (Japan); MONARPOP: Monitoring Network in the Alpine Region for Persistent and Other Organic Pollutants; NABEL: Swiss National Air Pollution Monitoring Network; NCU: National Central University; NOAA: National Oceanic and Atmospheric Administration; NPS: National Park Service; ONM: Déportent Météorologique Régional Sud of the Office National de la Météorologie; PSI: Paul Scherrer Institute; SHARE: Stations at High Altitude for Research on the Environment; SMNA: Servicio Meteorológico Nacional; TCCON: Total Column Carbon Observation Network; UBA: Federal Environmental Agency (Umweltbundesamt); UNEP: United Nations Environmental Programme; WACS 2010: Whistler Aerosol and Cloud Study 2010; WDCGG: World Data Centre for Greenhouse Gases; WMO: World Meteorological Organization

Acknowledgements

The authors thank the Acid Deposition Monitoring Network in East Asia (EANET, http://www.eanet.asia/), EBAS (http://ebas.nilu.no), the US Environmental Protection Agency Clean Air Markets Division Clean Air Status and Trends Network (CASTNET, http://www.epa.gov/castnet), the Interagency Monitoring of Protected Visual Environments (IMPROVE, http://vista.cira.colostate.edu/improve/), and the World Data Centre for Greenhouse Gases (WDCGG, http://ds.data.jma.go.jp/gmd/wdcgg/) for providing the observational data. The Office National de la Météorologie (ONM), Algeria and the Earth System Research Laboratory, NOAA (NOAA/ESRL) provided the Assekrem data. The Kenya Meteorological Department (KMD) provided the Mt. Kenya data. The Izana Atmospheric Research Center, Meteorological State Agency of Spain (AEMET) provided the Izaña data. The China Meteorological Administration (CMA) and NOAA/ESRL provided the Mt. Waliguan data. The Kyrgyz National University (KSNU) provided the Issyk-Kul data. The National Research Council, Institute of Atmospheric Sciences and Climate (ISAC) in collaboration with Department of Pure and Applied Science, University of Urbino provided the Pyramid and Monte Cimone data. Prof. Neng-Huei Lin at the National Central University, which is managed by the Taiwan Environmental Protection Administration, provided the Lulin data. EANET provided the Mondy data. Dr. Manish Naja at the Aryabhatta Research Institute for Observational Sciences and Dr. Shyam Lal at the Physical Research Laboratory provided the Mt. Abu data. Dr. Shungo Kato at Tokyo Metropolitan University and EANET provided the Mt. Happo data. The National Weather Service (SMNA) provided the La Quiaca Observatorio data. Environment and Climate Change Canada provided the Whistler Mountain data. Dr. Dan Jaffe at the University of Washington provided the Mt. Bachelor Observatory data. NOAA/ESRL provided the Niwot Ridge and Mauna Loa data. CASTNET and IMPROVE provided the Chiricahua NM, Great Basin NP, Lassen Volcanic NP, Yellowstone NP, and Yosemite NP data. The Malaysian Meteorological Department (MMD) provided the Tanah Rata data. The Environment Agency Austria provided the Sonnblick data. The Federal Environmental Agency Germany provided the Zugspitze data. The Slovenian Environment Agency (ARSO) provided the Krvavec data. Swiss Federal Laboratories for Materials Testing and Research (Empa) and the Laboratory of Atmospheric Chemistry of the Paul Scherrer Institute (PSI) provided the Jungfraujoch data. The data at Pico Mountain Observatory were collected in a collaboration among Michigan

Technological University, the University of Colorado and NOAA/ESRL. The NMHC sampling at Assekrem, Mt. Kenya, Izaña, and Mauna Loa is a component of the NOAA Global Greenhouse Gas Reference Network, with NMHC in the collected whole air samples being analyzed by the Institute of Arctic and Alpine Research at the University of Colorado. The authors also thank Dr. Edit Nagy-Tanaka, Dr. Hajime Akimoto of the National Institute of Environmental Studies (NIES), along with the editor and anonymous reviewers for providing valuable comments.

Funding
Not applicable.

Authors' contributions
SO reviewed the literature data and conducted the experimental study. SO and HT proposed the topic, conceived, and then designed the study. Both authors have read and approved the final manuscript.

Competing interests
The authors declare that they have no competing interests.

References

Acid Deposition Monitoring Network in East Asia (EANET). http://www.eanet.asia/. Accessed 15 Jan 2016

Ahlm L, Shakya KM, Russell LM, Schroder JC, Wong JPS, Sjostedt SJ, Hayden KL, Liggio J, Wentzell JJB, Wiebe HA, Mihele C, Leaitch WR, Macdonald AM (2013) Temperature-dependent accumulation mode particle and cloud nuclei concentrations from biogenic sources during WACS 2010. Atmos Chem Phys 13:3393_3407. doi:10.5194/acp-13-3393-2013

Alastuey A, Querol X, Castillo S, Escudero M, Avila A, Cuevas E, Torres C, Romero P-M, Exposito F, García O, Diaz JP, Van Dingenen R, Putaud JP (2005) Characterisation of TSP and PM2.5 at Izaña and Sta. Cruz de Tenerife (Canary Islands, Spain) during a Saharan Dust Episode (July 2002). Atmos Environ 39:4715_4728. doi:10.1016/j.atmosenv.2005.04.018

Ambrose JL, Reidmiller DR, Jaffe DA (2011) Causes of high O_3 in the lower free troposphere over the Pacific Northwest as observed at the Mt. Bachelor Observatory. Atmos Environ 45:5302–5315. doi:10.1016/j.atmosenv.2011.06.056

Andreae MO, Merlet P (2001) Emission of trace gases and aerosols from biomass burning. Glob Biogeochem Cycles 15:955–966. doi:10.1029/2000GB001382

Andrews E, Ogren JA, Bonasoni P, Marinoni A, Cuevas E, Rodríguez S, Sun JY, Jaffe DA, Fischer EV, Baltensperger U, Weingartner E, Collaud Coen M, Sharma S, Macdonald AM, Leaitch WR, Lin N-H, Laj P, Arsov T, Kalapov I, Jefferson A, Sheridan P (2011) Climatology of aerosol radiative properties in the free troposphere. Atmos Res 102:365–393. doi:10.1016/j.atmosres.2011.08.017

Angelbratt J, Mellqvist J, Blumenstock T, Borsdorff T, Brohede S, Duchatelet P, Forster F, Hase F, Mahieu E, Murtagh D, Petersen AK, Schneider M, Sussmann R, Urban J (2011a) A new method to detect long term trends of methane (CH_4) and nitrous oxide (N_2O) total columns measured within the NDACC ground-based high resolution solar FTIR network. Atmos Chem Phys 11:6167–6183. doi:10.5194/acp-11-6167-2011

Angelbratt J, Mellqvist J, Simpson D, Jonson JE, Blumenstock T, Borsdorff T, Duchatelet P, Forster F, Hase F, Mahieu E, De Mazière M, Notholt J, Petersen AK, Raffalski U, Servais C, Sussmann R, Warneke T, Vigouroux C (2011b) Carbon monoxide (CO) and ethane (C_2H_6) trends from ground-based solar FTIR measurements at six European stations, comparison and sensitivity analysis with the EMEP model. Atmos Chem Phys 11:9253-9269. doi:10.5194/acp-11-9253-2011

Aref'ev VN, Semenov VK (1994) Spectral transparency of the atmosphere in the center of the European-Asian continent. Atmos Environ 28:997–999. doi:10.1016/1352-2310(94)90259-3

Aref'ev VN, Kamenogradskiy NY, Semyonov VK, Sinyakov VP (1995) Ozone and nitrogen dioxide in the atmosphere above the Northern Tien Shan. Izv Atmos Ocean Phys 31:15–20

Aref'ev VN, Kashin FV, Krasnosel'tsev AV, Semenov VK, Sinyakov VP (2008) Structure of time variations in the atmospheric transparency in Central Eurasia (Issyk Kul monitoring station). Izv Atmos Ocean Phys 44:615–620. doi:10.1134/S0001433808050071

Aref'ev VN, Kashin FV, Semenov VK, Sinyakov VP (2009) Variations in nitrogen dioxide in the atmosphere over central Eurasia (Issyk Kul monitoring station). Izv Atmos Ocean Phys 45:575–582. doi:10.1134/S0001433809050053

Aref'ev VN, Kashin FV, Orozaliev MD, Sizov NI, Sinyakov VP, Sorokina LI (2013) Structure of carbon monoxide time variations in the atmospheric thickness over central Eurasia (Issyk Kul monitoring station). Izv Atmos Ocean Phys 49:148–153. doi:10.1134/S0001433813020047

Arimoto R, Duce RA, Ray BJ, Ellis WG Jr, Cullen JD, Merrill JT (1995) Trace elements in the atmosphere over the North Atlantic. J Geophys Res 100:1199–1213. doi:10.1029/94JD02618

Armerding W, Comes FJ, Crawack HJ, Forberich O, Gold G, Ruger C, Spiekermann M, Walter J, Cuevas E, Redondas A, Schmitt R, Matuska P (1997) Testing the daytime oxidizing capacity of the troposphere: 1994 OH field campaign at the Izaña observatory, Tenerife. J Geophys Res 102:10603–10611. doi:10.1029/96JD03714

Assonov SS, Brenninkmeijer CAM, Schuck TJ, Taylor P (2010) Analysis of ^{13}C and ^{18}O isotope data of CO_2 in CARIBIC aircraft samples as tracers of upper troposphere/lower stratosphere mixing and the global carbon cycle. Atmos Chem Phys 10:8575–8599. doi:10.5194/acp-10-8575-2010

Atkinson R (2000) Atmospheric chemistry of VOCs and NO_x. Atmos Environ 34:2063–2101. doi:10.1016/s1352-2310(99)00460-4

Atkinson R, Arey J (2003) Atmospheric degradation of volatile organic compounds. Chem Rev 103:4605–4638. doi:10.1021/cr0206420

Atlas E, Ridley B, Walega J, Greenberg J, Kok G, Staffelbach T, Schauffler S, Lind J, Hübler G, Norton R, PEM-West Science Team GTE, Dlugokeneky E, Elkins J, Oltmans S, Mackay G, Karecki D (1996) A comparison of aircraft and ground-based measurements at Mauna Loa Observatory, Hawaii, during GTE PEM-West and MLOPEX 2. J Geophys Res 101:14599–14612. doi:10.1029/96JD00213

Balkanski Y, Bauer SE, Van Dingenen R, Bonasoni P, Schultz M, Fischer H, Gobbi GP, Hanke M, Hauglustaine D, Putaud JP, Stohl A, Raes F (2003) The Mt Cimone, Italy, free tropospheric campaign: principal characteristics of the gaseous and aerosol composition from European pollution, Mediterranean influences and during African dust events. Atmos Chem Phys Discuss 3:1753–1776. doi:10.5194/acpd-3-1753-2003

Baltensperger U, Gäggeler HW, Jost DT, Lugauer M, Schwikowski M, Weingartner E, Seibert P (1997) Aerosol climatology at the high-alpine site Jungfraujoch, Switzerland. J Geophys Res 102:19707–19715. doi:10.1029/97JD00928

Balzani Lööv JM, Henne S, Legreid G, Staehelin J, Reimann S, Prévôt ASH, Steinbacher M, Vollmer MK (2008) Estimation of background concentrations of trace gases at the Swiss Alpine site Jungfraujoch (3580 m asl). J Geophys Res 113:D22305. doi:10.1029/2007JD009751

Bândă N, Krol M, van Noije T, van Weele M, Williams JE, Le Sager P, Niemeier U, Thomason L, Röckmann T (2015) The effect of stratospheric sulfur from Mount Pinatubo on tropospheric oxidizing capacity and methane. J Geophys Res 120:1202–1220. doi:10.1002/2014JD022137

Baray J-L, Courcoux Y, Keckhut P, Portafaix T, Tulet P, Cammas J-P, Hauchecorne A, Godin Beekmann S, De Mazière M, Hermans C, Desmet F, Sellegri K, Colomb A, Ramonet M, Sciare J, Vuillemin C, Hoareau C, Dionisi D, Duflot V, Vérèmes H, Porteneuve J, Gabarrot F, Gaudo T, Metzger J-M, Payen G, Leclair de Bellevue J, Barthe C, Posny F, Ricaud P, Abchiche A et al (2013) Maïdo observatory: a new high-altitude station facility at Reunion Island (21°S, 55°E) for long-term atmospheric remote sensing and in situ measurements. Atmos Meas Tech 6:2865–2877. doi:10.5194/amt-6-2865-2013

Barret B, De Mazière M, Demoulin P (2002) Retrieval and characterization of ozone profiles from solar infrared spectra at the Jungfraujoch. J Geophys Res 107:4788. doi:10.1029/2001JD001298

Barret B, De Mazière M, Demoulin P (2003a) Correction to "Retrieval and characterization of ozone profiles from solar infrared spectra at the Jungfraujoch." J Geophys Res 108:4372. doi:10.1029/2003JD003809

Barret B, De Mazière M, Mahieu E (2003b) Ground-based FTIR measurements of CO from the Jungfraujoch: characterisation and comparison with in situ surface and MOPITT data. Atmos Chem Phys 3:2217-2223. doi:10.5194/acp-3-2217-2003

Bauer SE, Balkanski Y, Schulz M, Hauglustaine DA, Dentener F (2004) Global modeling of heterogeneous chemistry on mineral aerosol surfaces: influence on tropospheric ozone chemistry and comparison to observations. J Geophys Res 109:D02304. doi:10.1029/2003JD003868

Beeston M, Grgić I, van Elteren JT, Iskra I, Kapun G, Močnik G (2012) Chemical and morphological characterization of aerosol particles at Mt. Krvavec, Slovenia, during the Eyjafjallajökull Icelandic volcanic eruption. Environ Sci Pollut Res 19:235–243. doi:10.1007/s11356-011-0563-8

Bey I, Jacob DJ, Logan JA, Yantosca RM (2001) Asian chemical outflow to the Pacific in spring: Origins, pathways, and budgets. J Geophys Res 106:23097–23113. doi:10.1029/2001JD000806

Bianchi F, Tröstl J, Junninen H, Frege C, Henne S, Hoyle CR, Molteni U, Herrmann E, Adamov A, Bukowiecki N, Chen X, Duplissy J, Gysel M, Hutterli M, Kangasluoma J, Kontkanen J, Kürten A, Manninen HE, Münch S, Peräkylä O, Petäjä T, Rondo L, Williamson C, Weingartner E, Curtius J, Worsnop DR, Kulmala M, Dommen J,

Baltensperger U (2016) New particle formation in the free troposphere: a question of chemistry and timing. Science 26. doi:10.1126/science.aad5456.

Bodhaine BA (1995) Aerosol absorption measurements at Barrow, Mauna Loa and the south pole. J Geophys Res 100:8967–8975. doi:10.1029/95JD00513

Bollasina M, Bertolani L, Tartari G (2002) Meteorological observations at high altitude in the Khumbu Valley, Nepal Himalayas, 1994-1999. Bull Glaciol Res 19:1–11

Bonasoni P, Calzolari F, Colombo T, Corazza E, Santaguida R, Tesi G (1997) Continuous CO and H_2 measurements at Mt. Cimone (Italy): preliminary results. Atmos Environ 31:959–967. doi:10.1016/S1352-2310(96)00259-2

Bonasoni P, Evangelisti F, Bonafe U, Ravegnani F, Calzolari F, Stohl A, Tositti L, Tubertini O, Colombo T (2000a) Stratospheric ozone intrusion episodes recorded at Mt. Cimone during the VOTALP project: case studies. Atmos Environ 34:1355-1365. doi:10.1016/S1352-2310(99)00280-0

Bonasoni P, Stohl A, Cristofanelli P, Calzolari F, Colombo T, Evangelisti F (2000b) Background ozone variations at Mt. Cimone Station. Atmos Environ 34:5183-5189. doi:10.1016/S1352-2310(00)00268-5

Bonasoni P, Cristofanelli P, Calzolari F, Bonafè U, Evangelisti F, Stohl A, Zauli Sajani S, van Dingenen R, Colombo T, Balkanski Y (2004) Aerosol-ozone correlations during dust transport episodes. Atmos Chem Phys 4:1201–1215. doi:10.5194/acp-4-1201-2004

Bonasoni P, Laj P, Angelini F, Arduini J, Bonafè U, Calzolari F, Cristofanelli P, Decesari S, Facchini MC, Fuzzi S, Gobbi GP, Maione M, Marinoni A, Petzold A, Roccato F, Roger JC, Sellegri K, Sprenger M, Venzac H, Verza GP, Villani P, Vuillermoz E (2008) The ABC-Pyramid Atmospheric Research Observatory in Himalaya for aerosol, ozone and halocarbon measurements. Sci Total Environ 391:252–261. doi:10.1016/j.scitotenv.2007.10.024

Bonasoni P, Laj P, Marinoni A, Sprenger M, Angelini F, Arduini J, Bonafè U, Calzolari F, Colombo T, Decesari S, Di Biagio C, di Sarra AG, Evangelisti F, Duchi R, Facchini MC, Fuzzi S, Gobbi GP, Maione M, Panday A, Roccato F, Sellegri K, Venzac H, Verza GP, Villani P, Vuillermoz E, Cristofanelli P (2010) Atmospheric Brown Clouds in the Himalayas: first two years of continuous observations at the Nepal Climate Observatory-Pyramid (5079 m). Atmos Chem Phys 10:7515–7531. doi:10.5194/acp-10-7515-2010

Bond TC, Bhardwaj E, Dong R, Jogani R, Jung S, Roden C, Streets DG, Trautmann NM (2007) Historical emissions of black and organic carbon aerosol from energy-related combustion, 1850-2000. Glob Biogeochem Cycles 21:GB2018. doi:10.1029/2006GB002840

Bowman KW, Shindell DT, Worden HM, Lamarque JF, Young PJ, Stevenson DS, Qu Z, de la Torre M, Bergmann D, Cameron-Smith PJ, Collins WJ, Doherty R, Dalsøren SB, Faluvegi G, Folberth G, Horowitz LW, Josse BM, Lee YH, MacKenzie IA, Myhre G, Nagashima T, Naik V, Plummer DA, Rumbold ST, Skeie RB, Strode SA, Sudo K, Szopa S, Voulgarakis A, Zeng G et al (2013) Evaluation of ACCMIP outgoing longwave radiation from tropospheric ozone using TES satellite observations. Atmos Chem Phys 13:4057–4072. doi:10.5194/acp-13-4057-2013

Bracci A, Cristofanelli P, Sprenger M, Bonafé U, Calzolari F, Duchi R, Laj P, Marinoni A, Roccato F, Vuillermoz E, Bonasoni P (2012) Transport of stratospheric air masses to the Nepal Climate Observatory-Pyramid (Himalaya; 5079 m MSL): a synoptic-scale investigation. J Appl Meteorol 51:1489–1507. doi:10.1175/jamc-d-11-0154.1

Brasseur GP, Hauglustaine DA, Walters S (1996) Chemical compounds in the remote Pacific troposphere: comparison between MLOPEX measurements and chemical transport model calculations. J Geophys Res 101:14795–14813. doi:10.1029/95JD03520

Brönnimann S, Buchmann B, Wanner H (2002) Trends in near-surface ozone concentrations in Switzerland: the 1990s. Atmos Environ 36:2841–2852. doi:10.1016/S1352-2310(02)00145-0

Brönnimann S, Cain JC, Staehelin J, Farmer SFG (2003) Total ozone observations prior to the IGY. II: data and quality. Q J R Meteorol Soc 129:2819–2843. doi:10.1256/qj.02.145

Bukowiecki N, Zieger P, Weingartner E, Jurányi Z, Gysel M, Neininger B, Schneider B, Hueglin C, Ulrich A, Wichser A, Henne S, Brunner D, Kaegi R, Schwikowski M, Tobler L, Wienhold FG, Engel I, Buchmann B, Peter T, Baltensperger U (2011) Ground-based and airborne in-situ measurements of the Eyjafjallajökull volcanic aerosol plume in Switzerland in spring 2010. Atmos Chem Phys 11:10011–10030. doi:10.5194/acp-11-10011-2011

Bukowiecki N, Weingartner E, Gysel M, Collaud Coen M, Zieger P, Herrmann E, Steinbacher M, Gäggeler HW, Baltensperger U (2016) A review of more than 20 years of aerosol observation at the High Altitude Research Station Jungfraujoch, Switzerland (3580 m asl). Aerosol Air Qual Res 16:764–788. doi:10.4209/aaqr.2015.05.0305

Calvert JG (1990) Glossary of atmospheric chemistry terms. Pure Appl Chem 62:2167–2219. doi:10.1351/pac199062112167

Campana M, Li Y, Staehelin J, Prevot ASH, Bonasoni P, Loetscher H, Peter T (2005) The influence of south foehn on the ozone mixing ratios at the high alpine site Arosa. Atmos Environ 39:2945–2955. doi:10.1016/j.atmosenv.2005.01.037

Carslaw N, Plane JMC, Coe H, Cuevas E (1997) Observations of the nitrate radical in the free troposphere at Izaña de Tenerife. J Geophys Res 102:10613–10622. doi:10.1029/96JD03512

Chalbot M-C, Kavouras IG, Dubois DW (2013) Assessment of the contribution of wildfires to ozone concentrations in the central US-Mexico border region. Aerosol Air Qual Res 13:1–11. doi:10.4209/aaqr.2012.08.0232

Chameides W, Walker JCG (1973) A photochemical theory of tropospheric ozone. J Geophys Res 78:8751–8760. doi:10.1029/JC078i036p08751

Chang YS, Carmichael GR, Kurita H, Ueda H (1989) The transport and formation of photochemical oxidants in central Japan. Atmos Environ 23:363–393. doi:10.1016/0004-6981(89)90584-2

Chang S-S, Lee W-J, Wang L-C, Lin N-H, Chang-Chien G-P (2013) Influence of the Southeast Asian biomass burnings on the atmospheric persistent organic pollutants observed at near sources and receptor site. Atmos Environ 78:184–194. doi:10.1016/j.atmosenv.2012.07.074

Chehade W, Weber M, Burrows JP (2014) Total ozone trends and variability during 1979-2012 from merged data sets of various satellites. Atmos Chem Phys 14:7059–7074. doi:10.5194/acp-14-7059-2014

Chen S-C, Hsu S-C, Tsai C-J, Chou CC-K, Lin N-H, Lee C-T, Roam G-D, Pui DYH (2013) Dynamic variations of ultrafine, fine and coarse particles at the Lu-Lin background site in East Asia. Atmos Environ 78:154–162. doi:10.1016/j.atmosenv.2012.05.029

Cheng H, Zhang G, Jiang JX, Li X, Liu X, Li J, Zhao Y (2007) Organochlorine pesticides, polybrominated biphenyl ethers and lead isotopes during the spring time at the Waliguan Baseline Observatory, northwest China: implication for long-range atmospheric transport. Atmos Environ 41:4734–4747. doi:10.1016/j.atmosenv.2007.03.023

Cheng F-Y, Yang Z-M, Ou-Yang C-F, Ngan F (2013) A numerical study of the dependence of long-range transport of CO to a mountain station in Taiwan on synoptic weather patterns during the Southeast Asia biomass-burning season. Atmos Environ 78:277–290. doi:10.1016/j.atmosenv.2013.03.020

Chevalier A, Gheusi F, Delmas R, Ordóñez C, Sarrat C, Zbinden R, Thouret V, Athier G, Cousin J-M (2007) Influence of altitude on ozone levels and variability in the lower troposphere: a ground-based study for western Europe over the period 2001–2004. Atmos Chem Phys 7:4311–4326. doi:10.5194/acp-7-4311-2007

Chi KH, Lin C-Y, Ou Yang C-F, Wang J-L, Lin N-H, Sheu G-R, Lee C-T (2010) PCDD/F measurement at a high-altitude station in central Taiwan: evaluation of long-range transport of PCDD/Fs during the Southeast Asia biomass burning event. Environ Sci Technol 44:2954–2960. doi:10.1021/es1000984

Chiapello I, Prospero JM, Herman JR, Hsu NC (1999) Detection of mineral dust over the North Atlantic Ocean and Africa with the Nimbus 7 TOMS. J Geophys Res 104:9277–9291. doi:10.1029/1998JD200083

Chou C, Stetzer O, Weingartner E, Jurányi Z, Kanji ZA, Lohmann U (2011) Ice nuclei properties within a Saharan dust event at the Jungfraujoch in the Swiss Alps. Atmos Chem Phys 11:4725–4738. doi:10.5194/acp-11-4725-2011

Clarke AD, Charlson RJ (1985) Radiative properties of the background aerosol: absorption component of extinction. Science 229:263–265. doi:10.1126/science.229.4710.263

Collaud Coen M, Weingartner E, Schaub D, Hueglin C, Corrigan C, Schwikowski M, Baltensperger U (2004) Saharan dust events at the Jungfraujoch: detection by wavelength dependence of the single scattering albedo and first climatology analysis. Atmos Chem Phys 4:2465–2480. doi:10.5194/acp-4-2465-2004

Collaud Coen M, Weingartner E, Nyeki S, Cozic J, Henning S, Verheggen B, Gehrig R, Baltensperger U (2007) Long-term trend analysis of aerosol variables at the high-alpine site Jungfraujoch. J Geophys Res 112:D13213. doi:10.1029/2006JD007995

Collaud Coen M, Weingartner E, Furger M, Nyeki S, Prévôt ASH, Steinbacher M, Baltensperger U (2011) Aerosol climatology and planetary boundary influence at the Jungfraujoch analyzed by synoptic weather types. Atmos Chem Phys 11:5931–5944. doi:10.5194/acp-11-5931-2011

Collaud Coen M, Andrews E, Asmi A, Baltensperger U, Bukowiecki N, Day D, Fiebig M, Fjaeraa AM, Flentje H, Hyvärinen A, Jefferson A, Jennings SG, Kouvarakis G, Lihavainen H, Lund Myhre C, Malm WC, Mihapopoulos N, Molenar JV, O'Dowd C, Ogren JA, Schichtel BA, Sheridan P, Virkkula A, Weingartner E, Weller R, Laj P (2013) Aerosol decadal trends-Part 1: in-situ optical measurements at GAW and IMPROVE stations. Atmos Chem Phys 13:869–894. doi:10.5194/acp-13-869-2013

Comes FJ, Armerding W, Spiekermann M, Walter J, Rüger C (1995) Point measurements of tropospheric trace gases at Tenerife by a laser absorption technique. Atmos Environ 29:169–174. doi:10.1016/1352-2310(94)00254-I

Comes FJ, Forberich O, Walter J (1997) OH field measurements: a critical input into model calculations on atmospheric chemistry. J Atmos Sci 54:1886–1894. doi:10.1175/1520-0469(1997)054<1886:OFMACI>2.0.CO;2

Cooper OR, Forster C, Parrish D, Trainer M, Dunlea E, Ryerson T, Hüber G, Fehsenfeld F, Nicks D, Holloway J, de Gouw J, Warneke C, Roberts JM, Flocke F, Moody J (2004) A case study transpacific warm conveyor belt transport: Influence of merging airstreams on trace gas import to North America. J Geophys Res 109:D23S08. doi:10.1029/2003JD003624

Cooper OR, Parrish DD, Ziemke J, Balashov NV, Cupeiro M, Galbally IE, Gilge S, Horowitz L, Jensen NR, Lamarque J-F, Naik V, Oltmans SJ, Schwab J, Shindell DT, Thompson AM, Thouret V, Wang Y, Zbinden RM (2014) Global distribution and trends of tropospheric ozone: an observation-based review. Elem Sci Anth 2:000029–28. doi:10.12952/journal.elementa.000029

Cozic J, Verheggen B, Weingartner E, Crosier J, Bower KN, Flynn M, Coe H, Henning S, Steinbacher M, Henne S, Collaud Coen M, Petzold A, Baltensperger U (2008) Chemical composition of free tropospheric aerosol for PM1 and coarse mode at the high alpine site Jungfraujoch. Atmos Chem Phys 8:407–423. doi:10.5194/acp-8-407-2008

Cristofanelli P, Bonasoni P (2009) Background ozone in the southern Europe and Mediterranean area: influence of the transport processes. Environ Pollut 157:1399–1406. doi:10.1016/j.envpol.2008.09.017

Cristofanelli P, Bonasoni P, Tositti L, Bonafè U, Calzolari F, Evangelisti F, Sandrini S, Stohl A (2006) A 6-year analysis of stratospheric intrusions and their influence on ozone at Mt. Cimone (2165 m above sea level). J Geophys Res 111:D03306. doi:10.1029/2005JD006553

Cristofanelli P, Bonasoni P, Bonafé U, Calzolari F, Duchi R, Marinoni A, Roccato F, Vuillermoz E, Sprenger M (2009a) Influence of lower stratosphere/upper troposphere transport events on surface ozone at the Everest-Pyramid GAW Station (Nepal): first year of analysis. Int J Remote Sens 30:4083-4097. doi:10.1080/01431160902821940

Cristofanelli P, Marinoni A, Arduini J, Bonafè U, Calzolari F, Colombo T, Decesari S, Duchi R, Facchini MC, Fierli F, Finessi E, Maione M, Chiari M, Calzolari F, Messina P, Orlandi E, Roccato F, Bonasoni P (2009b) Significant variations of trace gas composition and aerosol properties at Mt. Cimone during air mass transport from North Africa—contributions from wildfire emissions and mineral dust. Atmos Chem Phys 9:4603-4619. doi:10.5194/acp-9-4603-2009

Cristofanelli P, Fierli F, Marinoni A, Calzolari F, Duchi R, Burkhart J, Stohl A, Maione M, Arduini J, Bonasoni P (2013) Influence of biomass burning and anthropogenic emissions on ozone, carbon monoxide and black carbon at the Mt. Cimone GAW-WMO global station (Italy, 2165 m a.s.l.). Atmos Chem Phys 13:15–30. doi:10.5194/acp-13-15-2013

Cristofanelli P, Scheel H-E, Steinbacher M, Saliba M, Azzopardi F, Ellul R, Fröhlich M, Tositti L, Brattich E, Maione M, Calzolari F, Duchi R, Landi TC, Marinoni A, Bonasoni P (2015) Long-term surface ozone variability at Mt. Cimone WMO/GAW global station (2165 m a.s.l., Italy). Atmos Environ 101:23–33. doi:10.1016/j.atmosenv.2014.11.012

Crutzen P (1974) A review of upper atmospheric photochemistry. Can J Chem 52:1569–1581. doi:10.1016/S0140-6736(00)88507-4

Cuevas E, González Y, Rodríguez S, Guerra JC, Gómez-Peláez AJ, Alonso-Pérez S, Bustos J, Milford C (2013) Assessment of atmospheric processes driving ozone variations in the subtropical North Atlantic free troposphere. Atmos Chem Phys 13:1973–1998. doi:10.5194/acp-13-1973-2013

Cui J, Sprenger M, Staehelin J, Siegrist A, Kunz M, Henne S, Steinbacher M (2009) Impact of stratospheric intrusions and intercontinental transport on ozone at Jungfraujoch in 2005: comparison and validation of two Lagrangian approaches. Atmos Chem Phys 9:3371–3383. doi:10.5194/acp-9-3371-2009

Cui J, Pandey Deolal S, Sprenger M, Henne S, Staehelin J, Steinbacher M, Nédélec P (2011) Free tropospheric ozone changes over Europe as observed at Jungfraujoch (1990-2008): an analysis based on backward trajectories. J Geophys Res 116:D10304. doi:10.1029/2010JD015154

Danielsen EF (1968) Stratospheric-tropospheric exchange based on radioactivity, ozone and potential vorticity. J Atmos Sci 25:502–518. doi:10.1175/1520-0469(1968)025<0502:STEBOR>2.0.CO;2

Darzi M, Winchester JW (1982) Aerosol characteristics at Mauna Loa Observatory, Hawaii, after East Asian dust storm episodes. J Geophys Res 87:1251–1258. doi:10.1029/JC087iC02p01251

De Mazière M, Van Roozendael M, Hermans C, Simon PC, Demoulin P, Roland G, Zander R (1998) Quantitative evaluation of the post-Mount Pinatubo NO_2 reduction and recovery, based on 10 years of Fourier transform infrared and UV-visible spectroscopic measurements at Jungfraujoch. J Geophys Res 103:10849–10858. doi:10.1029/97JD03362

De Mazière M, Hennen O, Van Roozendael M, Demoulin P, De Backer H (1999) Daily ozone vertical profile model built on geophysical grounds, for column retrieval from atmospheric high-resolution infrared spectra. J Geophys Res 104:23855–23869. doi:10.1029/1999JD900347

de Reus M, Fischer H, Sander R, Gros V, Kormann R, Salisbury G, Van Dingenen R, Williams J, Zöllner M, Lelieveld J (2005) Observations and model calculations of trace gas scavenging in a dense Saharan dust plume during MINATROC. Atmos Chem Phys 5:1787–1803. doi:10.5194/acp-5-1787-2005

Decesari S, Facchini MC, Carbone C, Giulianelli L, Rinaldi M, Finessi E, Fuzzi S, Marinoni A, Cristofanelli P, Duchi R, Bonasoni P, Vuillermoz E, Cozic J, Jaffrezo JL, Laj P (2010) Chemical composition of PM_{10} and PM_1 at the high-altitude Himalayan station Nepal Climate Observatory-Pyramid (NCO-P) (5079 m a.s.l.). Atmos Chem Phys 10:4583–4596. doi:10.5194/acp-10-4583-2010

Deshler T (2008) A review of global stratospheric aerosol: measurements, importance, life cycle, and local stratospheric aerosol. Atmos Res 90:223–232. doi:10.1016/j.atmosres.2008.03.016

Díaz AM, Díaz JP, Expósito FJ, Hernández-Leal PA, Savoie D, Querol X (2006) Air masses and aerosols chemical components in the free troposphere at the subtropical Northeast Atlantic region. J Atmos Chem 53:63–90. doi:10.1007/s10874-006-2644-5

Dils B, Cui J, Henne S, Mahieu E, Steinbacher M, De Mazière M (2011) 1997-2007 CO trend at the high Alpine site Jungfraujoch: a comparison between NDIR surface in situ and FTIR remote sensing observations. Atmos Chem Phys 11:6735–6748. doi:10.5194/acp-11-6735-2011

Ding A, Wang T (2006) Influence of stratosphere-to-troposphere exchange on the seasonal cycle of surface ozone at Mount Waliguan in western China. Geophys Res Lett 33:L03803. doi:10.1029/2005GL024760

Duchatelet P, Mahieu E, Ruhnke R, Feng W, Chipperfield M, Demoulin P, Bernath P, Boone CD, Walker KA, Servais C, Flock O (2009) An approach to retrieve information on the carbonyl fluoride (COF_2) vertical distributions above Jungfraujoch by FTIR multi-spectrum multi-window fitting. Atmos Chem Phys 9:9027–9042. doi:10.5194/acp-9-9027-2009

Duchatelet P, Demoulin P, Hase F, Ruhnke R, Feng W, Chipperfield MP, Bernath PF, Boone CD, Walker KA, Mahieu E (2010) Hydrogen fluoride total and partial column time series above the Jungfraujoch from long-term FTIR measurements: impact of the line-shape model, characterization of the error budget and seasonal cycle, and comparison with satellite and model data. J Geophys Res 115:D22306. doi:10.1029/2010JD014677

Duncan BN, Martin RV, Staudt AC, Yevich R, Logan JA (2003) Interannual and seasonal variability of biomass burning emissions constrained by satellite observations. J Geophys Res 108:4100. doi:10.1029/2002JD002378

Duncan BN, Logan JA, Bey I, Megretskaia IA, Yantosca RM, Novelli PC, Jones NB, Rinsland CP (2007) Global budget of CO, 1988-1997: source estimates and validation with a global model. J Geophys Res 112:D22301. doi:10.1029/2007JD008459

Duncan BN, West JJ, Yoshida Y, Fiore AM, Ziemke JR (2008) The influence of European pollution on ozone in the Near East and northern Africa. Atmos Chem Phys 8:2267–2283. doi:10.5194/acp-8-2267-2008

EBAS. http://ebas.nilu.no. Accessed 9 Feb 2016

Elansky NF, Arabov AY, Elokhov AS, Makarov OV, Savastyuk VV, Senik IA (1995) Observations of minor species and UV radiation in atmosphere at high-mountain scientific station Kislovodsk. Izv Atmos Ocean Phys 31:5–14

Fahey DW, Hübler G, Parrsich DD, Williams EJ, Norton RB, Ridley BA, Singh HB, Liu SC, Fehsenfeld FC (1986) Reactive nitrogen species in the troposphere: measurements of NO, NO_2, HNO_3, particulate nitrate, peroxyacetyl nitrate (PAN), O_3, and total reactive odd nitrogen (NO_y) at Niwot Ridge, Colorado. J Geophys Res 91:9781–9793. doi:10.1029/JD091iD09p09781

Fehsenfeld FC, Dickerson RR, Hübler G, Luke WT, Nunnermacker LJ, Williams EJ, Roberts JM, Calvert JG, Curran CM, Delany AC, Eubank CS, Fahey DW, Fried A, Gandrud BW, Langford AO, Murphy PC, Norton RB, Pickering KE, Ridley BA (1987) A ground-based intercomparison of NO, NO_x, and NO_y measurement techniques. J Geophys Res 92:14710–14722. doi:10.1029/jd092id12p14710

Fenneteaux I, Colin P, Etienne A, Boudries H, Dutot AL, Perros PE, Toupance G (1999) Influence of continental sources on oceanic air composition at the eastern edge of the North Atlantic Ocean, TOR 1992-1995. J Atmos Chem 32:233–280. doi:10.1023/A:1006140223711Fierz-Schmidhauser R, Zieger P, Gysel M, Kammermann L, DeCarlo PF, Baltensperger U, Weingartner E (2010) Measured and predicted aerosol light scattering enhancement factors at the high alpine site Jungfraujoch. Atmos Chem Phys 10:2319–2333. doi:10.5194/acp-10-2319-2010

Finley BD, Swartzendruber PC, Jaffe DA (2009) Particulate mercury emissions in regional wildfire plumes observed at the Mount Bachelor Observatory. Atmos Environ 43:6074–6083. doi:10.1016/j.atmosenv.2009.08.046

Fiore AM, Dentener FJ, Wild O, Cuvelier C, Schultz MG, Hess P, Textor C, Schulz M, Doherty RM, Horowitz LW, MacKenzie IA, Sanderson MG, Shindell DT, Stevenson DS, Szopa S, Van Dingenen R, Zeng G, Atherton C, Bergmann D, Bey I, Carmichael G, Collins WJ, Duncan BN, Faluvegi G, Folberth G, Gauss M, Gong S, Hauglustaine D, Holloway T, Isaksen ISA et al (2009) Multimodel estimates of intercontinental source-receptor relationships for ozone pollution. J Geophys Res 114:D04301–21. doi:10.1029/2008JD010816

Fiore AM, Oberman JT, Lin MY, Zhang L, Clifton OE, Jacob DJ, Naik V, Horowitz LW, Pinto JP, Milly GP (2014) Estimating North American background ozone in U.S. surface air with two independent global models: variability, uncertainties, and recommendations. Atmos Environ 96:284–300. doi:10.1016/j.atmosenv.2014.07.045

Fischer H, Kormann R, Klüpfel T, Gurk C, Königstedt R, Parchatka U, Mühle J, Rhee TS, Brenninkmeijer CAM, Bonasoni P, Stohl A (2003) Ozone production and trace gas correlations during the June 2000 MINATROC intensive measurement campaign at Mt. Cimone. Atmos Chem Phys 3:725–738. doi:10.5194/acp-3-725-2003

Fischer EV, Jaffe DA, Marley NA, Gaffney JS, Marchany-Rivera A (2010a) Optical properties of aged Asian aerosols observed over the U.S. Pacific Northwest. J Geophys Res 115:D20209. doi:10.1029/2010JD013943

Fischer EV, Jaffe DA, Reidmiller DR, Jaeglé L (2010b) Meteorological controls on observed peroxyacetyl nitrate at Mount Bachelor during the spring of 2008. J Geophys Res 115:D03302. doi:10.1029/2009JD012776

Flentje H, Claude H, Elste T, Gilge S, Köhler U, Plass-Dülmer C, Steinbrecht W, Thomas W, Werner A, Fricke W (2010) The Eyjafjallajökull eruption in April 2010—detection of volcanic plume using in-situ measurements, ozone sondes and lidar-ceilometer profiles. Atmos Chem Phys 10:10085–10092. doi:10.5194/acp-10-10085-2010

Folini D, Kaufmann P, Ubl S, Henne S (2009) Region of influence of 13 remote European measurement sites based on modeled carbon monoxide mixing ratios. J Geophys Res 114:D08307. doi:10.1029/2008JD011125

Forrer J, Rüttimann R, Schneiter D, Fischer A, Buchmann B, Hofer P (2000) Variability of trace gases at the high-Alpine site Jungfraujoch caused by meteorological transport processes. J Geophys Res 105:12241–12251. doi:10.1029/1999JD901178

Francis T (2012) Temporal trends in ambient SO_2 at a high altitude site in semi-arid western India: observations versus chemical transport modeling. J Environ Prot 03:657–680. doi:10.4236/jep.2012.37079

Fu XW, Feng X, Liang P, Deliger ZH, Ji J, Liu P (2012) Temporal trend and sources of speciated atmospheric mercury at Waliguan GAW station, Northwestern China. Atmos Chem Phys 12:1951–1964. doi:10.5194/acp-12-1951-2012

Furger M, Dommen J, Graber WK, Poggio L, Prévôt ASH, Emeis S, Grell G, Trickl T, Gomiscek B, Neininger B, Wotawa G (2000) The VOTALP Mesolcina Valley Campaign 1996-concept, background and some highlights. Atmos Environ 34:1395–1412. doi:10.1016/S1352-2310(99)00377-5

Gaiero DM, Simonella L, Gassó S, Gili S, Stein AF, Sosa P, Becchio R, Arce J, Marelli H (2013) Ground/satellite observations and atmospheric modeling of dust storms originating in the high Puna-Altiplano deserts (South America): implications for the interpretation of paleo-climatic archives. J Geophys Res 118:3817–3831. doi:10.1002/jgrd.50036

Gallagher JP, McKendry IG, Macdonald AM, Leaitch WR (2011) Seasonal and diurnal variations in aerosol concentration on Whistler Mountain: boundary layer influence and synoptic-scale controls. J Appl Meteorol Climatol 50:2210–2222. doi:10.1175/JAMC-D-11-028.1

Gallagher JP, McKendry IG, Strawbridge K, Macdonald AM, Leaitch WR, Cottle PW (2012) Application of lidar data to assist airmass discrimination at the Whistler Mountaintop Observatory. J Appl Meteorol Climatol 51:1733–1739. doi:10.1175/JAMC-D-12-067.1

García OE, Schneider M, Redondas A, González Y, Hase F, Blumenstock T, Sepúlveda E (2012) Investigating the long-term evolution of subtropical ozone profiles applying ground-based FTIR spectrometry. Atmos Meas Tech 5:2917–2931. doi:10.5194/amt-5-2917-2012

Garreaud R, Vuille M, Clement AC (2003) The climate of the Altiplano: observed current conditions and mechanisms of past changes. Palaeogeogr Palaeoclimatol Palaeoecol 194:5–22. doi:10.1016/S0031-0182(03)00269-4

Genualdi SA, Massey Simonich SL, Primbs TK, Bidleman TF, Jantunen LM, Ryoo K-S, Zhu T (2009) Enantiomeric signatures of organochlorine pesticides in Asian, trans-Pacific, and western U.S. air masses. Environ Sci Technol 43:2806–2811. doi:10.1021/es803402q

Gil M, Puentedura O, Yela M, Cuevas E (2000) Behavior of NO_2 and O_3 columns during the eclipse of February 26, 1998, as measured by visible spectroscopy. J Geophys Res 105:3583–3593. doi:10.1029/1999JD900973

Gil M, Yela M, Gunn LN, Richter A, Alonso I, Chipperfield MP, Cuevas E, Iglesias J, Navarro M, Puentedura O, Rodríguez S (2008) NO_2 climatology in the northern subtropical region: diurnal, seasonal and interannual variability. Atmos Chem Phys 8:1635–1648. doi:10.5194/acp-8-1635-2008

Gilge S, Plass-Duelmer C, Fricke W, Kaiser A, Ries L, Buchmann B, Steinbacher M (2010) Ozone, carbon monoxide and nitrogen oxides time series at four alpine GAW mountain stations in central Europe. Atmos Chem Phys 10:12295–12316. doi:10.5194/acp-10-12295-2010

Ginoux P, Prospero JM, Torres O, Chin M (2004) Long-term simulation of global dust distribution with the GOCART model: correlation with North Atlantic Oscillation. Environ Model Softw 19:113–128. doi:10.1016/S1364-8152(03)00114-2

Gobbi GP, Barnaba F, Van Dingenen R, Putaud JP, Mircea M, Facchini MC (2003) Lidar and in situ observations of continental and Saharan aerosol: closure analysis of particles optical and physical properties. Atmos Chem Phys 3:2161–2172. doi:10.5194/acp-3-2161-2003

Gobbi GP, Angelini F, Bonasoni P, Verza GP, Marinoni A, Barnaba F (2010) Sunphotometry of the 2006–2007 aerosol optical/radiative properties at the Himalayan Nepal Climate Observatory-Pyramid (5079 m a.s.l.). Atmos Chem Phys 10:11209–11221. doi:10.5194/acp-10-11209-2010

Gomez-Pelaez AJ, Ramos R, Gomez-Trueba V, Novelli PC, Campo-Hernandez R (2013) A statistical approach to quantify uncertainty in carbon monoxide measurements at the Izaña global GAW station: 2008–2011. Atmos Meas Tech 6:787–799. doi:10.5194/amt-6-787-2013

Gratz LE, Jaffe DA, Hee JR (2015) Causes of increasing ozone and decreasing carbon monoxide in springtime at the Mt. Bachelor Observatory from 2004 to 2013. Atmos Environ 109:323–330. doi:10.1016/j.atmosenv.2014.05.076

Graustein WC, Turekian KK (1996) ^{7}Be and ^{210}Pb indicate an upper troposphere source for elevated ozone in the summertime subtropical free troposphere of the eastern North Atlantic. Geophys Res Lett 23:539–542. doi:10.1029/96GL00304

Greally BR, Manning AJ, Reimann S, McCulloch A, Huang J, Dunse BL, Simmonds PG, Prinn RG, Fraser PJ, Cunnold DM, O'Doherty S, Porter LW, Stemmler K, Vollmer MK, Lunder CR, Schmidbauer N, Hermansen O, Arduini J, Salameh PK, Krummel PB, Wang RHJ, Folini D, Weiss RF, Maione M, Nickless G, Stordal F, Derwent RG (2007) Observations of 1,1-difluoroethane (HFC-152a) at AGAGE and SOGE monitoring stations in 1994-2004 and derived global and regional emission estimates. J Geophys Res 112:D06308. doi:10.1029/2006JD007527

Greenberg JP, Helmig D, Zimmerman PR (1996) Seasonal measurements of nonmethane hydrocarbons and carbon monoxide at the Mauna Loa Observatory during the Mauna Loa Observatory Photochemistry Experiment 2. J Geophys Res 101:14581–14598. doi:10.1029/95JD01543

Gros V, Bräunlich M, Röckmann T, Jöckel P, Bergamaschi P, Brenninkmeijer CAM, Rom W, Kutschera W, Kaiser A, Scheel HE, Mandl M, van der Plicht J, Possnert G (2001) Detailed analysis of the isotopic composition of CO and characterization of the air masses arriving at Mount Sonnblick (Austrian Alps). J Geophys Res 106:3179–3193. doi:10.1029/2000JD900509

Guerova G, Bey I, Attié J-L, Martin RV, Cui J, Sprenger M (2006) Impact of transatlantic transport episodes on summertime ozone in Europe. Atmos Chem Phys 6:2057–2072. doi:10.5194/acp-6-2057-2006

Haagen-Smit AJ (1952) Chemistry and physiology of Los Angeles smog. Ind Eng Chem 44:1342–1346. doi:10.1021/ie50510a045

Haagenson PL (1979) Meteorological and climatological factors affecting Denver air quality. Atmos Environ 13:79–85. doi:10.1016/0004-6981(79)90247-6

Hahn CJ (1981) A study of the diurnal behavior of boundary-layer winds at the Boulder Atmospheric Observatory. Boundary-Layer Meteorol 21:231–245. doi:10.1007/BF02033941

Hahn CJ, Merrill JT, Mendonca BG (1992) Meteorological influences during MLOPEX. J Geophys Res 97:10291–10309. doi:10.1029/91JD02299

Hanisch F, Crowley JN (2003) Ozone decomposition on Saharan dust: an experimental investigation. Atmos Chem Phys 11:119–130. doi:10.5194/acp-3-119-2003

Hanke M, Umann B, Uecker J, Arnold F, Bunz H (2003) Atmospheric measurements of gas-phase HNO_3 and SO_2 using chemical ionization mass spectrometry during the MINATROC field campaign 2000 on Monte Cimone. Atmos Chem Phys 3:417–436. doi:10.5194/acp-3-417-2003

Hauglustaine DA, Madronich S, Ridley BA, Flocke SJ, Cantrell CA, Eisele FL, Shetter RE, Tanner DJ, Ginoux P, Atlas EL (1999) Photochemistry and budget of ozone during the Mauna Loa Observatory Photochemistry Experiment (MLOPEX 2). J Geophys Res 104:30275–30307. doi:10.1029/91JD02298

Heikkilä U, Beer J, Alfimov V (2008) Beryllium-10 and beryllium-7 in precipitation in Dübendorf (440 m) and at Jungfraujoch (3580 m), Switzerland (1998–2005). J Geophys Res 113:D11104. doi:10.1029/2007JD009160

Helmig D, Tanner DM, Honrath RE, Owen RC, Parrish DD (2008) Nonmethane hydrocarbons at Pico Mountain, Azores: 1. Oxidation chemistry in the North Atlantic region. J Geophys Res 113:D20S91. doi:10.1029/2007JD008930

Helmig D, Muñoz M, Hueber J, Mazzoleni C, Mazzoleni L, Owen RC, Val-Martin M, Fialho P, Plass-Duelmer C, Palmer PI Lewis AC, Pfister G (2015) Climatology and atmospheric chemistry of the non-methane hydrocarbons ethane and propane over the North Atlantic. Elem Sci Anth 3:000054. doi:10.12952/journal.elementa.000054

Hemispheric Transport of Air Pollution (2007) Hemispheric Transport of Air Pollution 2007, Air Pollution Studies No. 16. United Nations, New York and Geneva

Henne S, Furger M, Nyeki S, Steinbacher M, Neininger B, de Wekker SFJ, Dommen J, Spichtinger N, Stohl A, Prévôt ASH (2004) Quantification of topographic venting of boundary layer air to the free troposphere. Atmos Chem Phys 4:497–509. doi:10.5194/acp-4-497-2004

Henne S, Furger M, Prévôt ASH (2005) Climatology of mountain venting-induced elevated moisture layers in the lee of the Alps. J Appl Meteorol 44:620–633. doi:10.1175/JAM2217.1

Henne S, Klausen J, Junkermann W, Kariuki JM, Aseyo JO, Buchmann B (2008) Representativeness and climatology of carbon monoxide and ozone at the global GAW station Mt. Kenya in equatorial Africa. Atmos Chem Phys 8:3119–3139. doi:10.5194/acp-8-3119-2008

Henne S, Brunner D, Folini D, Solberg S, Klausen J, Buchmann B (2010) Assessment of parameters describing representativeness of air quality in-situ measurement sites. Atmos Chem Phys 10:3561–3581. doi:10.5194/acp-10-3561-2010

Henning S, Weingartner E, Schwikowski M, Gäggeler HW, Gehrig R, Hinz K-P, Trimborn A, Spengler B, Baltensperger U (2003) Seasonal variation of water-soluble ions of the aerosol at the high-alpine site Jungfraujoch (3580 m asl). J Geophys Res 108:4030. doi:10.1029/2002JD002439

Hoegger B, Levrat G, Staehelin J, Schill H, Ribordy P (1992) Recent developments of the Light Climatic Observatory-ozone measuring station of the Swiss Meteorological Institute (LKO) at Arosa. J Atmos Terr Phys 54:497–505. doi:10.1016/0021-9169(92)90093-Z

Hoerling MP, Schaack TK, Lenzen AJ (1991) Global objective tropopause analysis. Mon Weather Rev 119:1816–1831. doi:10.1175/1520-0493(1991)119<1816:GOTA>2.0.CO;2

Holmes J, Zoller W (1996) The elemental signature of transported Asian dust at Mauna Loa observatory. Tellus B 48:83–92. doi:10.3402/tellusb.v48i1.15669

Holzinger R, Kasper-Giebl A, Staudinger M, Schauer G, Röckmann T (2010) Analysis of the chemical composition of organic aerosol at the Mt. Sonnblick observatory using a novel high mass resolution thermal-desorption proton-transfer-reaction mass-spectrometer (hr-TD-PTR-MS). Atmos Chem Phys 10:10111–10128. doi:10.5194/acp-10-10111-2010

Honrath RE, Owen RC, Val Martín M, Reid JS, Lapina K, Fialho P, Dziobak MP, Kleissl J, Westphal DL (2004) Regional and hemispheric impacts of anthropogenic and biomass burning emissions on summertime CO and O_3 in the North Atlantic lower free troposphere. J Geophys Res 109:D24310. doi:10.1029/2004JD005147

Honrath RE, Helmig D, Owen RC, Parrish DD, Tanner DM (2008) Nonmethane hydrocarbons at Pico Mountain, Azores: 2. Event-specific analyses of the impacts of mixing and photochemistry on hydrocarbon ratios. J Geophys Res 113:D20S92. doi:10.1029/2008JD009832

Hoyle CR, Webster CS, Rieder HE, Nenes A, Hammer E, Herrmann E, Gysel M, Bukowiecki N, Weingartner E, Steinbacher M, Baltensperger U (2016) Chemical and physical influences on aerosol activation in liquid clouds: a study based on observations from the Jungfraujoch, Switzerland. Atmos Chem Phys 16:4043–4061. doi:10.5194/acp-16-4043-2016

Huebert BJ, Phillips CA, Zhuang L, Kjellström E, Rodhe H, Feichter J, Land C (2001) Long-term measurements of free-tropospheric sulfate at Mauna Loa: comparison with global model simulations. J Geophys Res 106:5479–5492. doi:10.1029/2000JD900627

Huntrieser H, Heland J, Schlager H, Forster C, Stohl A, Aufmhoff H, Arnold F, Scheel HE, Campana M, Gilge S, Eixmann R, Cooper O (2005) Intercontinental air pollution transport from North America to Europe: experimental evidence from airborne measurements and surface observations. J Geophys Res 110:D01305. doi:10.1029/2004JD005045

Hyslop NP, Trzepla K, Wallis CD, Matzoll AK, White WH (2013) Technical note: a 23-year record of twice-weekly aerosol composition measurements at Mauna Loa Observatory. Atmos Environ 80:259–263. doi:10.1016/j.atmosenv.2013.07.038

Igarashi Y, Sawa Y, Yoshida K, Matsueda H, Fujii K, Dokiya Y (2004) Monitoring the SO_2 concentration at the summit of Mt. Fuji and a comparison with other trace gases during winter. J Geophys Res 109:D17304. doi:10.1029/2003JD004428

Igarashi Y, Sawa Y, Yoshioka K, Takahashi H, Matsueda H, Dokiya Y (2006) Seasonal variations in SO_2 plume transport over Japan: observations at the summit of Mt. Fuji from winter to summer. Atmos Environ 40:7018–7033. doi:10.1016/j.atmosenv.2006.06.017

Interagency Monitoring of Protected Visual Environments (IMPROVE). http://vista.cira.colostate.edu/improve/. Accessed 8 Feb 2016

Intergovermental Panel on Climate Change (2013) Climate Change 2013: The Physical Science Basis, Working Group I Contribution to the Fifth Assesment Report of the Intergovermental Panel on Climate Change. Cambridge University Press, Cambridge

Ionov DV, Timofeyev YM, Sinyakov VP, Semenov VK, Goutail F, Pommereau J-P, Bucsela EJ, Celarier EA, Kroon M (2008) Ground-based validation of EOS-Aura OMI NO_2 vertical column data in the midlatitude mountain ranges of Tien Shan (Kyrgyzstan) and Alps (France). J Geophys Res 113:D15S08. doi:10.1029/2007JD008659

Iwasaka Y, Minoura H, Nagaya K (1983) The transport and spacial scale of Asian dust-storm clouds: a case study of the dust-storm event of April 1979. Tellus B 35:189–196. doi:10.1111/j.1600-0889.1983.tb00023.x

Jabbar T, Steier P, Wallner G, Priller A, Kandler N, Kaiser A (2012) Iodine isotopes (^{127}I and ^{129}I) in aerosols at high altitude Alp stations. Environ Sci Technol 46:8637–8644. doi:10.1021/es300948t

Jaffe D (2011) Relationship between surface and free tropospheric ozone in the western U.S. Environ Sci Technol 45:432–438. doi:10.1021/es1028102

Jaffe D, Ray J (2007) Increase in surface ozone at rural sites in the western US. Atmos Environ 41:5452–5463. doi:10.1016/j.atmosenv.2007.02.034

Jaffe D, Mahura A, Kelley J, Atkins J, Novelli PC, Merrill J (1997) Impact of Asian emissions on the remote North Pacific atmosphere: Interpretation of CO data from Shemya, Guam, Midway and Mauna Loa. J Geophys Res 102:28627–28635. doi:10.1029/96JD02750

Jaffe D, McKendry I, Anderson T, Price H (2003a) Six "new" episodes of trans-Pacific transport of air pollutants. Atmospheric Environment 37:391-404. doi:10.1016/s1352-2310(02)00862-2

Jaffe D, Price H, Parrish D, Goldstein A, Harris J (2003b) Increasing background ozone during spring on the west coast of North America. Geophys Res Lett 30:1613. doi:10.1029/2003gl017024

Jaffe D, Prestbo E, Swartzendruber P, Weiss-Penzias P, Kato S, Takami A, Hatakeyama S, Kajii Y (2005) Export of atmospheric mercury from Asia. Atmos Environ 39:3029–3038. doi:10.1016/j.atmosenv.2005.01.030

Jaffe D, Chand D, Hafner W, Westerling A, Spracklen D (2008) Influence of fires on O_3 concentrations in the western U.S. Environ Sci Technol 42:5885–5891. doi:10.1021/es800084k

Joly M, Peuch V-H (2012) Objective classification of air quality monitoring sites over Europe. Atmos Environ 47:111–123. doi:10.1016/j.atmosenv.2011.11.025

Jonson JE, Stohl A, Fiore AM, Hess P, Szopa S, Wild O, Zeng G, Dentener FJ, Lupu A, Schultz MG, Duncan BN, Sudo K, Wind P, Schulz M, Marmer E, Cuvelier C, Keating T, Zuber A, Valdebenito A, Dorokhov V, De Backer H, Davies J, Chen GH, Johnson B, Tarasick DW, Stübi R, Newchurch MJ, von der Gathen P, Steinbrecht W, Claude H (2010) A multi-model analysis of vertical ozone profiles. Atmos Chem Phys 10:5759–5783. doi:10.5194/acp-10-5759-2010

Junge CE (1962) Global ozone budget and exchange between stratosphere and troposphere. Tellus 14:363–377. doi:10.3402/tellusa.v14i4.9563

Kaiser A (2009) Origin of polluted air masses in the Alps. An overview and first results for MONARPOP. Environ Pollut 157:3232–3237. doi:10.1016/j.envpol.2009.05.042

Kaiser A, Scheifinger H, Spangl W, Weiss A, Gilge S, Fricke W, Ries L, Cemas D, Jesenovec B (2007) Transport of nitrogen oxides, carbon monoxide and ozone to the Alpine Global Atmosphere Watch stations Jungfraujoch (Switzerland), Zugspitze and Hohenpeissenberg (Germany), Sonnblick (Austria) and Mt. Krvavec (Slovenia). Atmos Environ 41:9273–9287. doi:10.1016/j.atmosenv.2007.09.027

Kajii Y, Someno K, Tanimoto H, Hirokawa J, Akimoto H, Katsuno T, Kawara J (1998) Evidence for the seasonal variation of photochemical activity of tropospheric ozone: continuous observation of ozone and CO at Happo, Japan. Geophys Res Lett 25:3505–3508. doi:10.1029/98GL02602

Kajino M, Ueda H, Satsumabayashi H, An J (2004) Impacts of the eruption of Miyakejima Volcano on air quality over far east Asia. J Geophys Res 109:D21204. doi:10.1029/2004JD004762

Kajino M, Ueda H, Satsumabayashi H, Han Z (2005) Increase in nitrate and chloride deposition in east Asia due to increased sulfate associated with the eruption of Miyakejima Volcano. J Geophys Res 110:D18203. doi:10.1029/2005JD005879

Kanakidou M, Crutzen PJ (1999) The photochemical source of carbon monoxide: importance, uncertainties and feedbacks. Chemosphere 1:91–109. doi:10.1016/S1465-9972(99)00022-7

Kanakidou M, Mihalopoulos N, Kindap T, Im U, Vrekoussis M, Gerasopoulos E, Dermitzaki E, Unal A, Koçak M, Markakis K, Melas D, Kouvarakis G, Youssef AF, Richter A, Hatzianastassiou N, Hilboll A, Ebojie F, Wittrock F, von Savigny C,

Burrows JP, Ladstaetter-Weissenmayer A, Moubasher H (2011) Megacities as hot spots of air pollution in the East Mediterranean. Atmos Environ 45:1223–1235. doi:10.1016/j.atmosenv.2010.11.048

Kandler K, Benker N, Bundke U, Cuevas E, Ebert M, Knippertz P, Rodríguez S, Schütz L, Weinbruch S (2007) Chemical composition and complex refractive index of Saharan Mineral Dust at Izaña, Tenerife (Spain) derived by electron microscopy. Atmos Environ 41:8058–8074. doi:10.1016/j.atmosenv.2007.06.047

Kaneyasu N, Igarashi Y, Sawa Y, Takahashi H, Takada H, Kumata H, Höller R (2007) Chemical and optical properties of 2003 Siberian forest fire smoke observed at the summit of Mt. Fuji, Japan. J Geophys Res Atmos 112:D13214. doi:10.1029/2007JD008544

Karl T, Crutzen PJ, Mandl M, Staudinger M, Guenther A, Jordan A, Fall R, Lindinger W (2001a) Variability-lifetime relationship of VOCs observed at the Sonnblick Observatory 1999—estimation of HO-densities. Atmos Environ 35:5287-5300. doi:10.1016/S1352-2310(01)00341-7

Karl T, Fall R, Crutzen PJ, Jordan A, Lindinger W (2001b) High concentrations of reactive biogenic VOCs at a high altitude site in late autumn. Geophys Res Lett 28:507-510. doi:10.1029/2000GL012255

Karl T, Hansel A, Märk T, Lindinger W, Hoffmann D (2003) Trace gas monitoring at the Mauna Loa Baseline Observatory using Proton-Transfer Reaction Mass Spectrometry. Int J Mass Spectrom 223–224:527–538. doi:10.1016/S1387-3806(02)00874-6

Kashin FV, Aref'ev VN, Kamenogradskii NE, Semenov VK, Sinyakov VP (2007) Carbon dioxide content in the atmospheric thickness over central Eurasia (Issyk Kul Monitoring Station). Izv Atmos Ocean Phys 43:480–489. doi:10.1134/S0001433807040093

Kashin FV, Aref'ev VN, Semenov VK, Sinyakov VP, Upenek LB (2008) Structure of time variations in carbon dioxide in the atmospheric thickness over central Eurasia (Issyk Kul Monitoring Station). Izv Atmos Ocean Phys 44:94–103. doi:10.1134/S0001433808010106

Kasper A, Puxbaum H (1998) Seasonal variation of SO_2, HNO_3, NH_3 and selected aerosol components at Sonnblick (3106 m a.s.l.). Atmos Environ 32:3925–3939. doi:10.1016/S1352-2310(97)00031-9

Kato S, Kajii Y, Akimoto H, Bräunlich M, Röckmann T, Brenninkmeijer CAM (2000) Observed and modeled seasonal variations of ^{13}C, ^{18}O, and ^{14}C of atmospheric CO at Happo, a remote site in Japan, and a comparison with other records. J Geophys Res 105:8891–8900. doi:10.1029/1999JD901144

Kato S, Pochanart P, Hirokawa J, Kajii Y, Akimoto H, Ozaki Y, Obi K, Katsuno T, Streets DG, Minko NP (2002) The influence of Siberian forest fires on carbon monoxide concentrations at Happo, Japan. Atmos Environ 36:385–390. doi:10.1016/S1352-2310(01)00158-3

Keeling CD, Whorf TP, Wahlen M, van der Plicht J (1995) Interannual extremes in the rate of rise of atmospheric carbon dioxide since 1980. Nature 375:666–670. doi:10.1038/375666a0

Kentarchos AS, Roelofs GJ, Lelieveld J, Cuevas E (2000) On the origin of elevated surface ozone concentrations at Izana Observatory, Tenerife during late March 1996. Geophys Res Lett 27:3699–3702. doi:10.1029/2000GL011518

Khalil MAK, Rasmussen RA (1994) Global decrease in atmospheric carbon monoxide concentration. Nature 370:639–641. doi:10.1038/370639a0

Kivekäs N, Sun J, Zhan M, Kerminen V-M, Hyvärinen A, Komppula M, Viisanen Y, Hong N, Zhang Y, Kulmala M, Zhang X-C, Deli-Geer, Lihavainen H (2009) Long term particle size distribution measurements at Mount Waliguan, a high-altitude site in inland China. Atmos Chem Phys 9:5461–5474. doi:10.5194/acp-9-5461-2009

Kleissl J, Honrath RE, Dziobak MP, Tanner D, Val Martín M, Owen RC, Helmig D (2007) Occurrence of upslope flows at the Pico mountaintop observatory: a case study of orographic flows on a small, volcanic island. J Geophys Res 112:D10S35. doi:10.1029/2006JD007565

Klock R, Mullock J (2001) The weather of British Columbia: graphic area forecast 31. NAV Canada, Ottawa

Koch D, Rind D (1998) Beryllium 10/beryllium 7 as a tracer of stratospheric transport. J Geophys Res 103:3907–3917. doi:10.1029/97JD03117

Kovač-Andrić E, Šorgo G, Kezele N, Cvitaš T, Klasinc L (2010) Photochemical pollution indicators—an analysis of 12 European monitoring stations. Environ Monit Assess 165:577–583. doi:10.1007/s10661-009-0969-7

Krieg J, Notholt J, Mahieu E, Rinsland CP, Zander R (2005) Sulphur hexafluoride (SF_6): comparison of FTIR-measurements at three sites and determination of its trend in the northern hemisphere. J Quant Spectrosc Radiat Transf 92:383–392. doi:10.1016/j.jqsrt.2004.08.005

Krzyścin JW (2000) Total ozone influence on the surface UV-B radiation in the late spring-summer 1963-1997: an analysis of multiple timescales. J Geophys Res 105:4993–5000. doi:10.1029/1999JD900265

Kuebler J, van den Bergh H, Russell AG (2001) Long-term trends of primary and secondary pollutant concentrations in Switzerland and their response to emission controls and economic changes. Atmos Environ 35:1351–1363. doi:10.1016/S1352-2310(00)00401-5

Kumar A, Sarin MM (2009) Mineral aerosols from western India: temporal variability of coarse and fine atmospheric dust and elemental characteristics. Atmos Environ 43:4005–4013. doi:10.1016/j.atmosenv.2009.05.014

Kumar A, Sarin MM (2010) Atmospheric water-soluble constituents in fine and coarse mode aerosols from high-altitude site in western India: long-range transport and seasonal variability. Atmos Environ 44:1245–1254. doi:10.1016/j.atmosenv.2009.12.035

Kumar A, Wu S, Weise MF, Honrath R, Owen RC, Helmig D, Kramer L, Val Martin M, Li Q (2013) Free-troposphere ozone and carbon monoxide over the North Atlantic for 2001-2011. Atmos Chem Phys 13:12537–12547. doi:10.5194/acp-13-12537-2013

Kurokawa J, Ohara T, Morikawa T, Hanayama S, Janssens-Maenhout G, Fukui T, Kawashima K, Akimoto H (2013) Emissions of air pollutants and greenhouse gases over Asian regions during 2000-2008: regional Emission inventory in ASia (REAS) version 2. Atmos Chem Phys 13:11019–11058. doi:10.5194/acp-13-11019-2013

Lal D, Peters B (1967) Cosmic ray produced radioactivity on the Earth. In: Sitte K (ed) Kosmische Strahlung II / Cosmic Rays II. Springer, Berlin Heidelberg

Lamarque J-F, Bond TC, Eyring V, Granier C, Heil A, Klimont Z, Lee D, Liousse C, Mieville A, Owen B, Schultz MG, Shindell D, Smith SJ, Stehfest E, Van Aardenne J, Cooper OR, Kainuma M, Mahowald N, McConnell JR, Naik V, Riahi K, van Vuuren DP (2010) Historical (1850-2000) gridded anthropogenic and biomass burning emissions of reactive gases and aerosols: methodology and application. Atmos Chem Phys 10:7017–7039. doi:10.5194/acp-10-7017-2010

Lanz VA, Henne S, Staehelin J, Hueglin C, Vollmer MK, Steinbacher M, Buchmann B, Reimann S (2009) Statistical analysis of anthropogenic non-methane VOC variability at a European background location (Jungfraujoch, Switzerland). Atmos Chem Phys 9:3445–3459. doi:10.5194/acp-9-3445-2009

Lanz VA, Prévôt ASH, Alfarra MR, Weimer S, Mohr C, DeCarlo PF, Gianini MFD, Hueglin C, Schneider J, Favez O, D'Anna B, George C, Baltensperger U (2010) Characterization of aerosol chemical composition with aerosol mass spectrometry in Central Europe: an overview. Atmos Chem Phys 10:10453–10471. doi:10.5194/acp-10-10453-2010

Lapina K, Honrath RE, Owen RC, Val Martín M, Hyer EJ, Fialho P (2008) Late summer changes in burning conditions in the boreal regions and their implications for NO_x and CO emissions from boreal fires. J Geophys Res 113:D11304. doi:10.1029/2007JD009421

Leaitch WR, Macdonald AM, Anlauf KG, Liu PSK, Toom-Sauntry D, Li S-M, Liggio J, Hayden K, Wasey MA, Russell LM, Takahama S, Liu S, van Donkelaar A, Duck T, Martin RV, Zhang Q, Sun Y, McKendry I, Shantz NC, Cubison M (2009) Evidence for Asian dust effects from aerosol plume measurements during INTEX-B 2006 near Whistler, BC. Atmos Chem Phys 9:3523–3546. doi:10.5194/acp-9-3523-2009

Leclair De Bellevue J, Réchou A, Baray JL, Ancellet G, Diab RD (2006) Signatures of stratosphere to troposphere transport near deep convective events in the southern subtropics. J Geophys Res 111:D24107. doi:10.1029/2005JD006947

Lee HN, Tositti L, Zheng X, Bonasoni P (2007) Analyses and comparisons of variations of ^{7}Be, ^{210}Pb, and $^{7}Be/^{210}Pb$ with ozone observations at two Global Atmosphere Watch stations from high mountains. J Geophys Res 112:D05303. doi:10.1029/2006JD007421

Lee C-T, Chuang M-T, Lin N-H, Wang J-L, Sheu G-R, Chang S-C, Wang S-H, Huang H, Chen H-W, Liu Y-L, Weng G-H, Lai H-Y, Hsu S-P (2011) The enhancement of $PM_{2.5}$ mass and water-soluble ions of biosmoke transported from Southeast Asia over the Mountain Lulin site in Taiwan. Atmos Environ 45:5784–5794. doi:10.1016/j.atmosenv.2011.07.020

Lee AKY, Hayden KL, Herckes P, Leaitch WR, Liggio J, Macdonald AM, Abbatt JPD (2012) Characterization of aerosol and cloud water at a mountain site during WACS 2010: secondary organic aerosol formation through oxidative cloud processing. Atmos Chem Phys 12:7103–7116. doi:10.5194/acp-12-7103-2012

Legreid G, Folini D, Staehelin J, Balzani Lööv J, Steinbacher M, Reimann S (2008) Measurements of organic trace gases including oxygenated volatile organic compounds at the high alpine site Jungfraujoch (Switzerland): seasonal variation and source allocations. J Geophys Res 113:D05307. doi:10.1029/2007JD008653

Lelieveld J, Berresheim H, Borrmann S, Crutzen PJ, Dentener FJ, Fischer H, Feichter J, Flatau PJ, Heland J, Holzinger R, Korrmann R, Lawrence MG, Levin Z, Markowicz KM, Mihalopoulos N, Minikin A, Ramanathan V, de Reus M, Roelofs GJ, Scheeren HA, Sciare J, Schlager H, Schultz M, Siegmund P, Steil B,

Stephanou EG, Stier P, Traub M, Warneke C, Williams J et al (2002) Global air pollution crossroads over the Mediterranean. Science 298:794–799. doi:10.1126/science.1075457

Levy II H (1971) Normal atmosphere: large radical and formaldehyde concentrations predicted. Science 173:141–143. doi:10.1126/science.173.3992.141

Levy II H, Moxim WJ (1989) Influence of long-range transport of combustion emissions on the chemical variability of the background atmosphere. Nature 338:326–328. doi: 10.1038/338326a0

Li Y, Campana M, Reimann S, Schaub D, Stemmler K, Staehelin J, Peter T (2005) Hydrocarbon concentrations at the Alpine mountain sites Jungfraujoch and Arosa. Atmos Environ 39:1113–1127. doi:10.1016/j.atmosenv.2004.09.084

Li J, Wang Z, Akimoto H, Tang J, Uno I (2009) Modeling of the impacts of China's anthropogenic pollutants on the surface ozone summer maximum on the northern Tibetan Plateau. Geophys Res Lett 36:L24802. doi:10.1029/2009GL041123

Liang Q, Jaeglé L, Jaffe DA, Weiss-Penzias P, Heckman A, Snow JA (2004) Long-range transport of Asian pollution to the northeast Pacific: seasonal variations and transport pathways of carbon monoxide. J Geophys Res 109:D23S07. doi:10.1029/2003JD004402

Liang M-C, Tang J, Chan C-Y, Zheng XD, Yung YL (2008) Signature of stratospheric air at the Tibetan Plateau. Geophys Res Lett 35:L20816. doi:10.1029/2008GL035246

Lin YC, Lin CY, Hsu WT (2010) Observations of carbon monoxide mixing ratios at a mountain site in central Taiwan during the Asian biomass burning season. Atmos Res 95:270–278. doi:10.1016/j.atmosres.2009.10.006

Liu SC, Trainer M, Fehsenfeld FC, Parrish DD, Williams EJ, Fahey DW, Hübler G, Murphy PC (1987) Ozone production in the rural troposphere and the implications for regional and global ozone distributions. J Geophys Res 92:4191–4207. doi:10.1029/JD092iD04p04191

Liu H, Jacob DJ, Bey I, Yantosca RM, Duncan BN, Sachse GW (2003a) Transport pathways for Asian pollution outflow over the Pacific: interannual and seasonal variations. J Geophys Res 108:D20. doi:10.1029/2002JD003102

Liu W, Hopke PK, VanCuren RA (2003a) Origins of fine aerosol mass in the western United States using positive matrix factorization. J Geophys Res 108:D23. doi:10.1029/2003JD003678

Liu X, Kondo Y, Ram K, Matsui H, Nakagomi K, Ikeda T, Oshima N, Verma RL, Takegawa N, Koike M, Kajino M (2013) Seasonal variations of black carbon observed at the remote mountain site Happo in Japan. J Geophys Res 118:3709–3722. doi:10.1002/jgrd.50317

Lobert JM, Scharffe DH, Hao WM, Kuhlbusch TA, Seuwan R, Warneck P, Crutzen PJ (1991) Experimental evaluation of biomass burning emissions: nitrogen and carbon containing compounds. In: Levine JS (ed) Global Biomass Burning: Atmospheric, Climatic and Biospheric Implications. MIT Press, Cambridge, pp 289–304

Logan JA (1983) Nitrogen oxides in the troposphere: global and regional budgets. J Geophys Res 88:10785–10807. doi:10.1029/JC088iC15p10785

Logan JA, Staehelin J, Megretskaia IA, Cammas J-P, Thouret V, Claude H, De Backer H, Steinbacher M, Scheel H-E, Stübi R, Fröhlich M, Derwent R (2012) Changes in ozone over Europe: analysis of ozone measurements from sondes, regular aircraft (MOZAIC) and alpine surface sites. J Geophys Res 117:D09301. doi:10.1029/2011JD016952

Lugauer M, Baltensperger U, Furger M, Gäggeler HW, Jost DT, Schwikowski M, Wanner H (1998) Aerosol transport to the high Alpine sites Jungfraujoch (3454 m asl) and Colle Gnifetti (4452 m asl). Tellus B 50:76–92. doi:10.3402/tellusb.v50i1.16025

Luria M, Boatman JF, Harris J, Ray J, Straube T, Chin J, Gunter RL, Herbert G, Gerlach TM, Van Valin CC (1992) Atmospheric sulfur dioxide at Mauna Loa, Hawaii. J Geophys Res 97:6011–6022. doi:10.1029/91JD03126

Ma J, Tang J, Zhou X, Zhang X (2002) Estimates of the chemical budget for ozone at Waliguan Observatory. J Atmos Chem 41:21–48. doi:10.1023/A:1013892308983

Ma J, Zheng X, Xu X (2005) Comment on "Why does surface ozone peak in summertime at Waliguan?" by Bin Zhu et al. Geophys Res Lett 32:L01805. doi:10.1029/2004GL021683

Macdonald AM, Anlauf KG, Leaitch WR, Chan E, Tarasick DW (2011) Interannual variability of ozone and carbon monoxide at the Whistler high elevation site: 2002–2006. Atmos Chem Phys 11:11431–11446. doi:10.5194/acp-11-11431-2011

Mahieu E, Zander R, Delbouille L, Demoulin P, Roland G, Servais C (1997) Observed trends in total vertical column abundances of atmospheric gases from IR solar spectra recorded at the Jungfraujoch. J Atmos Chem 28:227–243. doi:10.1023/A:1005854926740

Marcq S, Laj P, Roger JC, Villani P, Sellegri K, Bonasoni P, Marinoni A, Cristofanelli P, Verza GP, Bergin M (2010) Aerosol optical properties and radiative forcing in the high Himalaya based on measurements at the Nepal Climate Observatory-Pyramid site (5079 m a.s.l.). Atmos Chem Phys 10:5859–5872. doi:10.5194/acp-10-5859-2010

Marenco A, Gouget H, Nédélec P, Pagés JP, Karcher F (1994) Evidence of a long-term increase in tropospheric ozone from Pic du Midi data series: consequences: positive radiative forcing. J Geophys Res 20:16617–16632. doi:10.1029/94JD00021

Marenco F, Bonasoni P, Calzolari F, Ceriani M, Chiari M, Cristofanelli P, D'Alessandro A, Fermo P, Lucarelli F, Mazzei F, Nava S, Piazzalunga A, Prati P, Valli G, Vecchi R (2006) Characterization of atmospheric aerosols at Monte Cimone, Italy, during summer 2004: source apportionment and transport mechanisms. J Geophys Res 111:D24202. doi:10.1029/2006JD007145

Maring H, Savoie DL, Izaguirre MA, McCormick C, Arimoto R, Prospero JM, Pilinis C (2000) Aerosol physical and optical properties and their relationship to aerosol composition in the free troposphere at Izaña, Tenetire, Canary Islands, during July 1995. J Geophys Res 105:14677–14700. doi:10.1029/2000JD900106

Marinoni A, Cristofanelli P, Laj P, Duchi R, Calzolari F, Decesari S, Sellegri K, Vuillermoz E, Verza GP, Villani P, Bonasoni P (2010) Aerosol mass and black carbon concentrations, a two year record at NCO-P (5079 m, Southern Himalayas). Atmos Chem Phys 10:8551–8562. doi:10.5194/acp-10-8551-2010

McKendry IG, Strawbridge KB, O'Neill NT, Macdonald AM, Liu PSK, Leaitch WR, Anlauf KG, Jaegle L, Fairlie TD, Westphal DL (2007) Trans-Pacific transport of Saharan dust to western North America: a case study. J Geophys Res 112:D01103. doi:10.1029/2006JD007129

McKendry IG, Macdonald AM, Leaitch WR, van Donkelaar A, Zhang Q, Duck T, Martin RV (2008) Trans-Pacific dust events observed at Whistler, British Columbia during INTEX-B. Atmos Chem Phys 8:6297–6307. doi:10.5194/acp-8-6297-2008

McKendry IG, Gallagher J, Campuzano Jost P, Bertram A, Strawbridge K, Leaitch R, Macdonald AM (2010) Ground-based remote sensing of an elevated forest fire aerosol layer at Whistler, BC: implications for interpretation of mountaintop chemistry. Atmos Chem Phys 10:11921–11930. doi:10.5194/acp-10-11921-2010

McKendry I, Strawbridge K, Karumudi ML, O'Neill N, Macdonald AM, Leaitch R, Jaffe D, Cottle P, Sharma S, Sheridan P, Ogren J (2011) Californian forest fire plumes over Southwestern British Columbia: lidar, sunphotometry, and mountaintop chemistry observations. Atmos Chem Phys 11:465–477. doi:10.5194/acp-11-465-2011

Melen F, Mahieu E, Zander R, Rinsland CP, Demoulin P, Roland G, Delbouille L, Servais C (1998) Vertical column abundances of COF_2 above the Jungfraujoch station, derived from ground-based infrared solar observations. J Atmos Chem 29:119–134. doi:10.1023/A:1005847829686

Meng Z-Y, Xu X-B, Wang T, Zhang X-Y, Yu X-L, Wang S-F, Lin W-L, Chen Y-Z, Jiang Y-A, An X-Q (2010) Ambient sulfur dioxide, nitrogen dioxide, and ammonia at ten background and rural sites in China during 2007-2008. Atmos Environ 44:2625–2631. doi:10.1016/j.atmosenv.2010.04.008

Middleton NJ, Goudie AS (2001) Saharan dust: sources and trajectories. Trans Inst Br Geogr 26:165–181. doi:10.1111/1475-5661.00013

Monks PS (2000) A review of the observations and origins of the spring ozone maximum. Atmos Environ 34:3545–3561. doi:10.1016/S1352-2310(00)00129-1

Monks PS, Granier C, Fuzzi S, Stohl A, Williams ML, Akimoto H, Amann M, Baklanov A, Baltensperger U, Bey I, Blake N, Blake RS, Carslaw K, Cooper OR, Dentener F, Fowler D, Fragkou E, Frost GJ, Generoso S, Ginoux P, Grewe V, Guenther A, Hansson HC, Henne S, Hjorth J, Hofzumahaus A, Huntrieser H, Isaksen ISA, Jenkin ME, Kaiser J et al (2009) Atmospheric composition change—global and regional air quality. Atmos Environ 43:5268–5350. doi:10.1016/j.atmosenv.2009.08.021

Mu Y, Pang X, Quan J, Zhang X (2007) Atmospheric carbonyl compounds in Chinese background area: a remote mountain of the Qinghai-Tibetan Plateau. J Geophys Res 112:D22302. doi:10.1029/2006JD008211

Nagashima T, Ohara T, Sudo K, Akimoto H (2010) The relative importance of various source regions on East Asian surface ozone. Atmos Chem Phys 10:11305–11322. doi:10.5194/acp-10-11305-2010

Naja M, Lal S, Chand D (2003) Diurnal and seasonal variabilities in surface ozone at a high altitude site Mt Abu (24.6°N, 72.7°E, 1680m asl) in India. Atmos Environ 37:4205–4215. doi:10.1016/s1352-2310(03)00565-x

Nakazawa T, Aoki S, Fukabori M, Tanaka M (1984) The concentration of atmospheric carbon dioxide on the summit of Mt. Fuji (3776m), Japan. J Meteorol Soc Japan 62:688–695

Naoe H, Heintzenberg J, Okada K, Zaizen Y, Hayashi K, Tateishi T, Igarashi Y, Dokiya Y, Kinoshita K (2003) Composition and size distribution of submicrometer aerosol particles observed on Mt. Fuji in the volcanic plumes from Miyakejima. Atmos Environ 37:3047–3055. doi:10.1016/S1352-2310(03)00295-4

Nappo CJ, Caneill JY, Furman RW, Gifford FA, Kaimal JC, Kramer ML, Lockhart TJ, Pendergast MM, Pielke RA, Randerson D, Shreffler JH, Wyngaard JC (1982)

The Workshop on the respresentativeness of meteorological observations, June 1981, Boulder, Colo. Bull Am Meteorol Soc 63:761–764

Narita D, Pochanart P, Matsumoto J, Someno K, Tanimoto H, Hirokawa J, Kajii Y, Akimoto H, Nakao M, Katsuno T, Kinjo Y (1999) Seasonal variation of carbon monoxide at remote sites in Japan. Chemospher 1:137–144. doi:10.1016/S1465-9972(99)00023-9

Novelli PC, Masarie KA, Lang PM (1998) Distributions and recent changes of carbon monoxide in the lower troposphere. J Geophys Res 103:19015–19033. doi:10.1029/98JD01366

Novelli PC, Masarie KA, Lang PM, Hall BD, Myers RC, Elkins JW (2003) Reanalysis of tropospheric CO trends: effects of the 1997-1998 wildfires. J Geophys Res 108:4464. doi:10.1029/2002JD003031

Nyeki S, Baltensperger U, Colbeck I, Jost DT, Weingartner E, Gäggeler HW (1998a) The Jungfraujoch high-alpine research station (3454 m) as a background clean continental site for the measurement of aerosol parameters. J Geophys Res 103:6097-6107. doi:10.1029/97JD03123

Nyeki S, Li F, Weingartner E, Streit N, Colbeck I, Gäggeler HW, Baltensperger U (1998b) The background aerosol size distribution in the free troposphere: an analysis of the annual cycle at a high-alpine site. J Geophys Res 103:31749-31761. doi:10.1029/1998JD200029

Nyeki S, Halios CH, Baum W, Eleftheriadis K, Flentje H, Gröbner J, Vuilleumier L, Wehrli C (2012) Ground-based aerosol optical depth trends at three high-altitude sites in Switzerland and southern Germany from 1995 to 2010. J Geophys Res 117:D18202. doi:10.1029/2012JD017493

Oliveira TS, Pio CA, Alves CA, Silvestre AJD, Evtyugina M, Afonso JV, Fialho P, Legrand M, Puxbaum H, Gelencsér A (2007) Seasonal variation of particulate lipophilic organic compounds at nonurban sites in Europe. J Geophys Res 112:D23S09. doi:10.1029/2007JD008504

Oltmans SJ, Levy II H (1994) Surface ozone measurements from a global network. Atmos Environ 28:9–24. doi:10.1016/1352-2310(94)90019-1

Oltmans SJ, Hofmann DJ, Lathrop JA, Harris JM, Komhyr WD, Kuniyuki D (1996a) Tropospheric ozone during Mauna Loa Observatory Photochemistry Experiment 2 compared to long-term measurements from surface and ozonesonde observations. J Geophys Res 101:14569-14580. doi:10.1029/95JD03004

Oltmans SJ, Levy II H, Harris JM, Merrill JT, Moody JL, Lathrop JA, Cuevas E, Trainer M, O'Neill MS, Prospero JM, Vömel H, Johnson BJ (1996b) Summer and spring ozone profiles over the North Atlantic from ozonesonde measurements. J Geophys Res 101:29179-29200. doi:10.1029/96JD01713

Oltmans SJ, Johnson BJ, Harris JM, Thompson AM, Liu HY, Chan CY, Vömel H, Fujimoto T, Brackett VG, Chang WL, Chen J-P, Kim JH, Chan LY, Chang H-W (2004) Tropospheric ozone over the North Pacific from ozonesonde observations. J Geophys Res 109:D15S01. doi:10.1029/2003JD003466

Oltmans SJ, Lefohn AS, Harris JM, Galbally I, Scheel HE, Bodeker G, Brunke E, Claude H, Tarasick D, Johnson BJ, Simmonds P, Shadwick D, Anlauf K, Hayden K, Schmidlin F, Fujimoto T, Akagi K, Meyer C, Nichol S, Davies J, Redondas A, Cuevas E (2006) Long-term changes in tropospheric ozone. Atmos Environ 40:3156–3173. doi:10.1016/j.atmosenv.2006.01.029

Oltmans SJ, Lefohn AS, Shadwick D, Harris JM, Scheel HE, Galbally I, Tarasick DW, Johnson BJ, Brunke EG, Claude H, Zeng G, Nichol S, Schmidlin F, Davies J, Cuevas E, Redondas A, Naoe H, Nakano T, Kawasato T (2013) Recent tropospheric ozone changes—a pattern dominated by slow or no growth. Atmos Environ 67:331–351. doi:10.1016/j.atmosenv.2012.10.057

Ou Yang C-F, Lin N-H, Sheu G-R, Lee C-T, Wang J-L (2012) Seasonal and diurnal variations of ozone at a high-altitude mountain baseline station in East Asia. Atmos Environ 46:279–288. doi:10.1016/j.atmosenv.2011.09.060

Ou-Yang C-F, Lin N-H, Lin C-C, Wang S-H, Sheu G-R, Lee C-T, Schnell RC, Lang PM, Kawasato T, Wang J-L (2014) Characteristics of atmospheric carbon monoxide at a high-mountain background station in East Asia. Atmos Environ 89:613–622. doi:10.1016/j.atmosenv.2014.02.060

Owen RC, Cooper OR, Stohl A, Honrath RE (2006) An analysis of the mechanisms of North American pollutant transport to the central North Atlantic lower free troposphere. J Geophys Res 111:D23S58. doi:10.1029/2006JD007062

Pandey Deolal S, Brunner D, Steinbacher M, Weers U, Staehelin J (2012) Long-term in situ measurements of NO_x and NO_y at Jungfraujoch 1998-2009: time series analysis and evaluation. Atmos Chem Phys 12:2551–2566. doi:10.5194/acp-12-2551-2012

Pandey Deolal S, Staehelin J, Brunner D, Cui J, Steinbacher M, Zellweger C, Henne S, Vollmer MK (2013) Transport of PAN and NO_y from different source regions to the Swiss high alpine site Jungfraujoch. Atmos Environ 64:103–115. doi:10.1016/j.atmosenv.2012.08.021

Parrington JR, Zoller WH (1984) Diurnal and longer-term temporal changes in the composition of atmospheric particles at Mauna Loa, Hawaii. J Geophys Res 89:2522–2534. doi:10.1029/JD089iD02p02522

Parrington JR, Zoller WH, Aras NK (1983) Asian dust: seasonal transport to the Hawaiian islands. Science 220:195–197. doi:10.1126/science.220.4593.195

Parrish DD, Stohl A, Forster C, Atlas EL, Blake DR, Goldan PD, Kuster WC, de Gouw JA (2007) Effects of mixing on evolution of hydrocarbon ratios in the troposphere. J Geophys Res 112:D10S34. doi:10.1029/2006JD007583

Parrish DD, Law KS, Staehelin J, Derwent R, Cooper OR, Tanimoto H, Volz-Thomas A, Gilge S, Scheel H-E, Steinbacher M, Chan E (2012) Long-term changes in lower tropospheric baseline ozone concentrations at northern mid-latitudes. Atmos Chem Phys 12:11485–11504. doi:10.5194/acp-12-11485-2012

Parrish DD, Law KS, Staehelin J, Derwent R, Cooper OR, Tanimoto H, Volz-Thomas A, Gilge S, Scheel H-E, Steinbacher M, Chan E (2013) Lower tropospheric ozone at northern midlatitudes: changing seasonal cycle. Geophys Res Lett 40:1631–1636. doi:10.1002/grl.50303

Parrish DD, Lamarque J-F, Naik V, Horowitz L, Shindell DT, Staehelin J, Derwent R, Cooper OR, Tanimoto H, Voltz-Thomas A, Glige S, Scheel H-E, Steinbacher M, Fröhlich M (2014) Long-term changes in lower tropospheric baseline ozone concentrations: comparing chemistry-climate models and observations at northern midlatitudes. J Geophys Res 119:5719–5736. doi:10.1002/2013JD021435

Penkett SA, Brice KA (1986) The spring maximum in photo-oxidants in the Northern Hemisphere troposphere. Nature 319:655–657. doi:10.1038/319655a0

Perry KD, Cahill TA, Schnell RC, Harris JM (1999) Long-range transport of anthropogenic aerosols to the National Oceanic and Atmospheric Administration baseline station at Mauna Loa Observatory, Hawaii. J Geophys Res 104:18521–18533. doi:10.1029/1998JD100083

Peterson MC, Honrath RE, Parrish DD, Oltmans SJ (1998) Measurements of nitrogen oxides and a simple model of NO_y fate in the remote North Atlantic marine atmosphere. J Geophys Res 103:13489–13503. doi:10.1029/97JD02307

Pfister G, Hess PG, Emmons LK, Lamarque J-F, Wiedinmyer C, Edwards DP, Pétron G, Gille JC, Sachse GW (2005) Quantifying CO emissions from the 2004 Alaskan wildfires using MOPITT CO data. Geophys Res Lett 32:L11809. doi:10.1029/2005GL022995

Pfister GG, Emmons LK, Hess PG, Honrath R, Lamarque J-F, Val Martin M, Owen RC, Avery MA, Browell EV, Holloway JS, Nedelec P, Purvis R, Ryerson TB, Sachse GW, Schlager H (2006) Ozone production from the 2004 North American boreal fires. J Geophys Res 111:D24S07. doi:10.1029/2006JD007695

Pierce JR, Leaitch WR, Liggio J, Westervelt DM, Wainwright CD, Abbatt JPD, Ahlm L, Al-Basheer W, Cziczo DJ, Hayden KL, Lee AKY, Li S-M, Russell LM, Sjostedt SJ, Strawbridge KB, Travis M, Vlasenko A, Wentzell JJB, Wiebe HA, Wong JPS, Macdonald AM (2012) Nucleation and condensational growth to CCN sizes during a sustained pristine biogenic SOA event in a forested mountain valley. Atmos Chem Phys 12:3147–3163. doi:10.5194/acp-12-3147-2012

Pochanart P, Hirokawa J, Kajii Y, Akimoto H, Nakao M (1999) Influence of regional-scale anthropogenic activity in northeast Asia on seasonal variations of surface ozone and carbon monoxide observed at Oki, Japan. J Geophys Res 104:3621–3631. doi:10.1029/1998JD100071

Pochanart P, Akimoto H, Maksyutov S, Staehelin J (2001) Surface ozone at the Swiss Alpine site Arosa: the hemispheric background and the influence of large-scale anthropogenic emissions. Atmos Environ 35:5553–5566. doi:10.1016/S1352-2310(01)00236-9

Pochanart P, Akimoto H, Kajii Y, Potemkin VM, Khodzher TV (2003) Regional background ozone and carbon monoxide variations in remote Siberia/East Asia. J Geophys Res 108:4028. doi:10.1029/2001jd001412

Pochanart P, Kato S, Katsuno T, Akimoto H (2004) Eurasian continental background and regionally polluted levels of ozone and CO observed in northeast Asia. Atmos Environ 38:1325–1336. doi:10.1016/j.atmosenv.2003.11.014

Prather MJ (1996) Time scales in atmospheric chemistry: theory, GWPs for CH_4 and CO, and runaway growth. Geophys Res Lett 23:2597–2600. doi:10.1029/96GL02371

Primbs T, Piekarz A, Wilson G, Schmedding D, Higginbotham C, Field J, Simonich SM (2008a) Influence of Asian and western United States urban areas and fires on the atmospheric transport of polycyclic aromatic hydrocarbons, polychlorinated biphenyls, and fluorotelomer alcohols in the western United States. Environ Sci Technol 42:6385-6391. doi:10.1021/es702160d

Primbs T, Wilson G, Schmedding D, Higginbotham C, Simonich SM (2008b) Influence of Asian and western United States agricultural areas and fires on the atmospheric transport of pesticides in the western United States. Environ Sci Technol 42:6519-6525. doi:10.1021/es800511x

Prinn RG, Huang J, Weiss RF, Cunnold DM, Fraser PJ, Simmonds PG, McCulloch A, Harth C, Salameh P, O'Doherty S, Wang RHJ, Porter L, Miller BR (2001)

Evidence for substantial variations of atmospheric hydroxyl radicals in the past two decades. Science 292:1882–1888. doi:10.1126/science.1058673

Prospero JM, Schmitt R, Cuevas E, Savoie DL, Graustein WC, Turekian KK, Volz-Thomas A, Díaz A, Oltmans SJ, Levy II H (1995) Temporal variability of summer-time ozone and aerosols in the free troposphere over the eastern North Atlantic. Geophys Res Lett 22:2925–2928. doi:10.1029/95gl02791

Prospero JM, Ginoux P, Torres O, Nicholson S, Gill TE (2002) Environmental characterization of global sources of atmospheric soil dust identified with the Nimbus 7 Total Ozone Mapping Spectrometer (TOMS) absorbing aerosol product. Rev Geophys 40:1002. doi:10.1029/2000rg000095

Puentedura O, Gil M, Saiz-Lopez A, Hay T, Navarro-Comas M, Gómez-Pelaez A, Cuevas E, Iglesias J, Gomez L (2012) Iodine monoxide in the north subtropical free troposphere. Atmos Chem Phys 12:4909–4921. doi:10.5194/acp-12-4909-2012

Putaud J-P, Van Dingenen R, Dell'Acqua A, Raes F, Matta E, Decesari S, Facchini MC, Fuzzi S (2004) Size-segregated aerosol mass closure and chemical composition in Monte Cimone (I) during MINATROC. Atmos Chem Phys 4:889–902. doi:10.5194/acp-4-889-2004

Ram K, Sarin MM (2009) Absorption coefficient and site-specific mass absorption efficiency of elemental carbon in aerosols over urban, rural, and high-altitude sites in India. Environ Sci Technol 42:8233–8239. doi:10.1021/es9011542

Ram K, Sarin MM, Hegde P (2008) Atmospheric abundances of primary and secondary carbonaceous species at two high-altitude sites in India: Sources and temporal variability. Atmos Environ 42:6785–6796. doi:10.1016/j.atmosenv.2008.05.031

Ramanathan V, Crutzen PJ (2003) New directions: atmospheric brown "clouds". Atmos Environ 37:4033–4035. doi:10.1016/S1352-2310(03)00536-3

Rastogi N, Sarin MM (2005) Long-term characterization of ionic species in aerosols from urban and high-altitude sites in western India: role of mineral dust and anthropogenic sources. Atmos Environ 39:5541–5554. doi:10.1016/j.atmosenv.2005.06.011

Rastogi N, Sarin MM (2008) Atmospheric ^{210}Pb and ^{7}Be in ambient aerosols over low- and high-altitude sites in semiarid region: temporal variability and transport processes. J Geophys Res 113:D11103. doi:10.1029/2007jd009298

Reidmiller DR, Jaffe DA, Chand D, Strode S, Swartzendruber P, Wolfe GM, Thornton JA (2009) Interannual variability of long-range transport as seen at the Mt. Bachelor observatory. Atmos Chem Phys 9:557–572. doi:10.5194/acp-9-557-2009

Reidmiller DR, Jaffe DA, Fischer EV, Finley B (2010) Nitrogen oxides in the boundary layer and free troposphere at the Mt. Bachelor Observatory. Atmos Chem Phys 10:6043–6062. doi:10.5194/acp-10-6043-2010

Ridley BA, Robinson E (1992) The Mauna Loa Observatory photochemistry experiment. J Geophys Res 97:10285–10290. doi:10.1029/91jd00945

Ridley B, Walega J, Hübler G, Montzka D, Atlas E, Hauglustaine D, Grahek F, Lind J, Campos T, Norton R, Greenberg J, Schauffler S, Oltmans S, Whittlestone S (1998) Measurements of NO_x and PAN and estimates of O_3 production over the seasons during Mauna Loa Observatory Photochemistry Experiment 2. J Geophys Res 103:8323–8339. doi:10.1029/98JD00075

Rieder HE, Staehelin J, Maeder JA, Peter T, Ribatet M, Davison AC, Stübi R, Weihs P, Holawe F (2010a) Extreme events in total ozone over Arosa—part 1: application of extreme value theory. Atmos Chem Phys 10:10021-10031. doi:10.5194/acp-10-10021-2010

Rieder HE, Staehelin J, Maeder JA, Peter T, Ribatet M, Davison AC, Stübi R, Weihs P, Holawe F (2010b) Extreme events in total ozone over Arosa—part 2: fingerprints of atmospheric dynamics and chemistry and effects on mean values and long-term changes. Atmos Chem Phys 10:10033-10045. doi:10.5194/acp-10-10033-2010

Rinsland CP, Zander R, Demoulin P (1991) Ground-based infrared measurements of HNO_3 total column abundances: long-term trend and variability. J Geophys Res 96:9379–9389. doi:10.1029/91JD00609

Rinsland CP, Zander R, Mahieu E, Demoulin P, Goldman A, Ehhalt DH, Rudolph J (1992) Ground-based infrared measurements of carbonyl sulfide total column abundances: long-term trends and variability. J Geophys Res 97:5995–6002. doi:10.1029/92JD00040

Rinsland CP, Zander R, Demoulin P, Mahieu E (1996) $ClONO_2$ total vertical column abundances above the Jungfraujoch Station, 1986-1994: long-term trend and winter-spring enhancements. J Geophys Res 101:3891–3899. doi:10.1029/95JD03349

Rinsland CP, Mahieu E, Zander R, Demoulin P, Forrer J, Buchmann B (2000) Free tropospheric CO, C_2H_6, and HCN above central Europe: recent measurements from the Jungfraujoch station including the detection of elevated columns during 1998. J Geophys Res 105:24235–24249. doi:10.1029/2000JD900371

Rinsland CP, Mahieu E, Zander R, Jones NB, Chipperfield MP, Goldman A, Anderson J, Russell JM III, Demoulin P, Notholt J, Toon GC, Blavier J-F, Sen B, Sussmann R, Wood SW, Meier A, Griffith DWT, Chiou LS, Murcray FJ, Stephen TM, Hase F, Mikuteit S, Schulz A, Blumenstock T (2003) Long-term trends of inorganic chlorine from ground-based infrared solar spectra: past increases and evidence for stabilization. J Geophys Res 108:4252. doi:10.1029/2002JD003001

Rinsland CP, Chiou L, Mahieu E, Zander R, Boone CD, Bernath PF (2008) Measurements of long-term changes in atmospheric OCS (carbonyl sulfide) from infrared solar observations. J Quant Spectrosc Radiat Transf 109:2679–2686. doi:10.1016/j.jqsrt.2008.07.008

Rodríguez S, González Y, Cuevas E, Ramos R, Romero PM, Abreu-Afonso J, Redondas A (2009) Atmospheric nanoparticle observations in the low free troposphere during upward orographic flows at Izaña Mountain Observatory. Atmos Chem Phys 9:6319–6335. doi:10.5194/acp-9-6319-2009

Rodríguez S, Alastuey A, Alonso-Pérez S, Querol X, Cuevas E, Abreu-Afonso J, Viana M, Pérez N, Pandolfi M, de la Rosa J (2011) Transport of desert dust mixed with North African industrial pollutants in the subtropical Saharan Air Layer. Atmos Chem Phys 11:6663–6685. doi:10.5194/acp-11-6663-2011

Rodríguez S, Torres C, Guerra J-C, Cuevas E (2004) Transport pathways of ozone to marine and free-troposphere sites in Tenerife, Canary Islands. Atmos Environ 38:4733–4747. doi:10.1016/j.atmosenv.2004.05.021

Rose C, Sellegri K, Velarde F, Moreno I, Ramonet M, Weinhold K, Krejci R, Ginot P, Andrade M, Wiedensohler A, Laj P (2015) Frequent nucleation events at the high altitude station of Chacaltaya (5240 m a.s.l.), Bolivia. Atmos Environ 102:18–29. doi:10.1016/j.atmosenv.2014.11.015

Ruckstuhl AF, Henne S, Reimann S, Steinbacher M, Vollmer MK, O'Doherty S, Buchmann B, Hueglin C (2012) Robust extraction of baseline signal of atmospheric trace species using local regression. Atmos Meas Tech 5:2613–2624. doi:10.5194/amt-5-2613-2012

Salisbury G, Williams J, Gros V, Bartenbach S, Xu X, Fischer H, Kormann R, de Reus M, Zöllner M (2006) Assessing the effect of a Saharan dust storm on oxygenated organic compounds at Izaña, Tenerife (July-August 2002). J Geophys Res 111:D22303. doi:10.1029/2005JD006840

Sandrini S, Giulianelli L, Decesari S, Fuzzi S, Cristofanelli P, Marinoni A, Bonasoni P, Chiari M, Calzolai G, Canepari S, Perrino C, Facchini MC (2014) In situ physical and chemical characterisation of the Eyjafjallajökull aerosol plume in the free troposphere over Italy. Atmos Chem Phys 14:1075–1092. doi:10.5194/acp-14-1075-2014

Satsumabayashi H, Kawamura M, Katsuno T, Futaki K, Murano K, Carmichael GR, Kajino M, Horiguchi M, Ueda H (2004) Effects of Miyake volcanic effluents on airborne particles and precipitation in central Japan. J Geophys Res 109:D19202. doi:10.1029/2003JD004204

Scarnato B, Staehelin J, Stübi R, Schill H (2010) Long-term total ozone observations at Arosa (Switzerland) with Dobson and Brewer instruments (1988-2007). J Geophys Res 115:D13306. doi:10.1029/2009JD011908

Schäfer K, Thomas W, Peters A, Ries L, Obleitner F, Schnelle-Kreis J, Birmili W, Diemer J, Fricke W, Junkermann W, Pitz M, Emeis S, Forkel R, Suppan P, Flentje H, Gilge S, Wichmann HE, Meinhardt F, Zimmermann R, Weinhold K, Soentgen J, Münkel C, Freuer C, Cyrys J (2011) Influences of the 2010 Eyjafjallajökull volcanic plume on air quality in the northern Alpine region. Atmos Chem Phys 11:8555–8575. doi:10.5194/acp-11-8555-2011

Schaub D, Brunner D, Boersma KF, Keller J, Folini D, Buchmann B, Berresheim H, Staehelin J (2007) SCIAMACHY tropospheric NO_2 over Switzerland: estimates of NO_x lifetimes and impact of the complex Alpine topography on the retrieval. Atmos Chem Phys 7:5971–5987. doi:10.5194/acp-7-5971-2007

Scheel HE, Areskoug H, Geiß H, Gomiscek B, Granby K, Haszpra L, Klasinc L, Kley D, Laurila T, Lindskog A, Roemer M, Schmitt R, Simmonds P, Solberg S, Toupance G (1997) On the spatial distribution and seasonal variation of lower-troposphere ozone over Europe. J Atmos Chem 28:11–28. doi:10.1023/A:1005882922435

Schmitt R, Volz-Thomas A (1997) Climatology of ozone, PAN, CO, and NMHC in the free troposphere over the southern North Atlantic. J Atmos Chem 28:245–262. doi:10.1023/A:1005801515531

Schneider M, Blumenstock T, Chipperfield MP, Hase F, Kouker W, Reddmann T, Ruhnke R, Cuevas E, Fischer H (2005a) Subtropical trace gas profiles determined by ground-based FTIR spectroscopy at Izaña (28°N, 16°W): five-year record, error analysis, and comparison with 3-D CTMs. Atmos Chem Phys 5:153-167. doi:10.5194/acp-5-153-2005

Schneider M, Blumenstock T, Hase F, Höpfner M, Cuevas E, Redondas A, Sancho JM (2005b) Ozone profiles and total column amounts derived at Izaña,

Tenerife Island, from FTIR solar absorption spectra, and its validation by an intercomparison to ECC-sonde and Brewer spectrometer measurements. J Quant Spectrosc Radiat Transf 91:245-274. doi:10.1016/j.jqsrt.2004.05.067

Schneider M, Redondas A, Hase F, Guirado C, Blumenstock T, Cuevas E (2008) Comparison of ground-based Brewer and FTIR total O_3 monitoring techniques. Atmos Chem Phys 8:5535–5550. doi:10.5194/acp-8-5535-2008

Schöner W, Böhm R, Auer I (2012) 125 years of high-mountain research at Sonnblick Observatory (Austrian Alps)—from "the house above the clouds" to a unique research platform. Theor Appl Climatol 110:491–498. doi:10.1007/s00704-012-0689-8

Schuepbach E, Davies TD, Massacand AC (1999) An unusual springtime ozone episode at high elevation in the Swiss Alps: contributions both from cross-tropopause exchange and from the boundary layer. Atmos Environ 33:1735–1744. doi:10.1016/S1352-2310(98)00232-5

Schuepbach E, Friedli TK, Zanis P, Monks PS, Penkett SA (2001) State space analysis of changing seasonal ozone cycles (1988–1997) at Jungfraujoch (3580 m above sea level) in Switzerland. J Geophys Res 106:20413–20427. doi:10.1029/2000JD900591

Schultz M, Schmitt R, Thomas K, Volz-Thomas A (1998) Photochemical box modeling of long-range transport from North America to Tenerife during the North Atlantic Regional Experiment (NARE) 1993. J Geophys Res 103:13477–13488. doi:10.1029/97JD01481

Schwikowski M, Seibert P, Baltensperger U, Gäggeler HW (1995) A study of an outstanding Saharan dust event at the high-alpine site Jungfraujoch, Switzerland. Atmos Environ 29:1829–1842. doi:10.1016/1352-2310(95)00060-C

Seibert P (1990) South foehn studies since the ALPEX experiment. Meteorol Atmos Phys 43:91–103. doi:10.1007/BF01028112

Seibert P, Kromp-Kolb H, Kasper A, Kalina M, Puxbaum H, Jost DT, Schwikowski M, Baltensperger U (1998) Transport of polluted boundary layer air from the Po Valley to high-alpine sites. Atmos Environ 32:3953–3965. doi:10.1016/S1352-2310(97)00174-X

Seinfeld JH, Pandis SN (2006) Atmospheric chemistry and physics: from air pollution to climate change, 2nd edn. Wiley, Hoboken

Sellegri K, Laj P, Venzac H, Boulon J, Picard D, Villani P, Bonasoni P, Marinoni A, Cristofanelli P, Vuillermoz E (2010) Seasonal variations of aerosol size distributions based on long-term measurements at the high altitude Himalayan site of Nepal Climate Observatory-Pyramid (5079 m), Nepal. Atmos Chem Phys 10:10679–10690. doi:10.5194/acp-10-10679-2010

Semenov VK, Smirnov A, Aref'ev VN, Sinyakov VP, Sorokina LI, Ignatova NI (2005) Aerosol optical depth over the mountainous region in central Asia (Issyk-Kul Lake, Kyrgyzstan). Geophys Res Lett 32:L05807. doi:10.1029/2004GL021746

Senik IA, Elansky NF (2001) Surface ozone concentration measurements at the Kislovodsk high-altitude scientific station: temporal variations and trends. Izv Atmos Ocean Phys 37:110–119

Senik IA, Elansky NF, Belikov IB, Lisitsyna LV, Galaktionov VV, Kortunova ZV (2005) Main patterns of the temporal variability of surface ozone in the region of the town of Kislovodsk at 870 and 2070 m above sea level. Izv Atmos Ocean Phys 41:67–79

Sepúlveda E, Schneider M, Hase F, García OE, Gomez-Pelaez A, Dohe S, Blumenstock T, Guerra JC (2012) Long-term validation of tropospheric column-averaged CH_4 mole fractions obtained by mid-infrared ground-based FTIR spectrometry. Atmos Meas Tech 5:1425–1441. doi:10.5194/amt-5-1425-2012

Sharma UK, Kajii Y, Akimoto H (2000) Seasonal variation of C_2–C_6 NMHCs at Happo, a remote site in Japan. Atmos Environ 34:4447–4458. doi:10.1016/S1352-2310(00)00162-X

Sheu G-R, Lin N-H, Wang J-L, Lee C-T, Ou Yang C-F, Wang S-H (2010) Temporal distribution and potential sources of atmospheric mercury measured at a high-elevation background station in Taiwan. Atmos Environ 44:2393–2400. doi:10.1016/j.atmosenv.2010.04.009

Sholkovitz ER, Sedwick PN, Church TM (2009) Influence of anthropogenic combustion emissions on the deposition of soluble aerosol iron to the ocean: empirical estimates for island sites in the North Atlantic. Geochim Cosmochim Acta 73:3981–4003. doi:10.1016/j.gca.2009.04.029

Škerlak B, Sprenger M, Wernli H (2014) A global climatology of stratosphere-troposphere exchange using the ERA-Interim data set from 1979 to 2011. Atmos Chem Phys 14:913–937. doi:10.5194/acp-14-913-2014

Smirnov A, Holben BN, Slutsker I, Welton EJ, Formenti P (1998) Optical properties of Saharan dust during ACE 2. J Geophys Res 103:28079–28092. doi:10.1029/98JD01930

Sofen ED, Bowdalo D, Evans MJ (2016) How to most effectively expand the global surface ozone observing network. Atmos Chem Phys 16:1445–1457. doi:10.5194/acp-16-1445-2016

Solomon S (1999) Stratospheric ozone depletion: a review of concepts and history. Rev Geophys 37:275–316. doi:10.1029/1999RG900008

Solomon S, Daniel JS, Neely RR III, Vernier J-P, Dutton EG, Thomason LW (2011) The persistently variable "background" stratospheric aerosol layer and global climate change. Science 333:866–870. doi:10.1126/science.1206027

Spangl W, Schneider J, Moosmann L, Nagl C (2007) Representativeness and classification of air quality monitoring stations. Final report. Umweltbundesamt, Vienna

Spivakovsky CM, Yevich R, Logan JA, Wofsy SC, McElroy MB (1990) Tropospheric OH in a three-dimensional chemical tracer model: An assessment based on observations of CH_3CCl_3. J Geophys Res Atmos 95:18441–18471. doi:10.1029/JD095iD11p18441

Staehelin J, Thudium J, Buehler R, Volz-Thomas A, Graber W (1994) Trends in surface ozone concentrations at Arosa (Switzerland). Atmos Environ 28:75–87. doi:10.1016/1352-2310(94)90024-8

Staehelin J, Kegel R, Harris NRP (1998a) Trend analysis of the homogenized total ozone series of Arosa (Switzerland), 1926–1996. J Geophys Res 103:8389-8399. doi:10.1029/97JD03650

Staehelin J, Renaud A, Bader J, McPeters R, Viatte P, Hoegger B, Bugnion V, Giroud M, Schill H (1998b) Total ozone series at Arosa (Switzerland): homogenization and data comparison. J Geophys Res 103:5827-5841. doi:10.1029/97JD02402

Stevenson DS, Dentener FJ, Schultz MG, Ellingsen K, van Noije TPC, Wild O, Zeng G, Amann M, Atherton CS, Bell N, Bergmann DJ, Bey I, Butler T, Cofala J, Collins WJ, Derwent RG, Doherty RM, Drevet J, Eskes HJ, Fiore AM, Gauss M, Hauglustaine DA, Horowitz LW, Isaksen ISA, Krol MC, Lamarque J-F, Lawrence MG, Montanaro V, Müller J-F, Pitari G et al (2006) Multimodel ensemble simulations of present-day and near-future tropospheric ozone. J Geophys Res 111:D08301. doi:10.1029/2005JD006338

Stohl A, Eckhardt S, Forster C, James P, Spichtinger N, Seibert P (2002) A replacement for simple back trajectory calculations in the interpretation of atmospheric trace substance measurements. Atmos Environ 36:4635–4648. doi:10.1016/S1352-2310(02)00416-8

Streets DG, Bond TC, Carmichael GR, Fernandes SD, Fu Q, He D, Klimont Z, Nelson SM, Tsai NY, Wang MQ, Woo J-H, Yarber KF (2003) An inventory of gaseous and primary aerosol emissions in Asia in the year 2000. J Geophys Res 108:8809. doi:10.1029/2002JD003093

Streets DG, Canty T, Carmichael GR, de Foy B, Dickerson RR, Duncan BN, Edwards DP, Haynes JA, Henze DK, Houyoux MR, Jacob DJ, Krotkov NA, Lamsal LN, Liu Y, Lu Z, Martin RV, Pfister GC, Pinder RW, Salawitch RJ, Wecht KJ (2013) Emissions estimation from satellite retrievals: a review of current capability. Atmos Environ 77:1011–1042. doi:10.1016/j.atmosenv.2013.05.051

Strode SA, Jaeglé L, Jaffe DA, Swartzendruber PC, Selin NE, Holmes C, Yantosca RM (2008) Trans-Pacific transport of mercury. J Geophys Res 113:D15305. doi:10.1029/2007JD009428

Sudo K, Akimoto H (2007) Global source attribution of tropospheric ozone: long-range transport from various source regions. J Geophys Res 112:D12302. doi:10.1029/2006JD007992

Suzuki I, Hayashi K, Igarashi Y, Takahashi H, Sawa Y, Ogura N, Akagi T, Dokiya Y (2008) Seasonal variation of water-soluble ion species in the atmospheric aerosols at the summit of Mt. Fuji. Atmos Environ 42:8027–8035. doi:10.1016/j.atmosenv.2008.06.014

Suzuki I, Igarashi Y, Dokiya Y, Akagi T (2010) Two extreme types of mixing of dust with urban aerosols observed in Kosa particles: "after" mixing and "on-the-way" mixing. Atmos Environ 44:858–866. doi:10.1016/j.atmosenv.2009.11.030

Swartzendruber PC, Jaffe DA, Prestbo EM, Weiss-Penzias P, Selin NE, Park R, Jacob DJ, Strode S, Jaeglé L (2006) Observations of reactive gaseous mercury in the free troposphere at the Mount Bachelor Observatory. J Geophys Res 111:D24301. doi:10.1029/2006JD007415

Takahama S, Schwartz RE, Russell LM, Macdonald AM, Sharma S, Leaitch WR (2011) Organic functional groups in aerosol particles from burning and non-burning forest emissions at a high-elevation mountain site. Atmos Chem Phys 11:6367–6386. doi:10.5194/acp-11-6367-2011

Tanimoto H (2009) Increase in springtime tropospheric ozone at a mountainous site in Japan for the period 1998–2006. Atmos Environ 43:1358–1363. doi:10.1016/j.atmosenv.2008.12.006

Tanimoto H, Sawa Y, Matsueda H, Uno I, Ohara T, Yamaji K, Kurokawa J, Yonemura S (2005) Significant latitudinal gradient in the surface ozone spring maximum over East Asia. Geophys Res Lett 32:L21805. doi:10.1029/2005GL023514

Tanimoto H, Mukai H, Sawa Y, Matsueda H, Yonemura S, Wang T, Poon S, Wong A, Lee G, Jung JY, Kim KR, Lee MH, Lin NH, Wang JL, Ou-Yang CF, Wu CF, Akimoto H, Pochanart P, Tsuboi K, Doi H, Zellweger C, Klausen J (2007) Direct assessment of international consistency of standards for ground-level ozone: strategy and implementation toward metrological traceability network in Asia. J Environ Monit 9:1183–1193. doi:10.1039/b701230f

Tanimoto H, Ohara T, Uno I (2009) Asian anthropogenic emissions and decadal trends in springtime tropospheric ozone over Japan: 1998–2007. Geophys Res Lett 36:L23802. doi:10.1029/2009GL041382

Tanimoto H, Zbinden RM, Thouret V, Nédélec P (2015) Consistency of tropospheric ozone observations made by different platforms and techniques in the global databases. Tellus B 67:27073. doi:10.3402/tellusb.v67.27073

Tarasova OA, Elansky NF, Kuznetsov GI, Kuznetsova IN, Senik IA (2003) Impact of air transport on seasonal variations and trends of surface ozone at Kislovodsk high mountain station. J Atmos Chem 45:245–249. doi:10.5194/acp-9-4157-2009

Tarasova OA, Brenninkmeijer CAM, Jöckel P, Zvyagintsev AM, Kuznetsov GI (2007) A climatology of surface ozone in the extra tropics: cluster analysis of observations and model results. Atmos Chem Phys 7:6099–6117. doi:10.5194/acp-7-6099-2007

Tarasova OA, Senik IA, Sosonkin MG, Cui J, Staehelin J, Prévôt ASH (2009) Surface ozone at the Caucasian site Kislovodsk High Mountain Station and the Swiss Alpine site Jungfraujoch: data analysis and trends (1990–2006). Atmos Chem Phys 9:4157–4175. doi:10.5194/acp-9-4157-2009

Thompson AM, Cicerone RJ (1986) Possible perturbations to atmospheric CO, CH_4, and OH. J Geophys Res 91:10853–10864. doi:10.1029/JD091iD10p10853

Thoning KW, Tans PP, Komhyr WD (1989) Atmospheric carbon dioxide at Mauna Loa Observatory 2. Analysis of the NOAA GMCC data, 1974–1985. J Geophys Res 94:8549–8565. doi:10.1029/JD094iD06p08549

Timonen H, Ambrose JL, Jaffe DA (2013) Oxidation of elemental Hg in anthropogenic and marine airmasses. Atmos Chem Phys 13:2827–2836. doi:10.5194/acp-13-2827-2013

Toh YY, Lim SF, von Glasow R (2013) The influence of meteorological factors and biomass burning on surface ozone concentrations at Tanah Rata, Malaysia. Atmos Environ 70:435–446. doi:10.1016/j.atmosenv.2013.01.018

Tressol M, Ordonez C, Zbinden R, Brioude J, Thouret V, Mari C, Nedelec P, Cammas J-P, Smit H, Patz H-W, Volz-Thomas A (2008) Air pollution during the 2003 European heat wave as seen by MOZAIC airliners. Atmos Chem Phys 8:2133–2150. doi:10.5194/acp-8-2133-2008

Trickl T, Cooper OR, Eisele H, James P, Mücke R, Stohl A (2003) Intercontinental transport and its influence on the ozone concentrations over central Europe: three case studies. J Geophys Res 108:8530. doi:10.1029/2002JD002735

Tsutsumi Y, Zaizen Y, Makino Y (1994) Tropospheric ozone measurement at the top of Mt. Fuji. Geophys Res Lett 21:1727–1730. doi:10.1029/94GL01107

Tsutsumi Y, Igarashi Y, Zaizen Y, Makino Y (1998) Case studies of tropospheric ozone events observed at the summit of Mount Fuji. J Geophys Res 103:16935–16951. doi:10.1029/98JD01152

Turekian KK, Nozaki Y, Benninger LK (1977) Geochemistry of atmospheric radon and radon products. Annu Rev Earth Planet Sci 5:227–255. doi:10.1146/annurev.ea.05.050177.001303

Turnbull JC, Lehman SJ, Miller JB, Sparks RJ, Southon JR, Tans PP (2007) A new high precision $^{14}CO_2$ time series for North American continental air. J Geophys Res 112:D11310. doi:10.1029/2006JD008184

Turquety S, Logan JA, Jacob DJ, Hudman RC, Leung FY, Heald CL, Yantosca RM, Wu S, Emmons LK, Edwards DP, Sachse GW (2007) Inventory of boreal fire emissions for North America in 2004: importance of peat burning and pyroconvective injection. J Geophys Res 112:D12S03. doi:10.1029/2006JD007281

Tuzson B, Henne S, Brunner D, Steinbacher M, Mohn J, Buchmann B, Emmenegger L (2011) Continuous isotopic composition measurements of tropospheric CO_2 at Jungfraujoch (3580 m a.s.l.), Switzerland: real-time observation of regional pollution events. Atmos Chem Phys 11:1685–1696. doi:10.5194/acp-11-1685-2011

U.S. Environmental Protection Agency (2006) Air quality criteria for ozone and related photochemical oxidants (2006 Final). U.S. Environmental Protection Agency, Washington, DC

U.S. Environmental Protection Agency Clean Air Markets Division Clean Air Status and Trends Network (CASTNET) [Ozone - Hourly, Filter Pack Concentration -Weekly]. http://www.epa.gov/castnet. Accessed 14 Jan 2016

Umann B, Arnold F, Schaal C, Hanke M, Uecker J, Aufmhoff H, Balkanski Y, Van Dingenen R (2005) Interaction of mineral dust with gas phase nitric acid and sulfur dioxide during the MINATROC II field campaign: first estimate of the uptake coefficient γ_{HNO_3} from atmospheric data. J Geophys Res 110:D22306. doi:10.1029/2005JD005906

Usher CR, Michel AE, Grassian VH (2003) Reactions on mineral dust. Chem Rev 103:4883–4940. doi:10.1021/cr020657y

Val Martín M, Honrath RE, Owen RC, Pfister G, Fialho P, Barata F (2006) Significant enhancements of nitrogen oxides, black carbon, and ozone in the North Atlantic lower free troposphere resulting from North American boreal wildfires. J Geophys Res 111:D23S60. doi:10.1029/2006JD007530

Val Martin M, Honrath RE, Owen RC, Lapina K (2008a) Large-scale impacts of anthropogenic pollution and boreal wildfires on the nitrogen oxides over the central North Atlantic region. J Geophys Res 113:D17308. doi:10.1029/2007JD009689

Val Martin M, Honrath RE, Owen RC, Li QB (2008b) Seasonal variation of nitrogen oxides in the central North Atlantic lower free troposphere. J Geophys Res 113:D17307. doi:10.1029/2007JD009688

Van der A RJ, Eskes HJ, Boersma KF, van Noije TPC, Van Roozendae M, De Smedt I, Peters DHMU, Meijer EW (2008) Trends, seasonal variability and dominant NO_x source derived from a ten year record of NO_2 measured from space. J Geophys Res 113:D04302. doi:10.1029/2007JD009021

Van Dingenen R, Putaud J-P, Martins-Dos Santos S, Raes F (2005) Physical aerosol properties and their relation to air mass origin at Monte Cimone (Italy) during the first MINATROC campaign. Atmos Chem Phys 5:2203–2226. doi:10.5194/acp-5-2203-2005

VanCuren RA (2003) Asian aerosols in North America: extracting the chemical composition and mass concentration of the Asian continental aerosol plume from long-term aerosol records in the western United States. J Geophys Res 108:D20. doi:10.1029/2003JD003459

VanCuren RA, Cahill TA (2002) Asian aerosols in North America: frequency and concentration of fine dust. J Geophys Res 107:D24. doi:10.1029/2002JD002204

VanCuren RA, Cliff SS, Perry KD, Jimenez-Cruz M (2005) Asian continental aerosol persistence above the marine boundary layer over the eastern North Pacific: continuous aerosol measurements from Intercontinental Transport and Chemical Transformation 2002 (ITCT 2K2). J Geophys Res 110:D09S90. doi:10.1029/2004JD004973

Verheggen B, Cozic J, Weingartner E, Bower K, Mertes S, Connolly P, Gallagher M, Flynn M, Choularton T, Baltensperger U (2007) Aerosol partitioning between the interstitial and the condensed phase in mixed-phase clouds. J Geophys Res 112:D23202. doi:10.1029/2007JD008714

Verver G, Raes F, Vogelezang D, Johnson D (2000) The 2nd Aerosol Characterization Experiment (ACE-2): meteorological and chemical context. Tellus B 52:126–140. doi:10.1034/j.1600-0889.2000.00090.x

Viatte C, Schneider M, Redondas A, Hase F, Eremenko M, Chelin P, Flaud J-M, Blumenstock T, Orphal J (2011) Comparison of ground-based FTIR and Brewer O_3 total column with data from two different IASI algorithms and from OMI and GOME-2 satellite instruments. Atmos Meas Tech 4:535–546. doi:10.5194/amt-4-535-2011

Vigouroux C, De Mazière M, Errera Q, Mahieu E, Duchatelet P, Wood S, Smale D, Mikuteit S, Blumenstock T, Hase F, Jones N (2007) Comparisons between ground-based FTIR and MIPAS N_2O and HNO_3 profiles before and after assimilation in BASCOE. Atmos Chem Phys 7:377–396. doi:10.5194/acp-7-377-2007

Vigouroux C, De Mazière M, Demoulin P, Servais C, Hase F, Blumenstock T, Kramer I, Schneider M, Mellqvist J, Strandberg A, Velazco V, Notholt J, Sussmann R, Stremme W, Rockmann A, Gardiner T, Coleman M, Woods P (2008) Evaluation of tropospheric and stratospheric ozone trends over Western Europe from ground-based FTIR network observations. Atmos Chem Phys 8:6865–6886. doi:10.5194/acp-8-6865-2008

Vingarzan R (2004) A review of surface ozone background levels and trends. Atmos Environ 38:3431–3442. doi:10.1016/j.atmosenv.2004.03.030

Visheratin KN, Kamenogradskii NE, Kashin FV, Semenov VK, Sinyakov VP, Sorokina LI (2006) Spectral-temporal structure of variations in the atmospheric total ozone in central Eurasia. Izv Atmos Ocean Phys 42:184–202. doi:10.1134/S000143380602006X

Volz A, Kley D (1988) Evaluation of the Montsouris series of ozone measurements made in the nineteenth century. Nature 332:240–242. doi:10.1038/332240a0

Wai KM, Lin N-H, Wang S-H, Dokiya Y (2008) Rainwater chemistry at a high-altitude station, Mt. Lulin, Taiwan: comparison with a background station, Mt. Fuji. J Geophys Res 113:D06305. doi:10.1029/2006JD008248

Wainwright CD, Pierce JR, Liggio J, Strawbridge KB, Macdonald AM, Leaith RW (2012) The effect of model spatial resolution on Secondary Organic Aerosol predictions: a case study at Whistler, BC, Canada. Atmos Chem Phys 12:10911–10923. doi:10.5194/acp-12-10911-2012

Wang T, Wong HLA, Tang J, Ding A, Wu WS, Zhang XC (2006) On the origin of surface ozone and reactive nitrogen observed at a remote mountain site in the northeastern Qinghai-Tibetan Plateau, western China. J Geophys Res 111:D08303. doi:10.1029/2005JD006527

Weiss-Penzias P, Jaffe DA, Swartzendruber P, Dennison JB, Chand D, Hafner W, Prestbo E (2006) Observations of Asian air pollution in the free troposphere at Mount Bachelor Observatory during the spring of 2004. J Geophys Res 111:D10304. doi:10.1029/2005JD006522

Weiss-Penzias P, Jaffe D, Swartzendruber P, Hafner W, Chand D, Prestbo E (2007) Quantifying Asian and biomass burning sources of mercury using the Hg/CO ratio in pollution plumes observed at the Mount Bachelor observatory. Atmos Environ 41:4366–4379. doi:10.1016/j.atmosenv.2007.01.058

Welton EJ, Voss KJ, Gordon HR, Maring H, Smirnov A, Holben B, Schmid B, Livingston JM, Russell PB, Durkee PA, Formenti P, Andreae MO (2000) Ground-based lidar measurements of aerosols during ACE-2: instrument description, results, and comparisons with other ground-based and airborne measurements. Tellus B 52:636–651. doi:10.1034/j.1600-0889.2000.00025.x

Wild O, Pochanart P, Akimoto H (2004) Trans-Eurasian transport of ozone and its precursors. J Geophys Res 109:D11302. doi:10.1029/2003JD004501

Winkler P (1988) Surface ozone over the Atlantic ocean. J Atmos Chem 7:73–91. doi:10.1007/BF00048255

Wolfe GM, Thornton JA, McNeill VF, Jaffe DA, Reidmiller D, Chand D, Smith J, Swartzendruber P, Flocke F, Zheng W (2007) Influence of trans-Pacific pollution transport on acyl peroxy nitrate abundances and speciation at Mount Bachelor Observatory during INTEX-B. Atmos Chem Phys 7:5309–5325. doi:10.5194/acp-7-5309-2007

Wong JPS, Lee AKY, Slowik JG, Cziczo DJ, Leaitch WR, Macdonald A, Abbatt JPD (2011) Oxidation of ambient biogenic secondary organic aerosol by hydroxyl radicals: effects on cloud condensation nuclei activity. Geophys Res Lett 38:L22805. doi:10.1029/2011GL049351

Worden HM, Bowman KW, Worden JR, Eldering A, Beer R (2008) Satellite measurements of the clear-sky greenhouse effect from tropospheric ozone. Nat Geosci 1:305–308. doi:10.1038/ngeo182

Worden HM, Bowman KW, Kulawik SS, Aghedo AM (2011) Sensitivity of outgoing longwave radiative flux to the global vertical distribution of ozone characterized by instantaneous radiative kernels from Aura-TES. J Geophys Res 116:D14115. doi:10.1029/2010JD015101

Worden HM, Deeter MN, Frankenberg C, George M, Nichitiu F, Worden J, Aben I, Bowman KW, Clerbaux C, Coheur PF, de Laat ATJ, Detweiler R, Drummond JR, Edwards DP, Gille JC, Hurtmans D, Luo M, Martínez-Alonso S, Massie S, Pfister G, Warner JX (2013) Decadal record of satellite carbon monoxide observations. Atmos Chem Phys 13:837–850. doi:10.5194/acp-13-837-2013

World Data Centre for Greenhouse Gases, Tokyo. http://ds.data.jma.go.jp/gmd/wdcgg/. Accessed 15 Dec 2015

World Health Organization (2006) WHO Air quality guidelines for particulate matter, ozone, nitrogen dioxide and sulfur dioxide: global update 2005: summary of risk assessment. WHO Press, Geneva

World Meteorological Organization (2010) Guidelines for the measurement of atmospheric carbon monoxide, GAW Report No. 192. World Meteorological Organization, Geneva

World Meteorological Organization (2013) Guidelines for continuous measurements of ozone in the troposphere, GAW Report No. 209. World Meteorological Organization, Geneva

Wunch D, Toon GC, Blavier J-FL, Washenfelder RA, Notholt J, Connor BJ, Griffith DWT, Sherlock V, Wennberg PO (2011) The total carbon column observing network. Philos Trans R Soc A Math Phys Eng Sci 369:2087–2112. doi:10.1098/rsta.2010.0240

Xu WY, Lin WL, Xu XB, Tang J, Huang JQ, Wu H, Zhang XC (2015) Long-term trends of surface ozone and its influencing factors at the Mt. Waliguan GAW station, China—part 1: overall trends and characteristics. Atmos Chem Phys Discuss 15:30987–31024. doi:10.5194/acpd-15-30987-2015

Xue LK, Wang T, Zhang JM, Zhang XC, Poon CN, Ding AJ, Zhou XH, Wu WS, Tang J, Zhang QZ, Wang WX (2011) Source of surface ozone and reactive nitrogen speciation at Mount Waliguan in western China: new insights from the 2006 summer study. J Geophys Res 116:D07306. doi:10.1029/2010JD014735

Xue LK, Wang T, Guo H, Blake DR, Tang J, Zhang XC, Saunders SM, Wang WX (2013) Sources and photochemistry of volatile organic compounds in the remote atmosphere of western China: results from the Mt. Waliguan Observatory. Atmos Chem Phys 13:8551–8567. doi:10.5194/acp-13-8551-2013

Yienger J, Galanter M, Holloway TA, Phadnis MJ, Guttikunda SK, Carmichael GR, Moxim WJ, Levy II H (2000) The episodic nature of air pollution transport from Asia to North America. J Geophys Res 105:26931–26945. doi:10.1029/2000JD900309

Yonemura S, Tsuruta H, Kawashima S, Sudo S, Peng LC, Fook LS, Johar Z, Hayashi M (2002) Tropospheric ozone climatology over Peninsular Malaysia from 1992 to 1999. J Geophys Res 107:4229. doi:10.1023/A:1016152929037

Yurganov LN, Blumenstock T, Grechko EI, Hase F, Hyer EJ, Kasischke ES, Koike M, Kondo Y, Kramer I, Leung F-Y, Mahieu E, Mellqvist J, Notholt J, Novelli PC, Rinsland CP, Scheel HE, Schulz A, Strandberg A, Sussman R, Tanimoto H, Velazco V, Zander R, Zhao Y (2004) A quantitative assessment of the 1998 carbon monoxide emission anomaly in the Northern Hemisphere based on total column and surface concentration measurements. J Geophys Res 109:D15305. doi:10.1029/2004JD004559

Yurganov LN, Duchatelet P, Dzhola AV, Edwards DP, Hase F, Kramer I, Mahieu E, Mellqvist J, Notholt J, Novelli PC, Rockmann A, Scheel HE, Schneider M, Schulz A, Strandberg A, Sussmann R, Tanimoto H, Velazco V, Drummond JR, Gille JC (2005) Increased Northern Hemispheric carbon monoxide burden in the troposphere in 2002 and 2003 detected from the ground and from space. Atmos Chem Phys 5:563–573. doi:10.5194/acp-5-563-2005

Zander R, Demoulin P, Ehhalt DH, Schmidt U, Rinsland CP (1989) Secular increase of the total vertical column abundance of carbon monoxide above central Europe since 1950. J Geophys Res 94:11021–11028. doi:10.1029/JD094iD08p11021

Zander R, Mahieu E, Demoulin P, Duchatelet P, Roland G, Servais C, De Mazière M, Reimann S, Rinsland CP (2008) Our changing atmosphere: evidence based on long-term infrared solar observations at the Jungfraujoch since 1950. Sci Total Environ 391:184–195. doi:10.1016/j.scitotenv.2007.10.018

Zander R, Duchatelet P, Mahieu E, Demoulin P, Roland G, Servais C, Auwera JV, Perrin A, Rinsland CP, Crutzen PJ (2010) Formic acid above the Jungfraujoch during 1985-2007: observed variability, seasonality, but no long-term background evolution. Atmos Chem Phys 10:10047–10065. doi:10.5194/acp-10-10047-2010

Zanis P, Schuepbach E, Gäggeler HW, Hübener S, Tobler L (1999) Factors controlling beryllium-7 at Jungfraujoch in Switzerland. Tellus B 51:789–805. doi:10.1034/j.1600-0889.1999.t01-3-00004.x

Zanis P, Gerasopoulos E, Priller A, Schnabel C, Stohl A, Zerefos C, Gäggeler HW, Tobler L, Kubik PW, Kanter HJ, Scheel HE, Luterbacher J, Berger M (2003a) An estimate of the impact of stratosphere-to-troposphere transport (STT) on the lower free tropospheric ozone over the Alps using ^{10}Be and ^{7}Be measurements. J Geophys Res 108:8520. doi:10.1029/2002JD002604

Zanis P, Monks PS, Green TJ, Schuepbach E, Carpenter LJ, Mills GP, Rickard AR, Brough N, Penkett SA (2003b) Seasonal variation of peroxy radicals in the lower free troposphere based on observations from the FREE Tropospheric EXperiments in the Swiss Alps. Geophys Res Lett 30:1497. doi:10.1029/2003GL017122

Zanis P, Ganser A, Zellweger C, Henne S, Steinbacher M, Staehelin J (2007) Seasonal variability of measured ozone production efficiencies in the lower free troposphere of Central Europe. Atmos Chem Phys 7:223–236. doi:10.5194/acp-7-223-2007

Zellweger C, Ammann M, Buchmann B, Hofer P, Lugauer M, Rüttimann R, Streit N, Weingartner E, Baltensperger U (2000) Summertime NO_y speciation at the Jungfraujoch, 3580m above sea level, Switzerland. J Geophys Res 105:6655–6667. doi:10.1029/1999JD901126

Zellweger C, Forrer J, Hofer P, Nyeki S, Schwarzenbach B, Weingartner E, Ammann M, Baltensperger U (2003) Partitioning of reactive nitrogen (NO_y) and dependence on meteorological conditions in the lower free troposphere. Atmos Chem Phys 3:779–796. doi:10.5194/acp-3-779-2003

Zellweger C, Klausen J, Buchmann B (2007) System and performance audit for surface ozone and carbon monoxide at the Global GAW station Assekrem, Algeria, May 2007, WCC-Empa Report 07/2. Empa, Dübendorf

Zellweger C, Hüglin C, Klausen J, Steinbacher M, Vollmer M, Buchmann B (2009) Inter-comparison of four different carbon monoxide measurement techniques and evaluation of the long-term carbon monoxide time series of Jungfraujoch. Atmos Chem Phys 9:3491–3503. doi:10.5194/acp-9-3491-2009

Zenker T, Fischer H, Nikitas C, Parchatka U, Harris GW, Mihelcic D, Müsgen P, Pätz HW, Schultz M, Voltz-Thomas A, Schmitt R, Behmann T, Weißenmayer M, Burrows JP (1998) Intercomparison of NO, NO_2, NO_y, O_3, and RO_x measurements during the Oxidizing Capacity of the Tropospheric Atmosphere (OCTA) campaign 1993 at Izaña. J Geophys Res 103:13615–13634. doi:10.1029/97JD03739

Zhang L, Jacob DJ, Boersma KF, Jaffe DA, Olson JR, Bowman KW, Worden JR, Thompson AM, Avery MA, Cohen RC, Dibb JE, Flocke FM, Fuelberg HE, Huey LG, McMillan WW, Singh HB, Weinheimer AJ (2008) Transpacific transport of ozone pollution and the effect of recent Asian emission increases on air quality in North America: an integrated analysis using satellite, aircraft, ozonesonde, and surface observations. Atmos Chem Phys 8:6117–6136. doi:10.5194/acp-8-6117-2008

Zhang JM, Wang T, Ding AJ, Zhou XH, Xue LK, Poon CN, Wu WS, Gao J, Zuo HC, Chen JM, Zhang XC, Fan SJ (2009a) Continuous measurement of peroxyacetyl nitrate (PAN) in suburban and remote areas of western China. Atmos Environ 43:228-237. doi:10.1016/j.atmosenv.2008.09.070

Zhang L, Jacob DJ, Kopacz M, Henze DK, Singh K, Jaffe DA (2009b) Intercontinental source attribution of ozone pollution at western U.S. sites using an adjoint method. Geophys Res Lett 36:L11810. doi:10.1029/2009GL037950

Zhang F, Zhou LX, Novelli PC, Worthy DEJ, Zellweger C, Klausen J, Ernst M, Steinbacher M, Cai YX, Xu L, Fang SX, Yao B (2011) Evaluation of in situ measurements of atmospheric carbon monoxide at Mount Waliguan, China. Atmos Chem Phys 11:5195–5206. doi:10.5194/acp-11-5195-2011

Zhang B, Owen RC, Perlinger JA, Kumar A, Wu S, Val Martin M, Kramer L, Helmig D, Honrath RE (2014) A semi-Lagrangian view of ozone production tendency in North American outflow in the summers of 2009 and 2010. Atmos Chem Phys 14:2267–2287. doi:10.5194/acp-14-2267-2014

Zhu B, Akimoto H, Wang Z, Sudo K, Tang J, Uno I (2004) Why does surface ozone peak in summertime at Waliguan? Geophys Res Lett 31:L17104. doi:10.1029/2004GL020609

Zieger P, Kienast-Sjögren E, Starace M, Von Bismarck J, Bukowiecki N, Baltensperger U, Wienhold FG, Peter T, Ruhtz T, Collaud Coen M, Vuilleumier L, Maier O, Emili E, Popp C, Weingartner E (2012) Spatial variation of aerosol optical properties around the high-alpine site Jungfraujoch (3580 m a.s.l.). Atmos Chem Phys 12:7231–7249. doi:10.5194/acp-12-7231-2012

Zieman JJ, Holmes JL, Connor D, Jensen CR, Zoller WH, Hermann DM, Parrington JR, Gordon GE (1995) Atmospheric aerosol trace element chemistry at Mauna Loa Observatory 1. 1979–1985. J Geophys Res 100:25979–25994. doi:10.1029/93JD03316

Response of the mesosphere-thermosphere-ionosphere system to global change - CAWSES-II contribution

Jan Laštovička[1*], Gufran Beig[2] and Daniel R Marsh[3]

Abstract

Long-term trends in the mesosphere, thermosphere, and ionosphere are areas of research of increasing importance both because they are sensitive indicators of climatic change and because they affect satellite-based technologies which are increasingly important to modern life. Their study was an important part of CAWSES-II project, as they were a topic of Task Group 2 (TG-2) 'How Will Geospace Respond to Changing Climate'. Three individual projects of TG-2 were focused on important problems in trend investigations. Significant progress was reached in several areas such as understanding and quantifying the role of stratospheric ozone changes in trends in the upper atmosphere, reaching reasonable agreement between observed and simulated trends in mesospheric temperatures and polar mesospheric clouds, or understanding of why the thermospheric density trends are much stronger under solar cycle minimum conditions. The TG-2 progress that is reviewed in this paper together with some results reached outside CAWSES-II so as to have the full context of progress in trends in the upper atmosphere and ionosphere.

Keywords: Mesosphere; Thermosphere; Ionosphere; Long-term trends; Climatic change

Review

Introduction

The anthropogenic emissions of greenhouse gases carbon dioxide (CO_2), methane (CH_4), and nitrous oxide (N_2O) influence the troposphere, weather, and particularly climate. They appear to be the primary driver of a significant increase of surface temperature in recent decades, thus affecting directly our lives. Another anthropogenic pollutant of the atmosphere, ozone depleting substances (ODSs), is responsible for the depletion of the Earth's stratospheric ozone layer, particularly over the Antarctica, in the form of the spring ozone hole. Both greenhouse gases and ODS thus affect the environment and can adversely influence human health. However, these polluting substances affect not only the troposphere and stratosphere but also the upper atmosphere, including the mesosphere (approximately 50 to 90 km), the thermosphere (approximately 90 to 1000 km), and the ionosphere, which is embedded in the upper atmosphere (e.g., Rishbeth and Roble 1992; Laštovička et al. 2012). The thermosphere is the operating environment of many satellites, including the International Space Station, and thousands of pieces of dangerous space debris, the orbital lifetime of which depends on long-term changes of thermospheric density. Propagation of Global Positioning System (GPS) signals and radio communications are affected by the ionosphere. Thus, anthropogenic changes of the high-altitude regions, thermosphere, and ionosphere can also affect satellite-based technologies which are increasingly important to modern life. As with global change near the Earth's surface, the challenge faced by the upper atmosphere climate scientists is to detect long-term trends and understand their primary causes, so that society can mitigate against potential harmful changes.

The approximately 0.85°C increase in global surface air temperature since 1880 (IPCC 2013) has been attributed predominantly to the increasing atmospheric concentration of greenhouse gases. However, the greenhouse gas increase has an opposite, cooling effect in the upper atmosphere. Greenhouse gases in the troposphere are optically thick to outgoing long-wave (infrared) radiation, which they absorb. Then they emit radiation, corresponding to local

* Correspondence: jla@ufa.cas.cz
[1]Institute of Atmospheric Physics ASCR, Bocni II, Prague 14131, Czech Republic
Full list of author information is available at the end of the article

temperature, and part of this emission goes back to the surface to produce the heating effect. In contrast, greenhouse gases in the much lower density upper atmosphere are optically thin to outgoing infrared radiation. *In situ* collisional excitation results in atmospheric thermal energy easily lost to space via outgoing infrared radiation causing infrared radiative cooling, while the absorption of radiation emanating from the lower atmosphere plays only a secondary role in the energy balance. The net result is that the radiatively active greenhouse gases act as cooling agents, and their increasing concentrations enhance the cooling effect in the upper atmosphere. This effect of greenhouse gases may be called 'greenhouse cooling' (Cicerone 1990).

The global atmosphere is close to a hydrostatic equilibrium, which means that the height of a given pressure surface is determined by the average atmospheric temperature below. The cooling is expected to result in thermal contraction of the upper atmosphere, and we may expect a significant decline in thermospheric density at fixed heights, which is observed in long-term satellite drag data (e.g., Emmert et al. 2008). Downward displacement of ionospheric layers should accompany this contraction, and this has also been observed (e.g., Laštovička et al. 2012). The cooling also affects chemical reaction rates and, thus, the chemistry of minor constituents, resulting in further changes to the ionosphere.

The combination of the lower atmosphere heating and the upper atmosphere cooling is supported by a much stronger greenhouse effect that is observed on Venus where the 96% concentration of carbon dioxide in the atmosphere and a more dense atmosphere result in a troposphere that is more than twice as warm as the Earth's troposphere and a thermosphere that is 4 to 5 times colder than the Earth's thermosphere (e.g., Bougher and Roble 1991). While the Earth's troposphere is much colder than the thermosphere, the Venusian troposphere is substantially warmer than the thermosphere.

Long-term changes in the upper atmosphere and ionosphere have been of interest since the pioneering study of Roble and Dickinson (1989). They suggested that global cooling will occur in the upper atmosphere in conjunction with global warming in the troposphere due to long-term increase of greenhouse gas concentrations, particularly CO_2. Modeling studies by Rishbeth (1990) and Rishbeth and Roble (1992) broadened these results to the thermosphere-ionosphere system. Investigation of long-term changes in the upper atmosphere has become a significant topic in global change investigation, and many results have been published in the last more than 20 years (e.g., review by Laštovička et al. 2012 and references herein).

After years of rather chaotic and partly controversial observational results, a sufficient amount of information made it possible to create a first scenario of global change in the mesosphere, thermosphere, and ionosphere by an international team (Laštovička et al. 2006a, 2008a). Summarized in Figure 1, it reveals a cooling in the stratosphere, mesosphere, and thermosphere with no trend in the coolest part of the atmosphere, the mesopause, and some changes in the electron density profile in various ionospheric layers. Its main feature is cooling of the upper atmosphere, which results in thermal contraction of the mesosphere and thermosphere and changes in the ionosphere through changes in chemical composition and recombination rates. However, at that time, results concerning some variables were not consistent with the trend scenario and many variables were missing. This scenario has been continuously improved and updated (e.g., see reviews or progress reports by Qian et al. 2011; Laštovička et al. 2012; Danilov 2012; Laštovička 2013a). At present, the scenario is formed by essentially mutually consistent trends in the mesospheric temperature, electron concentration in the lower ionosphere, E region, F1 region, and F2 region; height of maximum electron concentration in the E region and F2 region; the thermospheric neutral density; F-region ion; and electron temperatures. These trends are basically mutually consistent and qualitatively agree with model simulations that show that they are the consequences of the enhanced greenhouse gases. However, there are also some problems with these trends. For example, ion temperature trends above approximately 350 km in middle latitudes are rather opposite to those at high latitudes (Laštovička 2013a). Electron temperature trends are still very preliminary. There are some discrepancies between electron density trends from different stations, and they seem to depend to some extent on the time of the day (e.g., Qian et al. 2009). Trend information about several quantities is missing or is not reliable.

It is useful to clarify what the term 'trend' means, as its understanding can differ among authors. Most authors use the term trend as a 'long-term trend', i.e., as a long-term tendency of essentially monotonic or quasi-linear change, either an increase or decrease of the values of a particular variable with time. If the long-term behavior is substantially unstable on medium time-scales, or oscillatory, the term 'long-term change' is usually used. Strictly speaking, trends are often not linear. However, in most cases, the linear trend approximation is sufficient; in other cases, we use the linear approximation to facilitate comparison with trends in other parameters. Ideally, a trend continues to infinity. In reality, a period of quasi-stable trends begins and ends in concrete years/periods, such as the Antarctic ozone hole, which appeared in the late 1970s, that has recently stabilized, and it is expected to disappear in the second half of this century (Eyring et al. 2010). Such behavior may be

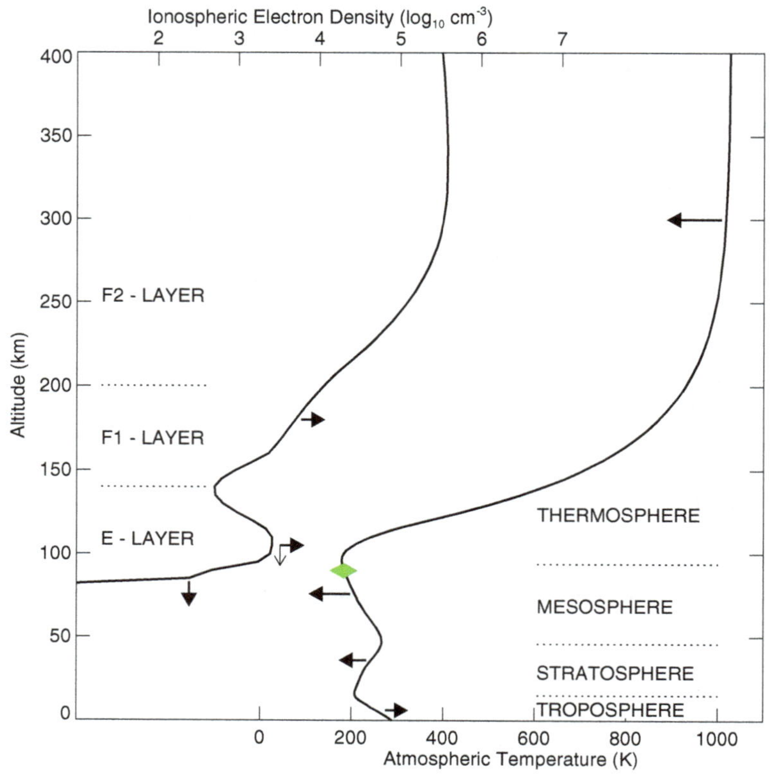

Figure 1 Trends in the Earth's atmosphere. The atmospheric layers are defined by the temperature profile. The ionospheric layers are defined by the electron density profile (midnight at equator). Arrows indicate the direction of change. Red - warming, blue - cooling, green - no change of temperature, black - changes in electron density (horizontal) and heights of ionospheric layers (vertical). After Laštovička et al. (2008a).

characterized by a sophisticated mathematical curve, but it is often approximated by a piecewise linear trend model (Weatherhead et al. 2000; Reinsel et al. 2005), wherein substantially different linear trends are fit to the data within different time intervals. The upper atmosphere and ionosphere linear trends should be understood in such a way, even though for some variables the period of quasi-stable trend may be identical with the period of available data for trend studies, i.e., such trends might appear to be monotonic.

It is also possible to use a slightly different and more formal mathematical approach. A truly linear trend in a time series, by definition, corresponds to the lowest frequencies that are not taken into account in a spectral decomposition. More precisely, quoting Kendall (1973), 'The essential idea of the trend is that it shall be smooth', which is equivalent to a continuous slowly varying change in a time series over long-time scales (Graigmile et al. 2004). Assuming that the trend is not a small temporal part of a major very long-period cycle or due to some temporally-coincident stochastic change, then it is possible to temporally and spectrally decompose the time series to separate the oscillatory and true trend elements using, for instance, the wavelet transform technique. However, practical application of such an approach has to be made with care, as the trend has a non-infinite period limited by the length of the dataset or by the limited length of the stable trend interval. This is probably the reason why the attempt of a Chilean team to use such an approach in foF2 trend analysis (Laštovička et al. 2006b) was unsuccessful (Laštovička et al. 2008b).

The international SCOSTEP project Climate and Weather of the Sun-Earth System (CAWSES-II), ran in 2009 to 2013. It consisted of several Task Groups (TGs), one of them being TG-2 'How Will Geospace Respond to Changing Climate'. Geospace, here, is understood to be the upper atmosphere (mesosphere and thermosphere) and ionosphere. TG-2 was organized around three projects that aimed to answer three questions:

Project 1. How do changes in tropospheric wave generation and their propagation through a changing atmosphere affect the dynamics of the MLT?

Project 2. By how much is the anthropogenic effect on the ionosphere/thermosphere enhanced during a quiet sun period?

Project 3. Are PMC/NLC frequency and brightness trending due to changes in temperature and water vapor? Do they differ between the hemispheres?

It should be mentioned that TG-2 also organized or co-organized international scientific meetings. Project 3organized the CAWSES-II workshop on modeling PMC trends. The workshop was held at LASP, University of Colorado, Boulder, USA on 3 to 4 May 2012 with the participation of 35 scientists and students. TG-2 co-organized two IAGA/ICMA/CAWSES workshops 'Long Term Changes and Trends in the Atmosphere' on 15 to 18 June 2010 at NCAR, Boulder, USA and on 11 to 14 September 2012 in Buenos Aires, Argentina. Participation at both workshops was approximately 70 scientists. A special issue of the Journal of Geophysical Research was published from each workshop. Both workshops provided a clear picture of where we are in studying long-term trends in the upper atmosphere and ionosphere and what are the goals for research in the near future.

The aim of this paper is to describe the main findings of each TG-2 project in the context of the general progress in the field of long-term trends in the mesosphere, thermosphere, and ionosphere. In recent years, it became quite clear that there are also other drivers of trends. This is described in the 'Other drivers of long-term trends' section. The 'Are there long-term changes in tropospheric wave generation and their propagation through a changing atmosphere?' and 'Are trends related to upward propagation of atmospheric waves observed in the mesosphere, lower thermosphere and ionosphere (MLTI)?' sections deal with the main results and activities of Project 1. Findings of Project 2 are in the 'By how much is the anthropogenic effect on the ionosphere/thermosphere enhanced during a quiet sun period?' section and those of Project 3 in the 'Are PMC/NLC frequency and brightness trending due to changes in temperature and water vapor?' section. The 'Important results reached outside of the CAWSES-II project TG-2' section deals with important results reached outside of these three projects; most of them were reached by scientists participating in TG-2 of CAWSES-II. The paper is closed by a brief 'Conclusions' section, which includes ideas for future investigations.

Other drivers of long-term trends

In recent years, it became increasingly clear that rising concentrations of greenhouse gases are the main, but not the only, driver of global change in the upper atmosphere (e.g., Laštovička et al. 2012). In the stratosphere, mesosphere, lower thermosphere, and related part of the ionosphere, some role is played by anthropogenic changes of the ozone layer (already mentioned in the introduction) and increasing concentration of water vapor. In the thermosphere and ionosphere, long-term variations of geomagnetic activity and secular changes of the Earth's magnetic field (both important particularly for the F2 region ionosphere) play a role. Due to correction for the solar cycle effect when calculating the trends, the direct impact of long-term changes of solar activity is largely removed. However, an indirect effect remains, at least in thermospheric density (see 'By how much is the anthropogenic effect on the ionosphere/thermosphere enhanced during a quiet sun period?' section). The poorly known trends in atmospheric circulation and particularly in atmospheric wave activity with predominantly tropospheric sources (one of potential trend drivers) are the key open question in the upper atmosphere and ionosphere trends. Temporal and spatial changes of long-term trends in the upper atmosphere and ionosphere are inevitable (e.g., Laštovička et al. 2012) because trends of some of these drivers change principally with time (e.g., ozone or geomagnetic activity) or location (e.g., Earth's magnetic field). Long-term changes in the relative role of greenhouse gases versus solar/geomagnetic activity in long-term trends in the atmosphere-ionosphere system are documented by Laštovička (2005). The long-term instability of trends is a problem in their investigation, simulation and interpretation, and points to the necessity of continuous monitoring and investigation of such trends.

An important driver of trends is the long-term change in geomagnetic activity. An increase of geomagnetic activity was observed throughout the twentieth century (e.g., Clilverd et al. 1998; Martini and Mursula, 2008) followed by a drop to rather low level at present (particularly 2008 to 2010). Geomagnetic activity might even be a dominant cause of ionospheric trends in the past. Bremer et al. (2009) reported the dominant role of the long-term change of geomagnetic activity on ionospheric trends around 1970 in the E region, in the mid-1990s in the F1 region, and around 2000 in the F2 region. Thus, it appears that at present, the long-term changes in geomagnetic activity are not the main driver of trends in any variable, contrary to the past, when they might have controlled ionospheric trends.

A regionally important driver is the secular change of the main magnetic field of the Earth. Its effect is insignificant in most regions but in the equatorial to southern mid-latitude western Atlantic Ocean and in related part of the South America it seems to play a substantial role in trends in hmF2 and foF2 (e.g., Cnossen and Richmond 2008, 2013). This effect is dominated by changes of inclination and to some extent also declination of the Earth's magnetic field, but changes of field strength cannot be neglected, as well.

In the mesosphere and lower thermosphere, important additional drivers appear to be stratospheric ozone, changes in CH_4, and mesospheric water vapor concentration. An important but unclear role is played by changes of the MLT region dynamics, including atmospheric wave activity. Figure 2 shows the results of model calculations of O_3 depletion (halved), H_2O increase (doubled), and CH_4 increase (doubled) effect on mesospheric and thermospheric temperature and neutral density (Qian et al. 2011). Model changes of O_3, H_2O, and CH_4 are much higher than the observed changes. Nevertheless, Figure 2 illustrates that effect of these trace gases at mesospheric and lower thermospheric altitudes cannot be neglected. Model calculations by Qian et al. (2013), with realistic values of concentrations of radiatively active trace gases in the middle atmosphere, confirm the dominant role of CO_2 in the thermosphere with minor but, even at 400 km, not quite negligible influence of middle atmospheric ozone, CH_4, and water vapor. Ionospheric evidence from the D and E regions supports the role of stratospheric ozone in long-term trends (e.g., Laštovička et al. 2012 and references herein).

An increase in mesospheric water vapor is expected from the 150% increase in CH_4 over preindustrial levels (IPCC 2013), and that water is produced from CH_4 oxidation in the middle atmosphere. However, observations show a recent drop in the stratospheric water vapor concentration in 2001 and again in 2008 (e.g., Solomon et al. 2010a), suggesting trends in the mesospheric water vapor concentration are far from linear. The role of stratospheric ozone in long-term trends in the MLTI region is crucial but it is treated in 'Are trends related to upward propagation of atmospheric waves observed in the mesosphere, lower thermosphere and ionosphere (MLTI)?' section.

Are there long-term changes in tropospheric wave generation and their propagation through a changing atmosphere?

It is very likely that trends in the lower atmosphere have an influence on the upper atmosphere via dynamical coupling through upward propagating waves. There is modeling evidence that fluxes of waves entering the mesosphere, lower thermosphere, and ionosphere (MLTI) have changed and will continue to change. Trends in the stratospheric Brewer-Dobson circulation were analyzed by Hardiman et al. (2013) in a number of historical and future simulations using 'high-top' climate models. They found annual mean tropical upwelling increased in the models, and that the width of upwelling region decreased below the 20 hPa pressure level and increased above that level. This is believed to be caused by changes in the propagation of planetary and gravity waves. Such changes in the tropical upwelling could affect the timing of the quasi-biennial oscillation and its imprint in the mesosphere (Ratnam et al. 2008), and also the speed of the tropical water vapor tape recorder. However, age of air data does not confirm a simple pattern of reduction of age of air as a consequence of the Brewer-Dobson circulation intensification (Engel et al. 2009; Stiller et al. 2012). There is not one positive or negative linear variation of age of air over the globe (Stiller et al. 2012). There are hemispheric differences and seasonal differences in the long-term behavior of age of air. A different long-term behavior of the lower and upper branches of the Brewer-Dobson circulation is quite possible.

New modeling studies using the Whole Atmospheric Community Climate Model (WACCM) (Smith et al. 2010) indicate a significant role of the development of the Antarctic stratospheric ozone hole in long-term trends in the MLTI region. Dramatic ozone loss in the lower stratosphere results in a temperature decrease in the polar lower stratosphere and a change in the zonal wind distribution. Smith et al. (2011) showed that this change modifies gravity wave propagation into the mesosphere during late spring so that the mesospheric residual mean circulation is weakened by the gravity wave drag. This leads to a reduction in polar ascent rates and thus anomalous adiabatic heating and more than a 4 K increase in temperature at 80 km between 1960 and 2006 over the summer pole, and a 6 K cooling above that level. In contrast, little change was seen over the summer pole in the Northern Hemisphere. The mesospheric warming over

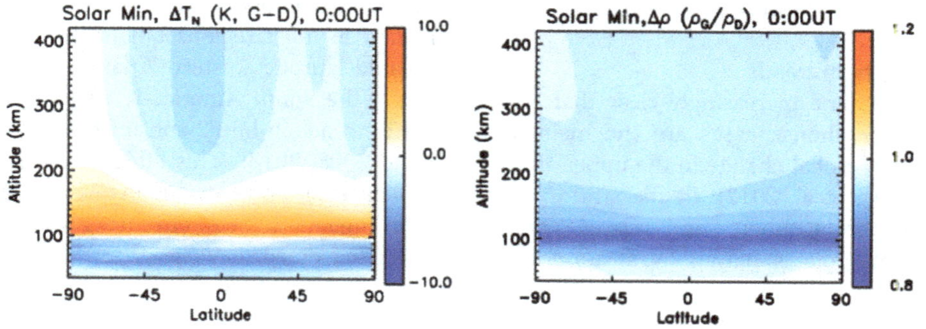

Figure 2 Changes of zonal mean. Changes of zonal mean neutral temperature (left panel) and neutral density (right panel) due to trace gases other than CO_2, such as O_3 depletion (halved), H_2O increase (doubled), and CH_4 increase (doubled) under solar minimum and geomagnetic quiet conditions, at 0000 LT, doubled CO_2, simulated by the NCAR TIMEGCM model. After Qian et al. (2011).

the Antarctic was not present in a simulation in which ODS concentrations were kept constant at pre-ozone-hole levels. This influence of the ozone hole in changing wave propagation into the MLTI was later confirmed in simulations using the Canadian Middle Atmosphere Model (CMAM) (Lossow et al. 2012), although their effect was less pronounced in December in CMAM than WACCM, likely due to too delayed breakup of the polar vortex in WACCM.

Another possible source of medium-to-long-term variability in wave propagation is the 11 year solar cycle. Ern et al. (2013) presented model simulations that suggest a small increase in planetary wave activity during solar maximum. There is also some observational evidence that gravity wave momentum fluxes vary out of phase with the solar cycle in the middle stratosphere (Ern et al. 2011); however, this finding is based on just nine years of data. The role of stratospheric ozone in long-term trends in the MLTI region is crucial, but it is treated in the next section.

Are trends related to upward propagation of atmospheric waves observed in the mesosphere, lower thermosphere, and ionosphere (MLTI)?

Two quantities, with some long-term data series of observations at MLT heights, are winds and temperatures; therefore, below we summarize trends in these two quantities. However, the problem is that the situations with trends in atmospheric wave activity at MLTI heights is not clear and, unfortunately, there are strong indications that these trends might differ substantially by region.

First results on long-term trends in zonal winds from multi-instrument measurements in the tropical mesosphere over India, 1978 to 2006, for heights approximately 70 to 80 km (Ratnam et al. 2013) reveal statistically significant negative trends in northern winter (December-March) and in September with maximum trends between 2 and 3 m/s/year in late winter (February-March), changing zonal wind from strong eastward during the 1970s to weak westward in recent years. This data series is the compilation of rocket measurements at Thumba, MST (mesosphere-stratosphere-troposphere) radar measurements at Gadanki and HRDI/UARS satellite measurements in the Indian sector near 10°N. Simulations with WACCM for 11.5°N, in winter, qualitatively confirm this result. They provide similar negative trends for all longitudes at heights of 70 to 80 km, whereas trends above 80 km are longitude dependent (Ratnam et al. 2013). Rao et al. (2012) analyzed approximately 80 to 100 km altitude winds at low latitudes based on measurements of four medium-frequency radars and three meteor radars. The analyzed data series were 20 years (1990 to 2010) long or shorter at individual stations, i.e., shorter than that used by Ratnam et al. (2013). Rao et al. (2012) did not find trends in the zonal wind; in the meridional wind, they found trends only for some stations, for other station they did not observe any trend.

At middle latitudes, winds at heights of approximately 90 to 95 km reveal considerable differences in long-term behavior between various regions. Figure 3 illustrates such differences for three stations near 52°N, Obninsk near Moscow in Russia, Collm in eastern Germany, and Saskatoon in western Canada in winter (December-February). For zonal wind (Figure 3a), each station behaves in its separate way, Obninsk and Saskatoon being almost out-of-phase. The meridional wind reveals more coherent pattern with some long-term (not short-term) similarity for all three stations. The most different pattern is provided by the semidiurnal tidal amplitudes (Figure 3c). Obninsk and Collm display large, out-of-phase changes, whereas no long-term change is seen for Saskatoon. Also, trends in semidiurnal tidal phases reveal evidently different behavior: negative trend in Saskatoon, no trend in Collm, and positive trend in Obninsk. However, such a diverse behavior of mid-latitudinal winds at heights approximately 90 to 95 km is supported by the results of model simulations by Qian et al. (2011), which reveal a different behavior of trends for different longitudinal regions even as for sign of the trend for zonal wind.

Jacobi and Beig (2013) reported a change of trends in winds in the 1990s. Laštovička et al. (2008a) reported a change of trends in annual prevailing MLT winds at northern higher middle latitudes near 1990. The change of trends around 1990 occurs not only in the MLT annual prevailing wind but also in the lower stratospheric winds at 100 hPa at 50°N to 70°N, in the occurrence frequency of small laminae in ozone profiles in Europe near 50°N (Laštovička et al. 2010), and in the stationary planetary wave with zonal wavenumber 1 near 50°N (Jacobi et al. 2009). This indicates the possibility of changes in trends in dynamics of the entire northern mid-latitude middle atmosphere.

Little progress on trends in atmospheric wave activity has been reached. This is the key problem of upper atmospheric trends. Jacobi et al. (2012) found regionally significantly different trends in tidal activity based on the mesopause region wind measurements at higher middle latitudes as mentioned above (Figure 3).

Hoffmann et al. (2011) analyzed winds measured at Juliusruh in the northernmost Germany. They observed a positive trend in activity of T = 3 to 6 h gravity waves in the height range of approximately 80 to 90 km over 1990 to 2010. However, they also mentioned that the gravity wave activity at middle latitudes tends to reveal trends at some stations but no trend at other stations in general agreement with previous results. One possible reason of even locally opposite trends in gravity wave activity at middle latitudes might be the shifts of storm tracks, one of mid-latitude gravity wave sources, towards

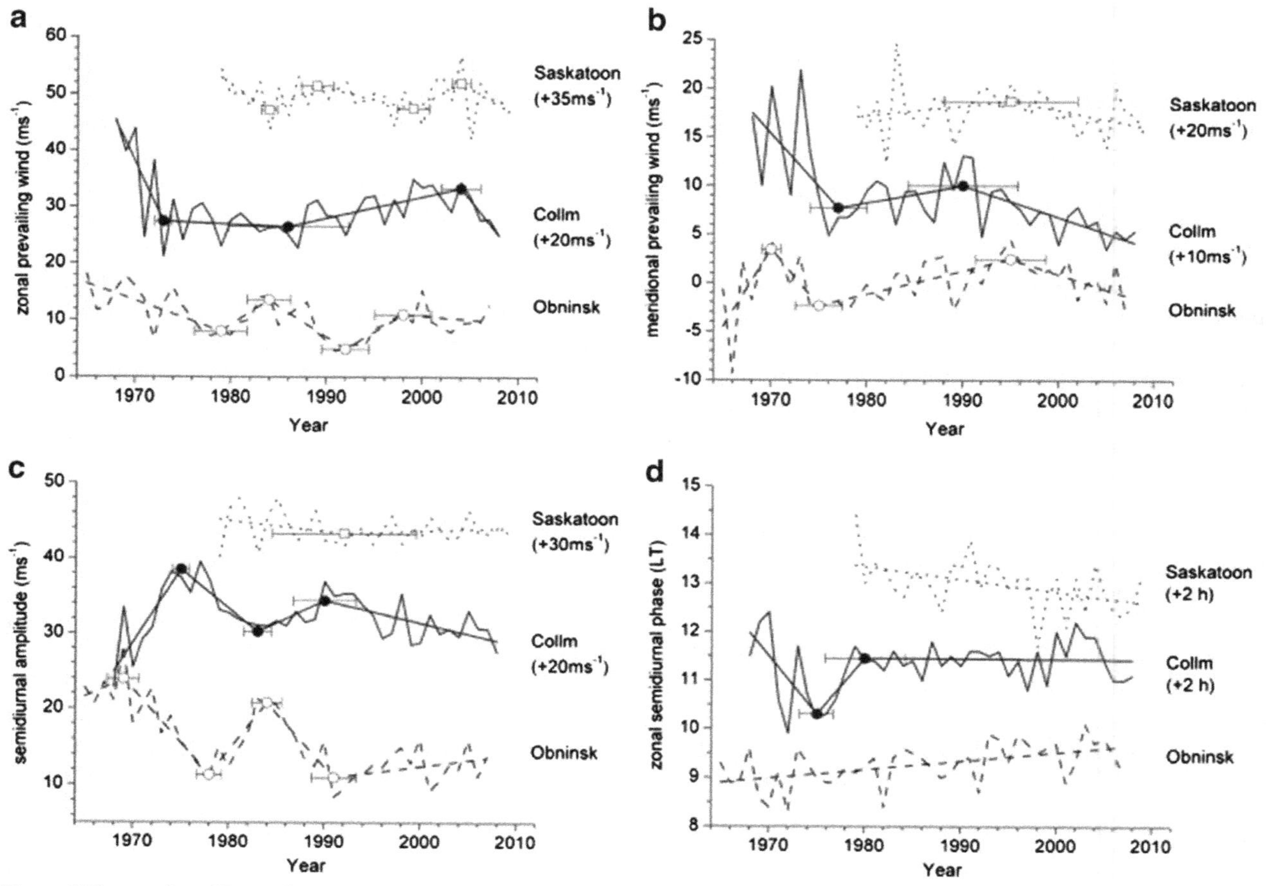

Figure 3 Time series of December to February. Time series of December to February averages of approximately 90 to 95 km winds at northern higher middle latitudes for Collm, Obninsk, and Saskatoon: **(a)** zonal prevailing winds, **(b)** meridional prevailing winds, **(c)** semidiurnal amplitude, and **(d)** zonal semidiurnal phase. After Jacobi et al. (2012).

or away from stations. Jacobi and Beig (2013) reported for Juliusruh and Collm (Germany) a positive trend in gravity wave activity in winter and no or a weak negative trend in summer. Tsuda et al. (2013) analyzed data of seven radars in the tropics since the 1990s; they found strengthening of shorter period (20 to 120 min) gravity wave activity and no trend for longer periods (2 to 10 h).

The planetary wave activity trends in the mesopause region over Central Europe are highly variable with period range and season, being generally positive in zonal wind and weaker and negative in meridional wind, and in total wind being in average slightly positive. The trends are more pronounced for short periods (T < 7 days) and for summer, whereas at longer periods and in other seasons, they are relatively small and statistically insignificant. Together with scarce information from other regions, they do not provide a clear pattern, as summarized by Laštovička et al. (2012).

Jacobi and Beig (2013) observed a tendency towards an earlier spring transition and a later fall transition in winds. Sonnemann et al. (2012) simulated the impact of dynamics on long-term trends in the MLT region with the LIMA model. They found that in summertime, middle and high latitude MLT region changes in dynamics can counteract the direct impact of anthropogenic increase of greenhouse gases.

Significant progress has been reached since the first review on trends in mesospheric and mesopause region temperatures by Beig et al. (2003). Temperature trend research in the mesopause region (80 to 100 km) witnessed dramatic ups and downs during the past decade (Beig 2011). The consensus of the trend community was that the mesopause region was not trending while the upper mesosphere and thermosphere were cooling. However, as the data series has become longer, it is observed that the trend in this region is showing the tendency towards cooling which is change from the earlier perception. The change of trends in stratospheric ozone in the mid-to-second-half of the 1990s had a noticeable impact and resulted in some changes in temperature trends as described in 'Other drivers of long-term trends' section.

Lübken et al. (2013) pointed out differences between temperature trends at fixed heights and fixed pressure levels. These differences are noticeable even though not principal. Temporal changes of mesospheric temperature are larger at fixed heights than at fixed pressure levels as shown in Figure 4. This figure also illustrates the changes in trend of mesospheric temperatures in relation to changes in ozone trends - steeper negative trend in 1980 to 1995 versus no or slightly positive trend after the mid-1990s. Simulations with model LIMA show that the largest CO_2 effect on temperatures is in the middle mesosphere (Lübken et al. 2013). These simulations also reveal thermal shrinking of the mesosphere as a consequence of cooling, which reaches near the height of 80 km and about 1 km from 1960 to mid-1990s, which coincides well with the shrinking revealed by the observations of the LF radio wave phase reflection heights in Germany (Lübken et al. 2013).

OH airglow-derived temperatures (approximately 87 km, mesopause region) above Wuppertal (Germany, 51°N) over 1988 to 2008 reveal a slight but statistically significant, overall cooling trend with substantially different trends in individual months which ranges between 0 and −0.6 K/yr, and systematic changes in the shape of seasonal variation (size of summer) of temperature (Offermann et al. 2010). An important finding of this study is an apparent break in trends around 1996, which coincides with the beginning of the recovery phase of stratospheric ozone. She et al. (2009) have carried out an assessment of temperature trends between 85 and 105 km based on 960 nights of Na lidar observation over a period of approximately 20 years over Fort Collins (41°N, 105°W). This study suggests that the impact of volcanic eruptions should be accounted for in trend calculations.

Mlynczak et al. (2010) have analyzed 7 years of observations of the radiative cooling in the thermosphere as observed by the SABER/TIMED. The observed weakening of the equator to the pole gradient in radiative cooling over these 7 years is a strong indication that the large scale thermospheric dynamics have also weakened in that period.

Tropical data over the past 2 decades indicates a cooling trend, but the magnitude is found to be smaller than the earlier results reported mainly from rocketsonde measurements carried out during the late 1960s to early 1990s (Beig 2011). This cooling trend increases with height in the mesosphere and becomes around 3 K/decade around 70 km, but with an uncertainty of about ±1 to 2 K. The cooling trends from the UARS HALOE data set (more than 14 years) are now found to be significant at most latitudes of the middle and lower mesosphere. They range from −1 K/decade at low latitudes to about −3 K/decade at the middle latitudes (Remsberg 2009). Values of the order of −1 K/decade are reasonably consistent with those reported from the lidar measurements at low latitudes (Batista et al. 2009; Sridharan et al. 2009). At middle latitudes, there is reasonable agreement in the middle and upper mesosphere with published values from rocketsonde and lidar measurements of the preceding 2 decades (e.g. Keckhut et al. 1999; She et al. 2009). However, the HALOE trends are smaller in the lower mesosphere probably be due to a reduction in the stratospheric ozone effect compared to earlier periods (Remsberg 2009).

Angot et al. (2012) studied the neutral temperature trends in the height range of 30 to 80 km above Haute Provence Observatory (OHP, France, 44°N). They found that for comparison with model simulations of the direct effect of anthropogenic changes of radiative forcing, the trends based on background temperatures (excluding

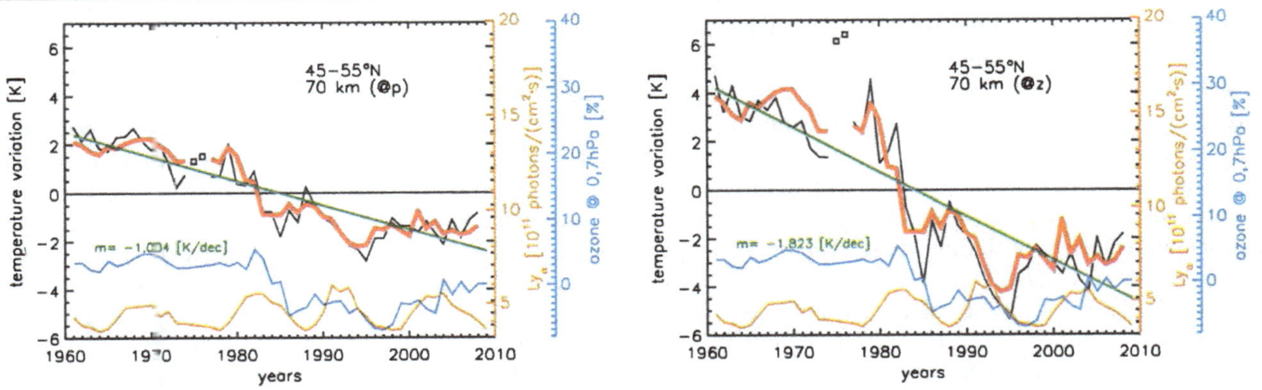

Figure 4 Temperature variations in the latitude band 45°N to 55°N. Temperature variations in the latitude band 45°N to 55°N on a (left) pressure and (right) geometrical altitude of 70 km. LIMA temperatures (black lines) are compared with a multivariate fit (red lines) consisting of carbon dioxide, ozone, and Ly-α. For comparison, a linear fit over the entire time period is also shown (green line). The orange curve shows the variation of Ly-α and the cyan curve shows the variation of ozone (right axis). The two squares mark temperatures in 1975 and 1976 which are ignored when fitting. After Lübken et al. (2013).

sudden stratospheric warming periods) should be considered, not total trends.

New observations of temperature trends in the high-latitude mesosphere near 87 to 90 km above Svalbard (78°N, 16°E) indicate a changing trend at this location. Older data based on OH-temperature measurements in winter (mostly inside the vortex) by Dyrland and Sigernes (2007) over 1983 to 2005 yield a positive, even though statistically insignificant, trend of +2 K/decade. On the other hand, year-round meteor radar measurements over 2002 to 2011 calibrated with satellite Aura data provide a relatively large negative trend of -4 ± 2 K/decade (Hall et al. 2012). More data are necessary to resolve this discrepancy.

The present status of mesopause region temperature trends can be described as follows: (i) the emergence of weak negative trends (a change from the earlier perception of no trend in this region), (ii) an indication of a break in trends at the mesopause region and a signal that trend is likely to strengthen (towards negative) in recent time as compared to the past which seems to be linked to the recovery of stratospheric ozone, (iii) trends in the entire MLT region need not be identical in all the seasons and likely have even monthly variability on long-term scales, and (iv) the importance of dynamics, particularly the role of gravity waves in understanding the variability in mesopause region trends, still remains an issue.

A growing number of model studies during the past decade have advanced to a level where they have started to reproduce a near zero temperature trend with an inclination towards negative trends near the mesopause region. While comparing various long term trend results, it should be noted that the period of measurements for which analysis is reported by various authors are different and it may have a strong bearing in comparisons. The seasonal differences near mesopause region with convincing interpretation still need to be understood sufficiently. The role of dynamics, particularly the role of gravity wave and its breaking, is very important to better understand the temperature trend features reported.

Current knowledge about trends in the ionosphere at heights corresponding to the mesosphere and lower thermosphere has been summarized in a review paper by Laštovička et al. (2012). There was little progress in this area during the CAWSES-II period, but it should be mentioned that the number of open questions here is smaller than it is for trends in the neutral mesosphere and lower thermosphere. Results of Ohya et al. (2011) reveal either no trend or a piecewise linear trend with a change in the second half of the 1990s for VLF tweek atmospherics observed at Kagoshima, Japan over 1976 to 2010. The trend is uncertain due to a large scatter of data.

The impact of trends in atmospheric wave activity on trends in the MLTI region is not yet clear enough, due in part to limited knowledge of trends in wave activity themselves. Still, as stated in the previous section, changes in greenhouse gases and stratospheric ozone are likely to affect wave propagation. Throughout the stratosphere, O_3 concentration has been decreasing since about 1979, reaching a minimum in the mid-1990s at northern middle latitudes (e.g., Harris et al. 2008) and later also in other regions. Future ozone changes are likely to affect trends in a way opposite to that in the 1980s. This should result in some modification of trends. In the mesosphere, HALOE trends are weaker than older rocket and lidar-based trends because the ozone effect over the period of HALOE observations was close to zero (Remsberg 2009). Temperature trends in the mesopause region in general changed from no trend to a slight negative trend (Beig 2011). Model results shown in Figure 5 reveal a clear effect of ozone on upper stratospheric and mesospheric temperatures; first, a strengthening of cooling after 1980 and then after the mid-1990s, a leveling-off of the trend, if not a slight statistically insignificant heating in the mesosphere. Lübken et al. (2013) estimated the share of CO_2 and O_3 on mesospheric temperature trends over 1962 to 2009 to be 2:1. This all means that the global scenario of trends remains qualitatively the same but changes in ozone trends somewhat changed it quantitatively. The effect of ozone on trends is temporary - it began near 1980 with the beginning of ozone depletion and will disappear after ozone reaches approximately the 1980 ozone level.

By how much is the anthropogenic effect on the ionosphere/thermosphere enhanced during a quiet sun period?

Long-term trends may be quantitatively different under solar maximum and minimum conditions (an indirect effect of solar cycle), such as in the case of thermospheric density (e.g., Emmert et al. 2008). The reason is the much larger relative role of CO_2 radiative cooling compared to NO radiative cooling under solar minimum conditions. The NO radiative power decreased by almost an order of magnitude from 2002 to 2009, whereas the CO_2 radiative power decreased only by approximately 35% according to SABER/TIMED measurements (Mlynczak et al. 2010). Emmert and Picone (2011) dealt with estimating statistical uncertainty of thermospheric density trends. They found that these trends are statistically significant at the 95% level, and that a sophisticated method of uncertainty estimation must be used because simple classical methods underestimate the uncertainty. Trends in thermospheric density are of high practical importance. Lewis et al. (2011) calculated the increase of space debris concentration due to a decreasing trend in thermospheric density.

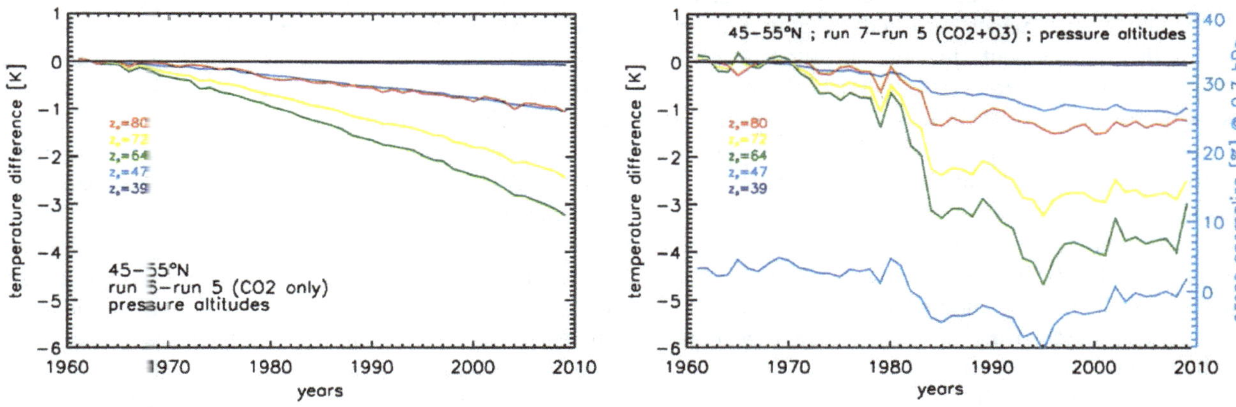

Figure 5 Temperature difference relative to the year 1961. Temperature difference relative to the year 1961 at pressure heights (in kilometers) caused by (left) CO_2 trends only and by (right) the combined effect of CO_2 and O_3. The cyan curve (right) shows the variation of ozone (right axis). After Lübken et al. (2013).

They found a remarkable increase in the concentration of space debris, particularly at low satellite orbits, leading to an increase in the number of damaging collisions of satellites with space debris. According to their calculations, the decline in thermospheric density results in a 74% increase in the LEO debris population growth rate over a 70 year projection (2009 to 2079).

The last solar activity minimum, particularly years 2008 to 2009, was significantly deeper than any other solar activity minimum during the era of ionospheric and thermospheric investigations, making trend estimations more difficult. Thermospheric densities were extremely low, and the ionosphere behaved partly in an unexpected way. An unexpectedly large decrease of thermospheric density in the recent extreme solar cycle minimum could make a problem in further thermospheric density trend estimates. Therefore, the focus of the CAWSES-II TG-2 project (2) was changed to clarifying the behavior of the ionosphere and thermosphere during this solar minimum. It is necessary to mention that we are not sure how deep the EUV flux decrease was in the last solar minimum. Standard solar indices, sunspot number and F10.7, are unable to describe this decrease. Solomon et al. (2010b) analyzed the extremely low thermospheric neutral densities in the solar minimum year 2008 compared to 1996 and concluded that the observed decrease of the solar EUV flux by 13% to 15% (SEM/SOHO observations) can account for this decrease of thermospheric density. Deng et al. (2012) carefully analyzed this density depletion and came to a conclusion that about 75% was caused by a decrease of the solar EUV irradiance and about 25% by very quiet geomagnetic activity. This would reduce the 1996 to 2008 EUV difference to some 10% to 12%. Laštovička (2013b) made an estimate based on noontime (10 to 14 UT) yearly average values of foF2 from very reliable station Juliusruh (the northernmost Germany, geomagnetic latitude 54.3°N). The estimated decrease of the EUV level from 1996 to 2008 is about 7% to 9% after considering the effect of changed geomagnetic activity. At the AGU Chapman conference in November 2012, T. Woods (private communication) estimated the difference between the 1996 and 2008 EUV irradiance levels to be probably close to 10% and within limits of 6% to 15% provided by various authors. Emmert et al. (2012), when interpreting the difference between the observed and modeled CO_2 concentration in the lower thermosphere, assumed some increase in eddy diffusion to get an agreement between model-based and observed CO_2 concentrations. This increase would contribute also to the anomalous thermospheric density decrease in the recent extreme solar minimum (Emmert et al. 2012).

As for ionosphere behavior, Klenzing et al. (2013) found that low latitude ExB drift as measured by C/NOFS satellite exhibit several differences compared with climatology based on previous solar activity minima including a downward drift in the early afternoon and a weak to non-existing pre-reversal enhancement, which results only can explain the observed post-sunset drop in hmF2. Solomon et al. (2013) fairly well simulated the observed anomalous decrease of electron density and hmF2 in the last solar minimum with model TIE-GCM. The simulation revealed a dominant role of the extreme solar EUV decrease in the last solar minimum together with minor contribution from lower geomagnetic activity and very minor contribution from anthropogenic increase of CO_2. Sunda et al. (2013) observed the effects of moderate geomagnetic storms in the equatorial and low latitude ionosphere during the extreme solar minimum to be more severe than under moderate and high solar activity. The same was found for middle latitudes by Buresova et al. (2014). Geomagnetic activity in years

2008 to 2009 was dominated by sequences of solar wind high speed streams, which resulted in only moderate geomagnetic storms but in a high-intensity, long-duration continuous auroral activity (HILDCAA) events. Verkhoglyadova et al. (2013) found surprisingly large enhancements of TEC at low latitudes during such events.

Summarizing, we can say that the ionosphere and thermosphere behaved in a different and partly unexpected way than in the previous solar activity minima. However, quantitatively, the difference is not yet determined enough, its understanding is only partial, and its impact on trend estimating is not sufficiently clear, even though there is some impact.

Zhang and Holt (2013) found the decrease of ion temperature above Millstone Hill to be largest under lower solar activity of F10.7 = 90 to 120 but not in the solar activity minimum. This, to some extent, contradicts the behavior of thermospheric density, where the highest decrease was found in the minimum of solar activity.

Are PMC/NLC frequency and brightness trending due to changes in temperature and water vapor?

The polar mesospheric clouds are observed in the summer polar upper mesosphere (approximately 82 to 85 km), i.e., in the region where the lowest atmospheric temperatures are observed. They consist of ice crystals. They are called polar mesospheric clouds (PMC) when observed from above by satellites and noctilucent clouds (NLC) when observed from below the surface. In the investigations of PMC/NLC, there were three open problems. First, PMC and NLC trends were apparently different; second, models did not capture trends in PMC/NLC; and third, it was not clear what the main driver of PMC/NLC trend was, if it water vapor content or temperature. WACCM-PMC model studies show that relatively dim clouds should have different trends than the brighter clouds. Thus comparison of time series of PMC brightness and frequency from different instruments need to consider this 'threshold effect'.

The first problem has been solved by Shettle et al. (2009). They analyzed 28 years of satellite PMC observations for latitudinal bands 50°N to 64°N (NLC observation latitudes), 64°N to 74°N, and 74°N to 82°N. They observed statistically significant increase only in 74°N to 82°N, whereas in 50°N to 64°N, it was statistically insignificant. At NLC observational latitudes, the PMC trend of Shettle et al. (2009) is +9.9%/decade, the NLC trend of Kirkwood et al. (2008) is +4.4%/decade for moderate and bright NLCs, and approximately 14%/decade for all NLCs; all trends being statistically insignificant. This means that the PMC and NLC trends are comparable and do not differ within accuracy of their determination in spite of partly different analyzed periods. Relatively large differences between PMC and NLC results were due to the fact that the occurrence frequency of PMC in the 74°N to 82°N band is four times as high as in the NLC latitudinal band and, therefore, the 74°N to 82°N band largely determines the overall trend in PMCs.

The second problem was basically solved recently with model LIMA/ICE. This model captures reasonably well (even quantitatively) the variability of PMC albedo (Lübken et al. 2012), as it is shown in Figure 6. The model simulates reasonably well the long-term behavior of PMCs both as to the solar cycle effect and as to the long-term trends in all three separated latitudinal zones of PMC albedo/brightness and supports the dominant role of high latitudes (74°N to 82°N) in both the solar cycle and trend effects in polar mesospheric clouds.

As for the third problem, Rong et al. (2012) analyzed the data of the Solar Occultation for Ice Experiment aboard the Aeronomy of Ice in the Mesosphere satellite (2007 to 2010, 4 years). Their results support the idea that the observational trends in PMC brightness (albedo) are primarily driven by changes in water vapor, not temperature. On the other hand, Remsberg's (2010) HALOE-based results support rather the role of temperature. Model results of Lübken et al. (2012) reveal important role of all three components, temperature, water vapor, and solar Lymna-alpha flux variability in changes of PMC brightness. Recent observations also suggest water vapor from space traffic exhaust may be sufficient to override the expected decrease in PMC as solar activity increases (Siskind et al. 2013). Thus, further research and more data are required to find the dominant driver of PMC trends. Temperature itself (T83 km) seems to be the main driver of trend in the height of PMCs, which is, however, very small (Lübken et al. 2012).

LIMA/ICE simulations reveal a change of trend of temperature at the NLC height of 83 km at high latitudes (69°N, ALOMAR) in the mid-1990s. The weak

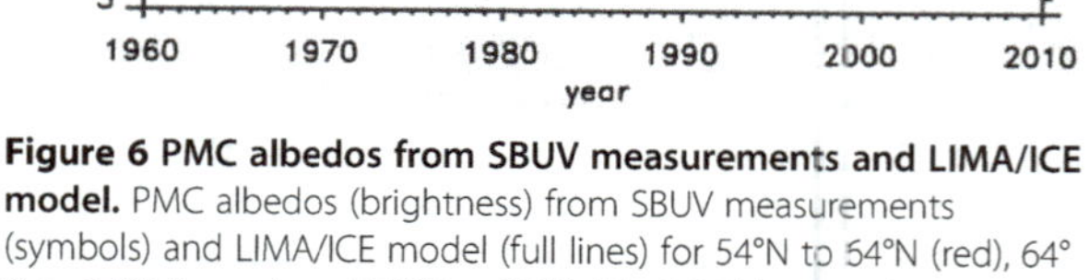

Figure 6 PMC albedos from SBUV measurements and LIMA/ICE model. PMC albedos (brightness) from SBUV measurements (symbols) and LIMA/ICE model (full lines) for 54°N to 54°N (red), 64° N to 74°N (green), and 74°N to 82°N (blue) (Lübken et al. 2012).

negative trend (−2 K per 35 years) is replaced by a similarly weak but positive trend just when the trends in the northern mid-latitude stratospheric ozone changed from depletion to recovery (Lübken et al. 2012).

A study by Benze et al. (2011) of Solar Backscatter Ultraviolet satellite instrument observations of PMC showed that the onset of the PMC season in the southern hemisphere (SH) is mostly influenced by the timing of the reversal of stratospheric winds in the same hemisphere from the winter to summer state. A secondary influence on SH PMC onset is the solar cycle. Combined, these factors appear to cause over a 40 day swing in the date of PMC onset. These authors, along with Smith et al. (2011) and Lossow et al. (2012), point out that any trend in timing of the reversal of the stratospheric jet could cause a trend in SH PMC onset.

Polar mesospheric summer echoes (PMSE) are phenomenon which correlates with PMCs. Radar observations at Andoya (69.3°N) over 1994 to 2008 reveal a positive trend in the PMSE occurrence frequency (Latteck and Bremer, 2013), which is qualitatively consistent with PMC trends.

Important results reached outside of the CAWSES-II project TG-2

The three CAWSES-II TG-2 projects were focused only on some specific important problems of trend investigations and, therefore, some progress was reached also out of the TG-2 activities. This progress is summarized in this section. It should be mentioned that many of these results were reached by CAWSES-II project participants.

One of the problems in the thermosphere and ionosphere is that the model simulations yield weaker trends than the observational data. Emmert et al. (2012) found quite recently that CO_2 concentration in the lower thermosphere increases substantially quicker than modelers expected. To repeat model simulations with new values of the CO_2 concentration in the lower thermosphere increases is an important challenge for modelers in the near future. It is very desirable to know to which extent these new CO_2 concentrations in the lower thermosphere help to reduce the quantitative difference between observed and simulated trends in the thermosphere and ionosphere.

First results about long-term trends in various quantities appeared during the CAWSES-II period but outside of the CAWSES-II projects. At thermospheric and ionospheric heights, the first information appeared on trends in ion and electron temperatures. Figure 7 displays trends of noontime Ti from Millstone Hill over a 100 to 550 km altitude range (Zhang et al. 2011). Above 200 km, the trends are negative; cooling reaches about 3%/decade (almost −4 K/year). Negative trends were expected because in the thermosphere (100 to 500 km), Ti and neutral temperature (T) are relatively closely related (e.g., Akmaev 2012); and both models (e.g., Roble and Dickinson 1989) and neutral density trends (Emmert et al. 2008) suggest thermospheric cooling above 200 km. Trends are apparently positive below 200 km if they are observed at fixed heights. As a consequence of the thermal contraction of the upper atmosphere together with the steep gradient of temperature in the height range approximately 100 to 200 km, the warm air from above replaces the air at a given height. The effect of local cooling is not sufficient to compensate for the effect of contraction, and therefore, we observe an apparent heating. This effect might be modulated to some extent by associated changes in chemistry and composition (e.g., the O/N_2 ratio). Donaldson et al. (2010) found at Saint Santin negative Ti trends of comparable magnitude above about 150 km and positive trends below for the period 1966 to 1987. The existence of apparent positive trends in temperature below approximately 150 km is in qualitative agreement with model simulations (e.g., Qian et al. 2011). The negative trends were increasing with height, and at 350 km, they reached an average of −3 K/year. However, they were much larger at noon than at midnight.

Zhang et al. (2011) observed larger trends of Ti for the period 1980 to 2006. Donaldson et al. (2010) found much weaker trends before than after 1979. This breakpoint coincides with the beginning of the period of strongly decreasing stratospheric ozone concentration at northern middle latitudes. This finding together with a large difference between daytime and nighttime trends in Ti prompted Walsh and Oliver (2011) to hypothesize that the ozone depletion may be the main cause of the accelerated cooling after 1979. However, Laštovička (2012) showed that the turnaround of ozone trends at northern middle latitudes (1995 to 1997) is reflected neither in Ti data from Millstone Hill nor in thermospheric density data series. Therefore, trend in Ti is not affected by ozone; the hypothesis of Walsh and Oliver (2011) was a consequence of too short of a data series at Saint Santin.

Zhang and Holt (2013) analyzed possible dependence of Ti above Millstone Hill on local time, season, solar, and geomagnetic activities. They found strong cooling in daytime but very weak cooling at nighttime. Therefore, the gross average cooling in 200 to 350 km (where Ti ~ Tn) is −4 K/decade, a trend close to estimates of Tn trends derived from thermospheric density trends. No consistent and substantial seasonal variation of cooling was found. A slight increase of cooling with increasing geomagnetic activity was observed.

Qian et al. (2011) simulated changes in Ti in response to doubling CO_2 during solar minimum conditions. Up to about an altitude of 350 km, the results of simulations

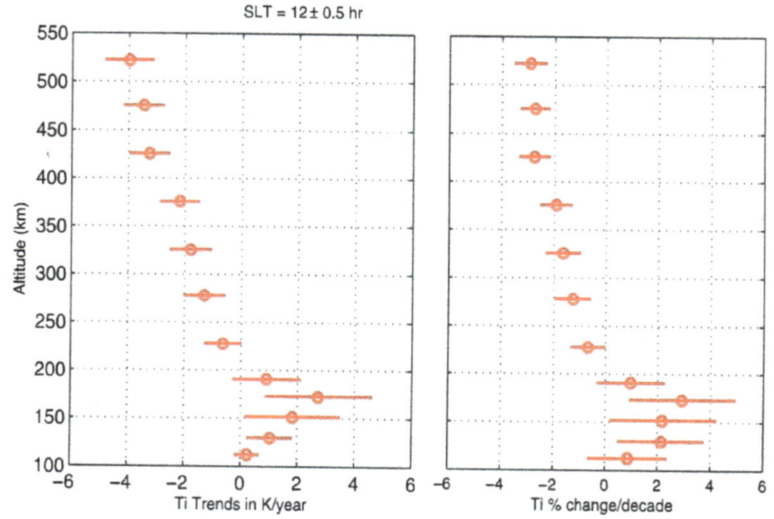

Figure 7 Trends in ion temperature as measured by the incoherent scatter radar. Trends in ion temperature as measured by the incoherent scatter radar at Millstone Hill, noontime, 1968 to 2006, heights 100 to 550 km (Zhang et al. 2011).

qualitatively agree with observations from Millstone Hill and Saint Santin, but above 350 km simulations provide a result opposite to the above observations, i.e., strong heating. The heating might be caused by atmospheric contraction in combination with Ti increasing with altitude in the uppermost thermosphere, a similar effect like that in 100 to 120 km altitudes. However, Akmaev (2012) claims that at a higher solar activity, the region of positive change in Ti might be shifted to much higher altitudes and, thus, not seen in radar data. This might result in the absence of the upper region of positive trends in Ti in the average trends presented in Figure 7.

High-latitude results by Ogawa et al. (2012) based on Ti data from EISCAT at Tromso differ from those of mid-latitude stations Millstone Hill and Saint Santin. A 30 year long (1981 to 2011) data series from Tromso yields for 11 to 12 LT a negative trend below 400 km, which is twice as strong as the trend simulated by Qian et al. (2011) but only of a half magnitude of Millstone Hill and Saint Santin trends. Upward, the trend quickly weakens and above 450 km it becomes positive. The trend in Ti somewhat depends on the solar proxy used (F10.7, F10.7aver, Mg II), but the differences are not large. There is also some dependence of trends at heights of about 400 to 450 km on local time. The trend in Ti is more negative near noon (11 to 14 LT) and more positive in the morning (08 to 11 LT).

The mid-latitude stations Millstone Hill and Saint Santin show at high altitudes negative trends, whereas the high latitude station Tromso and model simulations by Qian et al. (2011) show positive trends. Thus, there is an evident discrepancy, which needs to be analyzed and its origin looked for. Moreover, the absolute value of trends in Ti reported for mid-latitude stations Millstone Hill and Saint Santin is by about an order of magnitude higher than expected from model simulations of trends in neutral temperature (Qian and Solomon 2012) and trends in thermospheric densities derived from satellite drag (Emmert et al. 2008), as noticed by Akmaev (2012). The origin of this discrepancy is not yet known.

Oliver et al. (2013) claim that thermospheric cooling derived from ion temperature measurements above Millstone Hill is remarkably larger than the cooling necessary to explain the observed decrease of thermospheric density (however this is not confirmed by Zhang and Holt 2013). They try to explain this discrepancy by the effect of long-term changes in the gravity wave flux in the thermosphere initiated by a shift in climate regime in oceans in 1976/77, which launched a slow Rossby wave across oceans that continues to propagate until present. However, this hypothesis needs to be confirmed by observations and further model simulations.

As concerns the electron temperature Te, the first result on trends appeared based on Millstone Hill data (Zhang et al. 2011). It reveals positive trends in Te even though they are relatively little significant due to large scatter of data. At high altitudes, the positive trends may be due to the negative trends in electron density. The Te trend qualitatively agrees with model simulations by Qian et al. (2011). However, more data are needed to get statistically reliable trends as well as information from other measuring sites.

Available neutral temperature (Tn) data do not allow calculation of trends but based on trends in ion temperature and thermospheric neutral density we may expect trends qualitatively similar to those in Ti.

First results also appeared in long-term trends in the thermospheric neutral winds based on data series derived from Fabry-Perot nightglow measurements over Arecibo, 1980 to 2010 (Brum et al. 2012). These measurements are from heights near or slightly below the F2 region maximum. The trends reveal a complex pattern; they are strongly dependent (even as to sign) on the day of year and the time of night. However, if we take into account the behavior of long-term changes and trends in mesospheric and mesopause region winds as observed and also modeled (Qian et al. 2011), the existence of a complex pattern of trends in thermospheric wind is not surprising. Information on trends in thermospheric winds is required also from other sites, particularly if we take into account the behavior of MLT region wind trends, which are locally different.

The first information on trends also appeared for the total electron content (TEC) based on the International Geodetic Service (IGS) TEC maps over 1995 to 2010

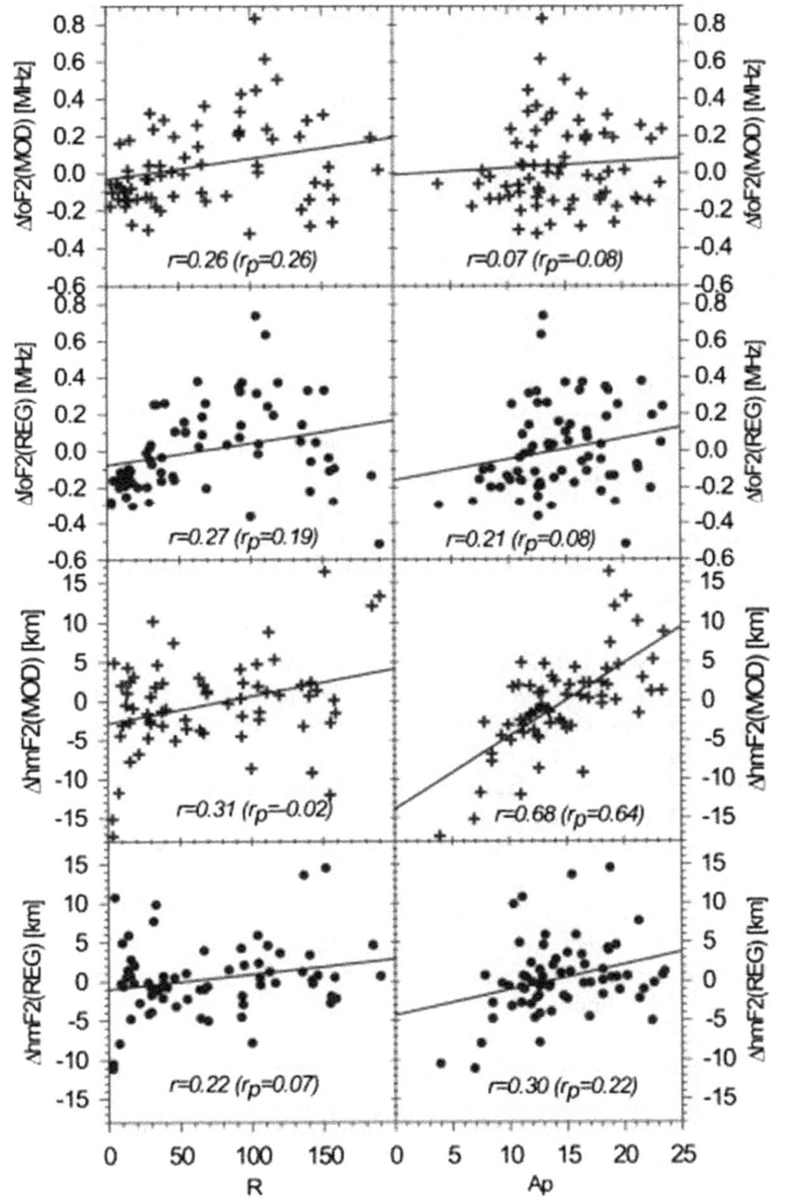

Figure 8 Long-term trends in hmF2 and foF2. Long-term trends in hmF2 (top panel) and foF2 (bottom panel) for Slough/Chilton (51.6N, 358.7E) with solar activity correction with F10.7 or with sunspot number R. ΔhmF2 and ΔfoF2 are differences between observed values and empirical model with F10.7 (R) and geomagnetic activity. After Bremer et al. (2012).

(Lean et al. 2011). These trends were strongly regionally dependent and predominantly positive. In some regions, like Europe, they are surprisingly strong and positive. Laštovička (2013b) showed that ionospheric trends in the period 1995 to 2010 were in general more positive than before, perhaps due to the extreme solar minimum 2008 to 2009, but TEC trends were more positive than foF2 trends in the past as well. Longer data series is necessary to clarify the potential impact of the last solar minimum on trends in TEC.

Danilov (2009) and Danilov and Vanina-Dart (2010) used behavior of ionospheric parameters foF2 and hmF2 including their scatter to deduce changes of trends in thermospheric dynamics. They found no trend before about 1980 but a systematic long-term trend after about 1980. They attributed this trend to the greenhouse cooling of the middle and upper atmosphere.

Bremer et al. (2012) and Mielich and Bremer (2013) used a new worldwide homogenized database of foF2 and M(3000)F2 to calculate long-term trends in foF2 and hmF2. They conclude that the global trends in yearly values of foF2 are very small and do not differ significantly from no trend. This is supported by model calculations of the greenhouse gas effects on foF2 because hmF2 closely follows the boundary between the increasing electron density below and decreasing electron density above, only a small trend in foF2 is expected (e.g., Qian et al. 2009). On the other hand, global trends in hmF2 are negative and statistically significant, again in qualitative agreement with model calculations of the greenhouse gas effect on hmF2. Bremer et al. (2012) and Mielich and Bremer (2013) also confirm and demonstrate for global data the necessity to use F10.7, not sunspot number for correction to solar activity (solar cycle), as it is illustrated by Figure 8. Mielich and Bremer (2013) shows that for F2 region trends F10.7 is even better than E10.7 index (Tobiska's ionospheric data-based index for description of solar EUV flux, which is close to F10.7) due to problems of the latter in 1957 to 1958. Trends based on the correction with F10.7 are larger and with the same or smaller uncertainty and stronger correlation than those based on the correction with sunspot numbers. The way how the correction for solar activity is made appears at least as important as the method how the trend is calculated, particularly for foF2. It is also necessary to mention that we cannot expect quite homogeneous trends globally, because in some areas, particularly in the low-latitude South America, secular changes in the Earth's magnetic field plays an important role (e.g., Cnossen and Richmond, 2008), and at high latitudes, the influence of long-term changes in geomagnetic activity has to play more important role than at middle and low latitudes. Historical values of trends in fofF2 and hmF2 as published by various author were different and sometimes even of opposite sign. A part of these differences was caused by using sunspot number versus F10.7 and in some cases also by data quality problems. However, recent results indicate also other reasons. Model calculations by Qian et al. (2009) show that on a global average basis, trends of foF2 are larger under solar minimum conditions and the global distribution of the trends tends to be more structured. The trends of foF2 are larger in the winter hemisphere than in the summer hemisphere, for both solar minimum and maximum conditions. Danilov (2012) claims that a strong influence of dynamical processes on foF2 at night compared to a weak influence during daytime may result in some difference between daytime and nighttime trends in foF2 and hmF2. As for trends of hmF2, they exhibit strong latitudinal, longitudinal, and diurnal variations; they are mostly negative but can be positive at night in some regions according to model calculations by Qian et al. (2009). Bremer et al. (2012) found stronger trends at lower than higher latitudes both in foF2 and hmF2, but the latitudinal dependence of trends was rather weak. All that can contribute to partly different results of various authors about trends in the F2 region.

Danilov and Konstantinova (2012) studied trends in foF2 over 1990 to 2010 based on data of 22 ionospheric stations. They found that trends in 1998 to 2010 were more negative than in 1990 to 1997. They interpret this finding as more important role of temperature (i.e., of greenhouse cooling) in the latter period. However, Laštovička (2013b) (found more positive trends in the period 1995 to 2010 than before, which contradicts the results of Danilov and Konstantinova (2012). Danilov and Konstantinova (2013) compared trends of foF2 based on different databases and found some, even though not principal, differences between them. They also reported twice as high negative trends in foF2 in winter compared to summer and remarkably larger standard deviations from the dependence of foF2 on F10.7 in 1998 to 2010 compared to 1958 to 1979; the latter maybe due to a larger variability of thermospheric winds. Scott et al. (2014) reported long-term changes in amplitude ratio of semiannual/annual variations of foF2 for Slough/Chilton (52°N) and Stanley (52°S). They attributed these changes to long-term changes in geomagnetic activity induced changes in thermospheric circulation and composition.

Another quantity which displays a long-term trend is the solar quiet (Sq) geomagnetic variation but this trend qualitatively agrees with the secular variation of the Earth magnetic field strength, so it is not of greenhouse gas origin (de Haro Barbas et al. 2013).

Conclusions

The TG-2 of CAWSES-II covered several important problems in the investigation of long-term trends in the upper atmosphere and ionosphere, but not the full spectrum of problems. Nevertheless, its contribution to the overall progress in the field of long-term trend was substantial. Some of these results are as follows:

1. The scenario of trends in the upper atmosphere and ionosphere became more complete and some discrepancies were removed or explained.
2. The role of stratospheric ozone and recovery or leveling off of its trend in the mid-1990s to later-1990s was proved to be remarkable and its impact on trends in the mesosphere, lower thermosphere and lower ionosphere was to a significant extent quantified.
3. A reasonable agreement between observations and simulations of trends in mesospheric temperatures and PMC albedo has been reached.
4. Generally, the progress in the field of PMC/NLC trends was substantial and several problems were solved.
5. Some progress in several other problems has also been reached including understanding of why the thermospheric density trends are much stronger under solar cycle minimum conditions, and of behavior of the upper atmosphere-ionosphere system in the extreme solar minimum 2008 to 2009 with impact on trends and their estimating.

Moreover, CAWSES-II TG-2 scientists made progress in the study trend topics outside the TG-2 itself.

Main tasks for future research on long-term trends are as follows:

1. The key open problem is trends in atmospheric dynamics, i.e., in atmospheric circulation and particularly in atmospheric wave activity. Atmospheric waves affect many parameters of the upper atmosphere and ionosphere. Relatively little progress was reached. However, now we are sure that trends in circulation are of a complex spatial pattern.
2. The agreement between observations and models is still only qualitative for many parameters. Reducing the quantitative discrepancy between observed and simulated trends is of high importance. It will result also in better understanding of mechanisms responsible for trends.
3. Trends are known mostly qualitatively, not quantitatively. Specifying and quantifying trends in various parameters and studying various finer details of trend behavior are tasks for future research. Some apparent discrepancies like those in ion temperature trends need to be explained.
4. The changing role of trend drivers induces spatial and particularly temporal changes of trends in various parameters. This has to be monitored and further studied.
5. On longer time scales, an important task is to try to join the upper atmospheric trends with long-term changes in the stratosphere into one scenario (if it is physically possible). As a first step, similarities and differences between upper atmospheric and stratospheric trends began to be studied.

Progress in studying the long-term trends in the mesosphere, thermosphere, and ionosphere over the period of CAWSES-II project are significant but still various questions remain open. Therefore in the CAWSES-II follow-up new SCOSTEP program VarSITI, project ROSMIC (Role Of the Sun and the Middle atmosphere/thermosphere/ionosphere In Climate), two of four working groups deal with long-term trends: WG-3 Trends in mesosphere and lower thermosphere, and WG-4 Trends and solar cycle effects in the thermosphere.

Abbreviations

AE: index of auroral zone geomagnetic activity; AGU: American Geophysical Union; ALOMAR: Arctic Lidar Observatory for Middle Atmospheric Research; CAWSES-II: Climate and Weather of the Sun-Earth System; C/NOFS: Communications/Navigation Outage Forecasting System satellite; EISCAT: European Incoherent Scatter Scientific Association; EUV: extreme ultraviolet radiation; GCM: global circulation model; GHG: greenhouse gases; GPS: Global Positioning System; HALOE: Halogen Occultation Experiment satellite; HILDCAA: high-intensity long-duration continuous AE activity; HRDI: high-resolution Doppler imager; IAGA: International Association of Geomagnetism and Aeronomy; ICMA: International Commission for the Middle Atmosphere; IPCC: Intergovernmental Panel on Climate Change; LASP: Laboratory for Atmospheric and Space Physics; LEO: low Earth orbiting; LIMA: Leibniz Institute Middle Atmosphere model; LIMA/ICE: Leibniz Institute Middle Atmosphere/Ice model; MLT: mesosphere-lower thermosphere; MLTI: mesosphere-lower thermosphere-ionosphere; NCAR: National Center for Atmospheric Research; NH: Northern Hemisphere; NLC: Noctilucent clouds; OHP: Observatory of Haute Provence; PMC: polar mesospheric clouds; ROSMIC: role of the sun and the middle atmosphere/thermosphere/ionosphere in climate; SABER: Sounding of the Atmosphere using Broadband Emission Radiometry instrument; SCOSTEP: Scientific Committee for Solar-Terrestrial Physics; SEM: solar EUV monitor instrument; SH: Southern Hemisphere; SOHO: Solar and Heliospheric Observatory; TEC: total electron content; TG: topical group; TIE: Thermosphere-Ionosphere-Electrodynamics model; Te: electron temperature; Ti: ion temperature; Tn: neutral temperature; TIMED: Thermosphere Ionosphere Mesosphere Energetics and Dynamics satellite; UARS: Upper Atmospheric Research Satellite; UCEMA: University Center of Macroeconomic Studies of Argentina; UT: Universal Time; VarSITI: Variability of the Sun and Its Terrestrial Impact; VLF: very low frequency; WACCM: Whole Atmospheric Community Climate Model.

Competing interests

The authors declare that they have no competing interests.

Authors' contributions

The three authors altogether chaired the CAWSES-II TG-2. JL wrote most of the paper. DM wrote the 'Are there long-term changes in tropospheric wave generation and their propagation through a changing atmosphere?' section, modified several other sections, particularly the 'By how much is the anthropogenic effect on the ionosphere/thermosphere enhanced during a quiet sun period?' section, and corrected the English grammar language of the paper. GB contributed substantially to the 'Are trends related to upward propagation of atmospheric waves observed in the mesosphere, lower thermosphere and ionosphere (MLTI)?' section. All authors read and approved the final manuscript.

Acknowledgements
The authors are grateful to many collaborators of CAWSES-II TG-2, particularly to the project leaders for their efforts. JL acknowledges the support by the Grant Agency of the Czech Republic, grant P209/10/1792, and the Ministry of Education, Youth and Sports of the Czech Republic, grant LD 12070. The National Center for Atmospheric Research is operated by the University Corporation for Atmospheric Research under the sponsorship of the National Science Foundation.

Author details
[1]Institute of Atmospheric Physics ASCR, Bocni II, Prague 14131, Czech Republic. [2]Indian Institute for Tropical Meteorology, Dr. Homi Bhabha Road, Pashan, Pune 411008, India. [3]National Center for Atmospheric Research, Boulder 80301CO, USA.

References
Akmaev RA (2012) On estimation and attribution of long-term temperature trends in the thermosphere. J Geophys Res 117:A09321, doi: 10.1029/2012JA018058

Angot G, Keckhut P, Hauchecorne A, Claud C (2012) Contribution of stratospheric warmings to temperature trends in the middle atmosphere from the lidar series obtained at Haute-Provence Observatory (44N). J Geophys Res 117: D21102, doi: 10.1029/2012JD017631

Batista PP, Clemesha BR, Simonich DM (2009) A 14-year monthly climatology and trend in the 35–65 km altitude range from Rayleigh Lidar temperature measurements at a low latitude station. J Atmos Solar-Terr Phys 71:1456–1462, doi: 10.1016/j.jastp.2008.03.005

Beig G (2011) Long-term trends in the temperature of the mesosphere/lower thermosphere region: 1. Anthropogenic influences. J Geophys Res 116: A00H11, doi: 10.1029/2011JA016646

Beig G, Keckhut P, Lowe RP, Roble RG, Mlynczak MG, Scheer J, Fomichev VI, Offermann D, French WJR, Shepherd MG, Semenov AI, Remsberg EE, She C-Y, Lübken F-J, Bremer J, Clemesha BR, Stegman J, Sigernes F, Fadnavis S (2003) Review of mesospheric temperature trends. Rev Geophys 41(4):1015, doi:10.1029/2002RG000121

Benze S, Randall CE, DeLand MT, Thomas GE, Bailey SM, Russell JM III, Merkel AW (2011) Evaluation of AIM CIPS measurements of polar mesospheric clouds by comparison with SBUV data. J Atmos Solar-Terr Phys 73:2065–2072, doi: 10.1016/j.jastp.2011.02.00

Bougher SW, Roble RG (1991) Comparative terrestrial planet thermospheres. 1. Solar cycle variations of global mean temperatures. J Geophys Res 96:11045–11055

Bremer J, Laštovička J, Mikhailov AV, Altadill D, Bencze P, Buresova D, De Franceschi G, Jacobi C, Kouris S, Perrone L, Turunen E (2009) Climate of the upper atmosphere. Annals Geophys 52:273–299

Bremer J, Damboldt T, Mielich J, Suessmann P (2012) Comparing long-term trends in the ionospheric F2-region with two different methods. J Atmos Solar-Terr Phys 77:174–185, doi: 10.1016/j.jastp.2011.12.017

Brum CGM, Tepley CA, Fentzke JT, Robles E, Terra dos Santos P, Gonzalez SA (2012) Long-term changes in the thermospheric neutral winds over Arecibo: climatology based on over three decades of Fabry-Perrot observations. J Geophys Res 117:A00H14, doi: 10.1029/2011JA016458

Buresova D, Laštovička J, Hejda P, Bochnicek J (2014) Ionospheric disturbances under low solar activity conditions. Adv Space Res http://dx.doi.org/10.1016/j.asr.2014.04.007

Cicerone RJ (1990) Greenhouse cooling up high. Nature 344:104–105

Clilverd MA, Clark TGC, Clarke E, Rishbeth H (1998) Increased magnetic storm activity from 1868 to 1995. J Atmos Solar-Terr Phys 60:1047–1056

Cnossen I, Richmond AD (2008) Modeling the effect of changes in the Earth's magnetic field from 1957 to 1997 on the ionospheric hmF2 and foF2 parameters. J Atmos Solar-Terr Phys 70:1512–1524

Cnossen I, Richmond AD (2013) Changes in the Earth's magnetic field over the past century: effects on the ionosphere-thermosphere system and solar quiet (Sq) magnetic variation. J Geophys Res 118:849–858

Danilov AD (2009) Scatter of hmF2 values as an indicator of trends in thermospheric dynamics. J Atmos Solar-Terr Phys 71:1586–1591, doi: 10.1016/j.jastp.2009.03.002

Danilov AD (2012) Long-term trends in the upper atmosphere and ionosphere (a review). Geomagn Aeron 52:271–291

Danilov AD, Konstantinova AV (2012) Variations in foF2 at the end of the 1990s and the beginning of the 2000s according to the SPIDR system data. Geomagn Aeron 52:350–355

Danilov AD, Konstantinova AV (2013) Trends in the F2 layer parameters at the end of the 1990s and the beginning of the 2000s. J Geophys Res Atmos 118:5947–5964, doi: 10.1002/jgrd.50501

Danilov AD, Vanina-Dart LB (2010) Parameters of the ionospheric F2 layer as a source of information on trends in thermospheric dynamics. Geomagn Aeron 50:187–200

de Haro Barbas BF, Elias AG, Cnossen I, Zossi de Artigas M (2013) Long-term changes in solar quiet (Sq) geomagnetic variations related to Earth's magnetic field secular variation. J Geophys Res 118:3712–3718, doi: 10.1002/jgra.50352

Deng Y, Huang Y, Solomon S, Qian L, Knipp D, Weimer DR, Wang J-S (2012) Anomalously low geomagnetic energy inputs during 2008 solar minimum. J Geophys Res 117:A09307, doi: 10.1029/2012JA018039

Donaldson JK, Wellman TJ, Oliver WL (2010) Long-term change in thermospheric temperature above Saint Santin. J Geophys Res 115:A11305, doi: 10.1029/2010JA015346

Dyrland ME, Sigernes F (2007) An update on the hydroxyl airglow temperature record from the Auroral Station in Adventdalen, Svalbard (1980–2005). Can J Phys 85:143–151

Emmert JT, Picone JM (2011) Statistical uncertainty of 1967–2005 thermospheric density trends derived from orbital drag. J Geophys Res 116:A00H09, doi:10.1029/2010JA016382

Emmert JT, Picone JM, Meier RR (2008) Thermospheric global average density trends, 1967–2007, derived from orbits of 5000 near-Earth objects. Geophys Res Lett 35:L05101, doi: 10.1029/2007GL032809

Emmert JT, Stevens MH, Bernath PF, Drob DP, Boone CD (2012) Observations of increasing carbon dioxide concentration in Earth's thermosphere. Nat Geosci 5:868–871, doi: 10.1038/NGEO1626

Engel A, Möbius T, Bönisch H, Schmidt U, Heinz R, Levin I, Atlas E, Aoki S, Nakazawa T, Sugawara S, Moore F, Hurst D, Elkins J, Schauffler S, Andrews A, Boering K (2009) Age of stratospheric air unchanged within uncertainties over the past 30 years. Nat Geosci 2:28–31, doi: 10.1038/ngeo388

Ern M, Preusse P, Gille JC, Hepplewhite CL, Mlynczak MG, Russell JM III, Riese M (2011) Implications for atmospheric dynamics derived from global observations of gravity wave momentum flux in stratosphere and mesosphere. J Geophys Res 116:D19107, doi: 10.1029/2011JD015821

Ern M, Arras C, Faber A, Froehlich K, Jacobi C, Kalisch S, Krebsbach M, Preusse P, Schmidt T, Wickert J (2013) Observations and ray tracing of gravity waves: implications for global modeling. In: Luebken F-J (ed) Climate and weather of the Sun-earth system (CAWSES): highlights from a priority program, 383–408. Springer, Dordrecht, The Netherlands, doi: 10.1007/978-94-007-4348-9_21

Eyring V, Cionni I, Bodeker GE, Charlton-Perez AJ, Kinnison DE, Scinocca JF, Waugh DW, Akiyoshi H, Bekki S, Chipperfield MP, Dameris M, Dhomse S, Frith SM, Garny H, Gettelman A, Kubin A, Langematz U, Mancini E, Marchand M, Nakamura T, Oman LD, Pawson S, Pitari G, Plummer DA, Rozanov E, Shepherd TG, Shibata K, Tian W, Braesicke P, Hardiman SC, et al (2010) Multi-model assessment of stratospheric ozone return dates and ozone recovery in CCMVal-2 models. Atmos Chem Phys 10:9451–9472, doi: 10.5194/acp-10-9451-2010

Graigmile PF, Guttorp P, Percival DB (2004) Trend assessment in a long memory dependence model using the discrete wavelet transform. Environmetrics 15:313–315

Hall CM, Dyrland ME, Tsutsumi M, Mulligan FJ (2012) Temperature trends at 90 km over Svalbard, Norway, (78°N, 16°E), seen in one decade of meteor radar observations. J Geophys Res 117:D08104, doi: 10.1029/2011JD017028

Hardiman SC, Butchart N, Calvo N (2013) The morphology of the Brewer–Dobson circulation and its response to climate change in CMIP5 simulations. Q J Roy Meteorol Soc doi: 10.1002/qj.2258

Harris NRP, Kyrö E, Staehelin J, Brunner D, Andersen SB, Godin-Beekmann S, Dhomse S, Hadjinicolaou P, Hansen G, Isaksen I, Jrrar A, Karpetchko A, Kivi R, Knudsen B, Krizan P, Lastovicka J, Maeder J, Orsolini Y, Pyle JA, Rex M, Vanicek K, Weber M, Wohltmann I, Zanis P, Zerefos C (2008) Ozone trends at northern mid and high latitudes - a European perspective. Ann Geophys 26:1207–1220

Hoffmann P, Rapp M, Singer W, Keuer D (2011) Trends in mesospheric gravity waves at northern middle latitudes during summer. J Geophys Res 116: D00P08, doi: 10.1029/2011JD015717

IPCC (2013) Summary for Policymakers. In: Stocker TF, Qin D, Plattner G-K, Tignor M, Allen SK, Boschung J, Nauels A, Xia Y, Bex V, Midgley PM (eds) Climate Change 2013: The Physical Science Basis. Contribution of Working Group I to the Fifth Assessment Report of the Intergovernmental Panel on Climate Change, Cambridge University Press, Cambridge, United Kingdom and New York, NY, USA

Jacobi C, Beig G (2013) Changes in MLT dynamics and composition. Invited paper presented at International CAWSES-II Symposium, Nagoya

Jacobi C, Hoffmann P, Liu RQ, Križan P, Laštovička J, Merzlyakov EG, Solovjova TV, Portnyagin YI (2009) Midlatitude mesopause region winds and wave sand comparison with stratospheric variability. J Atmos Solar-Terr Phys 71:1540–1546

Jacobi C, Hoffmann P, Liu RQ, Merzlyakov EG, Portnyagin YI, Manson AH, Meek CE (2012) Long-term trends, their changes, and interannual variability of Northern Hemisphere midlatitude MLT winds. J Atmos Solar-Terr Phys 75:81–91, doi: 10.1016/j.jastp.2011.03.016

Keckhut P, Schmidlin FJ, Hauchecorne A, Chanin ML (1999) Stratospheric and mesospheric cooling trend estimates from U.S. rocketsondes at low latitude stations (8°S–34°N), taking into account instrumental changes and natural variability. J Atmos Solar-Terr Phys 61:447–459, doi: 10.1016/S1364-6826(98)00139-4

Kirkwood S, Dalin P, Réchou A (2008) Noctilucent clouds observed from the UK and Denmark - trends and variations over 43 years. Ann Geophys 26:1243–1254

Klenzing J, Burrell AG, Heelis RA, Huba JD, Pfaff R, Simoes F (2013) Exploring the role of ionospheric drivers during the extreme solar minimum of 2008. Ann Geophys 31:2147–2156, www.ann-geophys.net/31/2147/2013/

Laštovička J (2005) On the role of solar and geomagnetic activity in long-term trends in the atmosphere–ionosphere system. J Atmos Solar-Terr Phys 67:83–92

Laštovička J (2012) On the role of ozone in the long-term trends in the upper atmosphere–ionosphere system. Ann Geophys 30:811–816, doi: 10.5194/angeo-30-811-2012

Laštovička J (2013a) Trends in the upper atmosphere and ionosphere: recent progress. J Geophys Res 118:3924–3935, doi: 10.1002/jgra.50341

Laštovička J (2013b) Are trends in total electron content (TEC) really positive? J Geophys Res 118:3831–3835, doi: 10.1002/jgra.50261

Laštovička J, Akmaev RA, Beig G, Bremer J, Emmert JT (2006a) Global change in the upper atmosphere. Science 314:1253–1254

Laštovička J, Mikhailov AV, Ulich T, Bremer J, Elias AG, Ortiz de Adler N, Jara V, Abarca del Rio R, Foppiano AJ, Ovalle E, Danilov AD (2006b) Long-term trends in foF2: a comparison of various methods. J Atmos Solar-Terr Phys 68:1854–1870

Laštovička J, Akmaev RA, Beig G, Bremer J, Emmert JT, Jacobi C, Jarvis MJ, Nedoluha G, Portnyagin YI, Ulich T (2008a) Emerging pattern of global change in the upper atmosphere and ionosphere. Ann Geophys 26:1255–1268

Laštovička J, Yue X, Wan W (2008b) Long-term trends in foF2: their estimating and origin. Ann Geophys 26:593–598

Laštovička J, Križan P, M. Kozubek M (2010) Long-term trends in the middle atmosphere dynamics at northern middle latitudes - one regime or two different regimes? Atmos Chem Phys Discuss 10:2631–2668, www.atmos-chem-phys-discuss.net/10/2633/2010/

Laštovička J, Solomon SC, Qian L (2012) Trends in the neutral and ionized upper atmosphere. Space Sci Rev 168:113–145, doi: 10.1007/s11214-011-9799-3

Latteck R, Bremer J (2013) Long-term changes of polar mesospheric summer echoes at 69°N. J Geophys Res Atmos 118:10441–10448, doi: 10.1002/jgrd.50787

Lean J, Emmert JT, Picone JM, Meier RR (2011) Global and regional trends in ionospheric electron content. J Geophys Res 116:A00H04, doi: 10.1029/2010JA016378

Lewis HG, Saunders A, Swinerd G, Newland RJ (2011) Effect of thermospheric contraction on remediation of the near-Earth space debris environment. J Geophys Res 116:A00H08, doi: 10.1029/2011JA016482

Lossow S, McLandress C, Jonsson AI, Shepherd TG (2012) Influence of the Antarctic ozone hole on the polar mesopause region as simulated by the Canadian Middle Atmosphere Model. J Atmos Solar-Terr Phys 74:111–123, doi: 10.1016/j.jastp.2011.10.010

Lübken F-J, Berger U, Kiliani J, Baumgarten G, Fiedler J (2012) Solar variability and trend effects in mesospheric ice layers. In: Lübken F-J (ed) Climate And Weather of the Sun-Earth System (CAWSES), 18th edn. Springer, Dordrecht, Netherlands, pp 317–338, doi:10.1007/978-94-007-4348-9

Lübken F-J, Berger U, Baumgarten U (2013) Temperature trends in the midlatitude summer mesosphere. J Geophys Res Atmos 118:13347–13360, doi:1002/2013JD020576

Martini D, Mursula K (2008) Centennial geomagnetic activity studied by a new, reliable long-term index. J Atmos Solar-Terr Phys 70:1074–1087

Mielich J, Bremer J (2013) Long term trends in the ionospheric F2 region with different solar activity indices. Ann Geophys 21:291–303

Mlynczak M, Hunt L, Marshall BT, Martin-Torres FJ, Mertens CJ, Russell JM III, Remsberg E, Lopez-Puertas M, Picard R, Winick J, Wintersteiner P, Thompson RE, Gordley LL (2010) Observations of infrared radiative cooling in the thermosphere on daily to multiyear timescales from the TIMED/SABER instrument. J Geophys Res 115:A03309, doi: 10.1029/2009JA014713

Offermann D, Hoffmann P, Knieling P, Koppmann R, Oberheide J, Steinbrecht W (2010) Long-term trends and solar cycle variations of mesospheric temperature and dynamics. J Geophys Res 115:D18127, doi: 10.1029/2009JD013363

Ohya H, Shiokawa K, Miyoshi Y (2011) Long-term variations in tweek reflection height in the D and lower E regions of the ionosphere. J Geophys Res 116: A10322, doi: 10.1029/2011JA016800

Oliver WL, Zhang S-R, Goncharenko LP (2013) Is thermospheric global cooling caused by gravity waves? J Geophys Res 118:3898–3908, doi: 10.1002/jgra.50370

Qian L, Solomon SC (2012) Thermospheric density: an overview of temporal and spatial variations. Space Sci Rev 168:147–173, doi: 10.1007/s11214-011-9810-z

Qian L, Burns AG, Solomon SC, Roble RG (2009) The effect of carbon dioxide cooling on trends in the F2-layer ionosphere. J Atmos Solar-Terr Phys 71:1592–1601, doi: 10.1016/j.jastp.2009.03.006

Qian L, Laštovička J, Solomon SC, Roble RG (2011) Progress in observations and simulations of global change in the upper atmosphere. J Geophys Res 116: A00H03, doi: 10.1029/2010JA016317

Qian L, Marsh D, Merkel A, Solomon SC, Roble RG (2013) Effects of trends of middle atmosphere gases on the mesosphere and thermosphere. J Geophys Res Space Phys 118:3846–3855, doi:10.102/jgra.50354

Rao NV, Tsuda T, Riggin DM, Gurubaran S, Reid IM, Vincent RA (2012) Long-term variability of mean winds in the mesosphere and lower thermosphere at low latitudes. J Geophys Res 117:A10312, doi: 10.1029/2012JA017850

Ratnam MV, Kumar GK, Murthy BVK, Patra AK, Rao VVMJ, Rao SVB, Kumar KK, Ramkumar G (2008) Long-term variability of the low latitude mesospheric SAO and QBO and their relation with stratospheric QBO. Geophys Res Lett 35:L21809, doi: 10.1029/2008GL035390

Ratnam MV, Kumar GK, Rao NV, Krishna Murthy BV, Laštovička J, Qian L (2013) Evidence of long-term change in zonal wind in the tropical lower mesosphere: observations and model simulations. Geophys Res Lett 40:397–401, doi: 10.1002/grl.50158

Reinsel GC, Miller AJ, Weatherhead EC, Flynn LE, Nagatani RM, Tiao GC, Wuebbles DJ (2005) Trend analysis of total ozone data for turnaround and dynamical contributions. J Geophys Res 110:D16306, doi: 10.1029/2004JD004662

Remsberg EE (2009) Trends and solar cycle effects in temperatures versus altitude from the Halogen Occultation Experiment for the mesosphere and upper stratosphere. J Geophys Res 114:D12303, doi: 10.1029/2009JD011897

Remsberg EE (2010) Observed seasonal to decadal scale responses in mesospheric water vapor. J Geophys Res 115:D06306, doi: 10.1029/2009JD012904

Rishbeth H (1990) A greenhouse effect in the ionosphere? Planet Space Sci 38:945–948

Rishbeth H, Roble RG (1992) Cooling of the upper atmosphere by enhanced greenhouse gases - modeling of thermospheric and ionospheric effects. Planet Space Sci 40:1011–1026

Roble RG, Dickinson DE RE (1989) How will changes in carbon dioxide and methane modify the mean structure of the mesosphere and lower thermosphere? Geophys Res Lett 16:1441–1444

Rong PP, Russell JM III, Hervig ME, Bailey SM (2012) The roles of temperature and water vapor at different stages of the polar mesospheric cloud season. J Geophys Res 117:D04208, doi: 10.1029/2011JD016464

Scott TJ, Stamper R, Rishbeth H (2014) Long-term changes in thermospheric composition inferred from a spectral analysis of ionospheric F-region data. Ann Geophys 32:113–119, www.ann-geophys.net/32/114/2014/

She C-Y, Krueger DA, Akmaev R, Schmidt H, Talaat E, Yee S (2009) Long-term variability in mesopause region temperatures over Fort Collins, Colorado (41Å°N, 105Å°W) based on lidar observations from 1990 through 2007. J Atmos Solar-Terr Phys 71:1558–1564, doi: 10.1016/j.jastp.2009.05.007

Shettle PE, DeLand MT, Thomas GE, Olivero JJ (2009) Long term variations in the frequency of polar mesospheric clouds in the Northern Hemisphere from SBUV. Geophys Res Lett 36:L02803, doi: 10.1029/2008GL036048

Siskind DE, Stevens MH, Hervig ME, Randall CE (2013) Recent observations of high mass density polar mesospheric clouds: a link to space traffic? Geophys Res Lett 40:2813–2817, doi: 10.1002/grl.50540

Smith AK, Garcia RR, Marsh DR, Kinnison DE, Richter JH (2010) Simulations of the response of mesospheric circulation and temperature to the Antarctic ozone hole. Geophys Res Lett 37:L22803, doi: 10.1029/2010GL045255

Solomon S, Rosenlof KH, Portmann RW, Daniel JS, Davis SM, Sanford TJ, Plattner G-K (2010a) Contributions of stratospheric water vapor to decadal changes in the rate of global warming. Science 327:1219–1223

Solomon SC, Woods TN, Didkovski LV, Emmert JT, Qian L (2010b) Anomalously low solar extreme-ultraviolet irradiance and thermospheric density during solar minimum. Geophys Res Lett 37:L16103, doi: 10.1029/2010GL044468

Solomon SC, Qian L, Burns AG (2013) The anomalous ionosphere between solar cycles 23 and 24. J Geophys Res 118:6524–6535, doi: 10.1002/jgra.50561

Sonnemann GR, Hartogh P, Berger U, Lübken F-J, Grygalashvyly M (2012) Anthropogenic effects on the distribution of minor chemical constituents in the mesosphere/lower thermosphere - a model study. Adv Space Res 50:598–618

Sridharan S, Prasanth PV, Kumar YB (2009) A report on long term trends and variabilities in middle atmospheric temperature over Gadanki (13.5°N, 79.2°E). J Atmos Solar-Terr Phys 71, doi: 10.1016/j.jastp.2008.09.017

Stiller GP, von Clarmann T, Haenel F, Funke B, Glatthor N, Grabowski U, Kellmann S, Kiefer M, Linden A, Lossow S, Lopez-Puertas M (2012) Observed temporal evolution of global mean age of stratospheric air for the 2002 to 2010 period. Atmos Chem Phys 12:3311–3331, www.atmos-chem-phys.net/12/3311/2012/

Sunda S, Vyas BM, Khekale PV (2013) Storm time spatial variations in TEC during moderate geomagnetic storms in extremely low solar activity conditions (2007–2009) over Indian region. Adv Space Res 52:158–176, http://dx.doi.org/10.1016/j.asr.2013.03.006

Tsuda T, Rao NV, Riggin D, Gurubaran S, Reid LM, Vincent RA (2013) Long-term variability of the mean winds in the mesosphere and lower thermosphere (MLT) observed with radars in the tropics. Paper presented at International CAWSES-II Symposium, Nagoya

Verkhoglyadova OP, Tsurutani BT, Mannucci AJ, Mlynczak MG, Hunt LA, Runge T (2013) Variability of ionospheric TEC during solar and geomagnetic minima (2008 and 2009): external high speed stream drivers. Ann Geophys 31:263–276

Walsh PL, Oliver WL (2011) Is thermospheric long-term cooling due to CO2 or O3? Ann Geophys 29:1779–1782, doi: 10.5194/angeo-29-1779-2011

Weatherhead EC, Reinsel GC, Tiao GC, Jackman CH, Bishop L, Hollandsworth Frith SM, DeLuisi J, Keller T, Oltmans SJ, Fleming EL, Wuebbles DJ, Kerr JB, Miller AJ, Herman J, McPeters R, Nagatani RM, Frederick JE (2000) Detecting the recovery of total column ozone. J Geophys Res 105:22201–22210

Zhang S-R, Holt J (2013) Long-term ionospheric cooling: dependency on local time, season, solar activity, and geomagnetic activity. J Geophys Res 118:3719–3730, doi: 10.1002/jgra.50306

Zhang S-R, Holt J, Kurdzo J (2011) Millstone Hill ISR observations of upper atmospheric long-term changes: height dependency. J Geophys Res 116:A00H05, doi:10.1029/2010JA016414

12

Cold surge event observed by radiosonde observation from the research vessel "Hakuho-maru" over the Philippine Sea in December 2012

Shin-Ya Ogino *, Peiming Wu[1], Miki Hattori[1], Nobuhiko Endo[1,2], Hisayuki Kubota[1,3], Tomoshige Inoue[1,4] and Jun Matsumoto[1,5]

Abstract

The thermal energy transfer from the sea surface to the atmosphere associated with a cold surge event was investigated with observations from radiosondes on the research vessel "Hakuho-maru" over the Philippine Sea in December 2012. These observations were analyzed, and the results were compared with those obtained from observations over the East China Sea in the Air Mass Transformation Experiment in 1974 (AMTEX '74). The horizontal advection of cold and dry air associated with the cold surge dominated at heights below 850 hPa. In spite of this strong advection, the local temporal variations in the temperature and moisture were small, because the advection was balanced by the transfer of heat and moisture from the sea surface, which is qualitatively the same behavior as observed during the cold surge event in AMTEX '74. The eddy transport of the total heat energy from the sea surface to the atmosphere was estimated at about 410 W/m^2, which is about half of the maximum value of 780 W/m^2 observed during AMTEX '74. This result shows the existence of considerable heat transfer from the sea surface to the atmosphere over the Philippine Sea, which is the downstream region of the cold surge, after it passed through the East China Sea.

Keywords: Cold surge, Air-sea interaction, Air mass transformation, Hakuho-maru, Philippine Sea, VPREX2012, AMY2007-2012

Introduction

Cold surges originating from the Siberian High are common during the Asian winter monsoon and can extend as far as the Southeast Asia, including the Philippines and the Indochina Peninsula. Two active propagation paths are known: the South China Sea (SCS) path and the Philippine Sea path (Compo et al. 1999). Several studies have shown that the surges along the SCS can result in heavy rainfall along the east coast of the Indochina Peninsula. The coupling of the cold surge and the westward-propagating disturbance in the SCS has been suggested as a key process of increased precipitation over central Vietnam (Yokoi and Matsumoto 2008; Wu et al. 2011; Chen et al. 2012). However, changes in the rainfall over the Philippine Sea area associated with a cold surge have not yet been carefully investigated.

Air mass transformation is one of the most important processes during a cold surge event, since it determines the thermodynamic features of the air mass that develops into convection and rainfall. However, a quantitative evaluation based on observations has not yet been conducted, except for the Air Mass Transformation Experiment in 1974 (AMTEX '74) (Ninomiya 1975). AMTEX '74 was conducted in February 1974 over the East China Sea, and the observation area is shown in Fig. 1. In the experiment, a cold surge and the associated air-mass transformation over the East China Sea were observed, when the cold surge encountered the sea surface after passing over land. Ninomiya (1975) showed that during the cold period of February 23–27, 1974, cold and dry advection dominated and was almost balanced with the eddy heat supply from the sea, which was confined below the 850 hPa level. The

*Correspondence: ogino-sy@jamstec.go.jp
[1] Japan Agency for Marine-Earth Science and Technology, 2-15 Natsushima-cho, Yokosuka 237-0061, Japan
Full list of author information is available at the end of the article

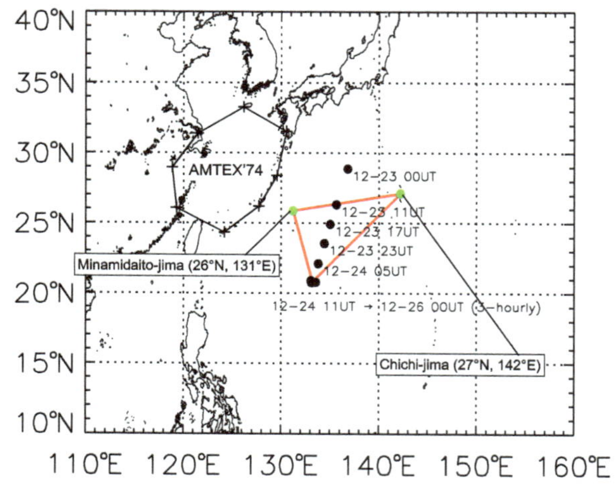

Fig. 1 The observation sites of the radiosondes used in this study. Black circles show the launch position of the radiosondes from the R/V Hakuho-maru. The radiosonde stations at Minamidaitojima and Chichijima are indicated by green circles. The AMTEX '74 observation area is shown by the black polygon

total heat supply from the sea to the atmosphere was estimated at about 580 W/m^2.

In December 2012, we succeeded in observing a cold surge event using radiosondes on the research vessel "Hakuho-maru" (hereafter, R/V Hakuho-maru) over the Philippine Sea. Since the observation area was located downstream of the AMTEX '74 observation area, we expected that the R/V Hakuho-maru observations would record the air mass that had already traveled over the East China Sea and had been influenced by the air-sea interaction. In this paper, we present a case study of this cold surge event to understand the air-sea interaction at locations far from land and to compare the results with those obtained from AMTEX '74.

Methods/Experimental

Observation and data

We conducted radiosonde observations from the R/V Hakuho-maru during December 23–26, 2012 as a part of the Vietnam Philippines Rainfall Experiment (VPREX) 2012 (http://www.jamstec.go.jp/rigc/j/tcvrp/mhcrt/vprex.html) under the Asian Monsoon Year 2007–2012 project (Wang et al. 2010). Figure 1 shows the observation sites and the launch times of the radiosondes. We launched radiosondes at 6- or 12-h intervals during December 23–24 between 29°N and 21°N along the cruise track from north to south. At the southernmost site (21°N, 133°E), we carried out a fixed-point observation at 3-h intervals for about 1.5 days during December 24–25. We then performed an energy budget analysis with the fixed-point observation data combined with the operational radiosonde data from the nearby Japan Meteorological Agency upper-air stations, Minamidaito-jima (26°N, 131°E) and Chichijima (27°N, 142°E). We also used the National Centers for Environmental Prediction (NCEP) reanalysis II (Kanamitsu et al. 2002) to examine the large-scale background fields and the Tropical Rainfall Measuring Mission (TRMM) 3B42 data set (e.g., Kummerow et al. 2000) to examine the rainfall intensity.

Thermal and moisture budget analysis

We performed a thermal and moisture budget analysis and compared the results with those obtained by Ninomiya (1975) from AMTEX '74. In order to compare these results, we used the same formula and vertical coordinates as those used by Ninomiya (1975), which are briefly described below.

The heat energy and moisture continuity equations averaged over a certain area are written as follows:

$$\frac{\delta T}{\delta t} + \frac{\partial}{\partial p}\overline{\omega' T'} = \frac{dT_R}{dt} + \frac{L}{C_p} m, \tag{1}$$

$$\frac{\delta q}{\delta t} + \frac{\partial}{\partial p}\overline{\omega' q'} = -m, \tag{2}$$

where T is temperature, p is pressure, ω is the vertical p-velocity, dT_R/dt is the heating rate due to net radiation, C_p is the specific heat of dry air at constant pressure, q is the water vapor mixing ratio, L is the latent heat of condensation, m is the net condensation, the overbar $\overline{X}$ denotes the area-averaged value of the variable X, and the prime X' denotes the deviation from $\overline{X}$. The individual changes, the changes in the area-averaged quantities following the motion, of temperature and moisture were calculated from the area-averaged thermodynamic and moisture budget equations as follows:

$$\frac{\delta T}{\delta t} = \frac{\partial \overline{T}}{\partial t} + \overline{\boldsymbol{v} \cdot \nabla T} + \overline{\omega}\left(\frac{\partial \overline{T}}{\partial p} - \frac{R}{C_p}\frac{\overline{T}}{p}\right), \tag{3}$$

$$\frac{\delta q}{\delta t} = \frac{\partial \overline{q}}{\partial t} + \overline{\boldsymbol{v} \cdot \nabla q} + \overline{\omega}\frac{\partial \overline{q}}{\partial p}, \tag{4}$$

where $\boldsymbol{v}$ is the horizontal wind and R is the gas constant. The individual change in the equivalent temperature is obtained by eliminating the net condensation m from Eqs. (1) and (2) as follows:

$$\frac{\delta}{\delta t}\left(T + \frac{L}{C_p} q\right) = -\frac{\partial}{\partial p}\overline{\omega'\left(T' + \frac{L}{C_p} q'\right)} + \frac{dT_R}{dt}. \tag{5}$$

By integrating Eq. (5) from the surface (p_{SFC}) to a given pressure level (p), the eddy transport of the total heat energy $F_c = -1/g\left(C_p\overline{\omega' T'} + L\overline{\omega' q'}\right)$ can be evaluated as follows:

$$F_c(p) = -\frac{1}{g}\int_{p}^{p_{SFC}} \frac{\delta}{\delta t}\left(C_pT + Lq\right) dp - \frac{1}{g}F_c(p_{SFC}) + \frac{C_p}{g}\int_{p}^{p_{SFC}} \frac{dT_R}{dt} dp, \tag{6}$$

where g is the gravitational acceleration, the second term on the right-hand side can be evaluated beforehand by integrating Eq. (5) from the surface to the top level under the assumption that the eddy heat and moisture transport vanish at the top level as described below, and the third term on the right-hand side can be evaluated by assuming that the radiative heating rate at each layer is as described below.

The area-average of the abovementioned thermodynamic quantities was calculated from the data obtained at R/V Hakuho-maru, Minamidaitojima, and Chichijima. The horizontal gradient and the horizontal divergence were calculated by fitting of the plane surface to the data from the three observation stations as in Davies-Jones (1993).

As in Ninomiya (1975), we evaluate the thermodynamic quantities in three layers: (1) between the surface and 850 hPa, (2) between 850 and 700 hPa, and (3) between 700 and 500 hPa. We assume that the eddy transport of heat and moisture vanish at 500 hPa.

The vertical p-velocity was estimated by accumulation of the horizontal divergence. The surface vertical p-velocity was calculated from the time derivative of the surface pressure. The original estimate of the vertical p-velocity ω_{org} was adjusted in order to reduce the estimate errors as follows:

$$\omega = \omega_{org} - \Delta\omega = \omega_{org} - (p - p_{sfc})\frac{\omega_{org,top} - \omega_{top}}{p_{top} - p_{sfc}},$$

where subscripts top and sfc denote the top level (500 hPa in this study) and the surface, respectively. The vertical p-velocity at the top level ω_{top} was estimated by

$$\omega_{top} = -\left(\frac{\partial \overline{T}}{\partial t} + \overline{\boldsymbol{v}\cdot\nabla T} - \frac{dT_R}{dt}\right) \Big/ \left(\frac{\partial \overline{T}}{\partial p} - \frac{R}{C_p}\frac{\overline{T}}{p}\right),$$

and assuming thermodynamic balance between the adiabatic motion and the radiative heating. This assumption is reasonable because condensation and evaporation did not seem to occur under the dry conditions observed at this level (the relative humidity was under 40%, not shown), and the eddy heat transport from the sea surface did not seem to penetrate the 700 hPa pressure level as discussed later.

For the radiative heating rate dT_R/dt, the climatological values near 20°N and 135°E taken from Katayama (1967) are used as in Ninomiya (1975). They are −0.042, −0.034, and −0.062 K/h for the layers between the surface and at 850 hPa, 850–700 hPa, and 700–500 hPa, respectively, and are insignificant compared with the other terms in the budget analysis of this study (e.g., the change in the equivalent temperature in the near-surface layer was estimated as about 0.5 K/hr).

The net condensation m was assumed to be zero, as we found no TRMM 3B42 precipitation over the analysis area during the analysis period (from 1200 UT on December 24 to 1200 UT on December 25).

Results

Synoptic situation

Figure 2 shows the distributions of the synoptic fields at 925 hPa and their temporal variations during the Hakuho-maru observation period. A tongue shaped ridge extending from the southeastern edge of the strengthened Siberian High appeared on December 21 (the upper panel of Fig. 2). On December 23 (the middle panel of Fig. 2), at the beginning of the observation period at the northernmost site (29°N), an anticyclone with cold air was separated from the southeastern edge of the Siberian High, which was weakened on the same day, and moved to the southern region of China. This anticyclone enhanced the northeasterly flow over the northern Indochina and SCS regions. These variations are typical characteristics of a SCS cold surge (e.g., Compo et al. 1999).

On the same day, a weak ridge was found in the East China Sea (around 30°N, 125°E) and the associated northwesterly flow reached the Hakuho-maru observation area after passing through the East China Sea. On December 26 (the lower panel of Fig. 2), the ridge became more pronounced and moved southeastward to reach the Hakuho-maru observation area, just as we finished the fixed-point observations at the southernmost site (21°N, 133°E). This ridge extension and the associated anticyclonic flow were considered to be a typical development of the Philippine Sea cold surge.

The downstream northeasterly flow reached the Philippines and seemed to have brought precipitation to the east coast of the Philippines on December 26. In the equatorial region between 0 and 15°N, a westward-propagating, cyclonic flow was observed near 150°E on December 21, near 140°E on December 23, and near 125°E on December 26. The easterly flow at the northern edge of the cyclonic disturbance and the northeasterly flow associated with the cold surge formed a convergence zone offshore of the Philippines. The formation of a similar convergence zone due to a cold surge and a westward-propagating disturbance is considered to be a key driver of the heavy rainfall that occurs in autumn along the east coast of the Indochina Peninsula (Yokoi and Matsumoto 2008; Wu et al. 2011; Chen et al. 2012).

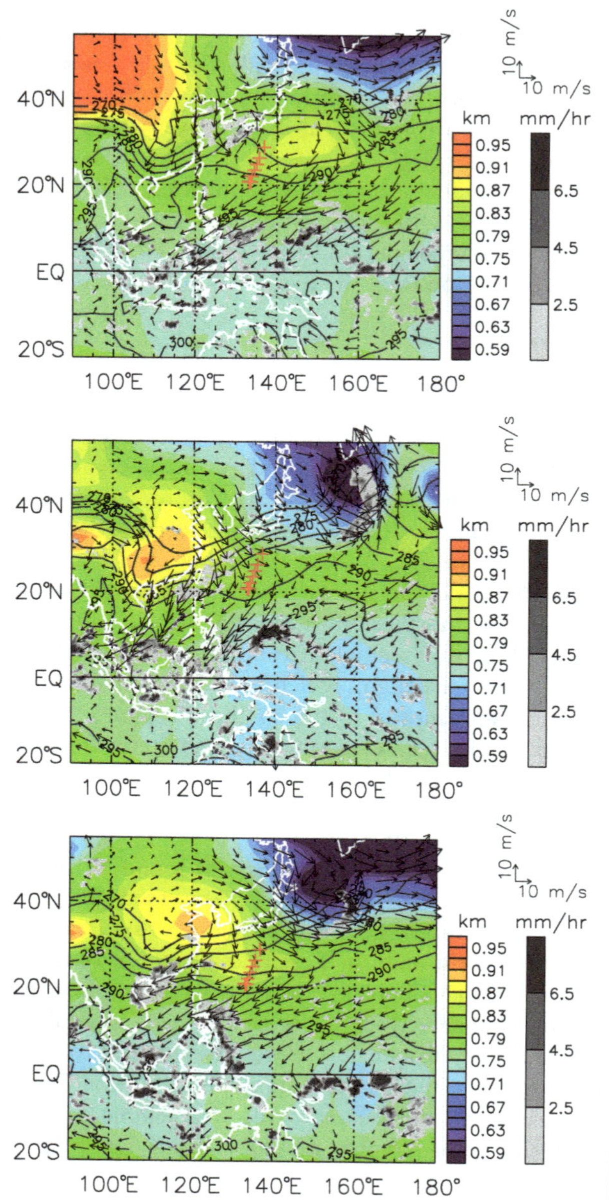

Fig. 2 Horizontal map of relevant meteorological variables during the Hakuho-maru observation period. The NCEP Reanalysis II geopotential height (color shading), temperature (Kelvin, black contours), horizontal wind (arrows), and TRMM 3B42 rainfall (gray shading) at the 925 hPa level at (upper) 1200 UT on December 21, (middle) 0600 UT on December 23, and (lower) 0000 UT on December 26, 2012. The launch sites of the radiosondes from the R/V Hakuho-maru are indicated by red pluses

Temperature and moisture variations obtained by the radiosondes

During the observation period between December 23–24 (upper panel of Fig. 3), a stable layer with a large vertical potential temperature gradient formed at a height of about 2 km. Below this layer, the temperature and humidity were higher in the southern area.

From the fixed-point observations (lower panel of Fig. 3), the stable layer at a height of around 2 km gradually intensified. Below the stable layer, both the potential temperature and the water vapor mixing ratio showed uniform vertical distributions, which were consistent with the well-mixed layer during a cold surge event.

Energy budget analysis

Using the radiosonde data from the fixed-point observations combined with the operational radiosonde data at Minamidaitojima and Chichijima, we performed a thermodynamic energy budget analysis to evaluate the thermodynamic energy transfer between the atmosphere and ocean.

The fixed-point observations were conducted during the period when the near-surface temperature reached its minimum as shown in Fig. 4. The near-surface temperatures at Minamidaitojima and Chichijima decreased before December 25 and increased after December 26 along with the cold surge passage. The near-surface temperature observed by the radiosondes from the R/V Hakuho-maru showed a temperature decrease between 1200 UT on December 24 and 1200 UT on December 25. We evaluated the thermodynamic and moisture budgets using the data obtained during this period. The area-average, near-surface temperature obtained from the radiosondes from the R/V Hakuho-maru, and at Minamidaitojima and Chichijima showed a slight decrease. Note that the result of the AMTEX '74 budget analysis showed the maximum heat and moisture transport from the ocean to the atmosphere near the end of the temperature decrease associated with the cold surge.

Figure 5 shows the results of the evaluation using Eqs. (3) and (4). In the layer below 850 hPa, the horizontal advection terms of temperature and moisture ($-\boldsymbol{v}\cdot\nabla T$, and $-\boldsymbol{v}\cdot\nabla q$, respectively) and the individual changes in the temperature and moisture ($\delta T/\delta t$ and $\delta q/\delta t$, respectively) were large and had opposite signs, while the local temperature and moisture changes ($\partial\overline{T}/\partial t$ and $\partial\overline{q}/\partial t$, respectively) and the vertical advection terms of temperature and moisture ($\overline{\omega}\left(\partial\overline{T}/\partial p - R/C_p\ \overline{T}/p\right)$ and $\overline{\omega}\ \partial\overline{q}/\partial p$, respectively) were small. This means the horizontal advection of cold and dry air was almost canceled by the vertical convergence of the eddy transport of temperature and moisture, resulting in the relatively small values of local changes in the temperature and moisture with time. This feature is consistent with the uniform vertical distribution of the potential temperature and water vapor mixing ratio observed with the R/V Hakuho-maru radiosondes shown in the previous section.

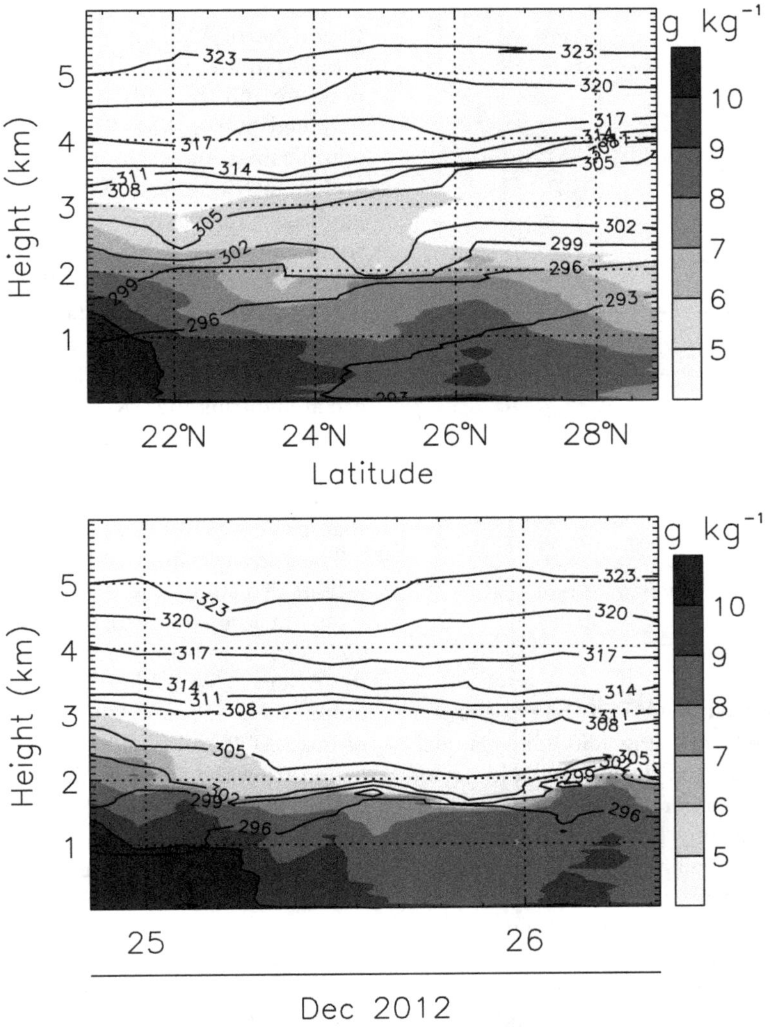

Fig. 3 The result of the radiosonde observation of the Hakuho-maru cruise. (Upper) Latitude-height section of water-vapor mixing ratio (gray shading) and potential temperature (black contours) from the moving observations from 29°N to 22°N. (Lower) The same as the upper panel, except for the time-height cross-section during the fixed-point observation during December 24–26, 2012

In the layer between 850 and 700 hPa, the cold horizontal advection had a similar magnitude to the layer below 850 hPa, and it seemed to balance the warm vertical advection. For the moisture budget in the same layer between 850 and 700 hPa, both horizontal and vertical advection of dry air were observed. Although the local moisture change was negative, it did not fully compensate for the horizontal and vertical advection of dry air, resulting in the large positive value of the individual change in moisture. Although it is possible that this large individual change of moisture was not only due to eddy moisture transport but also due to evaporation, this possibility will not be discussed further in this study, because the effects of evaporation (and condensation) canceled out in the evaluation of the equivalent temperature, which we will show later.

In the layer above 700 hPa, the cold horizontal advection balanced the warm vertical advection. All the moisture terms in this layer were small.

The individual change of the equivalent temperature (left panel of Fig. 6) shows a large value in the layer below 850 hPa, and the value in the layer between 700 and 850 hPa is comparable. The surface value of F_c, i.e., the total heat supply from the ocean to the atmosphere is estimated as 410 W/m^2. F_c at 850 hPa is estimated as 140 W/m^2. The results indicate that the thermal influence from the sea surface reached the 700–850 hPa layer through the 850 hPa level.

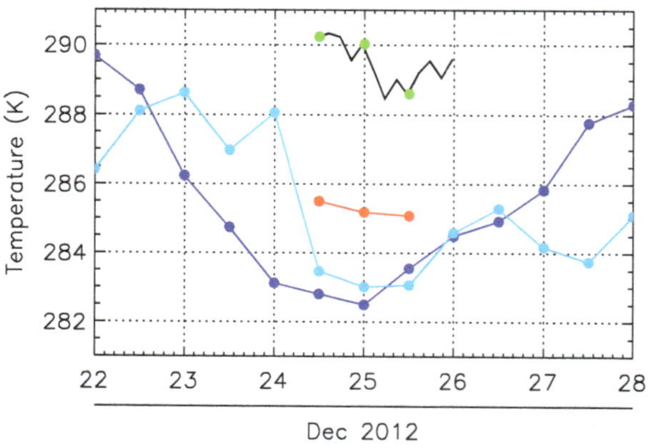

Fig. 4 Time series of the near-surface temperature during December 22–29, 2012. Values are the average between the surface and 850 hPa. (blue) Minamidaitojima, (light blue) Chichijima, (black) Hakuho-maru fixed-point observation data (3-hourly), (green filled circle) R/V Hakuho-maru data used in the budget analysis (12-hourly; 1200 UT on December 24, 0000 UT on December 25, and 1200 UT on December 25), (red) mean temperature obtained from the observations at the three site/stations (R/V Hakuho-maru, Minamidaitojima, and Chichijima)

Discussion: comparison with AMTEX '74 results

The energy budget in the layer below 850 hPa was qualitatively similar to that observed during the cold period in AMTEX '74 when the temperature decreased due to a cold surge; that is, the strong horizontal advection of cold and dry air was almost balanced by the heating and moistening due to eddy transport, and the local temperature and moisture changes were very small. Quantitatively, each term in the thermal budget (Eq. (3)) of this study had about half of the magnitude of that during the AMTEX '74 cold period. For the moisture terms, the magnitudes of the R/V Hakuho-maru observations were similar to or larger than those of AMTEX '74.

In Fig. 6, the individual change of the equivalent temperature and the eddy transport of the total heat energy obtained by the AMTEX '74 observations are indicated with dashed lines. The values were taken from Figs. 12 and 19 in Ninomiya (1975). The individual change of the equivalent temperature in the layer below 850 hPa of this study (about 0.5 K/hr) is about 60% of the magnitude of the averaged value of the north and south areas during the AMTEX '74 cold period (about 0.8 K/hr). On the other hand, in the layer between 850 and 700 hPa, the value obtained in this study is larger than that for the northern area during the AMTEX '74 cold period, and close to that of the southern area. The difference in the individual change in the equivalent temperature between the surface 850 hPa layer and the 850–700 hPa layer is smaller than that in the AMTEX '74 cold period.

The eddy transport of the total heat energy at the surface of this study was about 410 W/m^2, which is smaller than, but comparable to, those of the northern area (about 450 W/m^2) and southern area (about 570 W/m^2) during the AMTEX '74 cold period, and about half of the maximum value (about 780 W/m^2) observed near the end of the AMTEX '74 cold period. The eddy transport estimated from the R/V Hakuho-maru observations is much larger than that of the AMTEX '74 warm period during February 23–27 when no cold surge was observed. The result suggests that a significant amount of heat exchange occurred even over the open ocean far from land.

Conclusions

We succeeded in observing the air-sea interactions associated with a cold surge event using radiosondes on

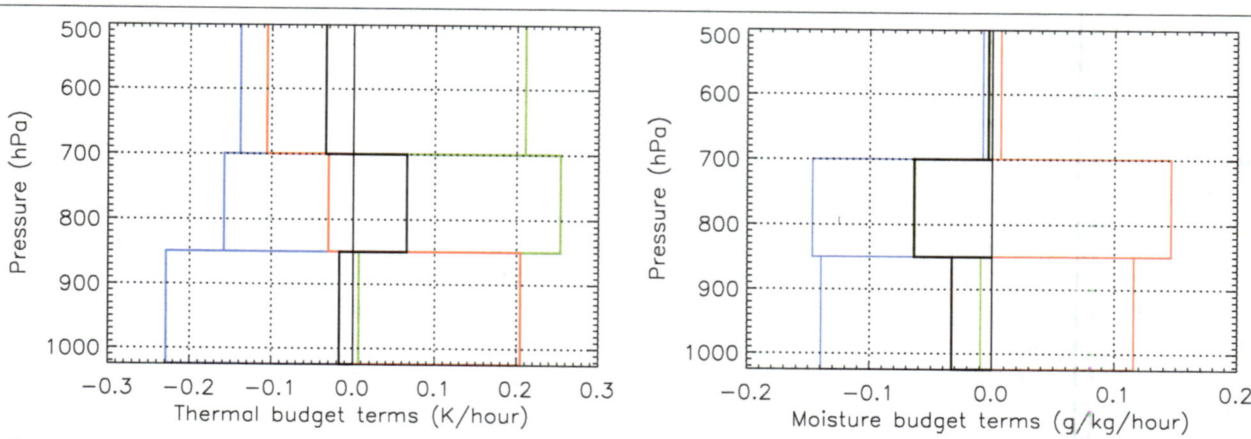

Fig. 5 Vertical profiles of the temperature and moisture budget. (Left) Local temperature change $\partial\overline{T}/\partial t$ (black), horizontal temperature advection $-\overline{\boldsymbol{v}\cdot\nabla T}$ (light blue), vertical temperature advection $-\overline{\omega}\left(\partial\overline{T}/\partial p - R/C_p\,\overline{T}/p\right)$ (green), and individual change in temperature $\delta T/\delta t$ (red). (Right) The same as the left panel, except for the moisture budget: local moisture change $\partial\overline{q}/\partial t$ (black), horizontal moisture advection $-\overline{\boldsymbol{v}\cdot\nabla q}$ (light blue), vertical moisture advection $-\overline{\omega}\,(\partial\overline{q}/\partial p)$ (green), and individual change in moisture $\delta q/\delta t$ (red)

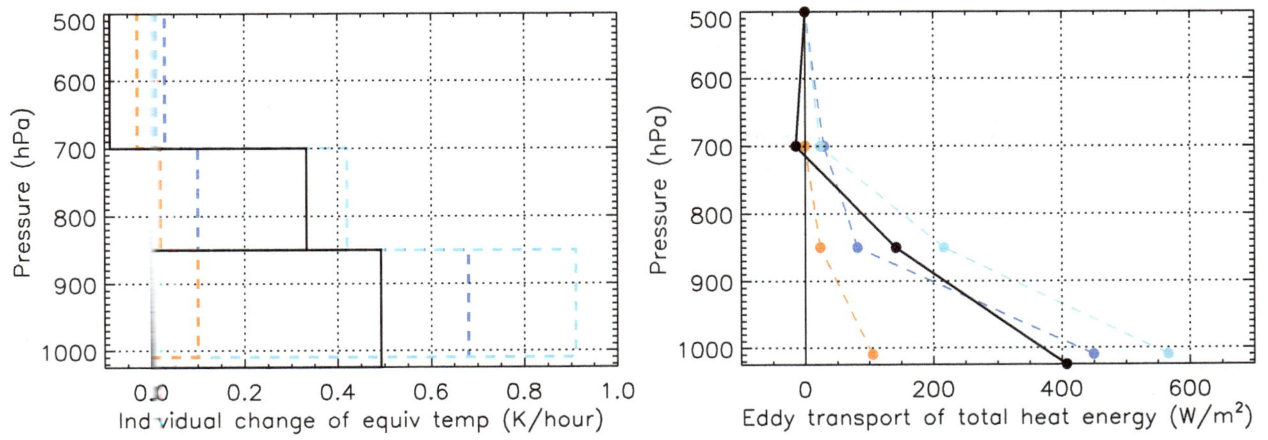

Fig. 6 Vertical profiles of results from R/V Hakuho-maru and AMTEX '74 observations. (Left) Individual change in the equivalent temperature. (Right) Eddy transport of the total heat energy. The results of the R/V Hakuho-maru observations are plotted with solid black lines. The results from AMTEX '74 are plotted with dashed lines: (blue) the results over the north area during the cold period, (light blue) the results over the south area during the cold period, and (orange) the results during the warm period

the R/V Hakuho-maru far from land in the Philippine Sea in late December 2012. We evaluated the thermal energy transport from the sea surface to the atmosphere, and compared it with the results obtained from the AMTEX '74 experiment conducted over the East China Sea.

The strong horizontal advection of cold and dry air due to the cold surge dominated in the layer below 850 hPa. In spite of this strong horizontal advection, the local temperature and moisture changes were small, since the horizontal advection was balanced by a supply of heat and moisture from the sea surface, which is qualitatively the same finding as observed in the AMTEX '74 cold period. The eddy transport of the total heat energy from the sea surface to the atmosphere was estimated as about 410 W/m^2, which is roughly similar to the mean value (about 510 W/m^2) and about half of the maximum value (about 780 W/m^2) observed during the AMTEX '74 cold period.

This result shows the existence of considerable heat transfer from the sea surface over the Philippine Sea, in the region downstream of the cold surge after it passed through the East China Sea. The horizontal advection of cold and dry air had an effect on decreasing the local temperature and humidity, i.e., to suppress the convective instability. However, this effect was compensated by the eddy heat and moisture transport from the sea surface. As a result, the convective instability tended to be maintained, which, in the case of this study, possibly led to the enhanced convection and precipitation observed along the east coast of the Philippines.

In this study, it is also indicated that the coupling of the northeasterly flow due to the cold surge and the westward-propagating cyclonic disturbance strengthened the convergent flow toward the convective area over the Philippines. This type of coupling has been highlighted as one of the key drivers of heavy rainfall over the east coast of the Indochina Peninsula in autumn. The indication of a similar development mechanism over the Philippines is a key result of this study.

We hope to estimate the thermal budget during different phases of a cold surge and further south of the region considered in this study in order to understand the effect of cold surges on tropical convection as soon as we have collected the necessary data.

Abbreviations

AMTEX: Air mass transformation experiment; AMY: Asian monsoon years; NCEP: National Centers for Environmental Prediction; TRMM: Tropical rainfall measuring mission; VPREX: Vietnam Philippines rainfall experiment

Acknowledgements

We would like to thank Dr. Hodaka Kawahata (the University of Tokyo), Mr. Katsura Kameo (the University of Tokyo), Mr. Ei Hatakeyama (Marine Works Japan LTD), and all of the crew of the R/V Hakuho-maru for their great help during the radiosonde observations on board the research vessel. We wish to express our gratitude to Drs. Kaoru Sato (the University of Tokyo), Naohiko Hirasawa, and Yoshihiro Tomikawa (National Institute of Polar Research) for permitting us to use their balloon launcher.

Funding

Part of this study was supported by the Green Network of Excellence (GRENE) program of the Japanese Ministry of Education, Culture, Sports, Science and Technology and by Japan Agency for Marine-Earth Science and Technology.

Authors' contributions

SYO and JM proposed the topic, conceived and designed the study. SYO and PW carried out the radiosonde observations on board the research vessel "Hakuho-maru" with support by MH who provided the meteorological information during the "Hakuho-maru" cruise. MH, HK, NE, and TI discussed the obtained results and provided comments on the manuscript. All authors read and approved the final manuscript.

Competing interests

The authors declare that they have no competing interest.

Author details
[1]Japan Agency for Marine-Earth Science and Technology, 2-15 Natsushima-cho, Yokosuka 237-0061, Japan. [2]Present address: National Institute for Agro-Environmental Sciences, 3-1-3 Kannondai, Tsukuba 305-8604, Japan. [3]Faculty of Science, Department of Earth and Planetary Sciences, Hokkaido University, Kita 10 Nishi 8, Kita-ku Sapporo 060-0810, Japan. [4]Present address: Tokyo Metropolitan University, 1-1, Minami-Osawa, Hachioji 192-0397, Japan. [5]Tokyo Metropolitan University, 1-1, Minami-Osawa, Hachioji 192-0397, Japan.

References

Chen T-C, Yen M-C, Tsay J-D, Tan Thanh NT, Alpert J (2012) Synoptic development of the Hanoi heavy rainfall event of 30–31 October 2008: Multiple-scale processes. Mon Wea Rev 140:1219–1240. https://doi.org/10.1175/MWR-D-11-00111.1

Compo GP, Kiladis GN, Webster PJ (1999) The horizontal vertical structure of east Asian winter monsoon pressure surges. Q J R Meteorol Soc 125:29–54

Davies-Jones R (1993) Useful formulas for computing divergence, vorticity, and their errors from three or more stations. Mon Wea Rev 121:713–725. https://doi.org/10.1175/1520-0493(1993)121<0713:UFFCDV>2.0.CO;2

Kanamitsu M, Ebisuzaki W, Woollen J, Yang S-K, Hnilo JJ, Fiorino M, Potter GL (2002) NCEP-DOE AMIP-II Reanalysis (R-2). Bull Amer Meteor Soc 83:1631–1643. https://doi.org/10.1175/BAMS-83-11-1631

Katayama A (1967) On the radiation budget of the troposphere over the northern hemisphere (iii) —Zonal cross-section and energy consideration—. J Met Soc Japan 45:26–39

Kummerow C, Simpson J, Thiele O, Barnes W, Chang ATC, Stocker E, Adler RF, Hou A, Kakar R, Wentz F, Ashcroft P, Kozu T, Hong Y, Okamoto K, Iguchi T, Kuroiwa H, Im E, Haddad Z, Huffman G, Ferrier B, Olson WS, Zipser E, Smith EA, Wilheit TT, North G, Krishnamurti T, Nakamura K (2000) The status of the tropical rainfall measuring mission (TRMM) after two years in orbit. J Appl Meteor 39:1965–1982. https://doi.org/10.1175/1520-0450(2001)040<1965:TSOTTR>2.0.CO;2

Ninomiya K (1975) Large-scale aspects of air-mass transformation over the East China Sea during AMTEX '74. J Met Soc Japan 53:285–303

Wang B, Matsumoto J, Wu G, Li J (2010) The Science Plan for Asian Monsoon Years (AMY 2007–2012), A Cross-cutting WCRP Initiative. China Meteorological Press, Beijing

Wu P, Fukutomi Y, Matsumoto J (2011) An observational study of the extremely heavy rain event in northern Vietnam during 30 October–1 November 2008. J Met Soc Japan 89A:331–344. https://doi.org/10.2151/jmsj.2011-A23

Yokoi S, Matsumoto J (2008) Collaborative effects of cold surge and tropical depression-type disturbance on heavy rainfall in central Vietnam. Mon Wea Rev 136:3275–3287

PERMISSIONS

All chapters in this book were first published in PEPS, by Springer; hereby published with permission under the Creative Commons Attribution License or equivalent. Every chapter published in this book has been scrutinized by our experts. Their significance has been extensively debated. The topics covered herein carry significant findings which will fuel the growth of the discipline. They may even be implemented as practical applications or may be referred to as a beginning point for another development.

The contributors of this book come from diverse backgrounds, making this book a truly international effort. This book will bring forth new frontiers with its revolutionizing research information and detailed analysis of the nascent developments around the world.

We would like to thank all the contributing authors for lending their expertise to make the book truly unique. They have played a crucial role in the development of this book. Without their invaluable contributions this book wouldn't have been possible. They have made vital efforts to compile up to date information on the varied aspects of this subject to make this book a valuable addition to the collection of many professionals and students.

This book was conceptualized with the vision of imparting up-to-date information and advanced data in this field. To ensure the same, a matchless editorial board was set up. Every individual on the board went through rigorous rounds of assessment to prove their worth. After which they invested a large part of their time researching and compiling the most relevant data for our readers.

The editorial board has been involved in producing this book since its inception. They have spent rigorous hours researching and exploring the diverse topics which have resulted in the successful publishing of this book. They have passed on their knowledge of decades through this book. To expedite this challenging task, the publisher supported the team at every step. A small team of assistant editors was also appointed to further simplify the editing procedure and attain best results for the readers.

Apart from the editorial board, the designing team has also invested a significant amount of their time in understanding the subject and creating the most relevant covers. They scrutinized every image to scout for the most suitable representation of the subject and create an appropriate cover for the book.

The publishing team has been an ardent support to the editorial, designing and production team. Their endless efforts to recruit the best for this project, has resulted in the accomplishment of this book. They are a veteran in the field of academics and their pool of knowledge is as vast as their experience in printing. Their expertise and guidance has proved useful at every step. Their uncompromising quality standards have made this book an exceptional effort. Their encouragement from time to time has been an inspiration for everyone.

The publisher and the editorial board hope that this book will prove to be a valuable piece of knowledge for researchers, students, practitioners and scholars across the globe.

LIST OF CONTRIBUTORS

Akiyoshi Wada
Meteorological Research Institute, Japan Meteorological Agency, 1-1 Nagamine, Tsukuba, Ibaraki 305-0052, Japan

Tomohiro Uehara and Shiro Ishizaki
Japan Meteorological Agency, 1-3-4 Otemachi, Chiyoda, Tokyo 100-8122, Japan

Enkhbat Erdenebat and Tomonori Sato
Graduate School of Environmental Science, Hokkaido University, Kita-10, Nishi-5, Sapporo 060-0810, Japan

Iain Murray Reid
Department of Physics, School of Physical Sciences, The University of Adelaide, Adelaide 5005, Australia
ATRAD Pty Ltd, 20 Phillips Street, Thebarton 5031, Australia

Junya Uchida, Shota Misawa and Masaki Satoh
Atmosphere and Ocean Research Institute (AORI), The University of Tokyo, 5-1-5 Kashiwanoha, Kashiwa, Chiba 277-8568, Japan

Teruyuki Nakajima
Atmosphere and Ocean Research Institute (AORI), The University of Tokyo, 5-1-5 Kashiwanoha, Kashiwa, Chiba 277-8568, Japan
Earth Observation Research Center (EORC), Japan Aerospace Exploration Agency (JAXA), 2-1-1 Sengen, Tsukuba, Ibaraki 305-8505, Japan

Haruo Tsuruta
Atmosphere and Ocean Research Institute (AORI), The University of Tokyo,5-1-5 Kashiwanoha, Kashiwa, Chiba 277-8568, Japan
Remote Sensing Technology Center of Japan, 3-17-1 Toranomon, Minato-ku, Tokyo 105-0001, Japan

Yu Morino, Toshimasa Ohara and Daisuke Goto
National Institute for Environmental Studies (NIES), 16-2 Onogawa, Tsukuba, Ibaraki 305-8506, Japan

Toshihiko Takemura
Research Institute for Applied Mechanics, Kyushu University, Kasuga Park 6-1, Kasuga, Fukuoka 816-8580, Japan

Yasuji Oura and Mitsuru Ebihara
Graduate School of Science and Engineering, Tokyo Metropolitan University, 1-1 Minami-Osawa, Hachioji, Tokyo 192-0397, Japan

Peiming Wu and Shuichi Mori
Japan Agency for Marine-Earth Science and Technology, 2-15 Natsushima-cho, Yokosuka-City, Kanagawa 237-0061, Japan

Fadli Syamsudin
Indonesian Agency for Assessment and Application of Technology, Jl. Raya Puspiptek, South Tangerang 15314, Indonesia

Robert A Vincent
Department of Physics, School of Physical Sciences, University of Adelaide, North Terrace., SA, 5005 Adelaide, Australia

Karen H. Rosenlof
NOAA Earth System Research Laboratory Chemical Sciences Division, 325 S. Broadway, Boulder, CO 80305, USA

Sean M. Davis
NOAA Earth System Research Laboratory Chemical Sciences Division, 325 S. Broadway, Boulder, CO 80305, USA
Cooperative Institute for Research in Environmental Sciences, University of Colorado at Boulder, Boulder, CO 80303, USA

Birgit Hassler
Deutsches Zentrum für Luft- und Raumfahrt (DLR), Institut für Physik der Atmosphäre, Oberpfaffenhofen, Germany

Daiki Tsurushima
IDEA Consultants, Inc., Hayabuchi 2-2-2 Tsuzuki, Yokohama 224-0025, Kanagawa, Japan

Kiyotaka Sakaida
Graduate School of Environmental Studies, TohokuUniversity, Aoba 6-3 Aramaki Aoba, Sendai 980-8678, Miyagi, Japan

Noriyasu Honma
R&D Center of Tohoku Electric Power Company, Inc, Nakayama 2-1 Aoba, Sendai 981-0952, Miyagi, Japan

Hiroshi G. Takahashi
Department of Geography, Tokyo Metropolitan University, 1-1 Minamioosawa, Hachioji, Tokyo 192-0397, Japan
Japan Agency for Marine-Earth Science and Technology (JAMSTEC), Yokohama, Japan

Shingo Watanabe
Japan Agency for Marine-Earth Science and Technology (JAMSTEC), Yokohama, Japan

Makiko Nakata
Kindai University, Higashiosaka, Japan

Toshihiko Takemura
Research Institute for Applied Mechanics, Kyushu University, Fukuoka, Japan

Sachiko Okamoto and Hiroshi Tanimoto
Center for Global Environmental Research, National Institute for Environmental Studies, 16-2 Onogawa, Tsukuba, Ibaraki 305-8506, Japan

Jan Laštovička
Institute of Atmospheric Physics ASCR, Bocni II, Prague 14131, Czech Republic

Gufran Beig
Indian Institute for Tropical Meteorology, Dr. Homi Bhabha Road, Pashan, Pune 411008, India

Daniel R Marsh
National Center for Atmospheric Research, Boulder 80301CO, USA

Shin-Ya Ogino and Miki Hattori
Japan Agency for Marine-Earth Science and Technology, 2-15 Natsushimacho, Yokosuka 237-0061, Japan

Nobuhiko Endo
Japan Agency for Marine-Earth Science and Technology, 2-15 Natsushimacho, Yokosuka 237-0061, Japan
National Institute for Agro-Environmental Sciences, 3-1-3 Kannondai, Tsukuba 305-8604, Japan

Hisayuki Kubota
Japan Agency for Marine-Earth Science and Technology, 2-15 Natsushimacho, Yokosuka 237-0061, Japan
Faculty of Science, Department of Earth and Planetary Sciences, Hokkaido University, Kita 10 Nishi 8, Kita-ku Sapporo 060-0810, Japan

Tomoshige Inoue
Japan Agency for Marine-Earth Science and Technology, 2-15 Natsushimacho, Yokosuka 237-0061, Japan
Tokyo Metropolitan University, 1-1, Minami-Osawa, Hachioji 192-0397, Japan

Jun Matsumoto
Japan Agency for Marine-Earth Science and Technology, 2-15 Natsushimacho, Yokosuka 237-0061, Japan
Tokyo Metropolitan University, 1-1, Minami-Osawa, Hachioji 192-0397, Japan

Index

www.ingramcontent.com/pod-product-compliance
Lightning Source LLC
LaVergne TN
LVHW080917210926
848485LV00004B/665